D0903970

Springer Series in Advanced Manufacturing

Other titles in this series

Assembly Line Design
B. Rekiek and A. Delchambre

Effective Resource Management in Manufacturing Systems:
Optimization Algorithms in Production Planning
Caramia and Dell'Olmo

Condition Monitoring and Control for Intelligent Manufacturing
Wang and Gao (Eds.)
Publication due: January 2006

Hoda A. ElMaraghy and Waguih H. ElMaraghy (Eds.)

Advances in Design

With 264 Figures

Hoda A. ElMaraghy, BSc, MEng, PhD, PEng, FSME, FCSME
Waguih H. ElMaraghy, BSc, MEng, PhD, PEng, FASME, FCSME
Intelligent Manufacturing Systems (IMS) Centre
Department of Industrial and Manufacturing Systems Engineering (IMSE)
Faculty of Engineering
University of Windsor
Windsor
Ontario
N9B 3P4
Canada

Series Editor:
Professor D. T. Pham
Intelligent Systems Laboratory
WDA Centre of Enterprise in
Manufacturing Engineering
University of Wales Cardiff
PO Box 688
Newport Road
Cardiff
CF2 3ET
UK

British Library Cataloguing in Publication Data
Advances in design. - (Springer series in advanced
manufacturing)
1. Engineering design 2. Engineering design - Data processing
I. ElMaraghy, H.A. II. ElMaraghy, W. H.
620'.0042
ISBN-10: 1846280044

Library of Congress Control Number: 2005933475

Springer Series in Advanced Manufacturing ISSN 1860-5168
ISBN-10: 1-84628-004-4 e-ISBN 1-84628-210-1 Printed on acid-free paper
ISBN-13: 978-1-84628-004-7

Printed in Germany

9 8 7 6 5 4 3 2 1

Springer Science+Business Media
springeronline.com

Preface

Engineering design encompasses a wide range of activities whose goal is to determine all attributes of a product before it is manufactured. Design research that produces a scientific foundation and helps in the development of new design methods and tools for consumer products as well as for the manufacturing systems that produce these products is a pre-requisite for industrial innovation.

Engineering design research and practice, particularly using systematic and tested methodologies contribute to the industrial innovation of nations by reducing reliance on outdated methods in design practice. It is important to educate a new generation of engineering graduates knowledgeable of good design methods and practices while fostering and developing their creativity. As such design engineering is an enabler for competitive advantage that can fuel the engine of economic growth and wealth creation.

A main theme of this book on Advances in Design emphasizes that the advances in computer technology and the exponential growth of the Internet have created opportunities for local and global communication which were never before possible. The book highlights contributions in the areas of design theories, principles and methodologies, their practical application, the development of computational tools to support distributed design and manufacturing and the whole product life cycle in a changing world.

In the current globally integrated industrial environment it is important to highlight the need for a holistic approach that deals with the entire system, as well as its components. Improvements in individual disciplines/methods are necessary, but not sufficient to effect the improvements in products and processes needed in the future. System level synthesis, analysis, and optimization tools are required. Important aspects for the foreseeable future are advances in collaborative design tools and techniques, functional design knowledge, design synthesis, analysis and optimization, human aspects, system integration tools, design frameworks, information support systems and integration with manufacturing activities. Greater demand for product efficiency, reliability, quality, compactness, variety and customization combined with life-cycle considerations and lower time to market, less cost, and better utilization of energy and natural resources are needed.

The discussion of leading edge research work is organized in this book on Advances in Design, in nine Chapters: Next Generation Design; Design Knowledge and Functional Design; Innovative and Conceptual Design; Design Frameworks; Design Management; Product Life Cycle; Collaborative Engineering Design; Design Intent and Tolerancing; and Modeling and Design for Manufacturing.

This book on Advances in Design is directed at three constituencies: researchers, design practitioners, and educators. Researchers will find latest research results in product design and design methodologies. Design professionals and practitioners will find the current state of engineering design practice, new approaches, methods and their applications. It is also important for educators to include the latest advances and methodologies for design curricula.

We acknowledge the contributions of all authors to this book, and those colleagues who assisted with the review of the original papers submitted and presented at the CIRP (International College of Production Engineering Research) Design Conference in May 2004. This book includes edited selected papers from that Conference.

Hoda ElMaraghy

Waguih ElMaraghy

London, Canada, April 2005

Contents

Part III Innovative and Conceptual Design

Part IV Design Frameworks

Part VII Collaborative Engineering Design

Part VIII Design Intent and Tolerancing

List of Contributors

S. Ahmed
Dept. of Mechanical Engineering,
Technical University of Denmark,
Building 404,
Lyngby, DK-2800, Denmark.
sah@mek.dtu.dk

A. Albers
MKL Institut fur
Maschinenkonstruktionslehre,
Universitat Karlsruhe
Geb/ 10.23 Raum 807
Kaiserstrasse 12, Karlsruhe, 76128,
Germany.
albers@mkl.uni-karlsruhe.de

D.C. Anghel
M3M Laboratory
Université de Technologie de
Belfort-Montbéliard
90010 Belfort Cedex, France.
daniel.constantin.anghel@utbm.fr

E. Arai
Department of Manufacturing
Science, Graduate School of
Engineering, Osaka University,
2-1, Yamadaoka, Osaka-shi, Osaka-
fu, 565-0871, Japan.
arai@mapse.eng.osaka-u.ac.jp

A. Barari
Intelligent Manufacturing Systems
(IMS) Centre, Faculty of
Engineering, University of Windsor,
204 Odette Bldg., 401 Sunset Ave.
Windsor, ON, N9B 3P4, Canada.
abararia@uwo.ca

M. Benassi
Politecnico di Milano
Via La Masa, 34
Milano, I-20158, Italy.
matteo.benassi@kaemart.it

A. Bernard
Institut de Recherche en
Communications et Cybernetique de
Nantes (IRCCyN), Ecole Centrale de
Nantes,
1, rue de la Noë, BP 92101
Nantes, Cedex 3, 44321, France.
Alain.Bernard@irccyn.ec-nantes.fr

M. Bordegoni
Politecnico di Milano
Via La Masa, 34
Milano, I-20158, Italy.
monica.bordegoni@kaemart.it

T. Boudouh
M3M Laboratory, Université de
Technologie de Belfort-Montbéliard,

Belfort, Belfort Cedex, 90010, France.
toufik.boudouh@utbm.fr

D. Brissaud
Laboratoire 3S (Sols, Solides, Structures), University of Grenoble, BP 53
Grenoble Cedex 9, 38041, France.
daniel.brissaud@hmg.inpg.fr

N. Burkardt
MKL Institut fur Maschinenkonstruktionslehre, Universitat Karlsruhe,
Geb/ 10.23 Raum 807
Kaiserstrasse 12, Karlsruhe, 76128, Germany.
burkardt@mkl.uni-karlsruhe.de

E. Caillaud
LICIA, INSA - Strasbourg
24 Boulevard de la Victoire
Strasbourg, 67084 Cedex, France.
emmanuel.caillaud@ipst-ulp.u-strasbg.fr

M. Cantamessa
Departimento di Sistemi di Produzione ed Economia dell'Azienda, Politecnico di Torino,
Corsco Duca degli Abruzzi 24-I,
Torino, 10129, Italy.
marco.cantamessa@polito.it

D. Cavallucci
Laboratoire d'Ingenierie de Conceptioin, Cognition, Intelligence Artificielle - Institut National des Sciences Appliquees (LICIA-INSA)
Strasbourg, 24 Boulevard de la Victoire, Strasbourg, 67084 Cedex, France.
cavallucci@insa-strasbourg.fr

H. Chen
Shandong University
1 Key CAD Lab., No73, Jingshi Rd.
Jinan, Shandong, 250061, China.
hongwuc@sdu.edu.cn

K.Z. Chen
Department of Mechanical Engineering, The University of Hong Kong
Pokfulam Road, Hong Kong, China.
kzchen@hkucc.hku.hk

X.C. Chen
School of Mechanical Engineering, Donghua University
1882, West Yan-An Road, Shanghai, 200051, China.
cimspaper@yeah.net

N. Chevassus
Corporate Research Center, European Aeronautic Defence and Space Company Corporate Research Centre (EADS CCR), 12 rue Pasteur BP 76,
Suresnes Cedex, 92152, France.
nicolas.chevassus@eads.net

D. Choulier
Laboratoire M3M (EA 3318), CID, Université de Technologie de Belfort-Montbéliard,
Belfort, Belfort Cedex, 90010, France.
denis.choulier@utbm.fr

P.J. Clarkson
Department of Engineering, University of Cambridge, Cambridge Engineering Design Centre, Trumpington St.
Cambridge, CB2 1PZ, U.K..

L. Clausson
Dept. of Production Engineering School of Industrial Engineering and Management, Royal Institute of Technology (KTH)

Brinellvagen 66
SE-100 44 Stockholm, Sweden.
staket@telia.com
leif.clausson@iip.kth.se

U. Cugini
Dipartimento di Meccanica,
Politecnico di Milano, Via La Masa
34, Milano, 20156, Italy.
umberto.cugini@kaemart.it

A. Deif
Intelligent Manufacturing Systems
(IMS) Centre, Faculty of
Engineering, University of Windsor,
204 Odette Bldg., 401 Sunset Ave.,
Windsor, ON, N9B 3P4, Canada.
deif@uwindsor.ca

A. Desrochers
Universite de Sherbrooke
2500, boul. de l'Universite
Sherbrooke, Quebec, J1K 2R1,
Canada.
Alain.Desrochers@USherbrooke.ca

G. Dragoi
Universitatea Politehnica din
Bucuresti, 313 Splaiul
Independentei,
Bucharest sector 6, 77206 Romania.
gdragoi@canad.ro

C. Eckert
Department of Engineering,
Cambridge Engineering Design
Centre, University of Cambridge,
Trumpington St.
Cambridge, CB2 1PZ, U.K.

E. ElBeheiry
Intelligent Manufacturing Systems
(IMS) Centre, Faculty of
Engineering, University of Windsor,
204 Odette Bldg., 401 Sunset Ave.,
Windsor, ON, N9B 3P4, Canada.
elbeheiry@yahoo.com

H. ElMaraghy
Intelligent Manufacturing Systems
(IMS) Centre, Faculty of
Engineering, University of Windsor,
204 Odette Bldg., 401 Sunset Ave.,
Windsor, ON, N9B 3P4, Canada.
hae@uwindsor.ca

W. ElMaraghy
Intelligent Manufacturing Systems
(IMS) Centre, Faculty of
Engineering, University of Windsor,
204 Odette Bldg., 401 Sunset Ave.,
Windsor, ON, N9B 3P4, Canada.
wem@uwindsor.ca

T. Eltzer
Laboratoire d'Ingenierie de
Conceptioin, Cognition, Intelligence
Artificielle - Institut National des
Sciences Appliquees (LICIA-INSA)
Strasbourg
24 Boulevard de la Victoire,
Strasbourg, 67084 Cedex, France.
eltzer@yahoo.fr

H. Falgarone
European Aeronautic Defence and
Space Company Corporate Research
Centre (EADS CCR), Corporate
Research Center,
12 rue Pasteur BP 76
Suresnes Cedex, 92152, France.
hugo.falgarone@eads.net

X.A. Feng
School of Mechanical Engineering,
Dalian University of Technology
Dalian
Liaoning Province, 116024, China.
xinanf@dlut.edu.cn

U. Frank
Heinz Nixdorf Institute,
Fürstenallee 11,
Dr. Roerig-Damm 98, Paderborn, D-
33102, Germany.

ursula.frank@hni.upb.de

R.Y.K. Fung
City University of Hong Kong
83 Tat Chee Avenue
Kowloon, Hong Kong, China.
richard.fung@cityu.edu.hk

G. Cascini
Dipartimento di Meccanica e Tecnologie Industriali, University of Florence, Universita degli Studi di Firenze, via S.Marta, 3
Firenze, FI, 50139, Italy.
gaetano.cascini@unifi.it

J. Gao
LSC Group, Cranfield University
Cranfield, Bedford, MK43 OAL, U.K.
x.gao@cranfield.ac.uk

O. Garro
M3M Laboratory, Université de Technologie de Belfort-Montbéliard
Belfort, Belfort Cedex, 90010, France.
olivier.garro@utbm.fr

J. Gausemeier
Heinz Nixdorf Institute,
Fürstenallee 11,
Dr. Roerig-Damm 98, Paderborn, D-33102, Germany.
jurgen.gausemeier@hni.upb.de

W. Ghie
Université du Québec à Trois-Rivières, 3351 Des Forges Blvd.
Trois-Rivieres, Quebec, G9A5H7, Canada.
walid_ghie@uqtr.ca

M. Giordano
Universite de Savoie
Laboratoire de Mécanique Appliquée (LMécA) ESIA, Domaine Universitaire, BP 806, Annecy Cedex, 74016, France.
max.giordano@esia.univ-savoie.fr

P. Girard
Laboratoire LAP GRAI, UMR 5131 CNRS, Universite Bordeaux I
Cours de la Liberation, 351, France.
girard@lap.u-bordeaux1.fr

P. Gu
Department of Mechanical & Manufacturing Engineering, The University of Calgary
Calgary, AB, T2N 1N4, Canada.
gu@enme.ucalgary.ca

J. Gustafsson
Royal Institute of Technology - KTH
Brinellvagen 66
Stockholm, SE 100 44, Sweden.
jg@iip.kth.se

A. Hamdi
Laboratoire Génie Industriel, Ecole Centrale Paris, Grande voie des vignes, F92295 Chatenay, Malabry Cedex, France.
abdelbasset.hamdi@lgi.ecp.fr

G. Harmel
Laboratoire Génie Industriel
Ecole Centrale Paris,
F92295 Chatenay, Malabry, France.
ghassen.harmel@lgi.ecp.fr

H. Hayka
Systems and Design Technology, Fraunhofer Institute for Production
Pascalstrasse 8-9
Berlin, D-10587, Germany.
haygazun.hayka@ipk.fhg.de

P. Hernandez
Universite de Savoie
LMecA, Ecole Superieure d'Ingenieurs d'Annecy, 5, Chemin de

Bellevue, Annecy Le Vieux, BP 806,
Annecy Cedex, 74016, France.
pascal.hernandez@esia.univ-savoie.fr

K.Z. Huang
Shandong University
1 Key CAD Lab., No73, Jingshi Rd.
Jinan, Shandong, 250061, China.
huangkz@sdu.edu.cn

T.A.W. Jarratt
Cambridge Engineering Design Centre, Department of Engineering, University of Cambridge
Trumpington St.
Cambridge, CB2 1PZ, U.K.
tawj2@eng.cam.ac.uk

C.I.V. Kerr
Enterprise Integration, School of Industrial and Manufacturing Science, Cranfield University
Building 53,
Cranfield, Bedford, England, MK43 0AL, U.K..
r.roy@cranfield.ac.uk

N. Khomenko
Laboratoire d'Ingenierie de Conceptioin, Cognition, Intelligence Artificielle - Institut National des Sciences Appliquees (LICIA-INSA) Strasbourg
24 Boulevard de la Victoire,
Strasbourg, 67084 Cedex, France.
jl-project@trizmink.org

A. Kjellberg
Royal Institute of Technology - KTH
Osterskarsvagen 84, Osterskar,
Stockholms Län, 18451, Sweden.
ann.kjellberg@iip.kth.se

T. Kjellberg
Royal Institute of Technology - KTH
Brinellvagen 66
Stockholm, SE 100 44, Sweden.
tk@iip.kth.se

J.A. Knowlton
Rensselaer Polytechnic Institute (RPI), USA
1213 Jefferson Pike
Knoxville, MD, 20817, USA.
knowlton@rpi.cat.edu

F.L. Krause
Systems and Design Technology, Fraunhofer Institute for Production,
Pascalstrasse 8-9
Berlin, D-10587, Germany.
frank-l.krause@ipk.fhg.de

E. Landel
Génie Industriel, Laboratoire Grande voie des vignes, Ecole Centrale Paris, F92295 Chatenay, Malabry Cedex, France.
eric.landel@lgi.ecp.fr

L. Laperrière
Université du Québec à Trois-Rivières, 3351 Des Forges Blvd
Trois-Rivieres, Quebec, G9A 5H7, Canada.
Luc_Laperriere@uqtr.ca

M. Larsson
Royal Institute of Technology - KTH
Brinellvagen 66
Stockholm, SE 100 44, Sweden.
ml@iip.kth.se

G. Legrais
Universite de Savoie
LMecA, Ecole Superieure d'Ingenieurs d'Annecy, 5, Chemin de Bellevue, Annecy Le Vieux, BP 806, Annecy Cedex, 74016, France.
geatan.legrais@esia.univ-savoie.fr

C. Lindfors
Royal Institute of Technology - KTH

Brinellvagen 66
Stockholm, SE 100 44, Sweden.
cl@iip.kth.se

M. Lombard
CRAN-UMR 7039
Faculy of Sciences and Techniques - BP 239, Vandoeuvre-les-Nancy, Lorraine, 54506, France.
muriel.lombard@cran.uhp-nancy.fr

P. Lonchampt
Laboratoire 3S (Sols, Solides, Structures), BP 53, University of Grenoble,
Grenoble Cedex 9, 38041, France.
pierre.lonchampt@hmg.inpg.fr

E. Lutters
University of Twente
CTW Z202, P.O. Box 217, AE Enschede, 7500, The Netherlands.
e.lutters@utwente.nl

P. Lutz
LAB, University Université de Franche Comté
24 Rue Alain Savary
Besancon, 25000, France.
plutz@ens2m.fr

L. Lv
Shandong University
1 Key CAD Lab., No73, Jingshi Rd.
Jinan, Shandong, 250061, China.
liangminl@sdu.edu.cn

M. Mauchand
Institut de Recherche en Communications et Cybernetique de Nantes (IRCCyN), Ecole Centrale de Nantes, 1, rue de la Noë, BP 92101
Nantes, Cedex 3, 44321, France.
magali.mauchand@irccyn.ec-nantes.fr

B.R. Meijer
Department of Mechanical Engineering, Delft University of Technology, Mekelweg 2,
Delft, CD Delft, 2628, The Netherlands.
b.r.meijer@wbmt.tudelft.nl

M. Milanesio
Departimento di Sistemi di Produzione ed Economia dell'Azienda, Politecnico di Torino
Corsco Duca degli Abruzzi 24-I,
Torino, 10129, Italy.
maurizio.milanesio@polito.it

M. Montero
Mechanical Engineering
Berkeley Manufacturing Institute,
University of California at Berkeley
2111 Etcheverry Hall
Berkeley, CA 94720 USA.
montero@kingkong.me.berkeley.edu

R. Movahed Khah
Laboratoire de Recherche M3M
Université de Technologie Belfort-Montébliard
Belfort, Belfort Cedex, 90010, France.
reza.movahedkhah@utbm.fr

J. Niemann
University of Stuttgart, Fraunhofer IPA, Nobelstrasse 12
Stuttgart, 70569, Germany.
jon@iff.uni-stuttgart.de

M. Ohmer
MKL Institut fur Maschinenkonstruktionslehre,
Universitat Karlsruhe, Geb/ 10.23
Raum 807
Kaiserstrasse 12, Karlsruhe, 76128, Germany.
ohmer@mkl.uni-karlsruhe.de

K. Ohtomi
Corporate Research & Development Center, Toshiba Corporation
1, Komukai-Toshiba-cho,
Saiwai-ku, Kawasaki, 212-8582,
Japan.
koichi.ootomi@toshiba.co.jp

E. Operti
Departimento di Sistemi di Produzione ed Economia dell'Azienda, Politecnico di Torino
Corsco Duca degli Abruzzi 24-I,
Torino, 10129, Italy.
elisa.operti@polito.it

E. Ostrosi
Laboratoire de Recherche M3M,
Université de Technologie Belfort-Montébliard,
Belfort, Belfort Cedex, 90010,
France.
egon.ostrosi@utbm.fr

M. Paouliquen
Institut de Recherche en Communications et Cybernetique de Nantes (IRCCyN), Ecole Centrale de Nantes, 1, rue de la Noë, BP 92101
Nantes, Cedex 3, 44321, France.
mamy.paouliquen@irccyn.ec-nantes.fr

M.W. Park
CADCAM Research Centre, Korea Institute of Science and Technology,
Hawolgok, Sungbuk,
Seoul, 136-791, Korea.
myon@kist.re.kr

B. Pasewaldt
Fraunhofer Institute for Production Systems and Design Technology
Pascalstrasse 8-9
Berlin, D-10587, Germany.
bernhard.paswealdt@ipk.fgh.de

H.J. Pels
Technische Universiteit Eindhoven
TM-I&T, Postus 513, 5600 MB
Eindhoven, The Netherlands.
H.J.Pels@tm.tue.nl

E. Pena
Laboratoire M3M (EA 3318), CID,
Université de Technologie de Belfort-Montbéliard
Belfort, Belfort Cedex, 90010,
France.
Ezio.Pena@UTBM.fr

N. Perry
Institut de Recherche en Communications et Cybernetique de Nantes (IRCCyN), Ecole Centrale de Nantes, 1, rue de la Noë, BP 92101
Nantes, Cedex 3, 44321, France.
nicolas.perry@irccyn.ec-nantes.fr

J.P. Petit
Laboratoire de Mécanique Appliquée (LMécA) ESIA,
Universite de Savoie, 806,
Annecy Cedex, 74016, France.
Jean-Philippe.Petit@univ-savoie.fr

K. Pinapunsri
Faculty Laboratoire 3S (Sols, Solides, Structures), Institut National Polytechnique de Grenoble,
Domaine Universitaire, BP 53
Grenoble Cedex 9, 38041, France.
kusol.pinapunsri@hmg.inpg.fr

G. Prudhomme
Laboratoire 3S (Sols, Solides, Structures), University of Grenoble,
BP 53
Grenoble Cedex 9, 38041, France.
guy.prudhomme@hmg.inpg.fr

B. Radulescu
Faculty Laboratoire 3S (Sols, Solides, Structures), Institut National

Polytechnique de Grenoble,
Domaine Universitaire, BP 53
Grenoble Cedex 9, 38041, France.
bruno.radulescu@hmg.inpg.fr

A. Ramelli
Dipartimento di Meccanica,
Politecnico di Milano
Via La Masa 34, Milano, 20156,
Italy.
andrea.ramelli@kaemart.it

C. Rizzi
Universita Bergamo
Viale A. Marloni, 5
24044 Dalmine (B6), Italy.
caterina.rizzi@unibg.it

V. Robin
Laboratoire LAP GRAI,
UMR 5131 CNRS, Universite
Bordeaux I
Cours de la Liberation, 351, France.
robin@lap.u-bordeaux1.fr

B. Rose
CRAN-UMR 7039
Faculy of Sciences and Techniques -
BP 239, Vandoeuvre-les-Nancy,
Lorraine, 54506, France.
bertrand.rose@cran.uhp-nancy.fr

R. Roy
Enterprise Integration, School of
Industrial and Manufacturing
Science, Cranfield University
Building 53,
Cranfield, Bedford, England, MK43
0AL, U.K..
r.roy@cranfield.ac.uk

P.J. Sackett
Enterprise Integration, School of
Industrial and Manufacturing
Science, Cranfield University
Building 53,
Cranfield, Bedford, England, MK43
0AL, U.K..
r.roy@cranfield.ac.uk

S. Samper
Laboratoire de Mécanique
Appliquée (LMécA) ESIA,
Universite de Savoie, Domaine
Universitaire, BP 806, Annecy
Cedex, 74016, France.
serge.samper@esia.univ-savoie.fr

J.C. Sand
VECO Canada Ltd.,
401 - 9th Avenue SW
Calgary, Alberta, T2P 3C5, Canada.
Jeff.Sand@veco.com

A. Schmidt
Heinz Nixdorf Institute,
Fürstenallee 11,
Dr. Roerig-Damm 98, Paderborn,
D-33102, Germany.
Andreas.Schmidt@hni.upb.de

J. Shah
Fulton School of Engineering
Arizona State University
Tempe, AZ, 96398-6106, USA.
jami.shah@asu.edu

M. Shahrokhi
Institut de Recherche en
Communications et Cybernetique de
Nantes (IRCCyN), Ecole Centrale de
Nantes, 1, rue de la Noë, BP 92101
Nantes, Cedex 3, 44321, France.
mahmoud.shahrokhi@irccyn.ec-nantes.fr

R. Sharma
LSC Group, Cranfield University
Cranfield, Bedford, MK43 OAL,
U.K.
lt_rohit_sharma@hotmail.com

K. Shirase
Department of Mechanical Engineering, Faculty of Engineering, Kobe University,
1-1,rokkoudai,nada-ku, Kobe-shi, Hyougo-ken, 657-8501, Japan.
shirase@mech.kobe-u.ac.jp

G. Sohlenius
Royal Institute of Technology - KTH
Osterskarsvagen 84, Osterskar, Stockholms Län, 18451, Sweden.
sohl@iip.kth.se

Y.T. Sohn
CADCAM Research Centre, Korea Institute of Science and Technology, Hawolgok, Sungbuk
Seoul, 136-791, Korea.
young@kist.re.kr

Z. Song
Shandong University
1 Key CAD Lab., No73, Jingshi Rd.
Jinan, Shandong, 250061, China.
zhenjuns@sdu.edu.cn

D. Steffen
Heinz Nixdorf Institute,
Fürstenallee 11,
Dr. Roerig-Damm 98, Paderborn,
D-33102, Germany.
daniel.steffen@hni.upb.de

K. Takeuchi
Fujitsu info Software technologies Ltd., The second development division, The second development department Project Director, Southpot Shizuoka Bldg,18-1,Minami-cyo,suruga-ku, Shizuoka-shi, Shizuoka-ken, 422-8572, Japan.
ktakeuchi@jp.fujitsu.com,
takeuchi@ist.fujitsu.com

S. Tichkiewitch
Faculty Laboratoire 3S (Sols, Solides, Structures), Institut National Polytechnique de Grenoble, Domaine Universitaire, BP 53
Grenoble Cedex 9, 38041, France.
serge.tichkiewitch@inpg.fr

M. Tideman
Laboratory of Design, Production & Management, Faculty of Engineering Technology, University of Twente, Drienerlolaan 5, 7522 NB, Enschede, The Netherlands.
m.tideman@utwente.nl

T. Tomiyama
Department of Mechanical Engineering and Marine Technology, Delft University of Technology, Mekelweg 2,
Delft, CD Delft, 2628, The Netherlands.
t.tomiyama@wbmt.tudelft.nl

A. Tsumaya
Department of Manufacturing Science, Graduate School of Engineering, Osaka University, 2-1,Yamadaoka
Osaka-shi, Osaka-fu, 565-0871, Japan.
tsumaya@mapse.eng.osaka-u.ac.jp

M. Ugolotti
Dipartimento di Meccanica, Politecnico di Milano,
Via La Masa 34
Milano, 20156, Italy.
marco.ugolotti@kaemart.it

R.J. Urbanic
Intelligent Manufacturing Systems (IMS) Centre, Faculty of Engineering, University of Windsor, 204 Odette Bldg., 401 Sunset Ave.

Windsor, ON, N9B 3P4, Canada.
rjurbanic@sympatico.ca

M.C. van der Voort
Laboratory of Design, Production & Management, Faculty of Engineering Technology, University of Twente, Drienerlolaan 5, 7522 NB, Enschede, The Netherlands.
m.c.vandervoort@utwente.nl

F.J.A.M. van Houten
Laboratory of Design, Production & Management, Faculty of Engineering Technology, University of Twente, Drienerlolaan 5, 7522 NB, Enschede, The Netherlands.
f.j.a.m.vanhouten@utwente.nl

N. Vargas
Fulton School of Engineering
Arizona State University
Tempe, AZ, 96398-6106, USA.
noe.vargas@asu.edu

H. Wakamatsu
Department of Manufacturing Science, Graduate School of Engineering, Osaka University, 2-1,Yamadaoka, Osaka-shi, Osaka-fu, 565-0871, Japan.
wakamatu@mapse.eng.osaka-u.ac.jp

K.M. Wallace
Engineering Design Centre, University of Cambridge, Trumpington Street, Cambridge, Cambridgeshire, CB3 0HH, U.K.

Y. Wang
Shandong University
1 Key CAD Lab., No73, Jingshi Rd.
Jinan, Shandong, 250061, China.
yandongw@sdu.edu.cn

E. Westkämper
Fraunhofer IPA, University of Stuttgart, Nobelstrasse 12, Stuttgart, 70569, Germany.
wke@ipa.fhg.de

M.J. Wozny
Rensselaer Polytechnic Institute (RPI), USA
1213 Jefferson Pike
Knoxville, MD, 20817, USA.
Wozny@rpi.cat.edu

P. Wright
Mechanical Engineering Dept., University of California, 5133 Etcheverry Hall
Berkeley, CA 94720-1740, USA.
pwright@me.berkeley.edu

B. Yannou
Laboratoire Génie Industriel
Grande voie des vignes, Ecole Centrale Paris,
F92295 Chatenay, Malabry Cedex, France.
bernard.yannou@lgi.ecp.fr

L. Yi
Shandong University
1 Key CAD Lab., No73, Jingshi Rd.
Jinan, Shandong, 250061, China.
yil@sdu.edu.cn

Part I

Next Generation Design

1

Economic Growth, Business Innovation and Engineering Design

Gunnar Sohlenius, Leif Clausson, and Ann Kjellberg

Abstract: Scientific knowledge of engineering within innovative industrial decision processes has a great potential to improve quality and productivity in industrial operations and hence improve profitability. This is a precondition for economic growth, which in turn is necessary to improve welfare. Innovative processes have to combine creativity with quality and productivity in order to achieve profitability and growth. The most important ways to improve profitability in industrial production are through an improved ability to meet more advanced requirements in new products and processes by using new knowledge and inventions and higher productivity through investments in more advanced and automatic tools. This is the fundamental mechanism behind industrial production seen as an engine of welfare. Besides the real world of the products and the production processes, the mechanisms for this development can be classified into three worlds. These are the *decision world*, the *human world* and the *model world*. In striving to obtain increased welfare through industrial production, fundamental knowledge about these worlds and about their relations to the products and processes has to be developed. This paper is a contribution to this understanding, which is necessary in order to combine Total Quality Management, (TQM) and Total Productivity Management (TPM) into Total Effective Management (TEM) by understanding Means.

Keywords: Economic growth, Innovation process, Decision theory

1.1 Wealth and Conditions for Stable Economic Growth

Economic growth is a pre-condition for increasing wealth and quality of life. However, growth, as it is usually defined, is not totally equivalent to wealth and quality of life. Focus should be placed on more sharing of what is produced, that is, to better distribution between the rich and the poor, while working to produce enough for human demands for a better balance with the Earth's carrying capacity. More attention should be given to qualitative rather than quantitative growth and to having more variety, complexity, diversity and customization. Therefore customized, order–based production is of central importance [1]. In order to obtain this, products and processes have to be developed concurrently [2].

However, economic growth cannot be taken for granted. The IT-economy, for example, which is based more upon expectations than profitability, has caused an enormous economic shrinkage because the pre-conditions for economic growth were not carefully considered.

Reflections on wealth and quality of life and their profound dependence on the effectiveness of industrial production, is necessary. People have different interests and preferences. Therefore, quality of life has, on the one hand, some common features and, on the other, many individual different components. Freedom to choose a place to live, education, profession and family, and to preserve a good outer and inner environment is a common ambition, which is fundamental to quality of life. Basic needs are also common, such as home, food, clean water, clothes, healthcare, safety and education. Above all, people are very individual, some have a profound interest in nature and hiking; others are sportsmen; some have humanistic interest in music, literature and arts. Some people want to travel and enjoy different environments, and others are collectors, some people are very materialistic and enjoy nice cars, boats and homes.

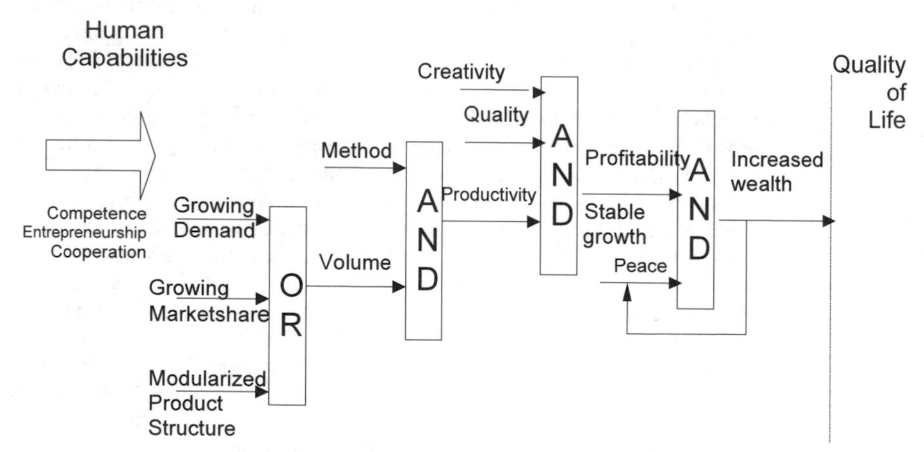

Figure 1.1 The logic of increased quality of life through economic growth

It is important to appreciate that all is dependent on having high–quality, productive technical systems in order to be obtainable and affordable. This includes systems for energy, transportation, telecommunication, information, health care, infrastructural and industrial production. These systems have to develop and produce products, which satisfy people functionally and aesthetically, provide work opportunities and an economic system that can act as an engine making everything that we appreciate, such as quality of life possible. In order to extend quality of life further, it is important to obtain economic growth through effectiveness in the industrial sector as well as in the public sector.

Consider the logic of increased quality of life through economic growth, as illustrated in Figure 1.1. Increased wealth is considered to increase quality of life. This is true if the resources we create meet our requirements and expectations and hence, add to our quality of life. This is exactly what is meant by *quality.* In other words, economic growth is connected to the ability to satisfy human needs. This requires ability to understand and to define relevant human needs. The resources to be spent in order to obtain quality also make sense. To achieve quality with minimal resources is fundamental. This is what is meant by *productivity.* Deviations from this are considered waste.

Therefore, the concepts of *quality and productivity* have to be used as guiding parameters in the management of product development for increased wealth. As stressed above, quality and productivity are of prime importance in view of the limited resources on earth. This is what is meant by *effectiveness.*

Quality and productivity are, however, not obtained automatically - they require *creativity* and engineering skills. A balanced combination of creativity and skills to obtain quality and productivity can result in profitability and stable economic growth. These conditions form the base for increased wealth in terms of quality of life. *Total Effectiveness Management* by understanding *Means* is therefore extremely important.

Increased wealth also requires peace. Generated wealth should not be destroyed again through war. Warfare is in general caused by threat and fear.

Productivity requires the development and choice of production methods, which provide reliable products with correct functions at lowest cost. Cost per product depends on volume - larger volume means better productivity. Volume in turn can be obtained through growing market demands, increased market shares and a smart modularized product structure.

H. Thomas Johnson and Anders Bröms [3] have pointed out that an industrial production system is a living system, which has to be *Managed by Means*. That is to say that management by means must be based upon an understanding of the fundamental conditions shown in Figure 1.1. To the left of the figure, we can see *Human capabilities (competence, entrepreneurship* and *cooperation).* To the right, we can obtain *Quality of life.* In between, there are the items to be designed, produced and managed in order to succeed. This understanding is fundamental to the concept of *Management by Means.*

There must be distinction between motivation for the growth of individual companies and that for the growth of a national or global economy. Without economic growth on the national and global level, increased wealth and welfare are impossible. At a company level, growth might not be necessary or desirable. Wealth and welfare mean a healthy community and a healthy environment. Hence, economic growth through effectiveness is fundamentally important.

In conclusion, profitability is a necessary condition for growth. The principles that guide natural systems, as Johnson and Bröms point out, are *self-definition, organization, interdependence* and *diversity* - these principles are mandatory. Skills in creativity, quality and productivity determine competitiveness and thus strength for survival and contribution to welfare.

The Swedish Industrial Fund, which financially supports the start-up of hundreds of new companies in Sweden, estimated in 2003 that 50% of them were based on innovation projects, 25% on new education and 25% on increased capital/employee.

Work, capital and *new technology* are the main economic factors. During a recent Swedish meeting in the Engineering Academy their relative importance was estimated to be 20%, 28% and 52%, respectively. The effect of new technology is considered to have the same weight on economic growth as the sum of both work and capital. This leads to the following conclusions:

- Increased wealth requires economic growth in the total economy.
- Profitability is more important than economic growth at the company level.
- It is necessary to adapt to new conditions in technology, markets and regulations in order to remain profitable.
- Products meeting more basic human needs have a more stable profitability but require high competence and skill in the supply, production and marketing processes using automation and IT.
- Products meeting high level human needs can be very profitable but they involve a higher risk and require high competence and skill for rapid change in many dimensions.

1.2 Total Effectiveness Management by Understanding Means

Profitability and economic growth require a management by means approach with:

- high competence in science–based engineering as a base for creativity,
- ability to understand and define market needs,
- ability to make decisions about products and processes in the innovation process leading to quality and productivity,
- ability to develop and manage competence.

These points are very important for *Total Effectiveness Management by Understanding Means*.

The development or innovation process is a fundamental sub-process within industrial production. This process is most clearly connected to decision-making. The decisions taken in the innovation process have a fundamental impact on profitability. This is the reason for examining this sub-process more thoroughly.

First, what we need and appreciate as customers and users of industrial products are considered and divided into the following points:

• Which functions do we need or appreciate?	*Functionality*
• How do we operate, *talk to,* the product?	*Human interface*
• How much do we have to spend in life-cycle cost?	*Purchase, operation, maintenance and recycling*

- How does the product communicate, (*talk to us),* *Styling emotionally*

When developing products technical possibilities have to be used to meet real users´ needs and expectations. This development can be technology driven or market driven – we need both. With the great new possibilities information technology provide the limits of technical possibilities cannot be foreseen. Information technology is the prime driver of new functions in almost all technical products, not only computers and telecommunication, but also cars, white wares, airplanes, homes and also business and manufacturing systems.

This development has been mostly technology driven by engineers. Sometimes the products contain more functions than needed. Very often these products are difficult to understand and operate. Modern mobile telephones, some digital cameras, video recorders and sometimes even cars are but few examples. In order to improve such products competence in cognitive psychology and ergonomics must be included in the industrial innovation processes as well as in engineering education.

It is also necessary to have good interaction with potential customers and users when developing new products. Much more effort and stringency have to be used in defining the functional requirements of products. The functional requirements have to be, as much as possible, specified independent of each other and with target values and allowed deviations (*tolerances*). When dealing with functionality, much more attention should be paid to the design of human interfaces. The communication must be designed according to natural cognitive human abilities.

Products are very often sold on emotions, especially when functions and prices are the same, the styling might be the most important selling factor. Functions are mostly considered self-evident and not especially noticed. Naturally, big dissatisfaction results due to functional failures. Unexpected functions can also create great emotional surprise. Styling and colour greatly affect our emotions. These factors are often crucial for the selling of the product. It is obvious that styling has to be developed in close contact with potential customers keeping in mind that the taste is different in different cultures and human generations, as well as from one individual to another.

This is natural also from a scientific point of view. Functional qualities can, as long as the requirements are carefully defined, be dealt with objectively and scientifically. It is only a matter of the features of the product itself. The effects of styling and colour, on the other hand, are emotional and subjective. This is a matter of the features of the product in combination with the reaction of the user. Styling and colour are of course objective features of the product, but their effect or function depends on and interplays with the emotional reactions of each individual user. This is an important distinction. Therefore, the following conclusions can be drawn:

- Functional requirements have to be defined in close contact with potential users and can be handled objectively and scientifically.
- The functional requirements of the human interface have to be defined based upon good knowledge of cognitive psychology and ergonomics.

- Styling of the products has a decisive importance for success. It is not only a matter of features of the product but also the users′ personal emotional character.

In spite of the importance of styling in innovation processes and engineering education, we now concentrate on functional engineering design. Many companies are still working in a *Build, Test and Fix* (BTF) mode, instead of working in a *Requirement, Concept and Improvement* (RCI) cycle. They are not successful in getting it right from the start, which results in deficient quality and low productivity in both the innovation process and the production process. In the long run this will also result in an inefficient engine of wealth and welfare [12].

A central issue within TEM should be to achieve the RCI cycle [22]. This requires a close co-operation, knowledge exchange and joint decisions between different occupational groups, traditionally belonging to different departments. Moreover, an understanding of the fundamental principles for decisions in the innovation and the production processes is required to provide transparency in the processes. This transparency, in combination with the right competences, could enable joint decision-making and result in quality and productivity in the processes.

With this background, the importance of a supporting system for good decision-making becomes obvious. This leads us to the following research question:

- Is it possible to define a theory for decision-making in the innovation process? A theory that especially addresses how to make decisions about the design of products and processes being developed.

1.3 Innovation and Production

After an ideation phase, all human activities have to consider preparation first then execution. In industrial processes, preparation is innovation and execution is production and distribution (Figure 1.2). In the innovation process, products and production systems are created. The production system is initiated from customer orders and cultivates material to carry functions appreciated by the customers.

This can also be more accurately presented with the SADT methodology, as shown in Figure 1.3. The overall functional requirements of the industrial process are:

- Innovate (develop product and production system) (FR1)
- Produce Data (to specify product and process in order to control production) (FR2)
- Produce Product (meeting market and customer requirements) (FR3)

The following three design parameters are required to meet these functional requirements:

- An Innovation System (DP1)
- A Planning System (Order-handling, CADCAM, Planning, Routing and Logistics) (DP2)

- A Manufacturing System (Machining, Forming, Heat treatment, Assembly, Painting, Delivery, Service, *etc.*.) (DP3)

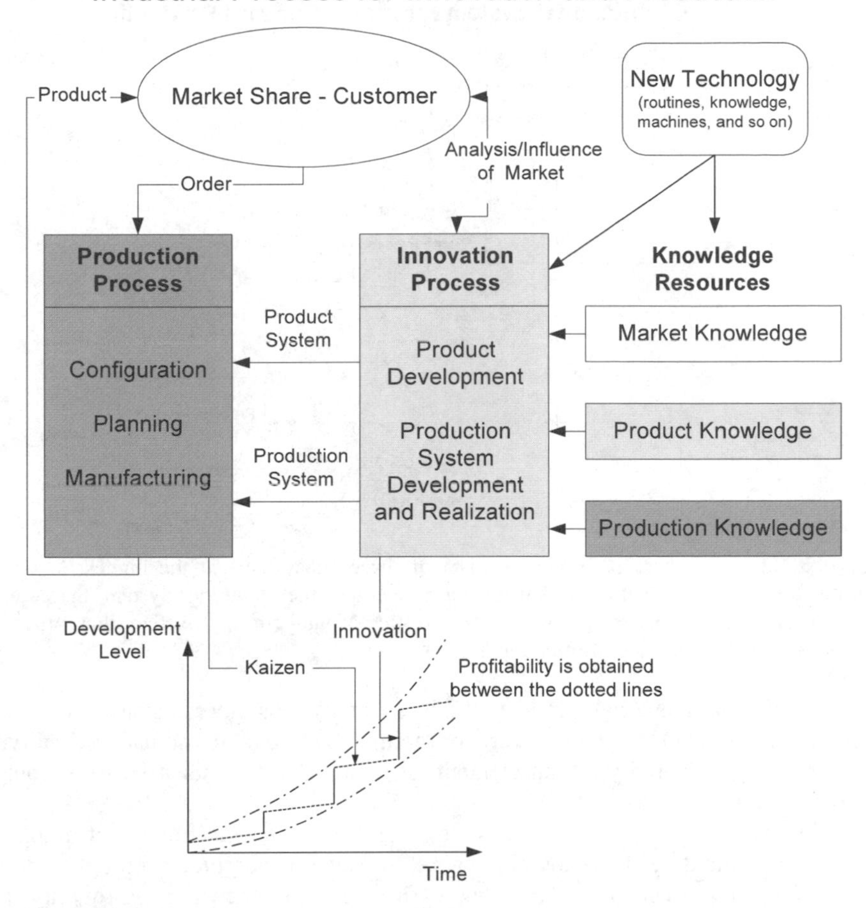

Figure 1.2 The Industrial Process contains the innovation and the production processes

The Manufacturing System together with the Planning System forms the Production System. The Product Model and the Prototype, the Process Model, the Planning System and the Manufacturing System have to be produced by the Innovation System. The Manufacturing System that meets Ordered Requirements and is controlled by the Planning System produces the products to be delivered to customers (Figure 1.3).

Innovation is primarily an information and knowledge development process, which employs the cognitive and visionary creative abilities. This requires specific competence and can be enhanced by problem solving tools. Production is an action that is very much dependent on human emotions; it has to be stimulated by leadership and company culture to develop competence, confidence, interest, belonging and joy. However, action without good and appropriate knowledge is waste and can even be dangerous. Knowledge without action is also wasted in this perspective.

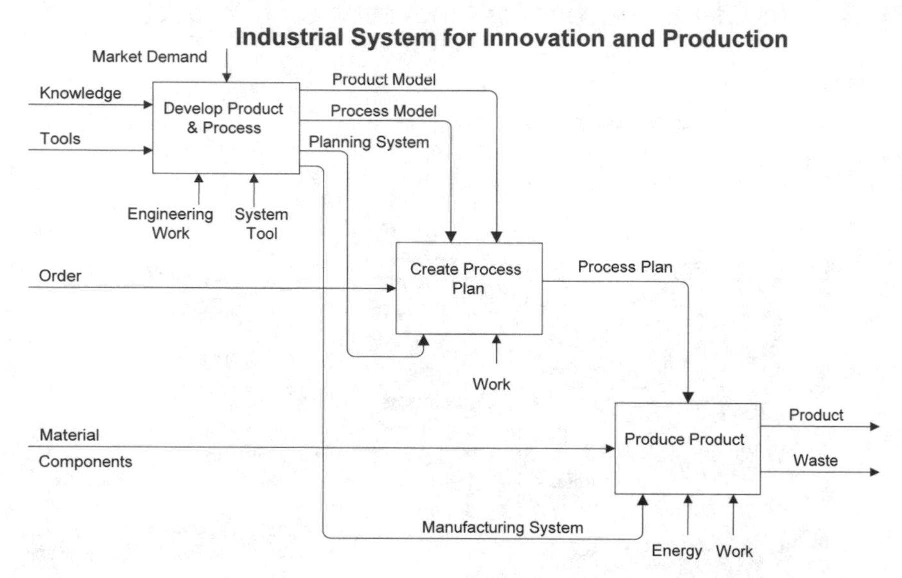

Figure 1.3 The Industrial System consists of three subsystems at the highest level: the Innovation system develops products and processes, the Planning System that creates process plans with orders as an input, and the Manufacturing System that processes materials, with energy and information, into products.

Sustainable industrial production must satisfy customers, shareholders and employees without harming nature or being hazardous to humans. Creativity, quality and productivity are all elements in a strategy for industrial development and production.

Understanding *Means* [3] has to start with an accurate definition of requirements and needs. Innovation processes have to be guided through accurate and competent decisions based upon firm criteria focusing on quality and productivity [4]. Competent decisions are demanding and require strict and strategic competence management.

Good decisions require defined goals. In industrial innovation processes, goals must be defined based upon knowledge about customers′ needs as well as expectations from shareholders and employees. If decisions can be based more on scientific theory and less on belief and heuristic experiences, it should be possible to increase creativity and effectiveness as well as the ability to meet defined requirements.

The Inherent Logic of the Innovation Process

Let us now try to find out the logical nature of innovation processes. In addition to the real world, there are three complementary worlds, which are essential in an innovation process. These worlds are the *model world*, the *decision world* and the *human competence world* (Figure 1.4).

The decisions in the innovation process directly influence the result. The decision criteria, therefore, have to be directly related to quality and productivity and as much as possible based upon criteria of an axiomatic nature, which provide a solid scientific basis.

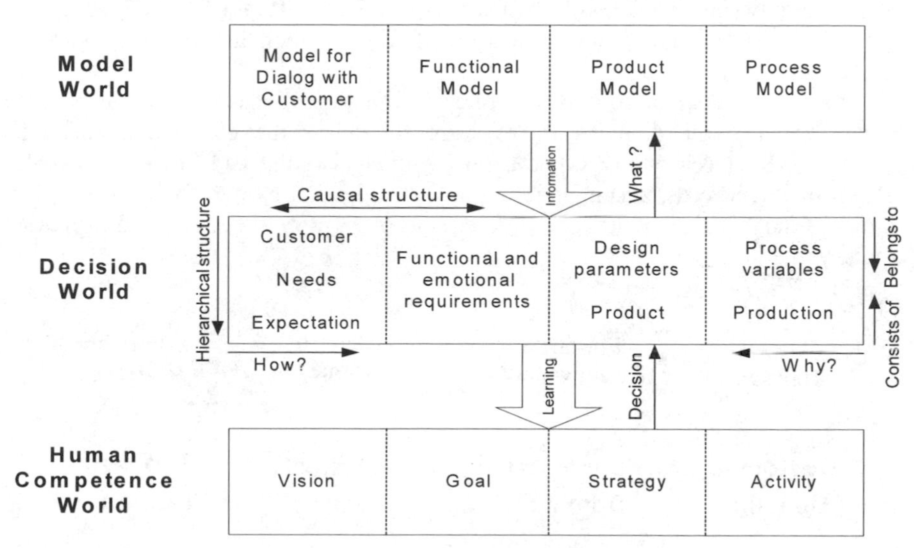

Figure 1.4 Systemic Map of the Innovation Process. The innovation process can be understood as a combination of the decision world with the human competence world and the model world.

Axiom: The Swedish National Encyclopaedia proposes the following definition: *"Axiom (Greek axioma appraisal, assessment, opinion, statement, which without proof is considered to be true). In everyday speech it means an obviously true statement. In science an axiom is considered a principle that in itself is not the subject of proof, but which is serving as the base for the proof of other statements."*

Decision World

Nam Suh proposed that the design process is a mapping between four domains, (Figure 1.5) [11]. This is a relevant and useful interpretation. The design process is principally a decision process, where the objectives are defined from the needs and expectations of the stakeholders, primarily customers, but also the employees and society as shareholders.

A closer look at the logical nature of this process reveals two important orthogonal structures: the hierarchical structure vertically and the causal structure horizontally, as shown in Figure 1.4. The hierarchical structure is related to the hierarchical nature of products. A product consists of modules, which consist of components, which consist of parts with features. The tree-structure is, in the ideal case, identical in the functional, design and process domains, and the structure is defined by the words *consists of* downwards and *belongs to* upwards. Decomposing and composing within the domains illustrate vertical and hierarchical relationships. The causal structure has to do with objectives and means and shows the connection between related positions in the hierarchical trees in adjacent domains. The words *how* and *why* are the horizontal guiding keys in this structure. Mapping between the domains exposes horizontal and causal relationships. In parallel the products and processes are modelled in the modelling world, answering the question *what?*

The existence of these two structures is the logical reason why we have to zigzag between the domains if we want to follow the connections between Functional Requirements (FRs), Design Parameters (DPs) and Process Variables (PVs). In other words, zigzagging, as shown in Figure 1.6, is to follow the logical nature of innovating products and processes in order to meet defined and expressed goals.

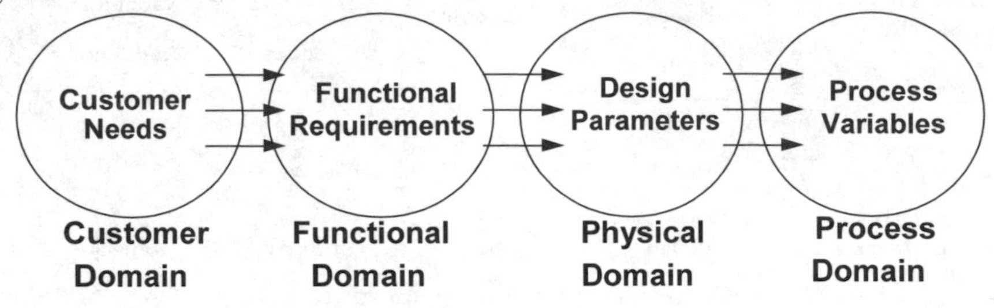

Figure 1.5 The design domains

Quality

The FRs are defined in a dialogue with customers to specify the minimum necessary functions of the product including allowed deviations from target values (tolerances). In design, the term quality means that the product DPs meet those defined FRs within specified tolerances. Decision criteria for this are defined in terms of two axioms. If the FRs are defined correctly, they specify requirements, which the customers need to control one at a time.

Axiom 1, *a design maintaining independence between functional requirements is superior,* defines a decision criterion for functionality, and axiom 2, *a design with higher probability to meet functional requirements is superior,* defines a decision criterion for success. Axiom 1 and axiom 2, therefore, are the correct and useful decision criteria for obtaining quality.

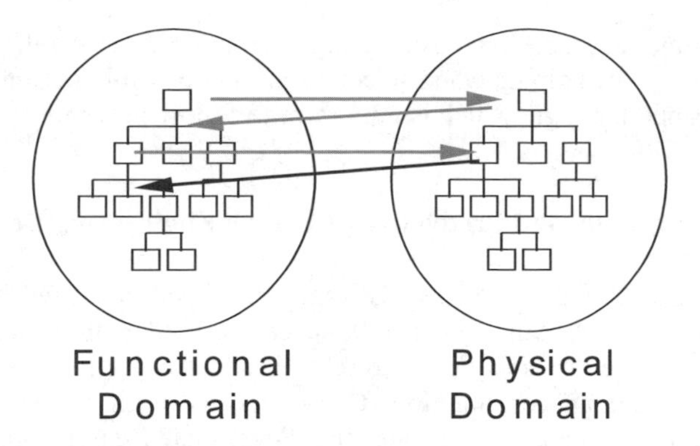

Figure 1.6 Mapping and decomposing by zigzagging

Productivity

In order to realize products, we extract materials from nature and process them with energy, which we control with information, which also requires energy to access and cultivate. Cost is a measure of the total energy needed in order to design and produce the products. However, cost is a very inaccurate measure of energy, as it is distorted due to market pricing, taxes and interest rates. From a practical point of view, however, it is correct and relevant to use cost as a measure of the energy used in order to design and produce the products.

After quality, therefore, low energy requirement is the next decision criterion, in engineering design. We can formally express it as an axiom 3 where: *Minimize energy in the selection of design parameters in order to meet the functional requirements* becomes the operational rule in following this axiom. It is a productivity axiom expressed as cost to meet a defined target, in this case defined in terms of quality.

Effectiveness requires a correct target (in this case quality) defined by axiom 1 and 2 together with axiom 3.

Productivity is both a matter of process design and a matter of waste. Waste occurs both as energy waste and as material waste. However, the cost for waste of material is also the cost for energy. Efficiency is generally expressed in % and defines the fraction of the input, which is the useful output: (input - waste) / input.

Time measures such as lead-time, Time To Market (TTM) and Time To Customer (TTC) are also important factors. Hence, a fourth axiom is proposed as a decision criterion for time, Axiom 4: *Minimize time in the selection of DP's meeting the FRs according to axiom 1 and 2* is the operational rule according to this axiom. In principle, those four decision criteria expressed as axioms 1 - 4 can and should guide decisions in any development work and especially in engineering design. They have to be related to each other in the order of numbering. This means that if there are difficulties in meeting requirements of an axiom of higher order, then the design decisions of the previous level should be reconsidered.

The order between decisions according to effort (cost) and time (axioms 3 and 4) is arbitrary. For example, in meeting short market opportunity windows, short time is more important than low cost. By zigzagging down in parallel through the domains and using the axioms when selecting among possible design alternatives, one can understand the logic which is important to follow in the design of products and processes.

Conclusions on Quality and Productivity through Engineering Design

Quality means Functionality, verified with the use of Axiom 1, and Probability to meet the requirements within agreed tolerances, verified with Axiom 2. This is important at each level in the design hierarchy. Productivity means to meet defined targets with low Energy consumption (Cost), verified with Axiom 3, and short Time, verified with Axiom 4. To meet the Functional Requirements with high Probability at low Energy and short Time is Effective. Quality and Productivity together define Effectiveness. Efficiency is a measure of a process and defines useful Output in relation to Input. Output is understood as Input minus Waste.

To be effective one has to be more conscious about waste:

- If a customer is buying products that do not satisfy his requirements, this causes waste
- If a marketing company succeeds in selling products, which the customer does not need, they are causing waste. There are marketing tricks used, which in fact have this effect.
- Insufficient quality causes waste.
- Low productivity causes waste. Observe that while high quality products are desired by the market in order to make the company is very profitable, high productivity is still important to minimize waste.

Waste limits the economic growth and the conditions for welfare, which is worth considering and being conscious about when developing products, processes and marketing strategies.

In conclusion, four axioms are the valid decision criteria for the choice of the best alternative solution at each level in the hierarchical function/design/process variable trees. The axioms we have found relevant for decisions in innovation processes are the following:

- Axiom 1: *A design maintaining the independence of functions is superior to coupled designs.*
- Axiom 2: *A design with higher probability to meet the functional requirements within specified tolerances is superior.*
- Axiom 3: *A design requiring less energy to be realized is superior.*
- Axiom 4: *A design requiring less time to be a realized is superior.*

The first two axioms define quality. They are the same as Nam Suh's design rules [9, 11]. The last two axioms define productivity and are proposed by Sohlenius. Quality has to be defined and met first then productivity. Whether cost or time has to be the next priority depends on the type product and market conditions. Decisions must also be based on proposed alternative design proposals. The decisions in this sense are choices between alternatives. The possibility to follow the logic of the innovation process also stimulates creativity through the use of the rationality of the real structures. Alternatively, cost and time can be defined as constraints. In this case, design solutions that meet axiom 1 and 2 have to be selected within these constraints. In this case, only two axioms are needed as decision criteria.

1.4 Sustainability through Learning and Competence – Human – Compentence World

The competitive development of the last decade has clearly shown that competence development is strategically crucial for industrial engineering. In order to increase competitiveness and use of new technology in industry, companies constantly have to improve products, services and working methods. This demands learning and competence development.

Decisions have to be made by humans who cooperate in the innovation process. It is also necessary to develop the human part of the industrial system, in its widest meaning, in parallel to and within the planning of the entire business process. The qualitative part of this human system is very much equal to the competence of the people involved. The ability/interest/desire to define goals and to make decisions is a prime core competence related to quality and productivity.

In this connection, it should be understood that competence is the ability of each individual to act correctly at the right time. This is naturally not easy. Therefore, competence management - a structured way of working, including a competence strategy - is needed. This perspective leads to the fact that the competence strategy has become an essential part of the total business strategy of the company. A competence strategy has to be developed integrated with and parallel to the other strategies of the company, such as market, product, process, finance, and environment strategies. By gathering all employees actively to break down targets and strategies into activities, from top management to the team level, all employees can be engaged in continuous improvements of the products, processes and organization in a consistent way.

In this work we are concerned with the following fundamental questions:

- Is the ability to stimulate competence development the most important possibility?
- Is our ability to know systematically and to define our competence in a planned way also needs a related primary possibility?

The following observations are also important and valid in connection to competence strategies:

- Modern tools for evaluation in organizational development in Sweden today emphasize leadership and competence issues (25 % of all criteria for analysis).
- A main goal is to see that one's own co-workers are more professionally skilled than those of the competitors. To make use of not yet discovered competence, to find it and understand its potential as a hidden asset - a treasure of the company - is a key question.

Innovative Competence

It might be a new requirement that every employee acquires an ability to take part in innovative development work. To which level this ability is developed will be more important as a measure than the number of participants in courses or numbers of proposals for improvements. Rather, it is of importance how many of the employees have been able to contribute to innovative work.

"The Knowledge Intensive Company" is an expression, which points at such a behaviour. A systematic re-education and training in new ways of cooperation for cross-disciplinary work are required [26]. It is also possible to envisage development of teams for innovative work - Teams of Excellence. These new teams would develop their own procedures, rules and ways of cooperation within individual development projects.

Competence depends on personal abilities and qualities, but it is also a result of the working conditions and leadership. To enhance competence development, the company has to provide a supportive environment. A problem in modern industry is the constantly changing market demands. This places a lot of stress on the organization. The consequences are often frustration and confusion, which hinder competence development and reduce work efficiency. The organization simply *has* to adapt to changes, but the changing demands must be handled in a way that does not take too much energy from the organization. The challenge is to create an organization able to strengthen already existing improvement work and credibly inspires further progress, in a world of constantly changing demands. To do this, the human need for stability has to be considered. For humans to accept changes, and even regard them as challenges, some conditions in the work situation must be sustained.

Sustainability is a word that has been in focus for many years in terms of the ecological aspects of industrialization. From the beginning, it mostly concerned the reuse of materials in the manufacturing processes. Recent discussions included the total production system [4]. Research in the field of “Reconfigurable Manufacturing” shows how machining systems can be designed, consisting of well thought out modules in combinations, which can be reconfigured to suit different customer demands. A reconfigurable system has a potential to combine the issues of sustainability and adaptation to changing market demands. The research and discussions about sustainability and reconfigurability have, however, yet not come to include the human aspects of working life in production systems. A balance between sustainability and renewal in working life is needed. This is approached from the viewpoint of competence development.

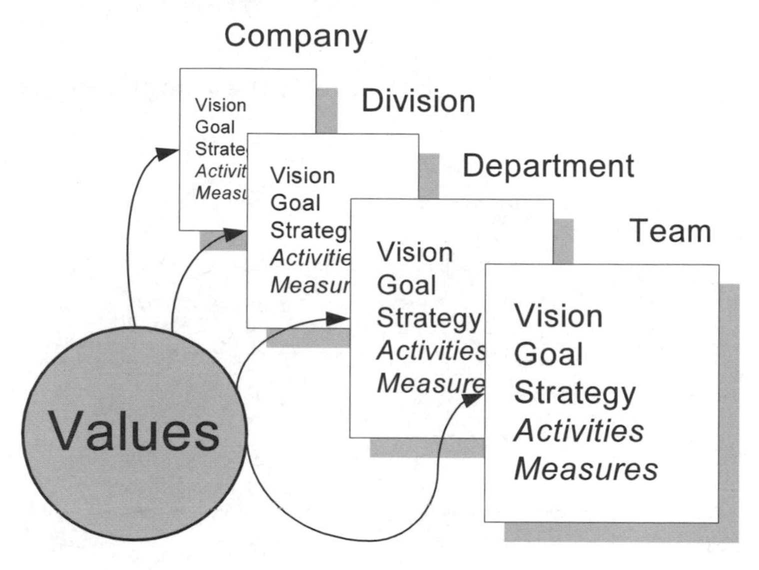

Figure 1.7 Coherence between the different organisational levels in the company; Source: Kjellberg, A.

A good way to help this development is to deal with concepts as *vision, goals, strategies, and activities* vertically in the company, (Figure 1.7). This dialogue is a good instrument to create coherence in the understanding of strategies between the different organizational levels in the company [27]. It is also useful and interesting to note that the concepts *vision, goal, strategy and activity* have the same conceptual meaning as *customer, function, design and process variable* in the decision world (Figure 1.4).

A successful business strategy has to be accompanied by a competence strategy. The competences needed for the new business processes around new products have to be defined and developed for and within the innovation process. The current competences also have to be defined as they form the starting point. The required new competencies have to be quantified as well as defined by competence levels. The differences are the competence gaps, which have to be filled through competence development activities, such as courses, as well as through mentorship, support from consultants, universities and alliances. The goals, strategies and activities for this have to be defined and planned (Figure 1.8) as a Competence Management Process. The CMP has to have its own strategy and procedures to be followed by all managers. Competence workshops can define competence gaps as a base for sourcing of competence. The gap defines functional requirements (FRs) in competence.

Some companies have involved all employees, from divisions to teams, in so–called competence workshops where competence gaps and activities to fill those gaps have been defined in relation to business demands. This should be combined with the vertical dialogue according to Figure 1.7. The gap defines the FRs on competence. The DPs have to be decided among possibilities such as those listed above.

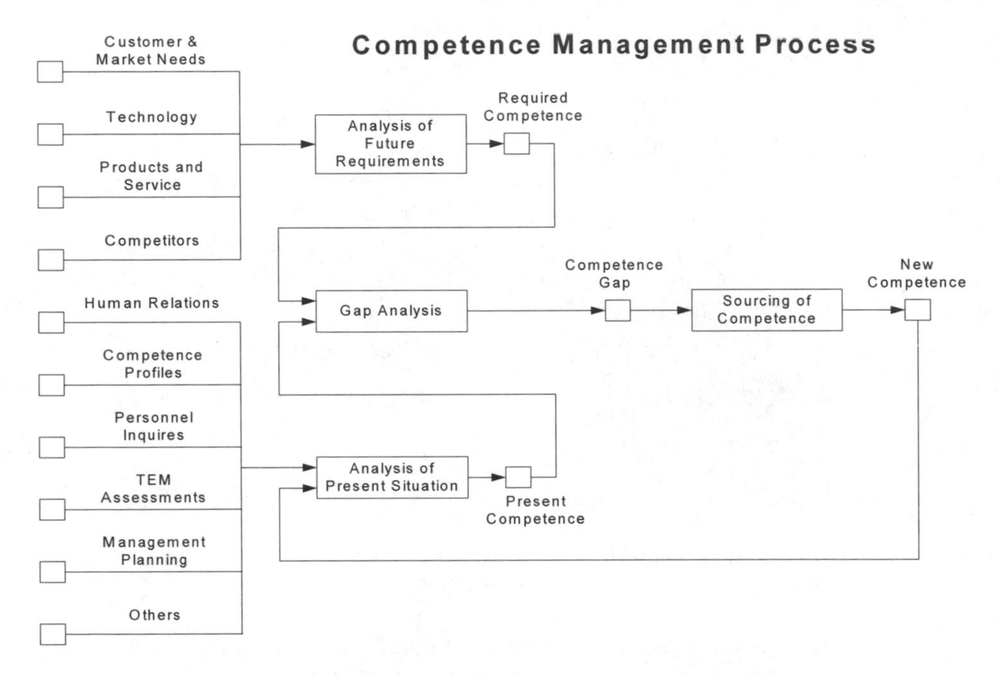

Figure 1.8 The Competence Management Process as a part of the innovation process, in itself also related to planned and desired business cases.

1.5 Industrial Company as a Business System

An industrial company can be regarded as a business system which is composed of sub-systems and processes. The business sub-systems represent different perspectives on the business inside and outside the company, and can be identified as market, innovation, supply and service. Most of the activities for business development are carried out in the sub-system for business innovation, henceforth called the innovation system. But development activities are also carried out in the other business sub-systems (market, supply and service). In the innovation system, therefore, concurrent activities from the three other business sub-systems are included. This concurrent engineering model is shown in the in Figure 1.9. The innovation system over-laps, to some extent, with the market system, supply system and service system.

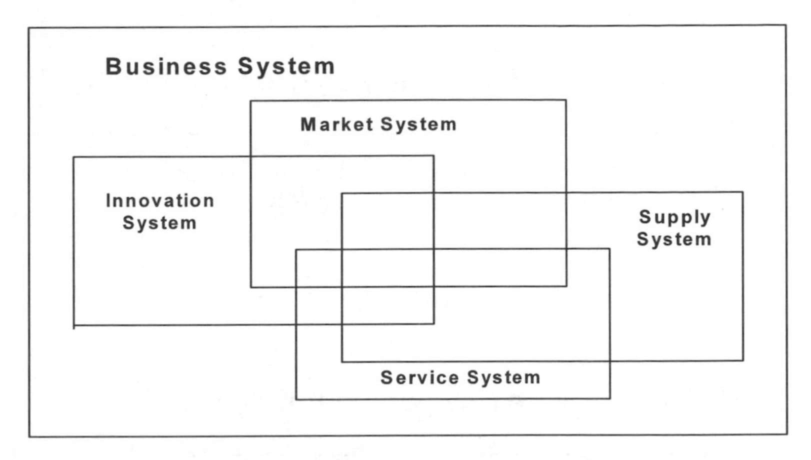

Figure 1.9 Model of business system with four sub-systems

The innovation system can be further decomposed into sub-systems, which are called product-market, product-function, product-design, product-supply and product-service. These sub-systems are regarded as different views of the innovation system, and handled as domains; hence, theories and methods from the engineering design area can be utilized for business innovation, including product development and realization.

Business Processes in an Industrial Company

In order to be competitive, an industrial company needs to have effective business processes. A business process is an assembly of connected activities designed to create value for customers. The business process is composed of four sub-processes, which are called pre-development (process-to-technology), main development (process-to-market), order and delivery (process-to-customer), and use and maintenance (process-to-service) according to the four phases in the product life cycle (Figure 1.10). The activities are executed within these processes.

Business innovation activities are executed in the business sub-processes for pre-development and main development. The innovation process is composed of these two sub-processes for business development (representing the product life phases: time-to-technology and time-to-market).

Business operation activities are executed in the business sub-processes for order and delivery (representing the product life phase time-to-customer) and for use and maintenance (representing the product life phase time-to-service). These two processes can be called the production process and the maintenance process.

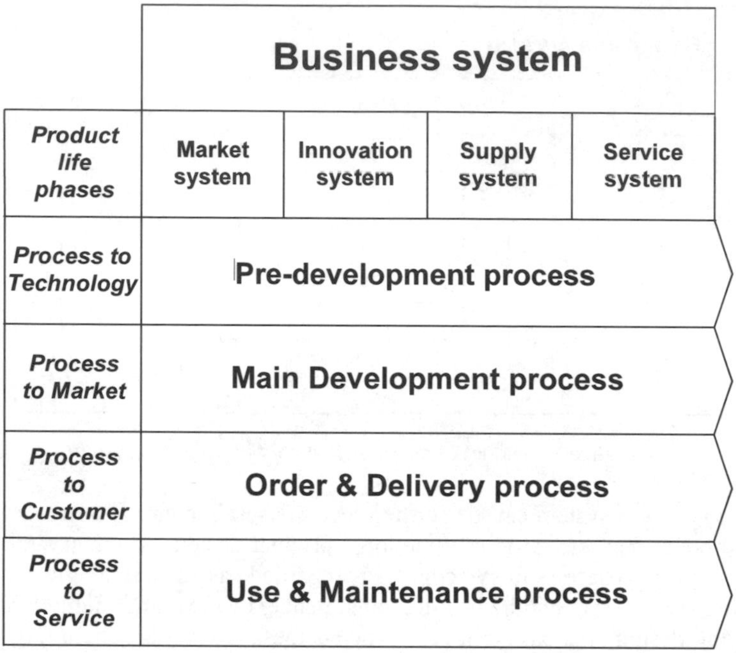

	Business system			
Product life phases	Market system	Innovation system	Supply system	Service system
Process to Technology	Pre-development process			
Process to Market	Main Development process			
Process to Customer	Order & Delivery process			
Process to Service	Use & Maintenance process			

Figure 1.10 Model of business system and business sub-processes

Interaction Mechanism

The Theory of Domains, by Andreasen [6, 7] describes the design task as navigation in relation to a basic pattern, which is composed of causal relationship between the domains. The domain model can be regarded as a basic map on which it is possible to chart the progress of the design task.

In the Axiomatic Design theory by Suh [9, 10, 11], the design process is seen as a mapping between four domains, customer domain, functional domain, physical domain and process domain, (Figure 1.5). Axiomatic design deals with the hierarchic nature of designs, which appears in the functional, physical and process domain as trees with more or less identical structures. The functional-, design- and process-trees grow through mapping between the domains and decomposing within the domains. The mapping between domains creates the branches on each level. One whole level has to be mapped over all domains before decomposition to the next level starts. This process, mapping between the domains and decomposing within the domains, is called zigzagging (Figure 1.6).

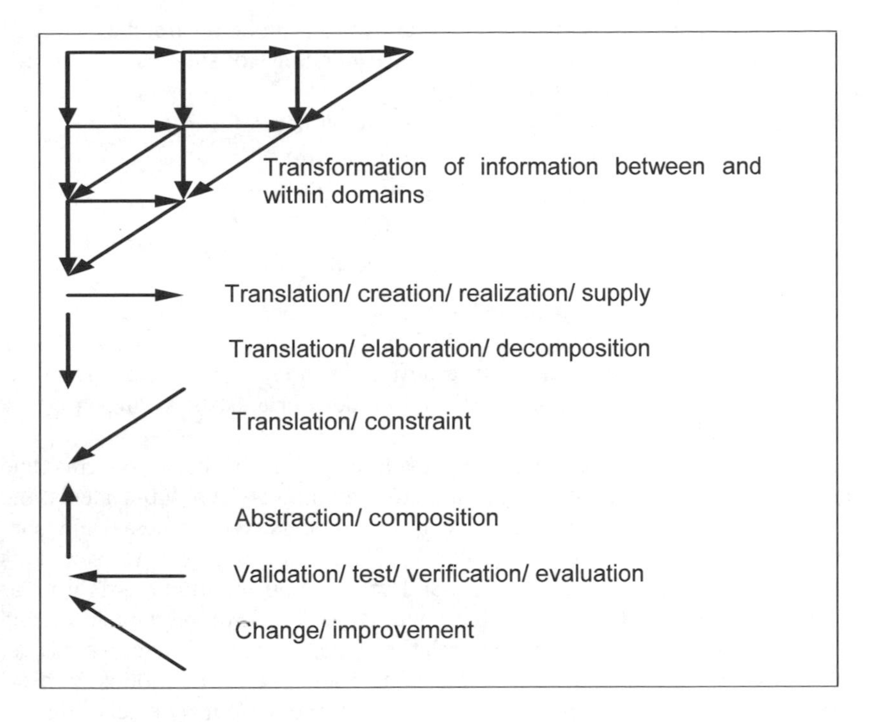

Figure 1.11 Mechanism for interactions between and within domains

For the navigation between and within the domains (sub-systems), a design tool called *Interaction Mechanism* is described and used in this paper. The Interaction Mechanism is based on the principles of the Theory of Domains and the theory of Axiomatic Design. The Mechanism can be used for innovation in the business system including transformation of information between the sub-systems and within the sub-systems. The transformations can be in the form of translation, creation, elaboration, realization, composition, decomposition, constraint, validation, verification, change or improvement. The mechanism can be regarded as a navigation tool comparable to a compass with interaction instructions [15, 23, 24], (Figure 1.11).

Business Innovation

Business Innovation is considered as innovation of business systems and processes. Activities for business innovation are performed in the innovation system and in its two sub-processes, pre-development and main development. The innovation system consists of sub-systems and by regarding these sub-systems as different views and handling them as domains, theories and methods from the engineering design area can be utilized for business innovation, sometimes called business (re) - engineering [5, 6, 9, 16, 17, 18].

The research approach is that Business Innovation can be regarded as a design task, and the following design objects, corresponding to sub-systems in the innovation system in Figure 1.10, are set up:

- Product-Market system (demand/package/offer structure)
- Product-Function system (function/system structure)
- Product-Design system (product/realization structure)
- Product-Supply system (production structure)
- Product-Service system (maintenance structure)

Design objects are also connected processes for business development and operation, which have to be designed, re-designed or re-used, (Figure 1.10). Design correspondences to decision-making in a design or innovation process using the decision world, human competence world and model world as shown in Figure 1.4 [19, 20, 21].

Design for business operation, *i.e.* development of business systems and processes, can be done by interactions/transformations between sub-systems and sub-processes in the matrix of business system/process. Sub-systems and sub-processes are interrelated; results (outputs) from sub-processes are stored as information in structures of sub-systems, and information in structures is used as input to sub-processes. Interactions horizontally have causal relationships, and interactions vertically have hierarchical relationships. Horizontal relations answer the questions *What, How* and *Why. 'What* is also addressing the model. Vertical relations answer the questions *What, Consists of* and *Belongs to* according to Figure 1.4 and Figure 1.10.

Business processes are designed by mapping (or zigzagging) between the subsystems (domains) and sub-processes with the Interaction Mechanism. By stating the interactions, business sub-processes can be developed and described with nouns and verbs for each process-step or activity (Figure 1.12). According to Figure 1.2 and Figure 1.3, the innovation process is designed first, which can be used for the development or innovation of the product and its support systems. The innovation process here consists of two sub-processes, pre-development process and main development process. First, the earliest sub-process, the PTT-process (Process-To-Technology) are developed. Thereafter, the other sub-process, the PTM-process (Process-To-Market), is developed based on the PTT-process.

An example for the development of the PTT- and PTM-processes with the use of the Interaction Mechanism in Figure 1.12 can be illustrated as follows:

- PTT/Scope/Need is translated into PTT/Function Requirement,
- which is created into / realized (virtually) by PTT/Digital product model,
- which is realized (virtually) by PTT/Production simulation.
- PTT/Scope/Need is elaborated into PTM/Concept/Requirement, which is also translated from/constrained by PTT/Function Requirement.
- PTM/Concept/Requirement is translated into PTM/Function specification, which is also elaborated from PTT/Function Requirement and constrained by PTT/Digital product model.

- PTM/Function specification is created into/realized by PTM/Prototype, which also is elaborated from PTT/Digital product model and constrained by PTT/Production simulation.
- PTM/Prototype is realized/produced by PTM/Prototype production, which is also elaborated from PTT/Production simulation.

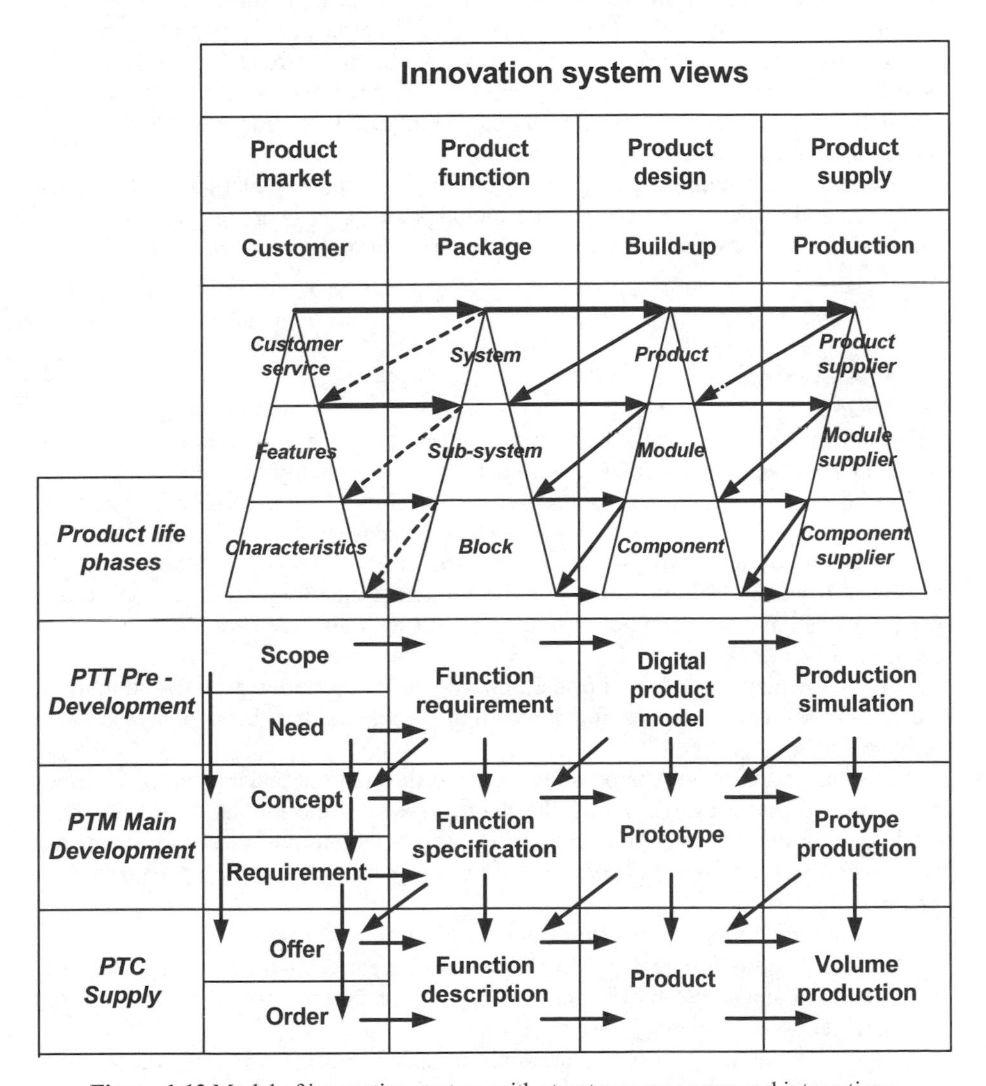

Figure 1.12 Model of innovation system with structures, processes and interactions

After developing the innovation process, we develop the production process, *i.e.* the PTC-process (Process-To-Customer), based on the PTM-process, including the product and the production system is developed. The maintenance process, the PTS-process (Process-To-Service), based on earlier processes and concurrently with the product and/or the production system can also be developed.

When the business processes are designed they can be used or re-used for the development of the sub-systems for product-market, product-function, product-design, product supply and product service. The Interaction Mechanism is then used for mapping between the domains (sub-systems) and decomposition within the domains. The results from these interactions or activities in the business processes can form the hierarchical information content for the product and its business as market offer, function structure, product structure, supply structure and service structure. The Product service view is not included in Figure 1.12. All business sub-processes have to be run through (mapped over all domains), and all hierarchical levels have to be decomposed/composed in order to develop and realize the product and its support systems.

Business innovation covers the area from business and product idea to business operation and product maintenance and includes customer demands and solutions. The main part of business innovation is development of the product and its support systems.

1.6 Conclusions

The use of fundamental knowledge in engineering science in innovative industrial decision processes has a great potential to improve quality and productivity in industrial operations and hence, improve profitability. This is a pre-condition for economic growth, which in turn is necessary to improve human welfare.

Based on a logic analysis of preconditions for increased welfare through profitability and economic growth, creativity, quality and productivity as necessary prerequisites have been defined. Developing these skills has to be the focus of a competence strategy.

The innovation process is of prime importance for economic growth and human welfare. A system structure for innovation processes has been analysed and defined.

The innovation process is possible to grasp through the understanding of three complementary worlds. These are the decision world, the human world and the model world. Fundamental knowledge about the functioning of these worlds opens up possibilities to improve *Total Effectiveness Management by understanding Means*.

The important mechanisms for economic growth are:

- New and more efficient product functions, created in innovation processes through human creativity and growing competence in the science of engineering.
- Increasing product quality and productivity can be realized through competence in defining needs, making decisions and using flexible automation and information technology in production processes.
- Based upon this overall strategic understanding, investments in **innovation** projects, **education** based upon science and practical experiences, improving skills and competence, and simple as well as advanced system **tools**, have the potential to yield profitability and economic growth.

1.7 References

[1] Sandell, P. Sohlenius, G. *et al.*, 1976, *Kundorderstyrd Produktion - PRODEVENT,* (Order Controlled, Customer Adapted Production – PRODEVENT) (in Swedish), Sveriges Mekanförbund.

[2] Sohlenius, G., 1992, "Concurrent Engineering," *CIRP Annals* Vol.41/2/1992.

[3] Johnson, H. T. Bröms, A., 2000, "Profit Beyond Measure," *The Free Press*.

[4] Sohlenius, G., 2000, "Productivity, Quality and Decision Theory Based upon Axiomatic Design," *Proceedings of ICAD 2000, First International Conference on Axiomatic Design,* 2000, Boston, MA, USA.

[5] Hubka, V., Eder, W.E., 1988, "Theory of Technical Systems: A Total Concept Theory for Engineering Design," *Springer Verlag,* ISBN 0-387-17451-6.

[6] Andreasen, M.M, 1992, "The Theory of Domains," *Workshop on Understanding Function and Function to Form Evolution*, 1991, Cambridge University, Cambridge, UK.

[7] Andreasen, M.M., 1992, *"Designing on a "Designer's Workbench" (DWB),"* *Proceedings of the 9th WDK Workshop*, 1992, Rigi, Switzerland.

[8] Aganovic, D., Nielsen, J., Fagerström, J., Clausson, L., Falkman, P., 2002, *"A Concurrent Engineering Information Model based on the STEP Standard and the Theory of Domains,"* *Proceedings of DESIGN 2002,* Dubrovnik, Croatia.

[9] Suh, N.P., 1990, *The Principles of Design*, Oxford University Press, ISBN 0-19-504345-6.

[10] Suh, N.P., 1995, "Design and Operation of Large Systems," *Journal of Manufacturing Systems*, Vol.14, No.3, pp. 203-213.

[11] Suh, N.P., 2001, *Axiomatic Design, Advances and Applications*, Oxford University Press, ISBN 0-19-513466-4.

[12] Sohlenius, G., Fagerström, J., Kjellberg, A., 2002, "The Innovation Process and the Principal Importance of Axiomatic Design," *Proceedings of ICAD2002, Second International Conference on Axiomatic Design*, 2002, Cambridge, MA, USA.

[13] Moestam Ahlström, L., Kjellberg, A., Sohlenius, G., 2002, "Principles and Experiences concerning Sustainability in Product Realisation," *XIIth World Productivity Congress,* 2002, Hong Kong.

[14] Fagerström, J., Moestam Ahlström, L., 2001, "Demands on Methods for Developing Work Focused on Concurrent Engineering," *Proceedings of ICPR-16*, 2001, Prague, Czech Republic.

[15] Clausson, L., Fagerström, J., Aganovic, D., Sahlin, M., 2002, "Business Process Engineering by Utilizing Design Theories and Methods," *Proceedings of CIRP 1st International Seminar on Digital Enterprise Technology, DET'02,* Durham, UK.

[16] Hubka, V., Eder, W.E., 1996, *Design Science: Introduction to the needs, scope and organization of engineering design knowledge,* Springer Verlag, ISBN 3-540-19997-7.

[17] Pahl, G., Beitz, W., 1996, *Engineering design: a systematic approach,* (original title, Konstruktionslehre), Springer, Berlin, ISBN 3-540-19917-9.

[18] Ulrich, K. T., Eppinger, S. D., 2000, *Product Design and Development*, The McGraw-Hill Companies Inc., ISBN 0-07-229647-X.
[19] Sahlin, M., Fagerström, J. Clausson, L. Aganovic, D. Sohlenius, G., 2002, "Concurrent Decision Making for High-tech Products and Supply Systems," *Design 2002*, Dubrovnik, Croatia.
[20] Fagerström, J., Aganovic, D., Nielsen, J., Falkman, P., 2002, "Multi-Viewpoint Modeling of the Innovation System: Using a Hermeneutic Method," *Proceedings of ICAD 2002*, Cambridge, MA, USA.
[21] Aganovic, D., Pandikow, A., 2002, "Towards Enabling an Innovation Process for Extended Manufacturing Enterprises," *Proceedings of CIRP 1st International Seminar on Digital Enterprise Tech., DET'02,* Durham, UK.
[22] Clausing, D. Fey, V., 2004, *Effective Innovation*, ASME Press.
[23] Clausson, L., 2003, "Innovation of Business System and Process," *2003 International CIRP Design Seminar*, Grenoble, France.
[24] Clausson, L., 2003, "Business Innovation by utilizing Engineering Design Theories and Methods," *Int. Working Conference, Total Quality Management – Advanced and Intelligent Approaches*, 2003, Kragujevac, Serbia.
[25] Sohlenius, G, Kjellberg, A., Clausson, L., 2003, "Economic Growth, Industrial Production and TQM," *International Working Conference, Total Quality Management – Advanced and Intelligent Approaches*, 2003, Kragujevac, Serbia.
[26] Kjellberg, A., 1998, "Teams - What´s Next?," *CIRP Annals,* Vol. 47, No.1
[27] Sohlenius, G. Kjellberg, A. Holmstedt, H., 1999, "Productivity, System Design and Competence Management," *World Xith Productivity Congress*, 1999, Edinburgh, U.K.

2

Directions of Next Generation Product Development

Tetsuo Tomiyama and Bart R. Meijer

Abstract: For the last 20 years, the focus has been on product development processes and developing tools to support them, addressing not only technological but also managerial issues. While these tools have been successfully supporting product development processes in a general sense, consensus on the direction of future developments seems to be lacking. In the paper, it is argued that horizontal seamless integration of product life cycle knowledge is the key toward the next generation product development. Knowledge fusion, rather than just knowledge integration, is considered crucial. In this paper, we will try to outline the directions of the next generation product development, its tools, and necessary research efforts.

Keywords: product development, integration, knowledge fusion

2.1 Introduction

Product development is a key process for manufacturing. This process includes product marketing, product planning, product design, prototyping and testing, production planning, production design, and product systems design. In the last two decades, product development focused on such issues as cost, quality, lead time, product variety, integration of various aspects, organization, and design for X. Companies that emphasized these issues, as well as lean production technologies, concurrent engineering technologies, have been successful. Research efforts in product development have resulted in enabling techniques and tools for efficient product development, such as digital design tools (CAD/CAM/CAE), rapid prototyping tools, and design and information management tools (PDM). Undoubtedly, these tools without a doubt contributed tremendously to achieving such difficult goals as cost reduction together with quality improvement and lead-time reduction while increasing product variety.

In this paper, we will try to outline the directions of the next generation product development and necessary research efforts. In Section 2.2, we will review the history of manufacturing and its influence on product development. We will argue that knowledge is the key, as the products become more multi-disciplinary and their boundaries expand to include a large variety of life cycle issues.

In Section 2.3, we will articulate the missing or insufficiently addressed issues within the current product development research trends. The first issue is "more horizontal integration" to include a wider range of engineering activities, and augment the current practices of overemphasizing shapes. The second issue is "seamless integration of activities" beyond data and knowledge-level integration. The activities within product development can include activities such as design, computation, procurement, prototyping, and testing. These might even be extended to supply chain management (SCM) and value chain management (VCM) based on product life cycle management (PLM), because products are now expanding their boundaries from a physical existence to a product-service system in a holistic product life cycle. The third issue is that product development still pursues "better quality, lower costs, more innovation, higher speed, and yet greener performance." Although concurrent engineering practices address these issues, these goals should be further emphasized.

In all of these three issues, knowledge integration plays a crucial role. Section 2.4 points out that knowledge fusion, rather than just knowledge integration, will be more important, because in order to be truly innovative, integrated knowledge is needed.

2.2 Analysis of the Production Development Paradigms

Historical Production Paradigms

Reviewing historical development always gives an insight into future directions. Manufacturing began with craftsmen and then became gradually organized first in the form of domestic industry and then as factory systems. Later, through the Industrial Revolution period, due to the necessity of capital investment of machinery, this factory system became dominant.

In the craft society, a craftsman designed and produced products all by himself. This single-man process was seamless, and there was no apparent division of labour. The knowledge needed for this process was undividable. Soon, however, product-wise specialization took place; goldsmiths used different processes and knowledge from those used by bookbinders. Nevertheless, the master was responsible for the final integration.

This was a qualitatively different division of labour compared with the factory concepts that appeared later in which the labour of many unskilled workers had to be organized and coordinated to achieve mass production. Improvements in productivity of this type of manufacturing, typically after the Industrial Revolution, was possible by employing more workers, by using mechanization that replaced human labour with mechanical operations, and by introducing more powerful and faster machines. This resulted in more specialized work in an organization.

In the early 20th century, two important concepts were introduced. One was the scientific management of production (Taylor system) and the other was the mass production (Ford system). The combination of these two resulted in powerful industrialization, particularly in North America and Western Europe, in the first half of the 20th century. This process was further accelerated by automation technologies (and later by computer technologies), replacing not only muscle work but also gradually the intellectual aspects of human workers.

Volume and productivity of production of mono-disciplinary products were achieved, at that time, by simplified mono-disciplinary division of labour, either component-wise or discipline-wise. The integration took place through the physical assembly of final products.

In the last three decades of the 20th century, quality goals replaced quantity goals. Production of many of the same was insufficient in the post-war competition among industrialized nations. First of all, batch size became smaller and smaller, while product variety increased. For instance, the concept of Flexible Manufacturing Systems (FMS), which aims at variety and small-batch-size production, gradually became popular. Secondly, quality became a more dominating factor in production than quantity, while lower costs were still pursued. Technologies such as lean production and concurrent engineering were developed to achieve these contradictory goals.

At the same time, products became increasingly more complex; a good example is mechatronics products that are typical of multi-disciplinary products. Computer based design support tools were critical to develop such complex products with higher quality. This further meant integration of data and information for product development, which was done at the systems level by assembling divided disciplines. Organizationally, efficiency achieved through the work of many-men by specialization and/or by organization was not enough. Competitive quality improvement and cost reduction were possible only through collaborative team work.

The important resources of production have been capital, labour, machines, production, and knowledge. First, in the craft society, all of them were distributed. Then the domestic production system gradually integrated production activities. This was an integration of distributed production. The factory production system started next, which meant that production, as well as machines and labour, was concentrated through capital investment.

As mass production started, capital, labour, machines, and production were all concentrated in one place. However, knowledge was distributed in many parts of the production system, largely in workers. The market system then played a crucial role in distributing those once concentrated production resources, because through the market, we can obtain these. Nowadays, capital is obtainable through the capital market, and even labour has its market. As a consequence of globalization and worldwide competition, except for knowledge, everything else is now obtainable through the market.

Analysis of the Product Development Paradigms

The analysis above outlines the current situation of manufacturing and further implies its influence on product development. The current production paradigm has the following features.

First, due to the ever-developing globalisation and universally developed market principles, we now see different forms of division in manufacturing. One is competence-based division of manufacturing and the other is geographical division. Both forms of specialization require integration through advanced information and communication technologies.

Second, products are becoming more multi-disciplinary. This has two meanings. Opto-mechatronics products, such as DVD players, which integrate more diverse disciplines, illustrate the first meaning. The second is the focus on product life cycle issues that include product life cycle management, end-of-life treatments, and different forms of added value generation such as service and product-service systems [1], due to increasing concerns about the global environment. This means that the shape and boundaries of products are further expanded. Classically, a product is a physical existence with material properties, physical shapes, performance, and functions. However, within the service-centred economy or production paradigm, a physical product is merely a device to deliver services that generate more added value. This means that in addition to PDM (Product Data Management), PLM (Product Life Cycle Management) should be developed to include SCM (Supply Chain Management), VCM (Value Chain Management), and CRM (Customer Relation Management).

These features of the current production paradigm have an impact on product development supported by a variety of tools that have become available over the last 40 years, since the Sketchpad system in 1963 [2]. During this period, as described in the previous section, product development had different goals to achieve, which resulted in a wide variety of technologies, methodologies, and tools. Among others, we can identify three important goals: improving productivity and quality through automation, and reducing costs. For instance, in the early phase of CAD (Computer Aided Design) in the 1960s and 1970s, the goal was to at least "automate of the drafting tasks" for better productivity, although "automation of designing tasks" was one of the aims.

Furthermore, CAD/CAM (Computer Aided Manufacturing) integration was pursued, so that CAM data is automatically generated on CAD. During the 1980s, together with the advances in 3D CAD technologies, product modelling became one of the goals. This further resulted in PDM, combined with the development of CAD data exchange formats such as STEP. In addition, CAE (Computer Aided Engineering) concepts using data defined on CAD systems in the engineering analysis stages have become interesting. There was a strong motivation to create a central database of products that can be used in every stage of product development to first improve productivity, then quality. However, the improvement of products quality has become very complicated during these periods which benefited the trend to reduce costs through automation.

In summary, information technology that has been applied to product development first identified automation as a primary goal, but then gradually turned to conversion of product development activities to an information-centred one as a dominant goal. In other words, data integration over various product development activities was the central issue targeting better productivity as well as quality and cost improvement.

During the 1980s, knowledge engineering approaches have established their role in product development. While the so-called expert systems approach quickly disappeared due to its technological immaturity and insufficient understanding of product development processes, the knowledge-centred view of product development has remained. Today, knowledge management for product development is simply an urgent issue. It aims at integration of knowledge for product development compared with the data integration view. This is considered essential for both quality and cost improvement, while productivity improvement is also addressed.

Concurrent engineering through improved communication among different product development participants has been highlighted from the late 1980s and 1990s, including such technologies as DfX (Design for X) and collaborative teamwork environment. As the products as well as product development processes became more and more complex, it was essential to have better collaboration among different product development participants. These not only shorten the product development lead time (*i.e.*, reduced costs as well as competitiveness) by having greater overlaps between processes that were formerly performed sequentially, but also help to identify design defects before production. Thus, the primary focus was on competitiveness, rather than productivity, through improved costs and quality.

In summary, the next generation product development should focus more on knowledge-centred integration due to the increasing importance of knowledge intensive products and product-service systems that require a wide variety of product life cycle knowledge.

2.3 Future Directions of Product Development

What are the Problems?

We have identified the trends in product development in the previous sections and concluded that products are becoming more multi-disciplinary with their boundaries expanded from just a physical existence to the whole life cycle system. A key to this is the knowledge-centred integration of product development.

In addition, the economy of advanced industrialized countries is increasingly becoming knowledge and service intensive, rather than energy and material intensive. This knowledge intensiveness can have two meanings; one is knowledge-intensive products, which have embedded intelligence for innovative features and functionalities, and the other is knowledge intensive product development. Due to the expanding product boundaries (*i.e.*, a product as a package to a product-service system with life cycle aspects), product development should include these life cycle aspects as well, addressing integration with such activities as SCM and VCM.

Of course, the ever-increasing pressures for product development with better quality, lower costs, and quicker delivery with greener performance still exist for product development. Achieving these goals is only possible by integration of different types of product life cycle knowledge.

Having said these, we recognize that product development processes should still address the knowledge integration about product's life cycle. Also, we should notice that the integration of different types of knowledge is the source of innovation [3]. This integrated knowledge should result in better integration of various activities over product's life cycle. These are summarized in the following sections.

Horizontal Integration

The first issue is "more horizontal integration" to include a wider range of engineering fields, because products are becoming more multi-disciplinary and their boundaries are expanding. Horizontal integration means taking more different fields into consideration during product development.

For example, it is not a surprising practice that during mechatronics products development, electronics and software design are totally separated from mechanical design. However, as concurrent engineering dictates, it is essential to promote collaboration among mechanical designers, electronics designers, software designers and production designers. If such collaboration does not exist, the easiness and cost of modifications at later stages of product development vary depending on the nature of the product. For example, a mechanical design error could be corrected by software modification with less cost. The differences in the degrees of component standardization among these disciplines also contribute to the differences in the development lead time, resulting in a tendency that software development receives extra pressure due to design changes from mechanical and electronics design. This never leads to good product development practices.

Seamless Integration of Activities

The second issue is "seamless integration of activities" based on information and knowledge level integration. The activities within product development can include such tasks as design, computation, procurement, prototyping, and testing. These activities do share a common PDM database, for instance. However, since the shared model is basically a product model, information about the product development process and these activities is not shared. In addition, a proper feedback or feed-forward mechanism that can prevent communication failures is missing from the current systems.

Additionally as product boundaries expand, we need further integration of SCM, VCM, and CRM with PDM to arrive at PLM. This will seamlessly integrate a variety of activities within a product life cycle.

Better, Cheaper, More Innovative, Speedier and yet Greener Product Development

The result of "horizontal" "seamless integration" of activities should help us to achieve product development with "better quality, lower costs, more innovation, higher speed, and yet greener performance." Concurrent engineering practices address these issues. However, the recent economic development in many areas in the globe demands further drastic improvements in achieving these goals.

It is worthwhile to examine the business practices exercised by the INCS Japanese Company [4]. This company, for example, promises delivery of a rapid prototyped component (using laser lithography technology) within four days after the order and of a moulding die within fourteen days. This kind of speed is possible not only through technology but also through an innovative management style, resulting in innovation in product development processes that improve both quality and cost. This not only signifies the importance of integration through computer-based tools but also addresses an important feature that is missing from the current research in product development; namely, speed.

2.4 From Knowledge Integration to Knowledge Fusion

In the previous chapter, we pointed out that "horizontal" "seamless integration of knowledge" about product's life cycle is the key to arriving at product development for better, more innovative, quicker, and still greener products.

A knowledge system that represents a mono-discipline required for developing a simple, mono-disciplinary product is shown in Figure 2.1 (a). We may then need a set of closely related knowledge systems for multi-disciplinary product development. Integrating these closely related knowledge systems requires defining at least interfaces (Figure 2.1 (b)). Such multi-disciplinary integration is a key for innovative product development [3].

However, to be more innovative, we may need to go one step further; knowledge fusion (Figure 2.1 (c)). Knowledge fusion is to create a new knowledge system that can be operated as a whole to develop truly multi-disciplinary products. Knowledge integration is still a collection of independent knowledge systems with clearly defined interfaces and describes common concepts among those integrated knowledge systems, while knowledge fusion is a situation in which these systems have been totally fused to create a new knowledge system.

Knowledge fusion is of course not automatically possible. Mechatronics is now considered to form an integrated knowledge system, but still we can see distinctions among mechanical technology, control technology, electronics, sensor technology, and software technology. This suggests that first, only a knowledge system based on knowledge fusion can arrive at better product development, and second, it is not just a matter of developing product development technologies but we need efforts to create a fused knowledge system. While details of knowledge fusion are yet subject to research, we may point out that knowledge-structuring efforts are considered useful [5, 6].

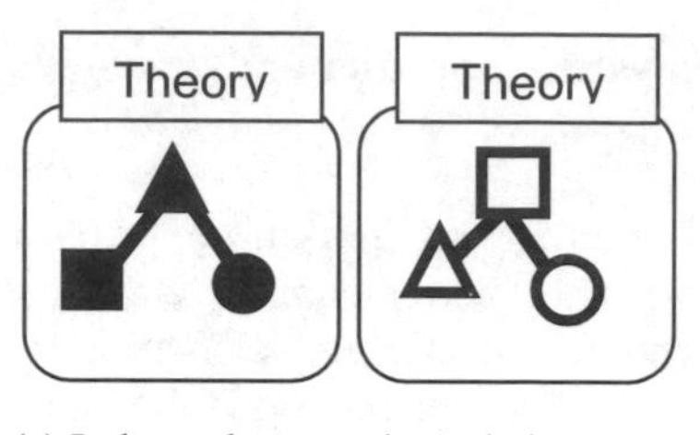

(a) Independent two knowledge systems

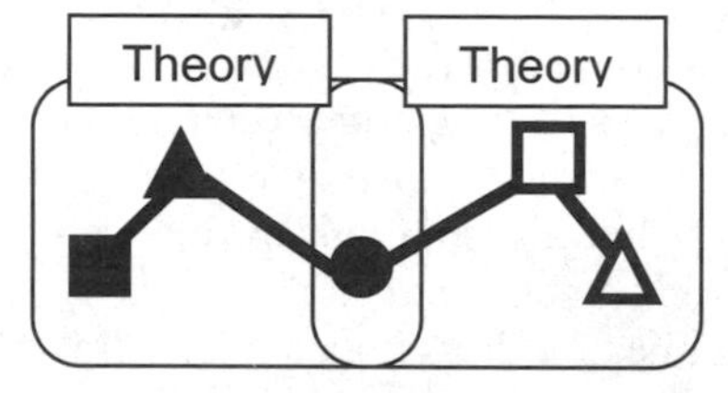

(b) Knowledge Integration

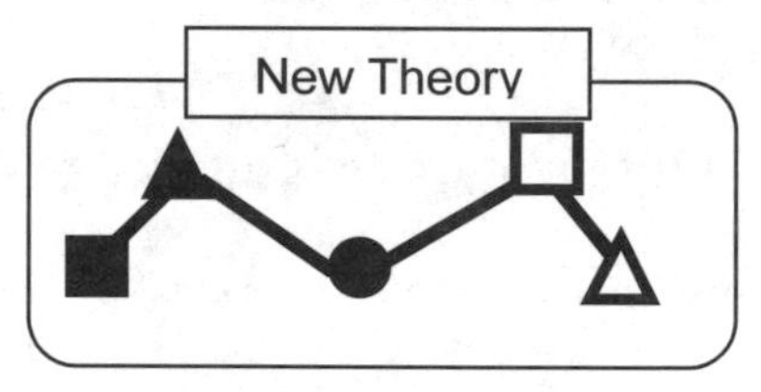

(c) Knowledge fusion

Figure 2.1 Knowledge integration and knowledge fusion

2.5 Conclusions

This paper is an attempt to identify the directions of the next generation product development and necessary research efforts. Three key issues were identified. The first issue is "more horizontal integration" to include a wider range of engineering activities. The second is "seamless integration of activities" beyond data and knowledge level integration. The activities within product development can include such tasks as design, computation, procurement, prototyping, and testing, and might even be extended to SCM and VCM based on PLM. The third is product development still pursues "better quality, lower costs, more innovation, higher speed, and yet greener performance."

For these three issues, knowledge integration plays a crucial role. However, since knowledge integration only arrives at a set of knowledge collection of which interfaces are clearly defined, we may need even another step; knowledge fusion. While knowledge fusion itself is not yet a clear concept, we have identified some research directions in knowledge structuring as being useful.

2.6 References

[1] Tomiyama, T., 2000, "Knowledge Intensive Engineering towards Sustainable Products with High Knowledge and Service Contents," *TMCE 2000, Third Int. Symp. on Tools and Methods of Competitive Engineering*, April 18-20, 2000, Delft University Press, Delft, The Netherlands, pp55-67.

[2] Sutherland, I.E., 1963, "SKETCHPAD – A Man-Machine Graphical Communication System," *Proc. of Spring Joint Computers Conf., IEEE/ACM*, pp. 329.

[3] Tomiyama, T., Takeda, H., Yoshioka M., and Shimomura, Y., 2003, "Abduction for Creative Design," *15th Int. Conf. on Design Theory and Methodology– DTM'03, Proc. 2003 ASME Design Eng. Tech. Conf. & Comp. and Info. in Eng. Conf.*, September 2-6, 2003, Chicago, IL, USA, Paper Number: DETC2003/DTM-48650, ISBN 0-7918-3698-3, 10 pages.

[4] http://www.incs.co.jp

[5] Tomiyama, T., 2003, "Knowledge Deployment: How to Use Design Knowledge," In: *Human Behaviour in Design –Individuals, Teams, Tools*, U. Lindemann (ed.), Springer, Berlin, Heidelberg, New York, ISBN 3-540-40632-8, pp. 261-271.

[6] Meijer, B.R., Tomiyama, T., van der Holst, B.H.A., and van der Werff, K., 2003, "Knowledge Structuring for Function Design," *CIRP Annals 2003*, Vol. 52/1, ISBN 3-905 277-39-5, pp. 89-92.

3

'What-if' Design as an Integrative Method in Product Design

Fred van Houten, and Eric Lutters

Abstract: In product development, many different aspects simultaneously influence the advancement of the process. Many specialists contribute to the specification of products, whilst in the meantime the consistency and mutual dependencies have to be preserved. Consequently, much effort is spent on mere routine tasks, which primarily distract members of the development team of their main task of creating the best solution for the design problem at hand. Many of these routine tasks can be translated into problems with a more or less tangible structure; often they are in fact an attempt to assess the consequences of a certain design decision on the rest of the product definition. Therefore, such questions can be formulated as: "what happens if....". The question is subsequently translated into a need for evolution of the information content determining the product definition. Based on this need for information, immediate workflow management processes can be triggered. This results in a 'train' of design and engineering processes that are carried out, leading to a viable answer to the question. As the structure of a 'what-if' question is independent of the domain under consideration, the 'what-if' questions can relate to any aspect in the information content at any level of aggregation. Consequently 'what-if' questions can range from anything between 'What if another machine tool is used' to 'What does this product look like if it is made from sheet metal'. Such a way of looking at products under development obviously strongly binds different domains and downstream processes under consideration, thus enabling a more integrated approach of the design process.

Keywords: Information management, design support system, 'what-if' design

3.1 Introduction

In product development, many different processes interact in order to determine the best manner to bring a product from an initial idea to the 'box' that ultimately arrives at the store shelves. In these processes, often a considerable number of designers, engineers and domain specialists are gathered in development (sub)teams to perform all the tasks that are related to the development of the product.

This implies that both the design team and the co-operating specialists are not only able to survey all aspects involved in the project, but also that they are able to weigh the consequences of all decisions made during product development at any given moment. In everyday practice, this may seem a clear infeasibility; nevertheless, development teams are able to cope with it because of their knowledge, experience and intuition, which is something that cannot be grasped by any computer. This, however, does not imply that computers are not able to support design teams in their efforts to arrive at better solutions for the problems they encounter.

The way in which the development processes can be supported is related to the 'maturity' of the product definition. The more a product shifts from conceptual design to actual production and assembly, the more orderly the processes involved can be described and governed. This is obviously related to the fact that during the initial phases, the lifeline for the product is established. Considered from a cost perspective: the actual expenses during manufacturing are largely related to downstream processes (*e.g.* production, assembly), whereas the establishment of costs that have to be made occurs during the upstream processes (*e.g.* design, engineering).

This implies that downstream phases have a decreasing freedom of action as more design aspects are established. Consequently, for these processes, more restraints are available, and therefore, many of these processes can be specified and controlled by means of distinct process models or scenario-like approaches. Following this train of thought, downstream processes can be considered as 'additive' processes appending information to what is generated by the upstream processes. This immediately emphasizes the synthesizing role of the design process in the entire design cycle [1]. Here, all aspects that are related to the entire product creation cycle have to be taken into account, to avoid flaws in the design that otherwise would only be recognized much later, as well as to use the downstream analyses to iteratively improve the design.

In positioning the design phases at the core of the product creation process, and as such interrelating, integrating and synthesizing the large majority of all other processes involved, it will be immediately clear that no overall and unequivocal process models or scenarios for the design phase can possibly exist. This is emphasized by the fact that the design process is, by definition, a creative process that cannot be captured by a comprehensive set of rules and constraints.

'What-if' Design

The best way to support designers in their work is to supply them with information on the hidden effects of their design decisions. Based on the simple axiom that a design decision leads to a change of the product definition, resulting in one or more updated downstream analyses that can be used to assess the consequences of the decision, a pattern for a structured way of working can be deduced. This structure hypothesizes that design decisions can often be rephrased as so-called 'what-if' questions. For example (see also Figure 3.1) "*What* is the change in the environmental impact of this part *if* the material is changed from aluminium to polyurethane?"

'What-if' Design is all about the management of information that underlies and determines the product definition. It deals with aspects of the life cycle of a product and will, in order to be able to meet its goal, present information on these phases during the design phase of a product. It has to gather, manipulate and present information on many different aspects of the product. This has to be done in a way that gives a user, who in most of the fields of interest is a non-expert useful information on the design at hand.

3.2 Outline of a System for 'What-if' Design

Posing a 'what-if' question, is actually asking for additional information. However, merely information (regardless of its quantity and quality) is insufficient to find adequate answers to questions. The cause for this is that relations that exist between the required information entities are probably more important than the entities themselves. In other words, each design decision requires a framework of information, depicted by structured information, context information and knowledge [2].

As indicated in Section 3.1, if a sheer process-oriented approach to product development is applied, the dependencies between the tasks and functions related to all the members of a development team understandably become entangled to such an extent that extremely rigid process management methods have to be employed. This not only distracts the members of the design team from their actual activities, but also causes an enormous overhead in communication and assimilation. More importantly, it hampers the activities that designers and engineers are best at: employing creativity to solve unusual problems. Thus, for adequate support, the 'what-if' approach is based on the information content that describes the evolution of the product definition during the development process.

Two approaches can be applied simultaneously:

- A generic top-down approach, focusing on the methods of answering structured 'what-if' questions, whilst disregarding any specific domain information, and avoiding any bias of solution routines.
- A bottom-up approach, contributing to understanding the application of a 'what-if' system and support systems in general.

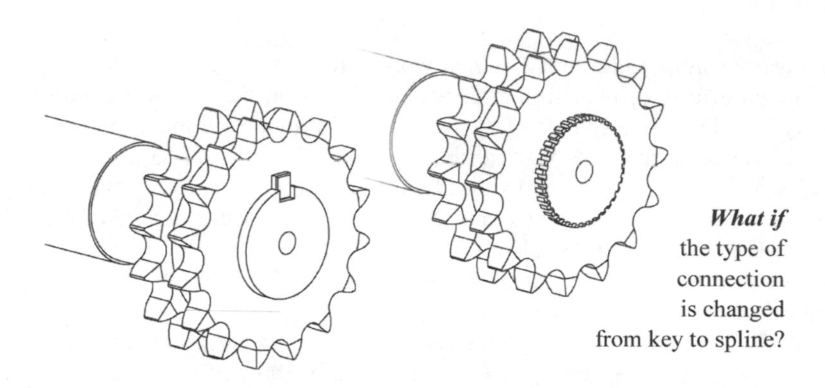

What if
the type of
connection
is changed
from key to spline?

Figure 3.1 Overly simplified example of a 'What-if' question

Consequently, this research focuses on combining both approaches in order to achieve a system that really enables synthesis in the design process.

In focusing on the information content instead of on all processes contributing to the development process, a different way of process control is required. Processes become means to achieve a required evolution of the information content, instead of a predefined next step towards an indistinctly phrased goal. Workflow management principles that are required to achieve this obviously need a different basis. If the evolution of the information content becomes the driving force in product development, it goes without saying that workflow management should conform to this evolution as well.

In a summary, 'what-if' design can be described as the information–based and workflow–driven, structured approach to chart the consequences of design decisions or changes in a design. The following sections deal with the different aspects of 'what-if' design in more detail.

3.3 Information as the Basis for the Development Product Life Cycle

There is no compelling foundation to regard the development cycle from a process-based point of view. This is especially true in recognising the fact that design and engineering activities need information as input and yield information as output. As an illustration, in everyday practice, most effort is indeed spent on the search, storage, retrieval, transformation, transportation, representation and interpretation of information. Moreover, in recognising the fact that each of the members of development teams in a company makes a myriad of decisions in order to generate required information, it is obvious that the reasoning behind all these decisions can hardly be transferred together with the information. Consequently, the need for feedback and interdepartmental communication increases, which may lead to extremely complex and uncontrollable flows of information between the separate departments.

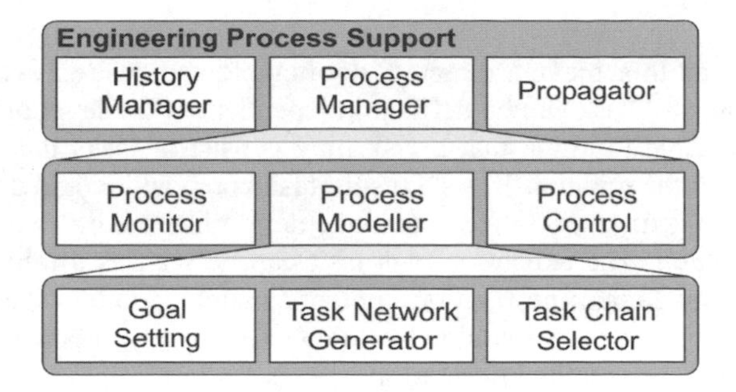

Figure 3.2 Functional components of engineering process support

However, providing that all the information generated by the different members and specialists of the design team during the development process is attached to an overall and widely accessible model, this situation may change considerably. In that case, instead of sending or rather 'pushing' information from one department to another, team members can 'pull' the information they require and are given access to on demand.

In this way, development cycles become information-based instead of process-based. Hence, in the design of manufacturing systems, the focus must be on information supply in support of an effective progress of the manufacturing process. Therefore, the course of the manufacturing processes should be governed by the need, supply and use of information.

3.4 Workflow Management

The engineering process support (EPS) concept that governs the workflow management is aimed at coordinating and supporting activities based on evolving information content [3]. The concept is divided into three main functional components addressing process management, decision support and knowledge management, respectively. There is a strong correlation between these three components, for example, between process management and knowledge management, where historic process information buttresses the construction of new process models (see Figure 3.2).

The process modelling methodology, as defined within the EPS concept, presents a novel approach to the management of engineering processes. The process modelling functionality is part of the 'process manager' component. The process modelling functionality itself is decomposed into 'Goal setting', 'Task network generation' and 'Task chain selector'. The methodology that has been developed for the generation of process models is based on linking task definitions to information content.

Tasks and Task Networks

Within the context of information management based systems, a task is defined as a systematic set of defined work to be done or undertaken in order to arrive at a specified result and to achieve a directed, predictable and desired evolution in the information content. A task can be defined independent of the level of aggregation; hence it is reasonable to argue that a task may consist of tasks that together can realize the work of the initial task. As a result, task composition and decomposition allow for the realization of process management functionality across different levels of aggregation. The execution of a task can be seen as a transition from a state established by certain information content to another state, established by an evolved information content with added value. Consequently, a task can be seen as a process that needs and yields information.

Individual task definitions can be linked to form task networks by mapping their respective inputs and outputs. In theory, when this is done for all tasks that have been defined within a manufacturing system, the resulting network will represent all possible process paths available to that system. As a practical application, this linking of tasks enables the generation of specific goal oriented task networks. Such a network is represented as a graph consisting of tasks and information entities, and it defines all possible paths from the process goal to the current state in information content.

The process goal is used as the starting point for the generation of a task network. The network is constructed by reasoning back towards the current information content of the product information structure, thus realizing an information pull methodology for process modelling.

In the resulting network, a set of optimal paths can be found that ensures the realization of the process goal. To enable the selection of optimal paths, the tasks in the network have to be valued and assigned 'cost' and preference parameters. Both operational information (such as resource availability) and historic information are used to define these parameters.

When tasks are executed, information is generated and, as a result, the state of the information content changes. Consequently, the current task chain will be evaluated to determine if it is still the optimal process solution. This illustrates how the mechanism reacts to the evolution of the information content. The process models are generated and evaluated at run-time.

3.5 Working Principles of 'What-if' Design

As mentioned above, two approaches for 'what-if' design can be applied simultaneously: an applied approach (bottom-up) and a generic approach (top-down). Both approaches are still in their infancy, but as a first onset for a design support system, the discussion of these rather abstract methods is a good indication for the proposed working methods.

Applied Approach

In order to understand the working methods related to 'what-if' design, it is important to not only sketch the big picture, but also especially to interpret the consequences of 'what-if' design on all levels of abstraction, starting with studies in a limited domain. For such a domain, existing correlations can be charted and rules, will later contribute to the composition and verification of the generic methodology can be deduced [4]. In one domain such as the mechanical domain, a design aspect hierarchy, based on an analysis of the design process, can be developed. Based on the hierarchy, the aspects involved are divided into two groups, driving and driven aspects (see Figure 3.3). Driving aspects (influencing all other aspects) contain the information that, together with aspect specific context information, is used to define the driven aspects (influenced by only the driving aspects). A 'what-if' question changes one of the driving aspects. This change will propagate through the design, affecting one or more driving and/or driven aspects. After the attenuation of this propagation, the product architecture changes into a new, valid state, based on which the 'what-if' question can be answered.

The relations between the set of driving aspects and the set of driven aspects are one-directional and downstream. The information needed to define the driven aspects can be deduced from the driving aspects and additional, not product specific, information. The relations between the four driving aspects are of a more complex nature. The mutual dependencies between these driving aspects can be simplified by assuming that the relations between function and process plan and between geometry and material are related via the intermediary aspects. If these relations (the dotted lines in Figure 3.3) are disregarded, then the risk of circular reasoning is also prevented.

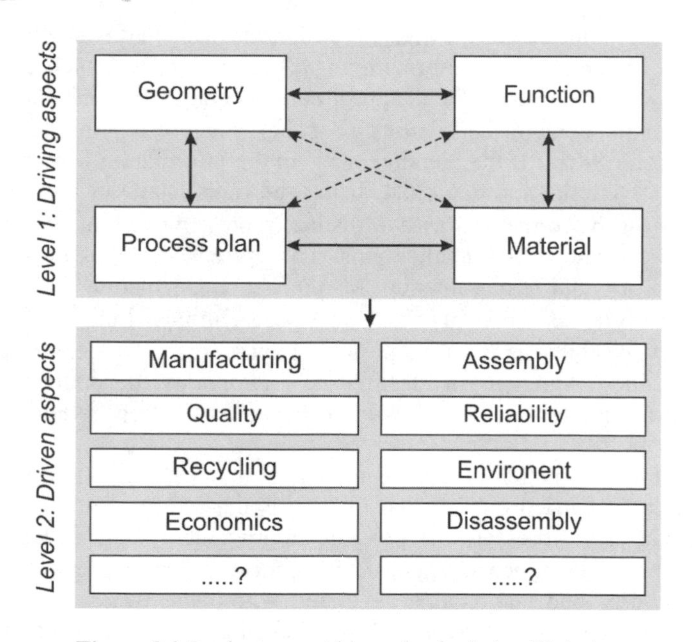

Figure 3.3 Product aspect hierarchy for 'what-if' design

Operating Procedure

In applied 'what-if' design, first the set of required information is obtained from information management functionality. When a designer considers a modification to an existing product, he has a certain improvement goal in mind. In order to reach this set goal, one of the four driving aspects will change. A new Product Information Structure (PRIS) will emerge comparing the new product configuration with the original [4]. A change in one of the driving aspects will initiate changes in other driving and driven aspects. If a check of the proposed change leads to conflicts within the other driving aspects, new non-conflicting variants of the design have to be generated.

Based on the information content, inconsistency detection functionality will indicate problems or possible flaws, leading to the generation of design variants. In following the arrows in level 1 of Figure 3.3, both clockwise and counter clockwise, two tree structures emerge, each representing a set of design variants, with more branches germinating at each level. For example, if another material is selected then more than one process is possible, each of which is capable of generating more than one geometry.

When many processes and materials are available to the designer, the tree structures will grow to a substantial size. As it is impossible to examine all the individual design variants, a ranking algorithm is needed in order to focus the designer's attention on the most promising design variants. Hence, design variants can be generated and assessed, whilst the validity of the solution is guaranteed.

Generic Approach

From a generic point of view, 'what-if' design can be seen as a request for the controlled evolvement of information. More precisely: 'What-if' design hypothesizes a certain status of the information content that deviates from the actual information content. This causes a Δ-state, which results in an ambiguity in the information content. If the 'derangement' in the information content is known, the two sets of information that exist (being the initial status S and the changed status S+Δ) can be formalised as two distinct representations of the information content or ones with different intentions. The current status of the information content S is a valid representation of the current situation in the product development cycle. As the status S+Δ is purposely imposed on the system, by a member of the design team, to test a possible improvement, S+Δ can be equated to a desired situation. Although this desire can be temporary (depending on whether or not the proposed change is really an improvement), it can be considered as a goal to focus on [5].

In using the information content as a basis, it is not important how and why a certain activity is carried out; all that matters is the information content it deals with. Thus, in order to assess the possible improvement suggested by the team member, S and S+Δ can be used to initiate Engineering Process Support. Based on the task networks and task chains generated within the workflow management procedures, a number of processes will be initiated to achieve a valid situation for S+Δ. This valid solution can probably be reached only by changing the related entities as well. Consequently, the related entities will have to accept a Δ-status as well. This again triggers EPS procedures, resulting in the initiation of design and engineering processes. At first sight, this leads to a stupefying chain reaction that instantaneously saturates any manufacturing system.

Manageability of 'What-if' Design

In addressing not only the information content, but also the denotation of the information (by means of an ontology) and its context, the mentioned chain reaction is confined. Using ontology constrains the chain reaction in two ways. Firstly, it values relations between entities, so dependencies are evaluated in a meaningful manner. Second, the ontology, together with workflow management, holds knowledge of the mutual dependencies between entities, thus stringently limiting the number of variations that have to be assessed. There is no theoretical proof that these measures are effective in achieving a practical environment for automated 'what-if' design. However, it goes without saying that indications that are usually employed to assess the quality of a design [6] remain unchanged if 'what-if' design is used.

There is another reason to believe that the disturbance caused by 'what-if' questions will be limited. After every change, the entire product description must still meet all the requirements and demands imposed by the designer. A 'what-if' based system can never leave the boundaries of what is permitted to 'return' to it in a subsequent step. This implies that after each change, the evolved value $S+\Delta$ is assessed, leading to early detection of impossibilities and indications for 'weak spots'. Therefore, the evolution of the information content can almost be guided along preferred routes, strongly limiting the number of variant situations that have to be compared.

3.6 Application and Prospects

A basis for a 'what-if' design method was introduced. A fully functional implementation is still an enormous amount of research and sheer hard work away. However, first impressions of how such a system might work are not as far away as it may seem. The modular constitution of the information–centred approach allows the implementation of relatively small applications. These applications by themselves cannot solve any question autonomously. However, because workflow management techniques can (re-) combine a multitude of different tasks in such a manner that the exact need for process capability can be compiled, 'what-if' design based on the information content can – in theory – address almost every conceivable question. Its main strength lies in the fact that all types of questions can be dealt with: from quotidian questions, to questions that have never been posed before. It is the combination of available tasks that determines the possibility of answering a question, not the focus or complexity of the question.

Prospects of 'What-if' Design

'What-if' design is constructed as a growing and flexible system: the more tasks are available for interfacing with the information content, the more complex questions can be answered. It is unimportant for the added tasks to add broader or more specific capabilities since workflow management deals with different levels of abstraction, and all types contribute to the system.

Once the set of available tasks is extensive enough, the system will become an efficacious and useful design support system. Not only can routine problems be solved by the system; but more importantly, members of product development teams can asses the consequences of decisions, or compare alternatives. In this manner, product developers can focus on the more creative jobs and limit the standard tasks to a minimum.

The 'what-if' approach has an important drawback: in order to arrive at an information content defined by S+Δ, the initial state S must be available and well defined. This contrasts with typical design problems where the initial state S itself is often somewhat blurry. Moreover, a 'what-if' system cannot assist in actually translating design constraints into design solutions. This is obvious, as no computer system will ever be able to take over the creative process done by designers. However, much of the constraints have a rather technical nature, requiring more good bookkeeping than bright ideas. This leads to a lot of routine work, which cannot be dealt with by the 'what-if' approach. Nevertheless, instinctively a solution for this is not impossible.

How-to Design

Providing that a 'what-if' system is extended with a module that can actually come up with valid descriptions of initial states S, the system can find a basis for questions itself. Referring to the notion of 'what-if' design, this added module can be indicated as how-to design. This shows that the scope is broadened, focussing on the fulfilment of constraints. Consequently, how-to design sets the context in which 'what-if' design is used to fully elaborate the problem. This obviously brings along an additional framework for the functioning of such an abstract system. Here the interaction between structured information and context information becomes even more important and entangled. Especially the role of knowledge and its application in design and engineering processes, as well as software applications supporting those processes, will change considerably. Additionally, the way in which knowledge can be extracted from experience, both of humans as well as these stored in design histories, must be studied, as this will contribute considerably to the understanding of generating design solutions in a more or less automatic mode.

3.7 Concluding Remarks

Based on information management and workflow management techniques, a design support system based on 'what-if' principles has been outlined. Although the realisation of such a system will take some doing (to say the least), the modular structure allows for the development of a growing system, which can contribute to the product development process (already in an early stage).

For the future, we foresee a design support system in which how-to and 'what-if' designs together aid the members of development teams to keep product designs consistent, to automatically perform a lot of routine work, and to actually come up with sensible proposals for design solutions. As all methods used in information management, workflow management, 'what-if' and how-to design are independent of the information content and the specific capability of each task, the design support system will not be limited to certain domains or areas of interest. As the used methods are generic, and the available tasks can be deployed in a flexible and combinatorial manner, the support system will become a learning system, adapting itself to the growing and changing situation in which it is employed in.

3.8 References

[1] Lutters, D., 2001, "Manufacturing Integration Based on Information Management," *PhD. Thesis*, University of Twente, ISBN 90-365-1583-1.
[2] Quine, W.V., "Ontological Relativity", In: *Ontological Relativity and Other Essays*, Columbia University Press, New York, ISBN 0231033079, pp. 26-68.
[3] Mentink, R.J., Van Houten, F.J.A.M., Kals, H.J.J., 2003, "Process Management for Engineering Environments Based on Dynamic Process Modelling," *Annals of the CIRP*, Vol. 52/1, pp. 351-354.
[4] Vaneker, T.H.J., Lutters, D., Hittorf, G., Van Houten, F.J.A.M., 2004, 'What-if-design; An illustration of applicability in the field of mechanical design," *Proc. of the Int. Conf. on Competitive Manufacturing*, Stellenbosch, 4-6 February 2004.
[5] Lutters, D., Vaneker, T.H.J., Van Houten, F.J.A.M., 2004, "'What-if design'; A Generic Approach Based on Information Management," *Proc. of the Int. Conf. on Competitive Manufacturing*, Stellenbosch, 4-6 February 2004.
[6] Suh, N.P., 1990, "The Principles of Design," *Oxford Series on Advanced Manufacturing*, Oxford University Press.

4

Self Organization in Design

Bart R. Meijer

Abstract: Principles of self organization are discussed as a frame of reference and a source of ideas for new design processes that can deal with more complexity in less time. It is demonstrated that set–based concurrent engineering makes effective use of these principles. Taking this idea one step further, an evolutionary organization for design processes is proposed.

Keywords: self organization, design processes, evolutionary problem solving

4.1 Introduction

Most academic institutions still teach structured design [1, 2] to their students. Not because it is the best method, guaranteed to lead to good designs, but merely because it addresses all the relevant areas of design processes in a comprehensible way to students unaware of their own early design experiences.

Axiomatic design [3] is not fundamentally different from structured design. The design phases are roughly identical. Axiomatic design is characterized by the design matrices, which represent an efficient data representation. They show where design decisions are complicated (coupled) and where they are not. The problem of developing a set of uncoupled or decoupled design matrices spanning the design space for our problem is as complex as solving the design problem by using structured design.

Following the principles of structured or axiomatic design, one could easily see a phased plan, perfectly fit for a work breakdown structure and presumably fit for effective and efficient development processes. Industrial practice shows that this approach often results either in risk-averse incremental development of a known concept, or in cyclic hard to finalize development processes in case a new concept was pursued. It is very hard to predict up-front what dependencies in which concepts are vital to a successful design. As a consequence, the product architectures of cars and aircrafts have not changed significantly over decades of their existence. Despite claims that technology developments are speeding up, the impact of new technology or new materials is often limited to redesign of sub-systems. The problems of introducing new and unknown relations are avoided as much as possible. The opportunities of new business models are the scope of implementing new technologies into existing product platforms.

Global competition has increased the required pace for product development and innovation. Dill and Pearson recognize the need for a focus on structure and communication mechanisms that enhance cross-functional and cross-disciplinary knowledge exchange as well as learning [4]. When new concepts do appear, prototyping and testing often takes a more prominent place in the development process. In other words, through trial and error, critical relations between sub-systems are discovered and dealt with or modified until working systems architecture is discovered. Although this approach is not really a self organization process, principles of self global change organization can serve as a reference for understanding how (sub)system boundaries are settled such that the interaction needs between (sub)systems relaxed to a level where the interaction becomes manageable under all operating conditions.

In this paper principles of self organization are being discussed. It is demonstrated that relatively novel design processes such as "set–based concurrent engineering" make use of these principles and have the potential to come up with new concepts without the risk of getting stuck in cyclic, non-convergent reasoning processes. Taking this idea one step further, an evolutionary organization for design processes is proposed.

4.2 Concepts of Self Organization

In this section concepts and notions of self organization will be discussed. Self organization is a term that has at least two interpretations. In the area of systems control and cybernetics, self organization refers to systems that are capable of changing their structure and their functionality on order to adapt to new environments. Another perspective on self organization originates from a systems perspective on understanding nature, life and organizations. This perspective, called autopoiesis (*self-production*), does not take adaptation as a response to changes in the environment as axiom, but it claims that living structures influence or adapt to their environment as a means to self-maintain and improve their chance of reproduction.

Both perspectives on self organization are relevant for understanding and improving design processes. The systems and cybernetics approach may be useful for developing design support, whereas the biology driven theories may be useful in developing a better understanding of agents that can act in self–organized systems. Through introducing an observer of the interactions of both types of systems, a complexity value can serve as a reference for an increase or decrease of the order or level of organization.

Autopoiesis

Maturana and Varela developed a theory of self organization for which they coined the term autopoiesis [5]. As biologists, their motivation was a desire to grasp the identity of living systems in terms of their autonomy as a phenomenon of their operation as unitary systems. They argue that living systems are organizationally closed, autonomous systems of interaction that make reference to themselves: *An autopoietic machine is a machine organized (defined as a unity) as a network of processes of production (transformation and destruction) of components that produce components which: (i) through their actions and transformations continuously regenerate and realize the network of processes (relations) that produced them; and (ii) constitute it (the machine) as a concrete unity in the space in which they (the components) exist by specifying the topological domain of its realization as such a network.*

Maturana and Varela state that autopoietic systems are purposeless in the sense that any purposeful interaction of autopoietic systems with other autopoietic or non-autopoietic systems is a construct of observations that belong only to the domain of observed actions. Since there are many structures and organizations possible and capable of generating these interactions, these interactions do not reveal the organization or internal structure of autopoietic systems. This principle is reversed in the theory through the reasoning that forcing interactions to change through changes in the environment, if at all possible, does not lead to changes in the elements and the internal structure. Living autopoietic systems shape their environment through selectively applying their potential for interactions. For living systems, this is a natural thing to do as long as the capacities for self-maintenance and reproduction benefit.

Sequential reproduction with the possibility of change in each reproductive step necessarily leads to evolution. In particular, autopoietic evolution is a consequence of self-reproduction. Evolution as a historical sequence of changes of autopoietic systems is the result of a coupling between the processes of change under perturbations (self maintenance) and changes during the processes of self-reproduction. Strictly speaking, autopoietic systems can exist only within a limited time frame of manageable external conditions. The required length of this period is determined by the maximum number of non-manageable changes that a species can absorb from one generation to the next.

Although the autopoiesis theory is an accepted system theory for living systems and although we cannot deny that living systems have a capacity for autonomy and self–organization, the theory is not really useful for self–organization in design. Autopoietic systems have self–organization capacities in their interactions, not in their internal structure or elements. Yet with respect to knowledge and cognition, the autopoiesis theory has implications for science in general as well as for design in particular. Knowledge generation and reasoning schemes may be considered as autopoietic systems. Theories, just like self-referred systems, span a closed space of possible experiments and observations. In fact, the abduction, induction, and deduction reasoning modes, express this as the importance of axioms. The autopoietic theory on observation implies that axioms form the basis that spans the space of existence for both theories and facts [6].

Self Organization and Complexity a Cybernetics View

In an attempt to define criteria for self organization, Gerhenson and Heylighen concluded that self organization is not a property of a system as such; it is the result of a system-perspective imposed by an observer [7]. This viewpoint is consistent with similar statements made by Materana and Varela. In support of this observation, they make use of the general concept of statistical entropy as a reference for the degree of order. For any system for which we can define a state space S and a probability distribution P for the occurrence of each state, the entropy for this system H(P) equals:

$$H(P) = -\sum_{s \in S} P(s) \log P(s) \tag{4.1}$$

This measure for entropy is also known as Shannon's information entropy and is often used as measure of the complexity of a state space [8]. In analogy with the entropy from thermodynamics, one may postulate that all such systems may evolve to their equilibrium, which is the case of maximum entropy or maximum disorder. One should note, however, that the observer makes the choice of observable states. The goal of observation and study determines what state variables to consider, how to map these variables onto states and how to map continuous variables onto a finite number of discrete states. If imposing structure means reducing the number of states, then entropy is also reduced. Yet, imposing a more fine-grained structure generally increases entropy. If imposing structures can go either way in terms of entropy development, what can one expect from self organization in terms of entropy?

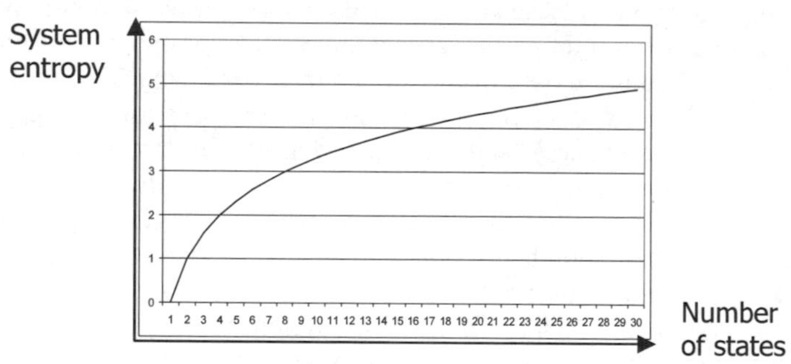

Figure 4.1 Entropy for systems with uniform probability density versus the number of states

Figure 4.1 shows the entropy for a uniform probability distribution of the states. This case is equivalent to the upper limit of the entropy for a system with n-states. Reducing the number of states means reducing that upper limit. If self organization can be associated with emergent order, then we assume that structure brings a reduction in entropy. Yet, the equilibrium of any system coincides with the entropy maximum for that system. Thus for lowering it, we need to impose a limiting condition on entropy. Such a condition allows a system to evolve to a structure with a certain level of entropy but not to break down any further into a disordered set of microstates. Fitness functions may be used for this purpose. In product design as well as in business process design, fitness functions are often productivity criteria that define the purpose of the system and favour state transitions that support this purpose over transitions that are more costly. Such fitness functions are usually based on customer value attributes [9, 10].

Sometimes the required functionality is two-fold; that is the fitness function has two attractors. To solve this problem, often two systems have to be built, each tuned to the specific requirements of one of the attractors. Thus an organizational split has occurred. This may sound trivial for functionalities without any commonality. In practice it often happens that the functional commonality in the requirements is large but that the required bandwidth is too large for a single system to handle. In such a case, a split that allows two specialized systems to be built is preferred operate together more profitably than one system that is no good at either of the tasks.

For a designer, choosing a perspective implies defining a fitness function for that particular aggregate. Such fitness functions may take the form of specifications and constraints. They may also be specified as assemblies of lower aggregate components or combinations of criteria, constraints and preferred assemblies. The mechanism of favouring some states over others is approximately equivalent to reducing the effective number of states taken into perspective. Choosing an aggregated perspective or taking the design discussion from a single design to a set of designs does the same thing, as argued in [11].

Evolutionary Problem Solving

Evolutionary problem solving is based on the structure of genetic algorithms. The basic structure of the genetic algorithm, originally developed by Holland is as follows [12]:

1. Initialize a starting population of physically feasible solutions.
2. Create a new generation through genetic operands such as mutation, crossover and reproduction.
3. Rank this population using the fitness function.
4. Select the top of this population and randomly select a couple of others to create a new starting population.
5. Repeat steps 2-4 until the top-member of a generation has a sufficient fitness score to be acceptable as a solution.

This scheme is sufficient for understanding how the algorithm works. An optimization problem is represented as a vector or a string of variables for which good or preferably optimal values have to be determined. One could call this a chromosome in genetic terminology. We can create a set of physically feasible but not yet optimal vectors. This is the initialization of the population.

Next, we create a new generation through applying genetic operands. Mutations can randomly change the value of variables or even replace the variables with others. Crossovers cut a part from the chromosome and try to replace this part with a similar part from another chromosome. Reproduction is simply creating a copy of the chromosome. Through these operands, a new generation is created.

Ranking and selection operation are highly non-linear and irreversible steps in the process. They are decisions over life and death. The chromosomes selected survive and will participate in the reproduction once more. The non-selected chromosomes are in fact dead and are removed from the process. The fitness function controls the chance of survival of the chromosomes with the best fit, so far. The random selection of some others is important for the process not to get stuck in a local minimum not yet good enough to be accepted as a solution. There needs to be new blood, so to speak, to create the diversity necessary to keep improving. The processes of creating a new generation, ranking and selection are repeated until the top member shows a sufficient fit to the ranking function to be accepted as a solution.

The success of this nature–inspired algorithm can be attributed to two properties that make it distinct from linear optimization techniques. The first property is redundancy and diversity. Rather than developing one solution, genetic algorithms develop and maintain multiple solutions concurrently. The resulting diversity is needed to maximize the probability to have solutions available at all times that can comply with all requirements and constraints. The second property is the non-linearity of the selection process. With linear optimization the fitness landscape is set from the start by the starting solution and the fitness function. Finding the optimum in this landscape could mean an exhaustive search through the entire landscape. Although the fitness landscape is set from the start, a genetic algorithm employs multiple starting points for the search, and the generation and selection steps cause the effective fitness landscape to be reshaped at the start of each generation.

For software implementation, a number of additional control parameters and heuristics are used to guide the creation of a new population. For example, one may use the ranking or fitness to steer how the genetic operands are used. If the selection of the operands is a random process, then control of the probabilities for each operand is used. The size of a generation and the size of the starting population are important variables that drive the diversity and the evolution speed of the algorithm. Also there are different strategies possible for selecting the next generation. All of these can be used to influence the efficiency of the algorithm if one can attach a meaningful interpretation to these controls in the context of the problem one is trying to solve [12].

Evolutionary Problem Solving and Self Organization

The genetic algorithm (GA) is a system model for the self–reproduction principles of the autopoiesis theorem. The solution patterns a GA may generate are predominantly the result of the initial set of solutions that were present at the start. The fitness function is the context within which structure changes may occur as long as survival as a unity or species is not at stake. Changing the fitness function will cause serious changes and may also cause death in case the present elements can not generate a sufficient fit (survival) to the new fitness function. In case of survival, biologists may recognize evolution, but they may also claim that the new organism is a different unity that is capable of a different set of interactions, fit for the new context. Thus the old species is declared extinct since it evolved into a new distinguishable organism.

Can we frame this as self–organization in terms of reducing or limiting entropy? How entropy emerges is mainly determined by the choice of chromosome variables, the fitness function and the operation of the genetic operands. We can take state transitions as variables, and the fitness function may attract certain patterns of interactions (structures) as being more attractive than others.

4.3 Set-based Concurrent Engineering

Ward and his co-authors argue that in concurrent engineering there are two fundamentally different approaches to be recognized: *point-based and set-based*. In case of point-based design, a single solution is synthesized first, then analyzed and changed accordingly. Even though the phases of the design process may be executed concurrently, all designers and specialists are investing their efforts in the pursuit of only one concept that is to be developed into a solution.

In set–based concurrent engineering, designers explicitly communicate and think about sets of design alternatives at both conceptual and parametric levels. The efficiency of set-based versus point-based design is that in communicating sets, implicitly or explicitly, all designers become more focused on relations and constraints between different aspects of the design than they would be when focusing at a point solution. All designers communicate their range of options rather than one preferred option. Sometimes to maintain focus, constraints for these sets can be set tighter than they would be in case of a point based design [13].

Ward and his co-authors found evidence that Japanese companies and, in particular, Toyota and Nippondenso deploy concurrent engineering practices that have much in common with the set-based concurrent engineering philosophy. In the next section the Toyota and Nippon-Denso concurrent engineering processes are exposed in more detail.

Set-based Concurrent Engineering: Toyota and Nippondenso

The set-based engineering processes of Toyota have the following characteristics [13]:

1. The team defines a set of solutions at the system level rather than a single solution.
2. It defines sets of possible solutions for various subsystems.
3. It explores these possible solutions in parallel, using analysis, design rules and experiments to characterize a set of possible solutions.
4. It uses analysis to gradually narrow the sets of solutions. In particular the team uses analysis of the set of possibilities for subsystems to determine appropriate specifications to impose on those subsystems.

5. Once the team establishes a single solution for any part of the design, it does not change it unless absolutely necessary.

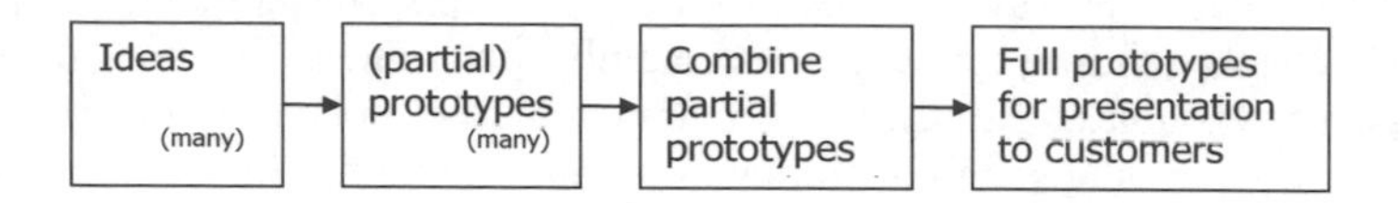

Figure 4.2 Nippondenso's R&D Process

Sets can be created by using design ranges that can be narrowed rationally once these areas have been explored. Toyota does this but is not limited to design ranges for defining sets. Toyota makes extensive use of so-called "lessons learned" books, not simply to record experience but also to define the manufacturable designs space. In the process, effort is made to avoid changes that expand the space of possible designs. In this way decisions remain valid throughout the project's life. In the early stages of conceptual car-body design, experts from all functional areas review all the alternatives, adding their manufacturability ranges to these designs and specifying possible conflict areas where existing capabilities may be insufficient. Sometimes these conflict areas can be resolved through making changes to the concept design; in other cases these areas give rise to capability enhancement projects to make the new design feasible. The lessons learned books also provide an opportunity for institutional learning. Documenting all explored solution areas and the starting point for each development provides possibilities for backtracking developments to their roots and maintains sight at built-in limitations that may not be so obvious after the concept has been reused and changed four times over.

Nippondenso, a partner of Toyota and a major automotive supplier of alternators and radiators, also applies a process that has characteristics of set-based concurrent engineering and extends this even to pre-design R&D. In this process, the degree of parallelism and redundancy is much higher than it typically is with Toyota.

As an automotive supplier, the demand for diversity is higher and their competitiveness is much affected by new technologies and new materials. In order to push the limits and to stay ahead of the competition, Nippondenso tests as many ideas as they can to create a platform (set) of solutions that is competitive and can be easily adapted to the specific interfacing requirements of different car makes. What may be a surprise is that the start of Nippondenso's development processes may be 3-5 years ahead of the start of the car development processes that adopt the new designs. Rather than pursuing rapid development once the outline of the specification from their customers is clear, Nippondenso pursues radical breakthrough designs that are ready before their customers ask for them. When they start working with their customers, the focus is on interfacing and not on the core technology, which enables them to aviod the major part of development risk.

4.4 Evolutionary Organization of Design Processes

It is possible to take the idea of set–based concurrent engineering one step further if we study the similarities between evolutionary problem solving and the practices of set–based concurrent engineering as implemented by Toyota and Nippondenso. The key to this development is, understanding the importance of the lessons learned book and the idea of combining partial prototypes as implemented by Nippondenso. Combining partial prototypes is like implementing crossings. The lessons learned book contains information on present and past fitness functions. This information is also important for assessing the fitness to the new design. The Nippondenso process starts with many concepts in parallel; another key element of genetic design processes includes parallelism and redundancy. Together these concepts contain the kernel of the genetic process; that is, how to start and create a population of promising designs.

We can make the process organizational by adopting the structure of a genetic algorithm and by taking measures to introduce and maintain the redundancy and parallelism in the process as well as in the organization. To do this, the following assumptions are needed. Most of these assumptions are general in nature; they apply to any project that is targeted and are confined with respect to time and resources:

- A target exists, and we can specify the target's requirements as a set of fitness functions to evaluate the results produced by this organization.
- There is a deadline. This means the process has to finalize before a certain date in order to meet the market window for introduction.
- All resources needed for the project are available.

Now the process can be as follows:

1. Divide the staff into n independent teams that are all capable of executing the entire project, and give all these teams an identical assignment and a deadline.
2. The teams will develop their concepts and solutions following set–based concurrent engineering practices, and they will record their achievements and findings in lessons learned books.
3. At regular intervals, a fair is organized where all teams present their progress and give insight in their lessons learned books.
4. At these fairs, team members look around for promising partial solutions with their "competitors".
5. After the fair, teams continue their own development, including ideas inspired by the last fair.
6. If a design with sufficient fitness has been achieved, stop; else, repeat steps 3-5.

The processes within the teams could also have the characteristics of a genetic algorithm if they apply brainstorming for finding and selecting ideas. However, the fair is really the place where crossovers and mutations occur. At the fair, everyone is looking for clever ideas that could fit to their own concept (crossings), and some ideas may also trigger new thoughts (mutations). Although a fitness function that could be used for ranking exists, the organizational form of a GA has the advantage that the ranking of partial ideas is fuzzy and not explicit. This means that ideas that may not be very successful in one context could be a perfect fit in another context. In case of explicit ranking, these ideas could have been lost. The implicit ranking also solves a social problem of working with a large engineering group force, where a dozen socially dominant engineers will monopolize the decision making at centralized meetings to a degree where a significant portion of the engineering staff effectively has no influence. Because the central meeting is now a fair where implicit recognition is the mechanism for the survival of ideas, good ideas, regardless of their source, stand a good chance of being inherited into the final concept. The process can be made more efficient if overlaid with a structured design process where the progress at the exchange moments (fairs) becomes synchronized.

4.5 Discussion and Conclusions

The genetic development process and evolutionary organization proposed in the previous section offer a recipe for self–organization in design. The process is self–organizing to the degree that the structure of the relations and the parts that provide the best fit to the target is not the result of careful causal sequential reasoning. Rather, it is the result of a process that provides focus through its fitness function and that, at the same time, allowed maximum freedom to copy, mutates and combines ideas from resources that may not even have been part of the design assets at the starting point.

The autopoiesis theory and the complexity/cybernetic perspective on self–organization share the concept of choosing an observer or perspective. If we try to include both in the context of design, the complexity measure offers a reference for the level of order or organization. Autopoiesis makes it explicit that we cannot expect behaviours from our agents other than the interaction patterns that were implemented through the internal elements and structure. Autopoietic systems can evolve only through evolution. The biggest challenge for design is to understand better how to deploy these principles of evolution. From the theory of evolutionary problem solving, it is learned that redundancy and concurrency are the most important ingredients for an efficient and effective optimization process.

4.6. References

[1] Pahl G., Beitz W., 1996, *Engineering Design; A Systematic Approach*, 2nd ed., Translated by Wallace K., London, Springer, ISBN 3-540-19917-9.

[2] Hubka V., Eder E., 1996, *Design Science*, London, Springer, ISBN 3-540-19997-7.

[3] Suh, N.P., 2001, *Axiomatic Design, Advances and Applications*, Oxford University Press, Oxford, New York.

[4] Dill D.D., Pearson A.W., 1991, "The Self Designing Organization: Structure, Learning, And The Management Of Technical Professionals," *IEEE Conference on Technology Management: The New International Language,* 1991, ISBN 0-7803-0161-7, pp. 33-36.

[5] Maturana H., Varela F., 1980, *Autopoiesis and Cognition: The Realization of the Living*, London, Reidl, ISBN 90-277-1015-5.

[6] Tomiyama, T., Takeda, H., Yoshioka, M., Shimomura Y., 2003, "Abduction For Creative Design," *Proceedings of ASME-DETC'03,* DETC2003/DTM-48650.

[7] Gershenson C., Heylighen F., 2003, "When Can We Call A System Self Organizing?" *Lecture Notes in Computer Science,* Vol. 2801, December 2003, pp. 606-614.

[8] Shannon, C.E., 1948, "A Mathematical Theory of Communication", *The Bell System Technical Journal*, Vol.27, July + October 1948, pp. 379-443, pp. 623-656.

[9] Kemperman J.E.B., Engelen J.M.L. van, 1999, "Operationalizing the Customer Value Concept", Competitive Paper, *28th EMAC Conf: Marketing And Competition In The Information Age*, 11-14 May 1999, Berlin, Germany.

[10] Meijer, B.R., Voûte, H.J., Tomiyama T., 2003, "Communicating Context And Strategy For Collaborative Design In Networks And Corporations," *CIRP Design Seminar,* 2003, Grenoble, France.

[11] Meijer B.R., 2002, "From Reducing Complexity To Adaptive Organizations", *Proceedings or the IEEE-IEMC 2002*, 18-20 August 2002, Cambridge, UK, ISBN 0-7803-7385-5, pp. 661-666.

[12] Banzhaf W., Nordin P., Keller R.E., Francone F.D., 1998, *Genetic Programming, An Introduction*, Morgan Kaufmann Publishers, ISBN 1-55860-510-X.

[13] Ward A.C., Liker J.K., Cristiano J.J., Sobek II D.K., "The Second Toyota Paradox: How Delaying Decisions Can Make Better Cars Faster", *Sloan Management Review*, Vol. 36, No.3, pp. 43-61.

5

Towards a Design Methodology for Self-optimizing Systems

Jürgen Gausemeier, Ursula Frank, Andreas Schmidt, and Daniel Steffen

Abstract: Self-optimizing systems will be able to react autonomously and flexibly to changing environments. They will learn and optimize their performance during their product life cycle. The key for the design of self-optimizing systems is to utilize reconfigurable system elements, communication structures and experienced knowledge. The concept of active principles of Self-optimization is an important starting point.

Keywords: Design Methodology, Mechatronics, Intelligent Systems, Self-optimization

5.1 Introduction

Information technology is increasingly penetrating the field of conventional mechanical engineering, and this offers considerable potential for innovation. Most modern mechanical engineering products already make use of the close interaction between classical mechanics, electronics, control engineering and software that is known as "mechatronics". The aim of mechatronics is to improve the behaviour of technical systems by using sensors to obtain information about the environment and the system itself. They process this information to enable the system to react optimally to its current situation.

The concept of self-optimization goes far beyond mechatronics includes systems with inherent "intelligence", which are able to adapt autonomously to varying environmental conditions. They open up fascinating prospects for mechanical engineering and related fields. To realize the vision of intelligent mechanical engineering products, there is a need for a novel design methodology. The remainder of this paper is organized as follows: First, the paradigm of self-optimizing systems is introduced. Then we show gaps that occur when trying to design those systems with the help of conventional design methods. Finally a process for the conceptual design of self-optimizing systems and a key element of the new philosophy, the active principle of Self-optimization, are presented.

5.2 Self-optimizing Systems

Future systems in the area of mechanical engineering will comprise configurations of intelligent system elements – also referred to as "solution elements"[1]. The communication and cooperation between intelligent system elements characterize the behaviour of the overall system (Figure 5.1). In terms of software engineering, this involves distributed systems of interacting agents:

> "An agent is an autonomous, proactive, cooperative and extremely adaptive function module. The term "autonomous" implies an independent control system, which it proactively initiates actions. Agents are regarded as function modules, which work in cooperation or competition with one another. "Adaptive" refers to a generic behaviour at run time, which may also, for example, include learning capabilities. A function module is taken to be a heterogeneous subsystem with electronic, mechanical and IT-related components." [7]

Combining the paradigm of intelligent agents with mechatronic structures makes it possible to construct self-optimizing mechanical engineering systems.

> "Self-optimization of a technical system refers to the endogenous modification of the target vector due to changing environmental conditions and the resulting target-compliant, autonomous adaptation of the structure, the behaviour and the parameters of this system. Self-optimization, therefore, far exceeds known control and adaptation strategies. Self-optimization enables empowered systems with inherent "intelligence," which are able to react autonomously and flexibly to changing environmental conditions" [7].

The examination of self-optimizing systems is based on four aspects: the target system (*e.g.* a hierarchical system of targets or a target vector), the structure (*i.e.* topology of mechanical components, sensors and actuators), the behaviour and the

---

[1] A solution element is a realized and proven solution to the fulfillment of a function. It will generally be a module or component that rests on an active principle. The computer representation of a solution element comprises various aspects such as behaviour and shape. Each of these aspects demonstrates a different concretization, and these correspond to the individual phases of the development process. The 'shape' aspect includes a rough specification for the determination of the principle solution and further specifications for determining the construction. In the case of software, the 'behaviour' aspect includes, among others, the abstract data types for the early development phases and code for the later ones [4]. A solution element may be a self-optimizing mechatronic function module (MFM) or an assembly of such modules, but it can also be a mechanical engineering element such as a hydraulic cylinder, or other components such as a control units or sensors [7].

parameters (Figure 5.2). Self-optimization is characterized accordingly by two features:

- The endogenous modification of the target system based on changing influences on the technical system, and
- The target-compliant autonomous adaptation of parameters, behaviour and structure.

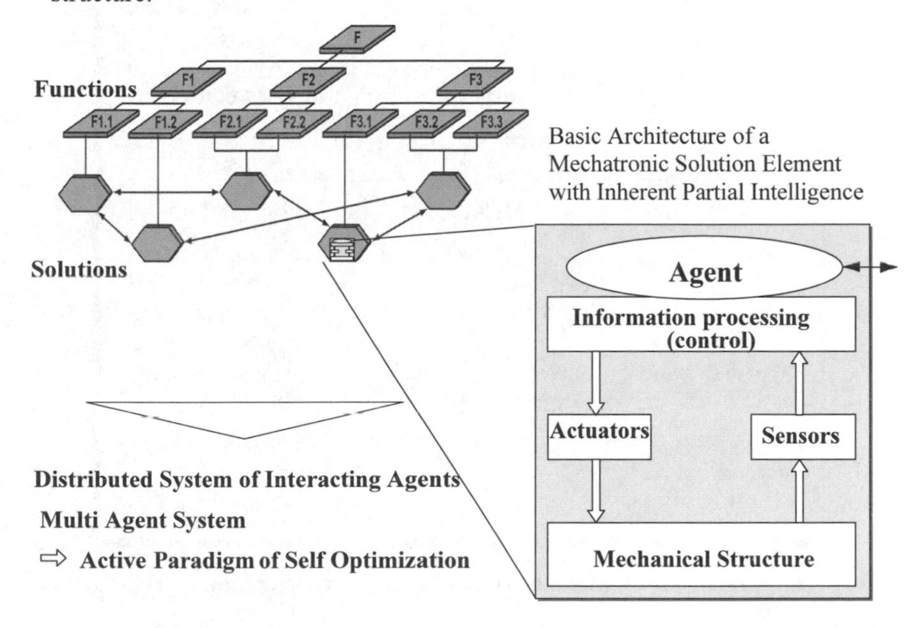

Figure 5.1 The basic idea – mechanical engineering products with inherent partial intelligence

The aim is to carry out the Self-optimization on the basis of mathematical models. The optimization process utilizes a realistic physical model of the controlled system. Whenever optimum parameters have been verified, these parameters are transferred to the controller. Oftentimes it will not be possible to calculate optimum parameters within acceptable time or with given resources. That is why behaviour-based Self-optimization is applied in combination with Man-based approaches. Behaviour-based Self-optimization acts cognitively, quasi-non-deterministically. Changes that occur during operation will be sensed and analyzed, and, as a result, either appropriate mathematical optimization models are loaded, or, if limitations of available models are exceeded, the system may revert to using past experience in the form of learned structure, behaviour and parameter settings from its knowledge base.

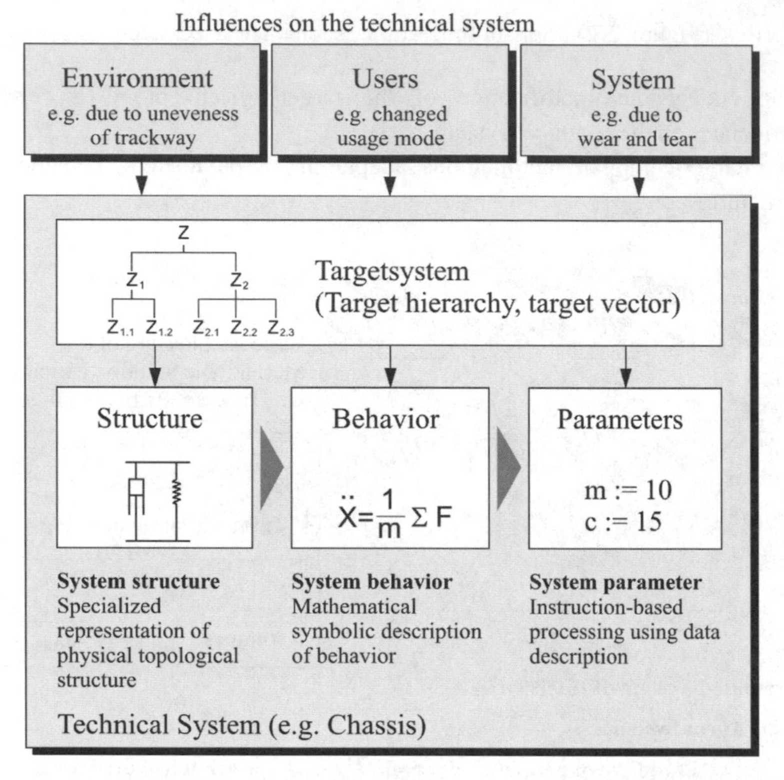

Figure 5.2 A self-optimizing system in terms of target, structure, behaviour and parameters

5.3 Design Methodology for Self-optimizing Systems

The guideline VDI 2206 "Design Methodology for Mechatronic Systems" [8] of the Association of German Engineers (VDI) suggests a systematic inter-domain design process for mechatronic systems. The fundamental structure of a mechatronic system consists of a control loop comprising a mechanical structure, sensors, online-information processing and actors. The active manipulation of the generally mechanical basic structure allows the system to compensate for interferences and adapt to changing environmental circumstances [6]. The system engineer lays down the principle adaptation behaviour of the mechatronic system at design time. The influences considered to be relevant have to be identified. A controller is designed for an acceptable behaviour under these circumstances. As a rule, it is a compromise between contradictory requirements. Adaptive controllers enable higher levels of adaptability, but remain still a compromise for anticipated situations.

In contrast to the design of mechatronic systems, the developer of self-optimizing systems is not anticipating every type of use implemented in the control system [3]. Moreover, the developer provides possibilities for independent adaptations and points out boundaries for Self-optimization processes. The restrictive definition of theses boundaries is made mainly through the target system.

The Process of Conceptual Design for Self-optimizing Systems

The development of self-optimizing systems represents an extension of the development of mechatronic systems. Therefore, the VDI guideline 2206 offers a useful starting point. In the system design process, all involved specialists elaborate a cross-domain concept of the aspired system, the principle solution. With the principle solution as a common basis, all individual domains start their simultaneous elaborations. Frequent adjustments and coordinated system integration ensure that all requirements are fulfilled.

Essential determinations of the future product are initiated in the phase of conceptual design. On that score the single domains need a sufficiently detailed and secured concept for their domain-specific elaborations. Within the conceptual design phase, predefinitions in form of an early assembly structure are made concerning the target systems of the self-optimizing system, the function structure and active structure, as well as a raw geometry. At the end of this phase, all domains need to have a clear definition of the system's structure and how functionalities are realized. From that every domain derives constraints for the system components, which they have to develop.

The conceptual design process can be shown in a phase diagram according to the one in systems engineering (Figure 5.3) [2].

1. Problem Analysis/Risk Analysis
   Starting from the development task (customer order), the expected core problems have to be defined. All influences on the system, possible sources of errors and risks, as well as their effects, are analyzed and requirements are derived.
2. Requirements Analysis
   Self-optimizing systems reach their functionality by pursuing a target system, which can be adapted to different situations. The target system is extracted out of the requirements. It is successively concretized during the design process.
3. Synthesis
   At first a function structure is created. Working with this structure, developers search for active principles and solution elements for the realisation of the functions. Geometrical and kinematic structures for the basic mechanical system are specified simultaneously. Then Self-optimization-scenarios are defined for typical situations of usage. Based on this, necessary communication structures and patterns of Self-optimization can be selected.
4. Analysis
   Alternating with the synthesis step, current results are analyzed and revised. The essential criterion is the fulfilment of the required functionality. This is examined by simulations (*e.g.* kinematics analysis) or by approximate calculations.

5. System Evaluation
   Alternative concepts are compared with one another and evaluated. This aims to save development resources by reducing the number of concepts pursued simultaneously.
6. System Decision
   The final result of the system design process is a detailed principle solution. At best, only one secured concept is released for the domain specific elaboration. The decision is made by all domains involved.

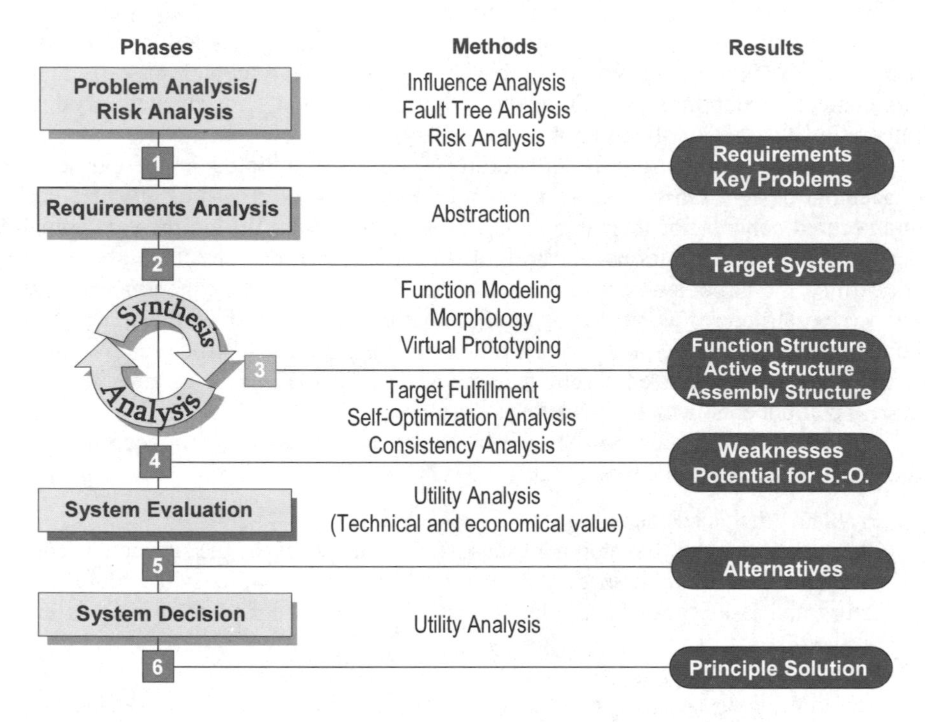

Figure 5.3 Conceptual design of self-optimizing systems

The design of self-optimizing systems requires that the six phases be passed several times. This is necessitated by strong dependences between characteristics of the design object (shape – kinematice – controller concept – communication system). Therefore, the developer can approach the goal only in an iterative, cyclic procedure. While passing the six phases, the emphasis is set on different aspects. First, for example, a functional structure is created followed by adding active and assembly structures until a comprehensive concept – the principle solution – is reached.

Particularities in Designing Self-optimizing Systems

The core phases within the design of self-optimizing systems are phases two to four. In these phases the targets of the system are defined and assigned to subsystems. This means that the most important influences in operating situations have to be considered and appropriate patterns of Self-optimization have to be chosen. The paradigm of independent adjustment at run-time means that at the end of the design cycle to identify and define the system adaptation potentials in which it may change its behaviour. The following principles determine Self-optimization:

- Reconfiguring system elements
  An adaptation to different environmental situations presupposes the presence of system elements which can be reconfigured or which can interact with other system elements in different combinations. In a chassis, for example, redundant actors (mechanical feather/spring, pneumatic spring, hydraulic cylinder) are used. They are used together in different ways (parallel/in series) to absorb different stimuli.
- Communication
  System elements behave like software agents. They pursue their targets according to the target system of the overall system. They achieve these targets by negotiations and co-operation with other system elements. For adjustment processes and negotiation principles, generic patterns are defined. Examples for communication relations are the chassis reconfiguration or an arrangement about the right of way between two vehicles.
- Experienced knowledge

In order to ensure the optimal behaviour in unknown operating situations or in situations that are not described in models, experienced knowledge embodied as cases is stored and used again in similar situations. It is shared with other systems, as well. So-called active principles of Self-optimization describe generic patterns of behaviour, which can be used in many situations [5]. Especially the use of active principles of Self-optimization creates greater opportunities and enables absolutely new functionalities. The concept of active principles of Self-optimization is described in the following section.

Active Principles of Self-optimization

Active principles of Self-optimization are meant to be a combination of a *technical system* and the *influences on the technical system* (the *environment*, the *user*, or other *system elements)* and *adaptation components*. The technical system consists of a *structure model,* in terms of the topology of mechanical components or the hierarchy of multi-agent systems, a *behaviour model*, such as differential equations or planning and learning systems, and the *parameterization* of the models. A *target system* prescribes the current goals which the technical system tries to achieve. In this way the active principle of Self-optimization allows for the endogenous modification of the technical system according to changing influences, as well as for target-compliant, autonomous adaptation of parameters, behaviour and structure. Adaptation strategies and adaptation tactics define the kind and process of modifications for long-term and medium- to short-term adaptation to application scenarios. Adaptation costs represent the effort of adaptation in terms of energy-consumption, time-delays, monetary payments and the like. Altogether the active principles reflect a structure of detailed or generalized behavioural patterns for a mechatronic system.

Figure 5.4 illustrates the process of applying the active principles of Self-optimization (APso) within the scenario of employing consistent target combinations in a multi-agent setting. The multi-agent system is composed of a drive-module agent, an engine agent and a battery agent. The technical system selects an active-principle of Self-optimization from the system's knowledge base based on the current influences from the environment, the user and other systems. The retrieval process operates according to the similarity paradigm of case-based reasoning [1]. Figure 5.4 identifies two active principles that are deemed to be appropriate in the given scenario – namely Autonomous Multi-Target Prosecution and Cooperative Multi-Target Conciliation. While the former follows the paradigm of individual agent optimization, the latter stresses the aspect of distributed problem solving. The latter is considered for the remainder of this section. Cooperative multi-target conciliation brings about a negotiation-based adaptation strategy. The involved agents negotiate for a common consistent target system by means of market auction mechanisms. Depending on individual characteristics, the agent executes different adaptation tactics at each time step of the negotiation process. A time-dependent tactic describes the agent behaviour under time constraints, *i.e.* when the time to close the negotiation runs short, an anxious agent will increase its offers with a higher gradient compared to an even-tempered agent. The resource-dependent tactic considers technical constraints such as energy levels of batteries. For example, if the energy level of a battery drops below a certain threshold, the drive-module will try to realize a behaviour which follows the target of loading the battery again. Behaviour-dependant adaptation tactics follow the goals of the other agents. Adaptation costs reflect the effort to adapt the system according to the selected adaptation strategy and tactic – in this case, costs of reconfiguring system controllers are affected.

The target system of the technical system is made up of consistent multi-target combinations. Consistent target combinations are developed in the course of system design according to the methodology of scenario-forecasting, techniques [4]. At first, the system designer develops a target-consistency matrix, which contains consistency values for all target tupels. The consistency values reflect whether two targets support each other (high consistency value) or contradict each other (low consistency value). A combinatorial recombination process results in a list of decreasing consistent target combinations depicted in the lower left of Figure 5.4. For example, the target system of the battery agent comprises the individual targets of safety, comfort and energy. According to target combination one, the safety target of agent one is judged consistent with the energy target of agent two and the comfort target of agent three. The structure of the technical system is reflected in a multi-agent setting. The behaviour of the multi-agent system realizes the selection of one of the target combinations by means of negotiation. The parameters, which guide multi-agent behaviour are made up of the above mentioned consistency values.

The technical system applies the active principle of "Cooperative Multi-Target Conciliation" as a negotiation-process for each agent. As a result of the negotiation, a consistent combination of targets is cooperatively determined at run time. All agents are then committed to realize those targets in the subsequent execution process.

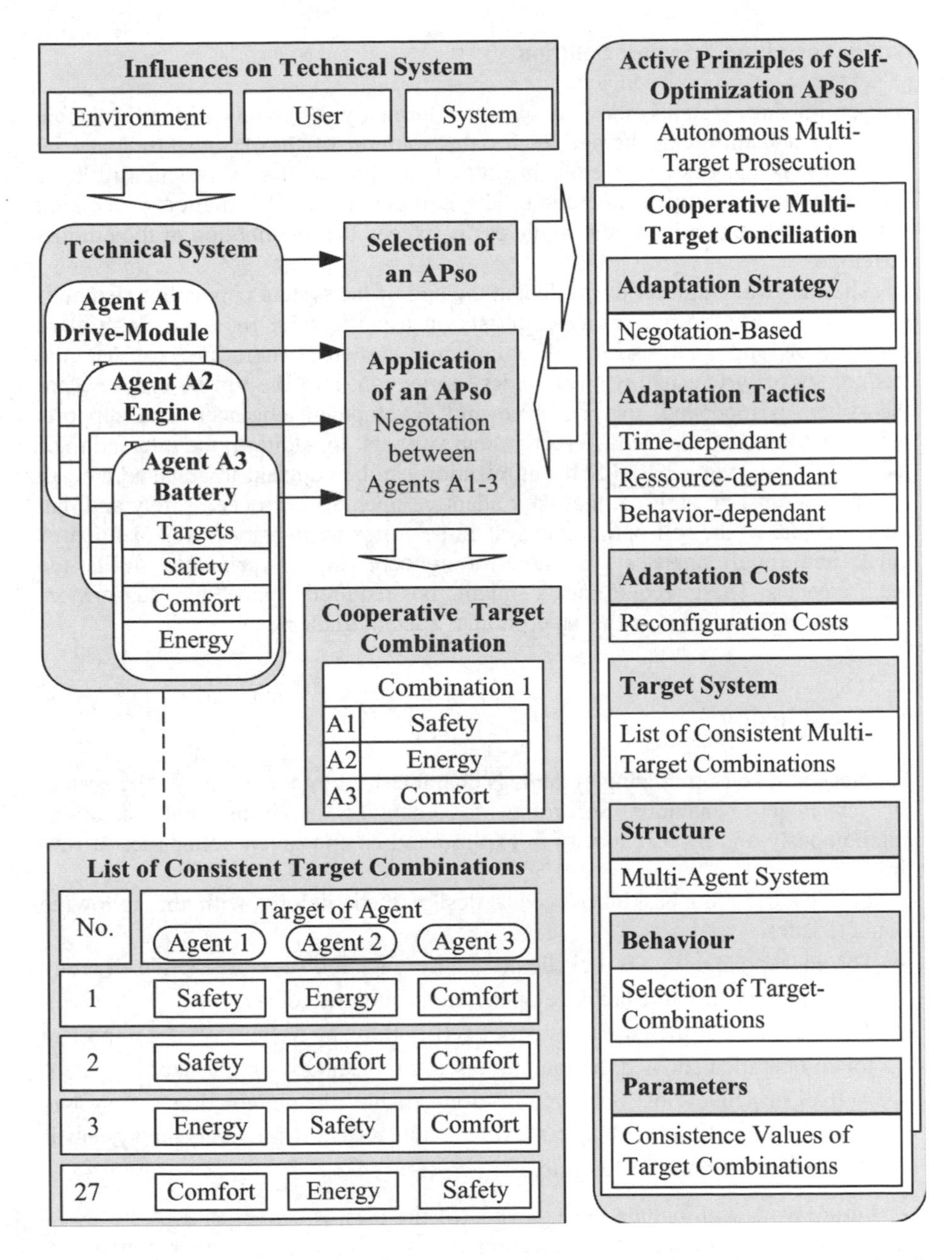

Figure 5.4 Active principles of self-optimization illustrated at the application scenario of cooperative multi-target conciliation

Validation of the Principle Solution

Self-optimizing systems show a quasi-evolutionary behaviour. Many possible situations and different alternatives for the solution of the design task must be tested during the design process, in order to guarantee that a system fulfils its functions later during its operation. The complexity of information processing within a self-optimizing system might easily exceed the imagination of the product developer.

The early validation of the system at the end of the system conception demands new approaches. We put strong emphasis on a particularly intuitive, descriptive composition and analysis of self-optimizing systems. Therefore, we use the technology of virtual reality: it provides immersion, into the virtual space where the system is modelled, for one or several development engineers and supports real-time testing of a lot of different system variants. In addition, the interaction of model- and behaviour-based Self-optimization can be adjusted in advance. For all concepts it must be verified, that their adaptive operations proceed safely and that the behaviour of the self-optimizing system performs in the way it was planned. A quick and rough analysis guarantees a concept that in principle fulfils the requirements. These requirements might be required space for suspension, admission of high forces or the absorption of special frequencies.

5.4 Conclusion

The design of self-optimizing systems is characterized by the ability of the system to select its behaviour according to changing environmental situations autonomously and by its capacity to explore and extend given boundaries at run-time.

This contribution has introduced a design methodology with the following characteristics.

- The conceptual design is a cyclic procedure: function structure, active structure and assembly structure are developed almost simultaneously.
- The definition of the target system of self-optimizing system is very important for co-operation and negotiation processes, so it is highly emphasized.
- Active principles of Self-optimization enable the interaction of system elements in several hierarchical levels of the system. Moreover, the system is able to learn during its life cycle.

Future work will include the extension of the idea of active principles of Self-optimization, in particular the evolutionary and evolving behaviour patterns of communicating and cooperating functional-module agents.

5.5 Acknowledgement

This contribution was developed in the course of the Collaborative Research Center "Self-Optimizing Concepts and Structures in Mechanical Engineering" (Speaker: Prof. Gausemeier), funded by the German Research Foundation (DFG), under grant number SFB614.

5.6 References

[1] Aamodt, A., Plaza, E., 1994, "Case-Based Reasoning: Foundational issues, methodological variations, and system approaches," *AI Communications*, Vol. 7(1), pp. 39-59.

[2] Daenzer, W.F., Huber, F., 1994, "Systems Engineering – Methoden und Praxis; 8. verbesserte Auflage;" *Verlag Industrielle Organisation*, Zürich, Germany.

[3] Gausemeier, J., 2002, "From Mechatronics to Self-Optimization," *Proc. 20th CAD-FEM Users Meeting, International Congress on FEM Technology*, 2002, Friedrichshafen, Germany.

[4] Gausemeier, J., Ebbesmeyer, P., Kallmeyer, F., 2001, *Produktinnovation – Strategische Planung und Entwicklung der Produkte von morgen*, Carl Hanser Verlag.

[5] Gausemeier, J, Schmidt, A., 2003, "Wirkprinzipien der Selbstoptimierung," *Proc. 14th Sym-posium Design for X*, 2003, Erlangen, Germany.

[6] Isermann, R., Lachmann, K.-H., Matko, D., 1992, *Adaptive Control Systems*, Prentice Hall, New York.

[7] SFB 614: *Einrichtungsantrag für den Sonderforschungsbereich 1799* (ab 1. Juli 2002: 614) "Selbstoptimierende Systeme des Maschinenbaus", Universität Paderborn, 2001.

[8] VDI 2206: *Entwicklungsmethodik für mechatronische Systeme*, Beuth-Verlag, Berlin, 2003.

Part II

Design Knowledge and Functional Design

6

Reusing Design Knowledge

Saeema Ahmed, and Ken Wallace

Abstract: The long-term aim of this research is to develop a method of indexing design knowledge that is intuitive to engineering designers and therefore assists the designers to retrieve relevant information. This paper describes the development and preliminary evaluation of a method of indexing design knowledge. The concepts for the method have been elicited from designers' descriptions of the design process. The method has been evaluated by indexing 92 reports related to one particular aero-engine.

Keywords: empirical studies, design support, design knowledge, indexing

6.1 Introduction

A recent report from the Department of Trade and Industry (UK) concluded that the most significant factor to improve innovation in the UK's manufacturing industry is to understand and transfer the knowledge possessed by engineering designers – along with the know-how they have about how to apply that knowledge [1]. In engineering design, a large amount of knowledge is generated during the design process. For example, in the aerospace industry, approximately 40,000 documents are produced in the design of a single aero-engine [2]. Some of this knowledge is captured in the form of memos, emails, sketches, reports, etc while some is retained in the head of the designers. Many systems propose methods of capturing and storing knowledge. Examples of such systems include DEKLARE, a methodology that supports engineering redesign, and PROSUS, which captures the rationale behind designs [3, 4].

There are many reasons why designers may wish to reuse knowledge, including enabling others to understand the original design process and the rationale behind the decisions made; and searching for past designs when working on a similar product or problem. The benefit of documents is related to the relevance of the knowledge captured within the documents. Recent research has shown that designers rarely access documents that they have not directly contributed to producing. A report is likely to be referred to only once a year and the most likely person to refer to it is the author [2]. Designers' awareness of the documents and their ability to retrieve them with ease contribute to the reuse of these documents.

Indexing design knowledge is one method to support the retrieval of knowledge from a system. The design knowledge, which may be in many formats including memos, emails, sketches, reports, etc, is either indexed manually when it is captured into a system or automatically, once captured. Current approaches to indexing design knowledge include automated indexing such as Dedal AI, which can improve the precision and recall of Boolean searches [5]. Dedal AI automatically indexes parts of a query by identifying generic design concepts. Another such method is the Precision Content Retrieval Method, which uses conceptual indexing to build a structured conceptual taxonomy of words and phrases extracted from the indexed material and uses specific passage retrieval to find specific passages and rank them according to relevance to the query [6].

The long-term aim of this research is to develop a method of indexing design knowledge that is intuitive to an engineering designer and therefore assisting the designer to retrieve relevant information. At the start of the research project it was unclear whether the knowledge should be manually indexed or automatically using parsing technologies for example. However, whether the indexes are available for the designer to see or form the underlying structure of a software tool is somewhat unimportant for the purpose of this research. It is the *root concepts* that form the indexing method that are considered important. The term *root concept* is used to refer to a group of terms that can be used for indexing. For example, the physical product may be a root concept. The root concept will have specific terms associated with it in the case of an automobile the terms associated with the root concept of the physical product could be door, steering wheel, wheel, engine, *etc.*. In order to identify these root concepts, an empirical research study was carried out in two aerospace companies.

This paper describes the development and preliminary evaluation of a method of indexing design knowledge that is based upon an empirical research study. The empirical research study aimed at understanding how designers described the process of designing a particular component or assembly and to use this understanding to identify the root concepts for a method of indexing design knowledge. The research method employed and the findings are described in the following sections; along with the proposed method of indexing and its subsequent evaluation.

6.2 Research Method

Interviews were carried out to understand how designers described the process of designing a particular component or assembly. The descriptions of the design processes were analysed and led to the development of a method of indexing design knowledge. The findings from this study are summarised here and are described in more detail in [7]. Prior to the interviews, it was hypothesised that how designers describe their processes of designing can be classified in four ways:

- the *process* itself, *i.e.* a description of the different tasks at each stage of the design process
- the physical *product* to be produced, *i.e.* the product, components, sub-assemblies and assemblies
- the *functions* that must be fulfilled by a particular component or assembly
- the *issues* whilst carrying out the design process there are several considerations the designer must make whilst designing, *i.e. issues*.

Participants

Eighteen engineering designers with different levels of experience and from two different companies were interviewed. Both of these companies were large aerospace companies based in the UK. The engineering designers interviewed were all graduated with degrees in mechanical engineering. Their experience within the aerospace industry ranged from 2 to 42 years. The designers were grouped into three different groups depending on their experience and also their design role. These groups were:

- designers with under 5 years of relevant experience
- experienced designers with between 11 and 23 years of relevant experience
- designers with between 28 and 42 years experience who had moved on to more managerial roles and were no longer directly designing.

A summary of the participants is presented in Table 6.1. Each row of the table describes the level of experience; the current team; and the assembly discussed. Designers 1-11 were all from the same company (referred to as company A) working on various assemblies of an aero-engine and designers 12-18 were from the second aerospace company (referred to as company B) working on various assemblies of an aircraft.

The eleven designers from company A were from three different teams within the aerospace company and each of these teams worked on a particular assembly of an aero-engine, for example turbines. The seven designers from company B were from five different teams, with the teams were divided on a role rather than product basis, for example engineering computation. Designers in company B would work across different teams on a project by project basis. The participants were selected from a number of teams to avoid overburdening any one team and also to ensure that the findings were not specific to one particular assembly or design task.

Methods

During the interviews the designers were asked to describe the process of designing a particular component or assembly that they were currently working on or had recently been working on (refer to Table 6.1 for a list). The designers were allowed to talk freely and care was taken not to communicate any of the expected results to the designers prior to the interviews.

In addition to collecting descriptions of the process of designing a particular component or assembly, the interviews also provided an opportunity to evaluate the suitability of two taxonomies for the purpose of indexing design knowledge. These two taxonomies were based upon the initial hypothesis and were: descriptions of functions using verbs and nouns; and a list of *issues*. The evaluation of these taxonomies was conducted after the designer had described their design process to avoid biasing their descriptions of the design process. Taxonomies for the design process and the physical product were not evaluated, as these were specific to the design task and are well understood.

Table 6.1 Level of experience of interviewees

Designer	Experience (years)	Company	Component or Assembly
1	11	Company A	Turbine Casing
2	18	Company A	Turbine Intermediate Pressure Casing
3	11	Company A	Turbine Internal Casing
4	20	Company A	High Pressure Compressor Casing
5	23	Company A	High Pressure Compressor Drum
6	2	Company A	Compressor Rotor Blade
7	4	Company A	Compressor Intermediate Annulus Line
8	4	Company A	Compressor Disc
9	2.5	Company A	Compressor Intermediate Pressure Rear Cone
10	2	Company A	Fans System Inner Ring
11	5	Company A	Fans System Inner Ring
12	28	Company B	Wing
13	27	Company B	Keel Post
14	39	Company B	Fin
15	30	Company B	Concept Aircraft
16	39	Company B	Refuel Door Panel
17	42	Company B	Aircraft Hydraulics
18	36	Company B	Foreplane

Method of Analysis

Each description of the design process was transcribed. The transcripts were broken down into small segments and each segment was analysed to identify any of the root concepts, *i.e.*:

- stages of the *design process*
- references to the *product,* including component, sub-assemblies, *etc.*.
- references to the *functions* to be fulfilled by the particular product
- references to *issues* that need to be considered.

An example of an analysed section of transcript is presented in Table 6.2, the first column is the designer's own words and each row represents a segment of the description, which are consecutive.

Table 6.2 Example of analysed description of design process

Designer's description	Design Process	Product	Functions	Issues
Define the material in order to assess weight and cost	Material selection, cost & weight assessment			Weight, unit cost
Think of assembly at a component and module level		Component /module		Assemble
Cost/function analysis, overall and individual costs	Cost/function analysis			Unit cost
Contain blade: calculation to check this	Calculations	Blade	Contain blade	
Pressure dilation, calculate if strong enough as a pressure vessel	Calculations		Withstand pressure	

6.3 Findings

The breakdown of all the descriptions of the design processes is shown in Figure 6.1. Each segment of a description referred to steps of the design process; components or assemblies; functions; or issues; or any combination of these. Therefore, the graph does not add up to one hundred percent, but instead represents the percentage of the description that referred to each of these.

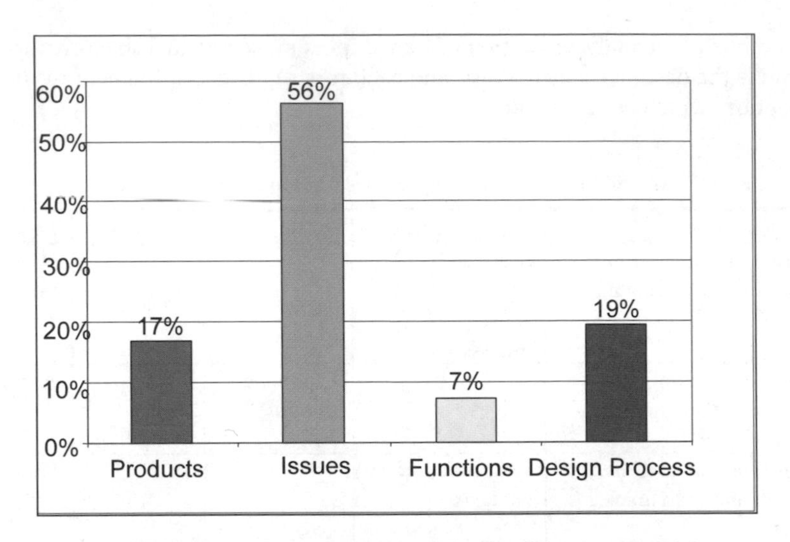

Figure 6.1 Breakdown of the descriptions of design processes

On average, 56% of a designer's description of the design process referred to issues that were considered; 19% referred to steps in the design process; 17% referred to products including the component or assembly being designed as well as surrounding components and assemblies; and 7% referred to the functions of the component or assembly. The descriptions of their processes varied with their level of experience. Figure 6.2 shows a breakdown of the descriptions of the design process against the level of experience of the designers. The designers with between 11 and 23 years of relevant experience referred to functions for 16% of their descriptions this was significantly higher (four times higher) than the designers with fewer than 5 years of experience and the designers with between 28 and 42 years, who were no longer designing.

The level of experience also influenced the number of references to steps of the design process; components and assemblies; function and issues. On average, the more experienced designers mentioned almost twice as many references (56) in their descriptions of the design process than the designers with under 5 years of experience (25).

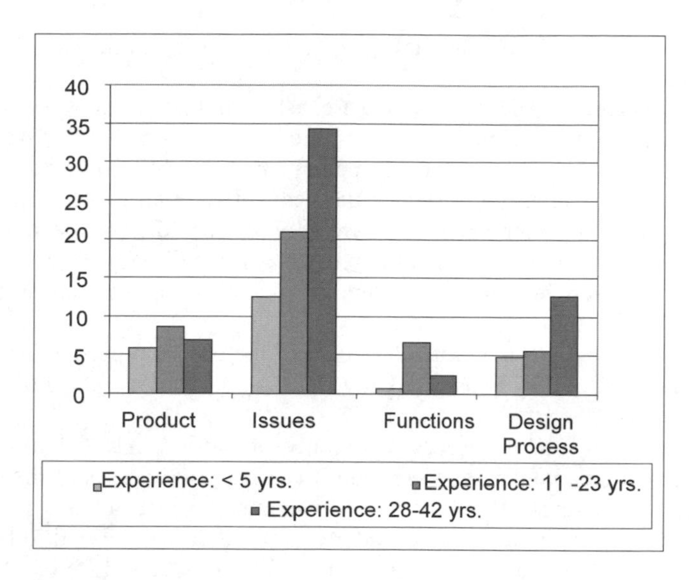

Figure 6.2 Breakdown of descriptions of design process against level of experience

Issues Taxonomy

Designers from Company A generated a list of sixty issues, specific to the aero-engine. The researcher then grouped these issues into four classes. These were issues that related to: 1) the lifecycle of the product; 2) the environment of the product and interfaces; 3) the functionality of the product; and 4) the characteristics of the product (refer to Figure 6.3). The four classes were identified from analysis of transcripts of think-aloud observation of designers working on design tasks, which was part of a separate research project [8].

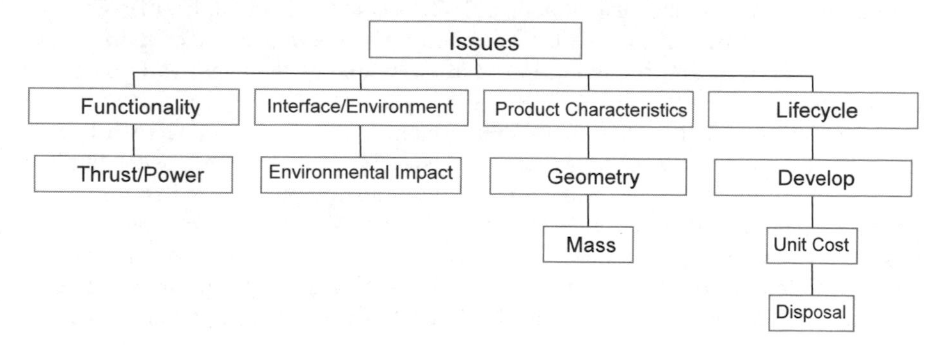

Figure 6.3 Issue taxonomy: generic classes with examples of issues

During the interviews the issue taxonomy was evaluated from three perspectives:

- Completeness: The designers were asked if they considered any issues whilst designing which were not on the list.
- Issues specific to particular components or assemblies: For each issue, the designers were asked to state if they considered that particular issue when designing the component or assembly discussed. The designers stated whether the issue was considered directly; indirectly (the issue was a consideration for that component or assembly but was considered by a different designer); or not at all.
- Issues specific to the particular stage of the design process. The designers were asked if they considered any issues at a particular stage of the design process.

The evaluation of the issues taxonomy identified one additional issue, corrosion. No additional classes were identified. The four classes of the taxonomy were found to be relevant to all the assemblies and sub-assemblies across the two companies. However, issues classed under functionality were found to be specific to a particular product, as the functionality of the products differed. The issues were of too high a granularity to identify if any issues were only relevant to certain stages of the design process.

Functions Taxonomy

The taxonomies of functions evaluated for their suitability of indexing design knowledge were those developed by Szykman *et al.,* and that of Hirtz *et al.,* and are applicable to a broad variety of engineering artefacts [9,10]. These taxonomies aim to facilitate the capture and exchange of function information. The taxonomy developed by Hirtz *et al.,* integrates the efforts of Szykman *et al.,* with those of Stone and Wood [11]. The resulting taxonomy is referred to as the functional basis with a set of functions (verbs) and flows (nouns). A function of a component or assembly can be described using the list of verbs combined with a list of nouns, for example, fasten material solid object rigid-body. Hirtz *et al.,* state the facilitation of indexing, searching and retrieving information as one of their motivations for the taxonomy.

In order to evaluate the function taxonomy, the designers were asked to describe the breakdown of the assembly or sub-assembly that they were familiar with. Each of the assemblies was broken down into components and features. The designers were asked to describe the function of each feature or component. The designers were shown examples of verbs and nouns that could be combined to describe functions. However, they were not asked to use any particular set of verbs or nouns and were able to describe the functions of each component or feature using their own words. The designers did not refer to the list of verbs and nouns. The evaluation of the functional taxonomies are summarised below and are described in more detail in [12].

In total, 207 descriptions of functions describing various sub-assemblies of an aero-engine and an aircraft design were collected. Eighty-six descriptions of functions were collected from designers 1–11 from company A and a further 121 descriptions of functions were collected from designers 12-18 from company B. The verbs from the descriptions were compared to two function taxonomies: a direct match, indirect match or no match was recorded. A direct match was defined if the verb used by the designer was the same as that from the taxonomy. An indirect match was recorded if a synonym was used or if it was possible to restate the description of a function with a combination of a verb and noun from the taxonomy. If a designer's description of the function could not be restated, no match was recorded. The nouns were also abstracted to be at the same level as the nouns from the taxonomy, for example, *blade* became *material: solid object: rigid-body*.

It was found that approximately 90% of the functions could be described using the functional basis proposed by Hirtz *et al.* The taxonomy of Hirtz was found to directly match 63% of the verbs that the designers used, which was a significant improvement on the earlier taxonomy of Szykman (refer to Table 6.3). However, 31% was not matched directly, *i.e.* a suitable alternative description of the function had to be found from the taxonomy. All of the cases that could not be matched were related to the verb seal, which was used several times by designers from both companies. In some cases this was used to describe sealing against a physical component rather than to enclose a material and was therefore difficult to describe in terms of the taxonomies.

Table 6.3 Evaluation of function taxonomy: matching of verbs

Taxonomy	Company	Direct match	Indirect match	Total matched	Not matched
Szykman *et al.*	A	26%	65%	91%	9%
Szykman *et al.*	B	31%	59%	89%	11%
Hirtz *et al.*	A	63%	31%	94%	6%
Hirtz *et al.*	B	50%	39%	89%	11%

The evaluation highlighted some of the issues that may be raised by introducing such a taxonomy to engineering designers in industry. The reasons for an indirect match of verb, extended further than the use of a synonym and highlights the need to consider both the verb and noun together when rephrasing a designer's description of a function to that of the evaluated taxonomy. The provision of a visible taxonomy when indexing knowledge may be sufficient for the designers to adapt their descriptions to the language of the taxonomy employed. The evaluation of the function taxonomy found that by abstracting a designer's description of functions to the verb and noun of the taxonomy resulted in a loss of information. However, these problems could be overcome by combining the function taxonomy with two of the other taxonomies from the indexing method: 1) issues; and 2) physical product. It was found that abstracting to the same level as the nouns from the function taxonomies resulted in *all* physical products becoming *material: solid object: rigid-body*, for example, *blade* became *material: solid object: rigid-body*. By combining with the product taxonomy, information that the knowledge indexed is about a blade is retained.

6.4 Indexing Method

The specific questions considered when developing the method to index design knowledge were: 1) How do designers wish to search for knowledge? and 2) Can the indexing terms encourage the use of knowledge that would not be ordinarily identified as appropriate. From the findings form the empirical research study and the evaluation of the two taxonomies, the following implications for the development of a method to index design knowledge were drawn. The four root concepts were identified in all of the descriptions and no additional root concepts were found, thereby confirming the initial hypothesis. Differences were found in the breakdown of descriptions between novice and experienced designers, particularly in the number of functions mentioned during the designers' descriptions of the design process. By including functions as one of the taxonomies, the visibility of the functions may encourage less experienced designers to think in terms of functions and access knowledge that they may have not considered searching for otherwise. The development of the method also needs to identify or develop the appropriate taxonomy for each of the root concepts. The evaluation of the taxonomies found that the part of the issues taxonomy related to the product functionality is product specific. The level of granularity of the terms on the issues taxonomy needs to be refined. The function taxonomies evaluated were found to cover 90% of the designers' descriptions, and therefore seem to be a good starting point.

The evaluation of the method is being carried out in three separate stages: 1) to evaluate suitable taxonomies for the indexing method; 2) to test the indexing to evaluate a sample set of documents; and 3) to evaluate the relevance of documents retrieved using this method. The first part of this evaluation has been completed for the function and issues taxonomy. This paper describes the second part of this evaluation. The third part of the evaluation will require the development of a computational software tool before it can be carried out. This is currently in progress and will form part of the future research. A total of 92 reports were indexed to evaluate the suitability of the indexing method. The reports varied in length from 25 to 250 words and are the first reports raised for a potential change on a product. The reports were indexed manually and classified under the headings of the four root concepts and against the individual terms. If more than one of the terms or root concepts was applicable, the report was indexed more then once. All 92 of these were indexed using the method. Table 6.4 shows the breakdown of the root concepts against which the reports were indexed. Almost 94% of the reports were indexed against *product,* and 94% against *issues*, although these are not necessarily the same reports. If indexing had been restricted to only one of the four taxonomies, then not all of the reports could have been indexed. This suggests that it is necessary to have more than one root concept as part of the indexing method.

Table 6.4 Number of reports indexed

	Product	Issues	Functions	Design Process
Number Indexed	86	86	11	51
Percentage Indexed	93.5%	93.5%	12.0%	55.4%

6.5 Key Conclusions

Structured interviews with engineering designers have been carried out to develop and test a proposed structure to index knowledge. The results from the interviews have supported the direction of the proposed indexing structure. A method of indexing design knowledge has been developed and a preliminary evaluation has been carried out. In addition, two *functions* taxonomies and an *issues* taxonomy were evaluated. The evaluation of the functions taxonomy suggested a need to combine the functions taxonomy with a product and issues taxonomy to avoid loss of information. The evaluation of the indexing taxonomy identified issues that were specific to the particular product. As part of the preliminary evaluation, over 90 reports were indexed and all the reports were successfully indexed. The evaluation indicated that it is necessary to index reports in more than one way, as not all of the reports could be indexed using only one of the four taxonomies. However, using all four taxonomies allowed all of the reports to be indexed. A further 300 reports will be indexed to evaluate the indexing method further. A separate evaluation to assess the ease of retrieving documents using the method will be carried out once software implementation has been completed.

6.6 Acknowledgements

This work was funded by the University Technology Partnership for Design, which is a collaboration between Rolls-Royce, BAE SYSTEMS and the Universities of Cambridge, Sheffield and Southampton. The authors acknowledge Dr. Michael Moss and Mr. Alastair Stewart for their specific assistance throughout the research and the designers who participated in the study.

6.7 References

[1] http://www.dti.gov.uk/insight_manufacturing.html

[2] Marsh, J. R., 1997, *The Capture and Utilisation of Experience in Engineering Design*, PhD. Thesis, Cambridge University, Cambridge, U.K.

[3] Arana, I., Ahriz, H., and Fothergill, P., 2000, "Redesign Knowledge Analysis, Representation and Reuse," In: *Industrial Knowledge Management: A Micro-Level Approach*, Ed. R. Roy, Springer-Verlag, London, pp.139-146.

[4] Blessing, L.T., 1996, "Design Process Capture and Support", *Proc. of the 2nd WDK Workshop on Product Structuring*, Delft, 1996, pp.109-121.
[5] Yang, M.C. and Cutkosky, M.R., 1997, "Automated Indexing Of Design Concepts", *Proc. of the Int. Conf. Eng. Design, ICED1997,* Vol. 2, Tampere, 1997, pp.191-196.
[6] Woods, W.A., 1997, *Conceptual Indexing: A Better Way To Organize Knowledge*. SMLI TR-97-61, April 1997, Sun Microsystems Laboratories, Mountain View, CA, pp. 91.
[7] Ahmed, S. and Wallace, K., 2003, "Indexing Design Knowledge Based Upon Descriptions of Design Processes," In: *Int. Conf. on Eng. Design '03*, Stockholm, Sweden.
[8] Ahmed, S., Wallace K.M., and Blessing, L.S., 2003, "Understanding The Differences Between How Novice And Experienced Designers Approach Design Tasks", *Research in Engineering Design*, Vol. 14(1), pp.1-11.
[9] Szykman, S., W.Racz, J., and Sriram, R.D., 1999, "The Representation of Function in Computer-Based Design," *Proceedings of the Design Theory and Methodology, ASME*, 1999, Las Vegas, Nevada.
[10] Hirtz, J.M., *et al.*, 2001, "Evolving a Functional Basis for Engineering Design," *Proc. of the ASME Design Eng. Technical Conf.: DETC2001*, 2001, Pittsburgh, PA, DTM-21688.
[11] Stone, R.B. and Wood, K.L., 1999, "Development of a Functional Basis for Design," *Proc. of the Design Theory and Methodology, ASME*, 1999, Las Vegas, Nevada.
[12] Ahmed S. and Wallace K., 2003, "Evaluating a Functional Basis," In: (eds.) *Design Theory and Methodology, Int. Design Eng. Technical Conf. & Comp. and Info. in Eng. Conf., ASME*, 2003, Chicago, Illinois, DTM-48685.

7

Structural and Functional Analysis for Assemblies

Hugo Falgarone, and Nicolas Chevassus

Abstract: This article presents a systemic method for designing assemblies. It is based on generic concepts such as modeling of assemblies using assembly nested graphs which reflect the product design breakdown, the interfaces between components. The proposed method enables to assess the product producibility and the robustness of the assembly process. It eases impact analysis following changes of modified product functions or features.

A software tool, called GAIA, has been developed to support this method; based on a user-friendly interface. It enables specifying assemblies through interfaces and performing a functional and structural analysis of assemblies. Interoperable with the Digital Mock-up and Product Management Systems, it speeds up design changes and impact analysis. Finally, it is useful to grasp the design intents and to capitalize and reuse this design knowledge.

The adoption of this advanced modeling technique in support of the engineering assembly process improves the quality of designed products and reduces the cost of change management, customization and fault rectification by solving assembly issues at the design stage.

Keywords: Assembly modeling, Design through interfaces, Assembly process analysis, Change management, Systems engineering, Structural and Functional Analysis, Computer Aided Design, Process Aided Design

7.1 Industrial Background

The design and manufacturing targets for aerospace products are derived from performance, quality and cost requirements. These products are made of a high number of parts, with various assembly levels. The prime aircraft manufacturers rely on an extensive manufacturing organization including a plurality of suppliers.

One way to monitor these targets is to perform as soon as possible in the design cycle, and next, to refined and maintain them, structural and functional analysis of assemblies for assessing alternative design and process solutions according to performance, producibility and affordability requirements.

Traditional Computer Aided Design tools (CAD) help designers to set up product geometrical definitions. However, these tools do not easily capture designers' intent as it should be needed in order to record the functional specification cascade with respect to the product's breakdown. For complex assemblies, the main hurdle that prevents designers from understanding the results of systemic analysis deals with the lack of representation over the product 3D geometry of both functional requirements and interfaces between components.

The modelling of functions and assembly interfaces is also needed to perform change and impact analysis over the product. A key question is: what are the effects on functions of a change in a product characteristic, and conversely, what are the consequences on product features of a functional modification. The effective identification of all the impacts based on lists of potentially concerned components as given by Product Data Manager (PDM) systems is a heavy task for which the Digital Mock-Up (DMU) and CAD tools provide weak support limited to clash detections.

Considering the foregoing, EADS Corporate Research Centre has developed a generic design method, supported by a software tool, for systemic analysis of assemblies. This software is capable of managing in liaison with the DMU the assembly requirements and interfaces through product breakdown and throughout the product life cycle.

7.2 Method for Systemic Analysis of Structural Assemblies

The need for systemic analysis of assemblies is mainly associated with the design stage of each assembly level, when the product design principles and the sub-structure breakdown have been defined.

The proposed method can be seen as a transposition to the specific field of structural assembly of the systems engineering approach as described in [4] where all subsystems are first defined from the interactions with other systems and the functions it shall fulfil.

Starting from the results of requirements analysis, the first step is to allocate requirements on the components of the product for each assembly level (requirement loop). In a second step, once functional requirements are allocated, the interfaces of subsystems are specified and the subsystems are designed (design loop). Finally, the resulting assembly design is validated in order to check initial product requirements. The overall systems engineering process loop is illustrated in Figure 7.1.

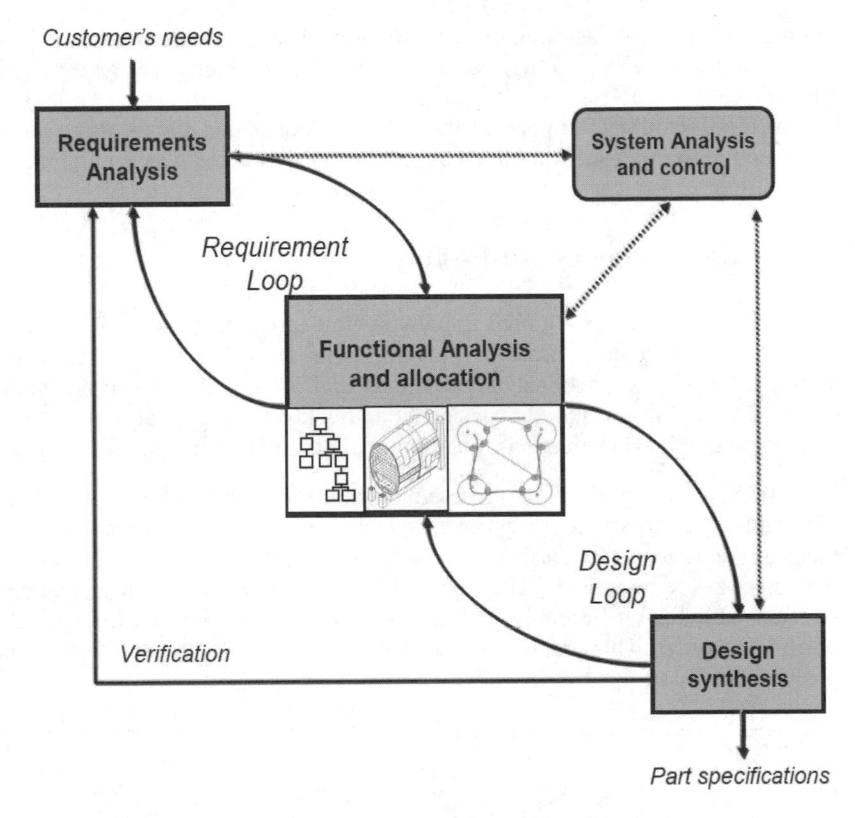

Figure 7.1 Assemblies engineering

The proposed method specifically considers the three following activities:

1. Identification of Product Key Characteristics (KCs), Assembly KCs, and finally Process KCs.

   The aim of this step is to identify geometrical KCs from functional analysis and to cascade those KCs through product breakdown.

2. Process design.

   This step leads to the definition and comparison of alternative assembly sequences according to different manufacturing solutions. The evaluation of sequences is based on various criteria like cost, lead-time, quality and accessibility indicators.

3. Specification of interfaces.

   This step deals with the allocation of geometric requirements between parts, the refinement of assembly principles and the exact tolerance distribution through the interfaces.

These activities are recursively applied to all refined components of the considered assembly.

Based on a model of assemblies with functional and structural features, the proposed method enables achieving a systemic and recursive analysis of assemblies.

This method is fully IT-supported by a design tool called GAIA that can be used in liaison with CAD software like CATIA or PDM systems like Windchill.

7.3 Assembly Modeling and Analysis

The proposed method offers a global framework to express and model design data representing a multi-layered assembly including its components, interfaces and assembly relationships, functional requirements and KCs. This structured model grasps the design intents and the design specifications. It is analysed through qualitative and graphical evaluations, quantitative computation or specific requests.

- The method is based on a representation through assembly graphs that highlights interfaces between parts as set by [2]. The circles represent parts, their grey ears represent surfaces of parts, the straight lines are interfaces and the doted arcs requirements. The Figure 7.2 shows an assembly graph made of five assembled parts assembly with one KC. The orientations of links set the positioning order. This graph shows that part 1 and part 5 are involved in the positioning of part 2.

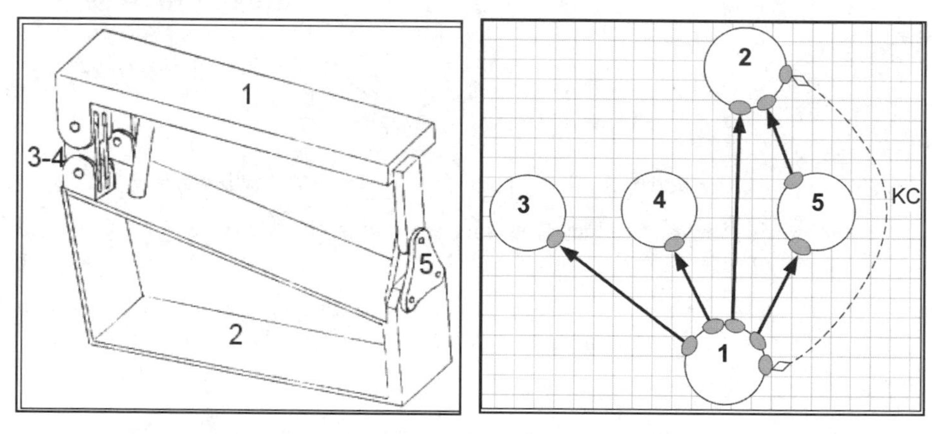

Figure 7.2 Interface principles of wing-pylon assembly

- Multi-level assembly structures can be defined with assembly graphs nested inside each other following the product breakdown. Figure 7.3 shows an assembly made of 2 assembly levels. At the upper level, there are 3 components, Subassembly 1, Subassembly 2 and Part 3, which are linked together with 3 global links, L1, L2 and L3. The two former subassemblies break down into parts, which are also linked together at a lower assembly level. The Global link L3 models the interface between the two subassemblies that are assembled together. As we know how these two subassemblies are refined, we can identify which parts and surfaces of each subassembly are involved in the implementation of the global link L3. A parent-child relationship is established between the global link L3 and the links L31 and L32. According to this example, the method helps to elaborate a complete tree structure of all the interfaces based on product breakdown.

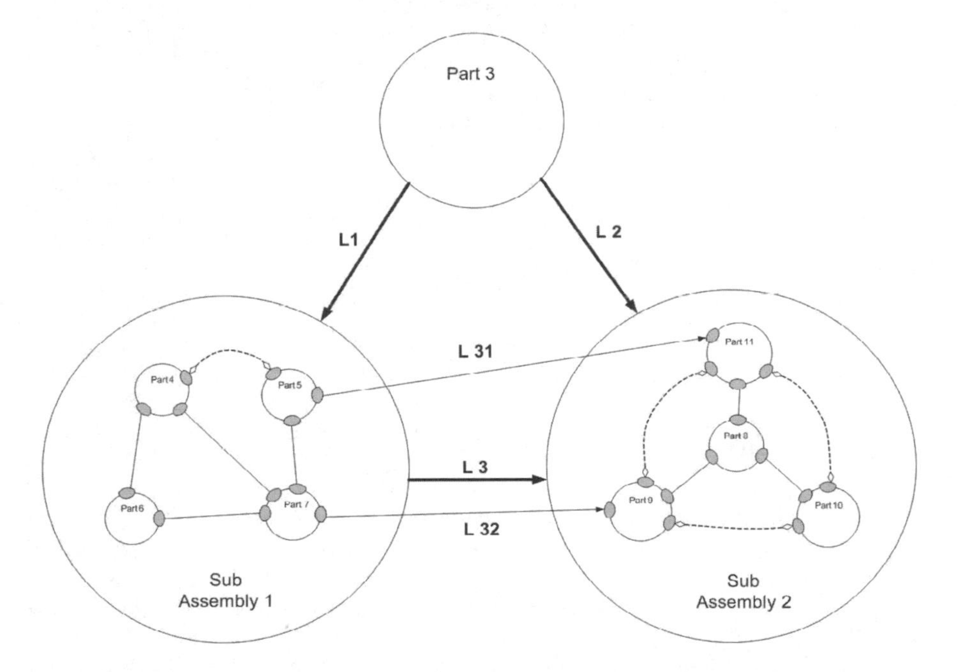

Figure 7.3 Assembly nested graph

- The proposed method makes it possible to perform a product functional analysis, to identify corresponding KCs and to cascade those KCs through the different assembly levels. The functional analysis starts with the identification of functions and constraints from external environment of the considered product or system. Each function, once refined, leads to several geometric requirements. They are assessed according to a risk analysis, which enables the selection of KCs. Each KC is also split into several requirements in assembly sub-levels and finally into geometrical requirements on functional surfaces. Finally, we obtain a cascade of product requirements that links each geometrical requirement to the main product functions.
- A global link represents an interface between two parts. It corresponds to the physical realization of mechanical joints and can embed various fastening technology [5]. At each assembly level, several assembly graphs can be set in order to compare alternative technological solutions or assembly sequences. In early process design stages, the method enables to compare alternative manufacturing solutions by focusing on cost, time and quality criteria without the availability of the full geometric definition.

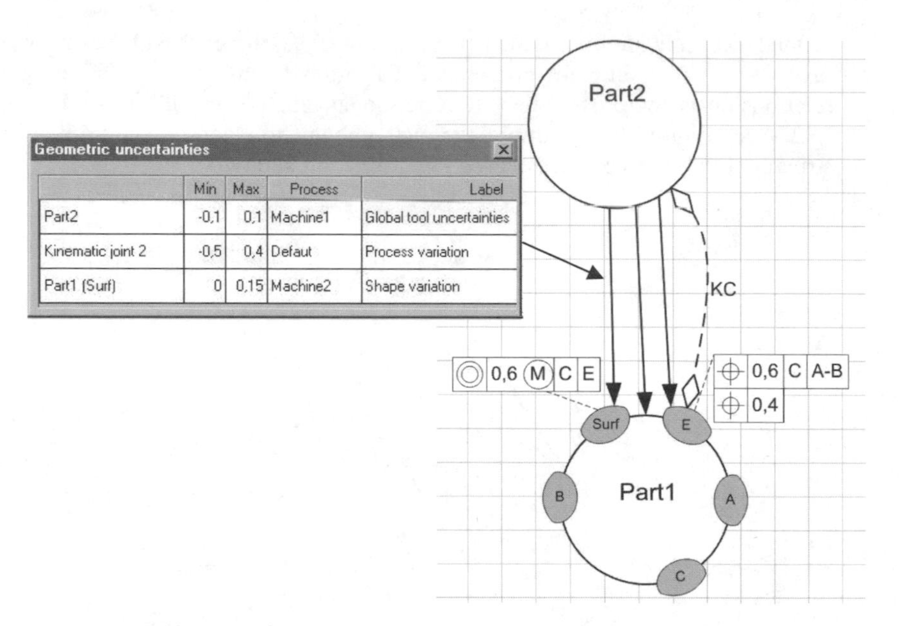

Figure 7.4 Kinematical liaisons with geometrical uncertainties

- The assembly graph model is a good starting point for setting up a geometric variation management plan. Each global link between two parts is detailed in terms of kinematical liaisons. As shown in Figure 7.4, the entire assembly graph can be seen as a kinematical scheme. Without any kinematical over-constraint, it provides input data for tolerance analysis studies. Refined analysis considering over-constrained liaisons can also be performed taking into account the flexibility of components. Finally, according to the assembly sequence, the links are oriented and the datum scheme is set.
- For each functional surface of components from the assembly oriented graph, it is possible to defined, using the support of the tolerance specification method provided in [3], the Geometrical Dimensioning & Tolerances (GDT) scheme.
- Once all the interfaces specifications have been established, some qualitative or quantitative analysis can be achieved. The best example is datum flow chain analysis for each KC. As explained in [1], the flow chain represents all the liaisons of the assembly that influence a specific KC. This list of liaisons is computed using a tolerance analysis solver. A sensitivity analysis offers a better understanding of all the liaisons that contribute to the KC variation. The variation management consists in selecting the best assembly sequences that minimize the geometric variations all over the KCs. In Figure 7.5, both chains represent the datum flow chain of two different KCs. The liaisons involved in the influencing chain are tagged with the contributing values. This helps to identify the most influential liaisons.
- Another use of the functional and structural assembly graph is to export relevant data into CAD tool in order to provide the designer with functional

and interface specification data while working with 3D complex shapes. These product specifications are useful within the DMU to elaborate interface-based design and beyond the DMU to perform direct and reverse impact analyses, which are straight forward as they rely on a dedicated structural and functional model.

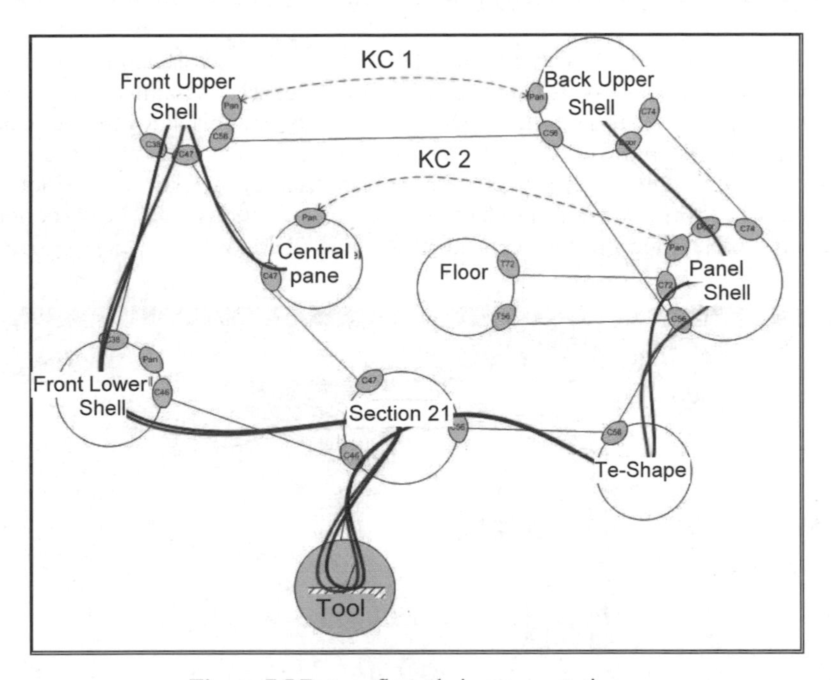

Figure 7.5 Datum flow chain representation

- The management of engineering changes at the design stage of a product life cycle is a key issue in the aircraft industry. The engineering change process suffers from numerous limitations, which restrict the efficiency of actors and the responsiveness of organization. Changes can have important consequences on both product and organization so they must be kept under control [6]. Engineering change information relative to the product is not modeled in the product architecture in current PDM systems. This weakness does not enable knowledge of the changed behaviour of components and assembly within a system to be used. With current PDM tool, it is fairly difficult to identify the consequences of a change in components on other components, because this propagation goes through items using links that are more complex than just "uses/used by" links. To analyze this propagation it is necessary to use a model of interfaces between components. In fact, as we have just seen before, the proposed model gathers the assembly requirement cascade, the interface breakdown and the datum flow chain of each requirement. Hence, it becomes possible to perform powerful multi-layered impact analysis of any local changes of an elementary part on the product's main requirements and conversely. These impacts analysis are especially relevant for large and complex assemblies if it can be driven within PDM systems.

7.4 GAIA Software for Systemic Analysis of Assemblies

The EADS Corporate Research Centre has developed a new design tool to support the method presented. This innovative piece of software is called GAIA which means Graphical Analysis of Interfaces for Assemblies. This software tool enables to grasp the design intent, the product structural and functional interfaces and the manufacturing process decisions with a user-friendly graphical user interface (GUI). The corresponding product-process specifications can be exported to various product lifecycle management (PLM) and PDM systems.

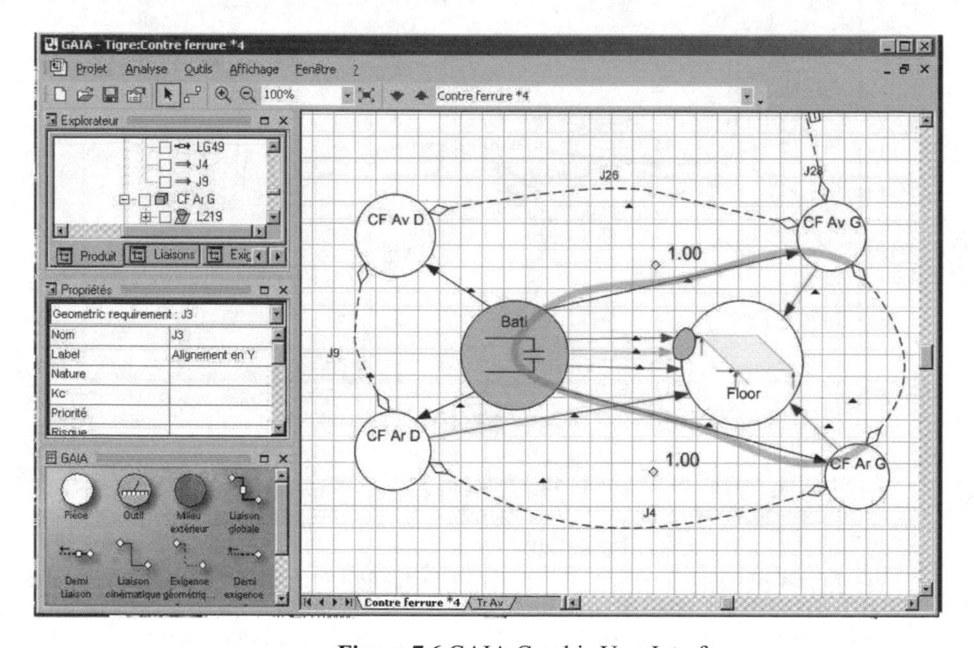

Figure 7.6 GAIA Graphic User Interface

GAIA is based on a MS Visio user interface and looks like an office tool. Its main features are:

- Support of the presented design method,
- Easy to handle,
- Adaptability to many engineering applications and
- Client-server architecture with a multi-user database.

7.5 Benefits

The table below highlights the benefits of the proposed method and tool. It stresses the relationships with other existing methods and tools for product-process assembled design.

GAIA provides a common framework and repository throughout the product-process design phases for supporting various methodologies and tools. Its main advantages are:

- GAIA is a visual tool for specifying assemblies through interfaces.
- GAIA supports the functional and structural analysis for assemblies.
- GAIA speeds up design changes and impact analysis.
- GAIA, coupled with CAD and Computer Aided Process tools (CAP), enabled iterative design from GAIA specifications to CAD/CAP definition and back.
- GAIA makes it possible to capitalize and reuse design knowledge about assemblies.

Table 7.1 Method and Tool comparison

<table>
<tr><th>Scenarios and applications</th><th colspan="2">Methods and Tools</th><th>Benefits of GAIA</th></tr>
<tr><td>Product specification</td><td>Functional analysis
System engineering</td><td rowspan="5">Systemic analysis & design of assemblies GAIA</td><td>Functions and product characteristics traceability.
Requirement cascade through product breakdown.</td></tr>
<tr><td>Interface specification & assembly design</td><td>CAD design
CATIA</td><td>Assembly modeling focused on interfaces.</td></tr>
<tr><td>Process Design</td><td>Process planning
Accessibility, cycle, cost analysis
DELMIA</td><td>Early process design.
Multi-disciplinary optimization without 3D data.</td></tr>
<tr><td>Variation management</td><td>Tolerance analysis
CeTol, 3DCS, Anatole</td><td>Visual datum flow chain.
Assembly sequence comparison.</td></tr>
<tr><td>Impact analysis</td><td>Product Data Management
WindChill</td><td>Liaison between function and product features.</td></tr>
</table>

7.6 Conclusion

In order to improve the design process of complex products, EADS Corporate Research Centre has developed a generic method for performing systemic analysis of large assemblies. This method is supported by GAIA, a piece of software that can be used stand-alone or in liaison with CAD/CAP packages.

The adoption of this advanced modelling technique in support of the engineering of assemblies improves the quality of designed products and reduces the cost of change management, customization and fault rectification by solving assembly issues at the design stage.

7.7 References

[1] Mantripragada, R., Whitney, D.E., 1998, "The Datum Flow Chain: A Systemic Approach to Assembly Design and Modeling," *Research in Engineering Design*, Vol. 10, pp. 150-165.

[2] Ballu, A., Mathieu, L., 1999, "Choice Of Functional Specifications Using Graphs Within The Framework Of Education," *Proceedings of the 6th CIRP Computer Aided Tolerancing Seminar*.

[3] Sellakh, R., Riviere, A., Chevassus, N., Marguet, B., 2001, "An Assisted Method For Specifying ISO Tolerances Applied To Structural Assemblies," *Proceedings of the 7th CIRP Computer Aided Tolerancing Seminar*.

[4] Sheard, S., *et al.*, 2002, "Systems Engineering Beyond Capability Models," *Proceedings of INCOSE*, August 2002.

[5] "Feature Based Assembly Modeling, FEAST," *Synthesis Report, Contract BRE2-CT94-1015*, Project funded by the European Community under the Brite/Euram programme, 1994.

[6] Riviere, A., Feru, F., Tollenaere, M., 2003, "Controlling Product Related Engineering Changes In The Aircraft Industry," *IDEC 2003*.

8

Knowledge Management for a Cooperative Design System

Serge Tichkiewitch, Bruno Radulescu, George Dragoï, and Kusol Pimapunsri

Abstract: Every five years, the French Ministry of Industry launches a study about the key technologies for the next five years. Knowledge capitalization was one of the mentioned technologies in 2000. This paper starts with the description of some problems forecasted at that time and the actual situation since. In this context, a definition for knowledge management is presented, and some related concepts are proposed.

Finally, it is shown how the expert system technology associated with a cooperative design modeler allows the implementation of the knowledge management concepts.

Keywords: Co-operative Design, Knowledge Management, Ontologies

8.1 Introduction

Production systems have been under constant change over the past 20 years because they have had to adapt to two major factors, the globalisation of the economy and the need for industrial innovation.

Concerning the first factor, and due to the higher labor costs, European factories in the domain of the traditional manufacturing processes have moved to some Asian countries: Vietnam, China, and Thailand. In the absence of any action, the knowledge of associated technologies will be lost in the not too distant future. In actuality, the "candidate countries" from the European Union have tended to specialize in low-cost production – a move reflected in limited transfers of production from the current Member States to the "candidate countries", which made it possible to retain activities in Europe.

In relation to the second factor, the innovation of products lies mainly in technology transfer, using unusual materials, new technologies or non-traditional manufacturing processes to obtain the parts which constitute the product. In innovation that utilizes new processes; large companies must seek partners, by using their competences in specific manufacturing processes and by integrating these competences during the design process. For example, in France when a car manufacturer wanted to introduce the "mono-space", the quantities of the launch did not allow the use of a traditional body made in steel, and it was necessary to find specialists in composite manufacture. Here, innovation is the result of integration of new partners in the design chain, in order to share their knowledge.

In order to keep the key production competences, it is imperative that the specialists are associated in a network, which allows them to share the competences and the knowledge that they still hold, and gives them the means of integrating these competences in new design systems. The CoDeMo system is an answer to this problem, as it allows the users to capitalize on their factual knowledge in the form of features, and their temporal knowledge in the form of production rules or algorithms.

One of the main objectives of CoDeMo is to create a collaborative integrated design platform allowing the different members of the network to participate either in a synchronous or asynchronous mode in collective design projects. Each member will bring in the knowledge related to his or her own expertise as part of a larger whole. Therefore, each member has to be connected to a common database and has to be able to understand in detail the part of the content, which he/she needs to use, as well as the scope of the knowledge which can be delivered by other partners involved in the network. The sharing of information in the right context needs a transformation of information into knowledge, in order to disseminate the same meaning to the different actors.

Knowledge may be universal, vehicular or vernacular. All people normally share universal knowledge. This is for example the case with geometrical knowledge. A specific actor who is only concerned with his or her own job only uses vernacular knowledge. It does not need to be shared. Vehicular knowledge is the type of knowledge which can be exchanged between two or more actors, allowing them for instance to perform collaborative design based on a common understanding. Therefore, the latter type of knowledge is very important for establishing a dialog between two partners. For example, a threaded plug in a product requires a manufacturing process, an assembly process (screwing phase) and finally a spatial structure to enable adequate access when screwing. At least, three different people are concerned with the decision to put a thread plug in a product during design; they all need to use about the same information, and apply it in their own context.

Every five years, the French Ministry of Industry launches a study about the key technologies for the five next years. Knowledge capitalization was one of the technologies raised in 2000. In this context, a definition for the knowledge management and related concepts are given in the next section. Finally, in a third part, we show how the expert system technology associated with a cooperative design modeler can give some answers to implementing knowledge management concepts.

8.2 One of the Key Technologies for 2005

Among the various technologies raised in the study about the key technologies for 2005 [1], managed by the French Ministry of Industry, the capitalization of knowledge obviously ranks high. Two problems were already presented in the 1995 study [2] and have since remained unanswered: How and what (knowledge) does one capitalize in?

These two problems are however, in 2000, considered in a context of integrated management. It is noted that: "the organization of the companies will have to adapt to an additional dimension, that of the rehabilitation of multiple steps, often taken into account in parallel: different actors in fact are today brought to work all at the same time to promote total quality, the respect of the standards, the innovation, an effective management of the processes, the environmental protection, ... A whole of behaviors, transverse to the company, is thus developed without there are necessarily coordinated or even there are dialogues between these parallel steps. However, the experiment shows that these efforts on different fields take part of the same logic and can call upon a joint base of practices. A reconciliation of these steps is likely to emerge in order to generate what some already recognize like a "integrated management", *i.e.* a coordinated management of knowledge and practices."

To tend towards a design of a product from the point of view of total service to the customer, by apprehending the better request, by integrating at the same time better human dimensions into the level of the end-users, as with that of the operators who will be brought to manufacture a product... multiplies the complexity of the environment in which the designers work.

Three major stakes emerged that requires to be translated into technological solutions.

Managing in a Complex Environment

The systems of management and production of companies (the concepts of chain logistic, capitalization of knowledge, and organization of production according to the request), induce problems of increasingly complex data management.

A Complex Environment

There is a need for developing the capacity to take into account qualitatively and quantitatively various parameters, corresponding to the needs for a "multi-actor" and "multitask" environment, in the context of increasing integration of these actors and activities. Thus, within the concept of integrated logistical support, the design of a product must include the concepts of costs, after-sales service, life cycle of the product, necessary materials, and the manufacturing chain, *etc*..... The integration of these very heterogeneous data necessarily challenges.

Volume of Information to be Treated, and thus to be Selected, Classified and Filed for Use

An existing problem of companies is the fact that the volume of information available is constantly growing with the size of the databases. The tools to assist creativity, and the capitalization of knowledge, as well as the computerized decision-making systems or integrated management, continue evolve with the development of "intelligent" software. This allows for a more effective use of information, meeting needs clearly expressed a priori. The management of the "rules of trades" gives a good example of the challenges involved. Thus, work of formalization (definition of the need: on what one will capitalize, and under which form?), and of management of the rules (how they evolve and progress?) is of considerable importance. There is also a paradox to be surmounted: the safeguarding of know-how in time, while avoiding the risks of obsolescence of any part of the data.

Means of Communication of Information

The problem of interface between systems is the principal glue for the divided information. The heterogeneous data to be integrated are very often from various types and sources (various systems of management, various computer set-ups, and various data formats). Data processing regulations and protocols are elements for the resolution of the technical difficulties. It remains that the concept of compatibility between the systems is a technological side challenge.

Restitution of Information

To fulfill the requirements of all involved individualized answers for each need to be formulated. For example, a designer will not require the same representation of a vehicle as the engineer charged with the design of its engine. Nevertheless, as they are working on the same product, their actions must be coordinated. The multi-representation of the same object, albeit from various perspectives, useful for the various actors, is a major technological objective. In an international and multicultural environment, the surmounting of mutual comprehension barriers is also important. In addition to the representation of objects, the simulation of human behaviors is also crucial, in particular the understanding of human perceptions. For example how to set-up the machines after sales instructions manuals which will be easily understood by all the users from several cultures? How to capture the individual needs and to deduce some useful information for the products information feedback? How to ensure a comprehensible utilization of knowledge for the future users of information? Linguistic engineering, work on the perception of the consumers are technologies that integrate this human dimension in the processes. It is therefore clearly seen that taking into account various points of view implies important considerations at the onset of the product design process.

Adapting the Tools, Parts and Materials Used

The requirements for speed and responsiveness faced by industry, in the fields of design, production and management, have a direct impact on the tools and materials used. The accentuated use of information systems and the development of "virtual" technologies (information systems, processes of digitization) make it possible to consider today largely digitized processes of design, automated production equipment, and a smoothly flowing and reactive management of companies.

Design
The increasingly reliable and complete representation of objects makes it possible for the designers to numerically consider the whole of the properties of a product. Thus fast "prototyping" evolves to the simultaneous creation of adapted tools very close to the pre-production.

Production
The impact of virtual technologies on production is to be noted. The digitization of the various processes of a factory is considered to facilitate the optimization of the entire flow in the factory such as scheduling of the lines of production, and management of the orders of production. In the same way, intelligent production machine tools are able to manage their rate of production. The remote monitoring of production systems (monitoring of routine and diagnoses of dysfunction), and the related decision systems (mean of action in the event of problems) are directly related to the technological solutions being implemented in this field. Sensors and actuators are the essential elements of this type of mechanism: the challenge relates in particular to their reliability and their capacity of resistance in constraining environments.

Forming / Evolving the Organization

Finally, it is in the "soft" field of sciences that great changes can be expected, although it is not still possible to specify completely what the target organization will look like in ten years.

Parallel concepts will evolve for integrated management, from integrated logistical support, customer service in social contexts, increasing demand for environmental laws, and with the development of the electronic commerce and information technologies. The technological developments will result in new industrial organizations structures that are flexible and reactive, without barriers to communication.

It remains to define the methods of evolution success and the defining of the changing role of each individual in the changing processes. This is important to ensure the smooth transition between the organizations of today and those of the future.

8.3 Knowledge Management

In order to define and to give some characteristics of knowledge management (KM), let us have a look at the proposition of Y. Malhotra in [4]:
"Knowledge management caters to the critical issues of organizational adaptation, survival and competence in face of increasingly discontinuous environmental change. Essentially, it embodies organizational processes that seek synergistic combination of data and information processing capacity of information technologies and the creative and innovative capacity of human beings."

This is a strategic view of KM that considers the synergy between technological and behavioral issues as necessary for survival in "turbulent environments". The need for synergy of technological and human capabilities is based on the distinction between the "old world of business" and the "new world of business."

Within this view, Malhotra defines the old world of business as characterized by predictable environments in which focus is on prediction and optimization based efficiencies. This is the world of competence based on "information" as the strategic asset, and the emphasis is on controlling the behavior of organizational agents toward fulfillment of pre-specified organizational goals and objectives. Information and control systems are used in this world for achieving the alignment of the organizational actors with pre-defined "best practices." The assumption is that such best practices retain their effectiveness over time.

In contrast, high levels of uncertainty and inability to predict the future characterize the new world of business. Use of the information and control systems and compliance with the pre-defined goals, objectives and best practices may not necessarily achieve long-term organizational competence. This is the world of "re-everything", which challenges the assumptions underlying the "accepted way of doing things". This world needs the capability to understand the problems afresh given the changing environmental conditions. The focus is not only on finding the right answers but also on finding the right questions. This world is differentiated from the "old world" by its emphasis on "doing the right thing" rather than "doing things right".

KM is a framework within which the organization views all its processes as knowledge processes. According to this view, all business processes involve creation, dissemination, renewal and application of knowledge toward organizational sustenance and survival.

This concept embodies a transition from the recently popular concept of "information value chain" to a "knowledge value chain". What is the difference? The information value chain, considers technological systems as key components guiding the organization's business processes, while treating humans as relatively passive processors that implement "best practices" archived in information databases. In contrast, the knowledge value chain treats human systems as key component that engage in continuous assessment of information archived in the technological system. In this view, the human actors do not implement best practices without active inquiry. Human actors engage in an active process of sense making to continuously assess the effectiveness of best practices. The underlying premise is that the best practices of yesterday may not be taken for granted as best practices of today or tomorrow. Hence double loop learning, unlearning and relearning processes need to be designed into the organizational business processes.

KM is necessary for companies because what worked yesterday may or may not work tomorrow. Considering a simplistic example, companies that were manufacturing the best quality of carbon paper became obsolete regardless of the efficiency of their process since their product definition did not keep up with the changing needs of the market. The same holds for assumptions about the optimal organization structure, the control and organization systems, the motivation and incentive schemes, and so forth. To remain aligned with the dynamically changing needs of the business environment, organizations need to continuously assess their internal theories of business for ongoing effectiveness. That is the only viable means for ensuring that today's "core competencies" do not become the "core rigidities" of tomorrow.

In the previous definition, KM embodies organizational processes that seek a synergistic combination of capacities of information technologies and human beings. So, if we want to adapt tools and organization in order to realize such synergy, we have to associate knowledge capitalization with the use of expert systems and concurrent engineering with the use of integrated design.

Some Tools for Knowledge

Expert systems have been introduced in the 1980's in order to address the issue of the gap between generations:

- At that time, older engineers had a lot of technical expertise in machine elements such as bearings, gears, etc... In France, Henriot was the "pope" of gears and has written three volumes about them [5] where it is described in the detail how to choose and to dimension the wheels. He was working at "Engrenages Citroën Messian" and was in competition with one of his German counterpart, Dr. Durand. The tools used by Henriot were descriptive geometry and logarithm tables. At the same time, Chamouard did similar work with the stamping and forging industry and has studied at least 4,000 different studies of rough forgings. Their background was mathematics, materials and technology.
- At the same time, new engineers progressed academically with developments of new computer applications simulating by numerical analysis highly non-linear problems. They manipulate plasticity theory, failure algorithms and obtain numerical results, that they cannot always verify, and they generally also lack the technological knowledge and details of machine elements.
- This gives a conflict between the newer and the older engineers. The former may lack the practical technical approach, while the latter are in conflict with their over confidence.

Artificial Intelligence was a new technique which permitted the computer not only to solve equations but also to reason as an intelligent actor in order to solve problems or to give diagnoses. Prolog, Frames, Production Rules, and Case-based reasoning are the new language used for the description of Expert Systems. New specialists in cognition were engaged in order to interview the older engineers, to extract their knowledge and to build virtual experts in different fields.

COPEST [6], an expert system to transform a desired manufactured part from a rough forging part had been developed. This expert tool included production rules expressed in natural language and a dictionary of terms and attributes able to be utilized by the production rules. An inference engine did pattern matching between the predicates of the rules and the information database in order to choose the best rule to be applied.

A new form of knowledge is introduced here in CoDeMo with several modules included in a translation file. The example of module given in Figure 8.1 concerns the possibility to replace a *relation* of type "*pivot_link*" whose name will be associated to *name_rel_0*, relation pointing the *link_0* of the *component_0* and the *link_1* of the *component_1*. This relation can be found in the *technological* view. The proposed solution is named 2_*ball_bearings*.

If a designer chooses such a solution, a mechanism of substitution and decomposition is induced from the module. The substitution replaces the relation, deleting the relation element and the previous links, and adding some components, links and relations at the same level of description of the concerned components 0 and 1. The decomposition does the same at a lower level of granularity. In this work, the names of the initial relation, links and components are introduced at the place of the generic names of the module, so that the same module can be applied for two different relations.

In order to use such knowledge in a KM context, diverse solutions will be proposed to the designer as possible solutions. The state of the product model at the time of the choice can bring to the designer some constraints, reducing the number of possible solutions.

Some Tools for Knowledge Management

In order to profit from the possible synergy of their work, we propose to the different teams to use the collaborative design modeler, CoDeMo, already discussed in [7] and [8]. With the sharing of a unique product model, the teams can work on the same product design, and provide an emerging initial solution, all the while taking into account their individual constraints. With such a system, each actor can share his/her own knowledge when he or she considers giving a solution to a problem raised by other actors. For example, if a proposed part may be obtained either by the forging process or by the extrusion process, the two different actors can give their own solution and the design team or the project leader can chose between the two. The merger of the different knowledge increases the quality of the design of the part and decrease the time within it can be accomplished.

CoDeMo is a client server system and gives access to multiple clients who have to work on the same project. The server manages the product model and delivers the wanted information to the actors depending of the trade they represent. The product model is made of components, links and relations, themselves instantiation of features or characteristics of features. The knowledge object is mainly based on ontologies, giving to the features some quality for sharing knowledge with universal, vehicular and vernacular features.

An actor can take into account what has previously been done and can add some information in order to advance the detailed project information or to constraint the system. To do this, each trade has access to the product model, can use general tools such as an inference engine, a features engine or a geometric kernel, and possesses specific software. When an actor wants to substitute a relation, he/she asks the system about the relation editor and can choose among the different solutions described by a module in the translation file.

```
Relation Technology
   L_Pivot name_rel_0 link_0 component_0 link_1 component_1
   Solution 2_Ball_Bearings
   Substitution
   DeleteRel name_rel_0
   DeleteLink link_0 component_0
   DeleteLink link_1 component_1
   CreateLink component_0 axe component_0_axe_0
   CreateLink component_0 axe component_0_axe_1
   CreateLink component_1 axe component_1_axe_0
   CreateLink component_1 axe component_1_axe_1
   CreateComponent Bearing Technol name_rel_0_Bearing_0
   CreateLink name_rel_0_Bearing_0 axe
                 name_rel_0_ earing_0_axe_0
   CreateLink name_rel_0_Bearing_0 axe
                 name_rel_0_Bearing_0_axe_1
   CreateComponent Bearing Technol name_rel_0_Bearing_1
   CreateLink name_rel_0_Bearing_1 axe
                 name_rel_0_Bearing_1_axe_0
   CreateLink name_rel_0_Bearing_1 axe
                 name_rel_0_Bearing _1_axe_1
   CreateRelation identity component_0_axe_0 component_0
     name_rel_0_Bearing_0_axe_0 name_rel_0_ Bearing_0
   CreateRelation identity component_1_axe_0 component_1
       name_rel_0_Bearing_0_axe_1 name_rel_0_Bearing_0
   CreateRelation identity component_0_axe_1 component_0
     name_rel_0_Bearing_0_axe_0 name_rel_0_Bearing_1
   CreateRelation identity component_1_axe_1 component_1
     name_rel_0_Bearing_0_axe_1 name_rel_0_Bearing_1
   Decomposition
     CreateComponent SkinShaft Frame component_0
       component_0_SkinShaft_0
   CreateLink component_0_SkinShaft_0 axe
           component_0_SkinShaft_0_axe_0
   CreateComponent SkinShaft Frame component_0
                 component_0_SkinShaft_1
   CreateLink component_0_SkinShaft_1 axe
                 component_0_SkinShaft_1_axe_0
   CreateRelation coaxility component_0_SkinShaft_0_axe_0
                 component_0_SkinShaft_0
                 component_0_SkinShaft_1_axe_0
                 component_0_SkinShaft_1
     …
   @
```

Figure 8.1 Module of knowledge for translation

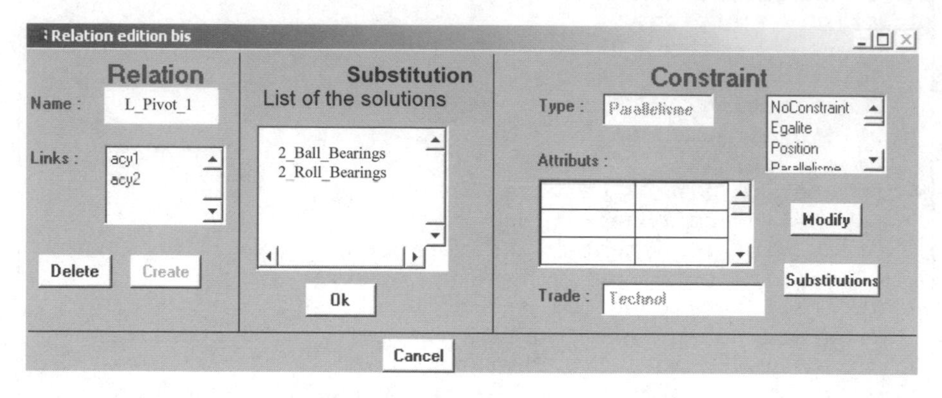

Figure 8.2 The window editor of a relation

8.4 Conclusion

As we saw in the study by the French Ministry of industry, a goal for the future of manufacturing is not only to capitalize on the knowledge of the different processes, and to use the best tool of the Information Processing Technology, but also to be able to react to external change as fast as possible. It is not the intention to provide a tool for automatic design, but rather to give to the designer, the expert, the tools in order to be ready to provide the right solutions in a multi-actors integrated system design environment.

8.5 References

[1] "Key technologies for 2005," 2000, *Les éditions de l'industrie*, Paris, ISBN 2 11 091986 8.

[2] "100 Technologies clés pour la France," 1995, *Les éditions de l'ind.*, Paris.

[3] Working document for the Manufuture 2003 Conference, *European Manufacturing of the Future: Role of Research and Education for European Leadership*, 1-2 December 2003, Milano, Italy, European Commission., http://europa.eu.int/comm/research/industrial_technologies/lists/list_112_en.html

[4] Malhotra, Y., 1998, "Knowledge Management, Knowledge Organizations & Knowledge Workers: A View from the Frontlines," *Maeil Business Newspaper*, February 19, 1998.

[5] *Henriot, Traité pratique sur les engrenages,* Dunot, France, 1978, 1980, 1982.

[6] Tichkiewitch, S., Boujut, J.F., 1990, "Fast Quotation Of Tri-Dimensional Stamped Part Using CAD System," *J. of Mat. Process Technology,* Vol. 24, pp. 127-135.

[7] Tichkiewitch, S., Véron, M., 1997, "Methodology And Product Model For Integrated Design Using A Multi-View System," *Annals of CIRP*, Vol. 46/1, pp. 81-84.
[8] Tichkiewitch S., Brissaud D., 2000, "Co-Ordination Between Product And Process Definitions In A Concurrent Engineering Environment," *Annals of the CIRP*, Vol. 49/1, pp. 75-78.

Part III

Innovative and Conceptual Design

9

AdaptEx: Extending Product Life Cycles through Strategic Product Upgrades

Jeff C. Sand, and Peihua Gu

Abstract: Increasing competition for better product functionality, quality, features, customization, environmental friendliness, lower cost and shorter delivery time will require that product-oriented manufacturing and engineering enterprises optimize the entire product life cycle and become more responsive in developing products. For manufacturing of relatively long life and one of a kind products such as power stations or ships, the manufacturing and construction of such products are influenced by the state of the art technology and knowledge as well as other related issues. To maintain or even enhance such engineering systems performance in their life cycles, technical upgrading is necessary. Therefore, it requires a new design thinking process as well as methodology to address these challenges. This paper proposes a new design approach using Adaptive Design Extension (AdaptEx) that incorporates key design information throughout the entire life cycle of the engineering systems. This helps ensure that the original function and design specifications are not lost or altered due to the operation, maintenance or upgrades made to the system during its life cycle. As the speed of technological change will be continuously increasing, this new methodology will allow design engineers to accommodate for this radical change in technology and be able to implement it into the design. AdaptEx will therefore focus on allowing design enhancements to continue throughout the product life cycle. This paper will reveal the need for this type of design engineering development and summarizes some of the potential benefits of implementing the AdaptEx process.

Keywords: adaptive, design, modularity, life cycle, extension, enhancement

9.1 Introduction

Increasing competition for better product functionality, quality, features, customization, environmental friendliness, lower cost and shorter delivery time will require that product-oriented manufacturing and engineering enterprises optimize the entire product life cycle and become more responsive in developing products [1-5]. For manufacturing of relatively long life and one of a kind products such as power stations or ships, the manufacturing and construction of such products are influenced by the state of the art technology and knowledge as well as other related issues. As current market demands require companies to develop new products and product lines very quickly, additional pressure is created to speed up the design process, which is a further challenge to the optimization of the entire product life cycle.

This paper provides an outline of the Adaptive Design Extension (AdaptEx) methodology. AdaptEx was used to describe this new process since it incorporates the two underlying principles. The first of these is Adaptive Design, or the ability of a product to better adjust to its operational environment. The second being Design Extension, represents the general extension to the product life cycle as well as the extension of design information into the operation phase of the product. Therefore, AdaptEx goes beyond the design process and can be thought of as the management of the entire product life cycle. AdaptEx focuses on the optimization of the product life cycle and implementation of strategic upgrades to extend the life cycle as long as possible. This is accomplished with the dissemination of design information throughout the life of the product under development. When looking at operational characteristics such as maintenance and upgrades it is crucial to fully understand the initial design intent and the information that went into designing the product.

The ultimate goal of AdaptEx is to enhance the overall life cycle of a product and allow for optimization to take place through future upgrades and enhancements during the operational phase. In theory this process will allow the design process to continue throughout the entire life cycle of the product from initial design concept through to product decomposition. Thus, it is expected that AdaptEx could lead to the development of a new methodology to design complex large-scale engineering systems.

9.2 The Need for Adaptive Design

At this time it is important to reiterate that AdaptEx will focus on two key processes. The first of which is extension, to both the product life cycle and to design knowledge into the operational phase. The second is the enhancement of the original design, which will be enhanced through the use of strategic upgrades that will be planned to improve the initial design as well as extend the overall project life cycle.

Both of these will be accomplished through the use of technological upgrades including those that are planned during the initial design phase of the project (type

1), and upgrades that are developed and implemented during the operational phase of the project (type 2).

With the rapid changes in technology that take place, today's designs are required to be much more robust than what was required in the past. Design historically consisted of a design phase in which everything for the project was completed. The project was then put into operation and the upkeep was left to the maintenance department. These models proved satisfactory when the rate of technological change for the product remained relatively slow without significant need for environmental concern. Therefore, system maintenance was used to try and withhold the design at its initial design criterion for as long as possible.

Figure 9.1 is a representation of the traditional design model. The design phase continues until the initial design is completed. The product then enters the operational phase and slowly begins to deteriorate. Product maintenance is implemented to try and keep the product at its original functionality, represented as a maintenance limit. However, as the product progresses through its life cycle it begins to degrade and lose operational functionality.

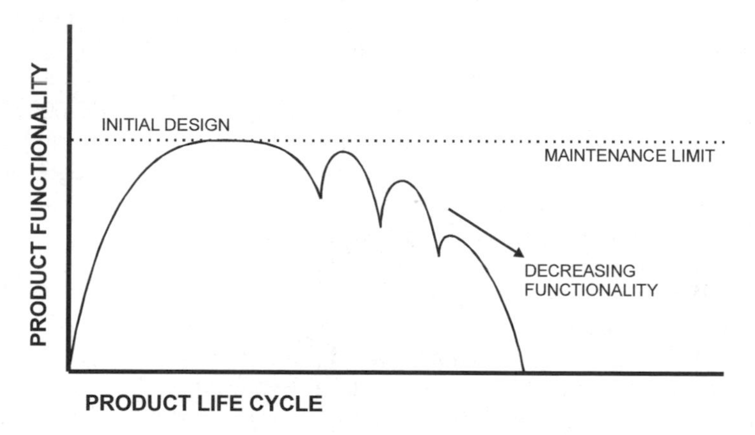

Figure 9.1 The traditional design model

When major changes and maintenance are required to the system, the changes would usually be implemented in the operational phase without the complete understanding of the initial design intent. This often led to band-aid patches that only focused on repairing a specific area and might be detrimental to other areas of the design and the overall design function. The design will go through a number of maintenance iterations until the product eventually reaches its end of life.

A type of "adaptive" design began to appear with the development of the personal computer (PC). Initially the PC was created to outlast its owner with an endless array of interchangeable and upgradeable modules and systems. However the downfall to this model was the weaknesses of the core platform on which this adaptive structure was based. In the case of PC's the core system, the motherboard, has one of the shortest life cycles of all of the systems within the computer. Therefore, the potential life of the computer is limited based on these core restrictions that make up the product platform. The usability of the computer is based on the motherboard and processor, which in many cases leads to the early retirement of other components that still have a useful life remaining.

Computers are currently changing to better accommodate these new trends. Computer manufacturers are beginning to move away from the traditional PCI bus system to a new ExpressCard system and newer technologies such as USB 2.0. This will enable easier transition of components from one machine to the next and allow for more successful upgrades to take place. Once in place these new design methods will better accommodate the rapid growth taking place in this industry and allow for the adaptation of new technologies to be implemented.

9.3 AdaptEx for Large Scale Engineering Systems

As discussed earlier the computer industry has embraced the need for adaptable and robust designs. However, this thinking is far from reality in the engineering, procurement and construction (EPC) industry that designs and constructs complex and large scale systems and infrastructure. The AdaptEx process will initially focus on the design of complex large-scale engineering systems (CLSES) for achieving both economical and environmental benefits.

The main goal of this research is that the newly developed AdaptEx process will become a new way of designing complex large-scale engineering systems (CLSES). It will include the use of strategically planned type 1 upgrades in the design phase as well as type 2 upgrades that will be developed and implemented within the operational phase of the project. The potential benefits of applying adaptable and robust design principles to large-scale problems are significant. Incorporating advancing technologies and environmental demands into CLSES can lead to more balanced and extended life cycles. Several examples of which include improved operations, design functionality and product recycleability.

The complexity within these one of a kind CLSES will allow for a complete core platform development. AdaptEx will then systematically proceed through the design phase, identifying key items such as the core platform as well as modules and modular relationships. In addition to these design identifications AdaptEx will also track and manage design weaknesses or deficiencies that exist due to economical and technological constraints. These weaknesses will be taken into consideration when upgrades are planned and implemented into the systems during the operational phase.

Within the AdaptEx process it is very important to identify and manage the technological and environmental deficiencies that exist within the design. Once the deficiencies have properly been identified and tracked it is possible to plan the implementation of strategic upgrades that will enhance the design as well as extend the product life cycle.

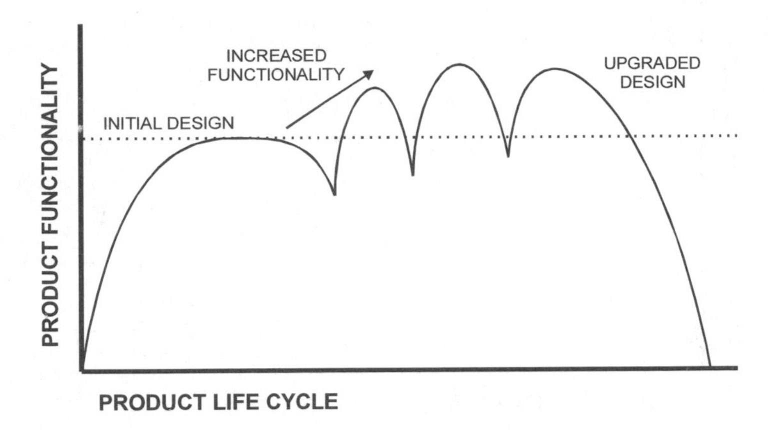

Figure 9.2 The AdaptEx design model

Figure 9.2 represents the AdaptEx method and how design upgrades will play a critical role in the fulfilment of the design enhancement and life cycle extension. The first upgrades that would be implemented would be type 1 upgrades that were planned during the initial design of the product. During the operational phase type 2 upgrades would be used to satisfy new technological and environmental requirements.

A major difference between the AdaptEx model and the traditional model shown in Figure 9.1 is how the upgrades are able to enhance the design function and extend the product life cycle.

When managing the life cycle of complex large scale engineering systems such as oil refineries and nuclear reactors, the AdaptEx methodology has enormous potential to improve and optimize their life cycle.

9.4 Establishing the AdaptEx Process

When a comparison is made of the traditional and AdaptEx design processes shown in Figure 9.1 and Figure 9.2 the potential benefits can clearly be seen. Figure 9.3 illustrates the differences between the two methods, mainly the enhanced functionality and extended life cycle.

When the AdaptEx process is implemented it can be seen that the upgrades enhance the initial design. When a type 1 upgrade is implemented it is based on information from the design phase of the project. The design information that was used in the design phase will be reused in the operational phase to help optimize the upgrade. This reusable engineering is very important to the AdaptEx process since it allows the upgrades to be implemented efficiently while remaining timely and cost effective.

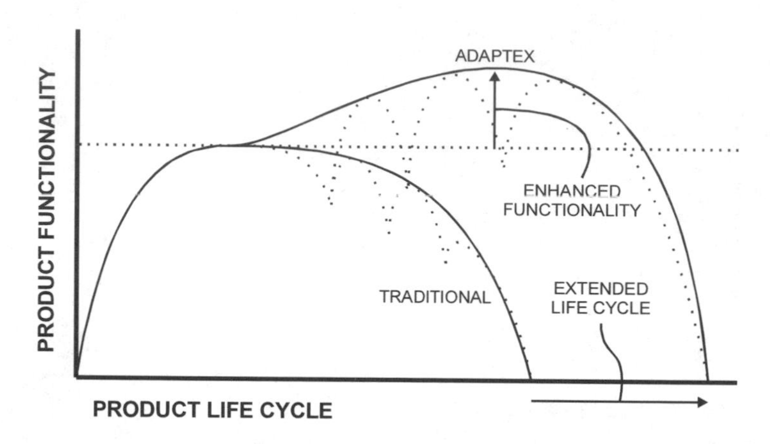

Figure 9.3 Comparing the traditional and AdaptEx design methods

AdaptEx will use information that exists from the design and development phase to plan for future requirements later on in the product life cycle. Often in design there is no ultimate design solution that will always satisfy the initial design requirements. Therefore by using design information to plan and accommodate future changes the resulting design will be more adaptable and robust. The AdaptEx method will focus on the two major phases of the product life cycle - the design phase and operational phase as shown in Figure 9.4.

In the design phase specification management and relationships will be used to manage the design process. The design process does not have to be drastically different from the way it is currently completed. What does have to change during the design process is the identification and understanding of the designs potential, including the identification of the designs benefits and deficiencies. Identifying a products strengths and weaknesses (deficiencies) along with incorporating information about technology and rate of technological change allows for efficient planning of future upgrades and design improvements. This structure will be based on the modular foundation developed within the House Of Modular Enhancement (HOME) methodology [6, 7].

Within the operational phase upgrades will be used to enhance the project. Two major types of upgrades will be implemented. The type 1 upgrades are planned upgrades, which were developed during the design phase of the project. This type of upgrade will be used to fix weaknesses that existed in the original design. The major focus of this type of upgrade will be based on technology, and most of the upgrade will have been defined during the design phase. New considerations will be made to include new features that came to realization during the operation of the facility.

Type 2 upgrades represent new upgrades developed within the operational phase of the project. It is critical that these new upgrades maintain the initial design function and strive to enhance the design and extend the product life.

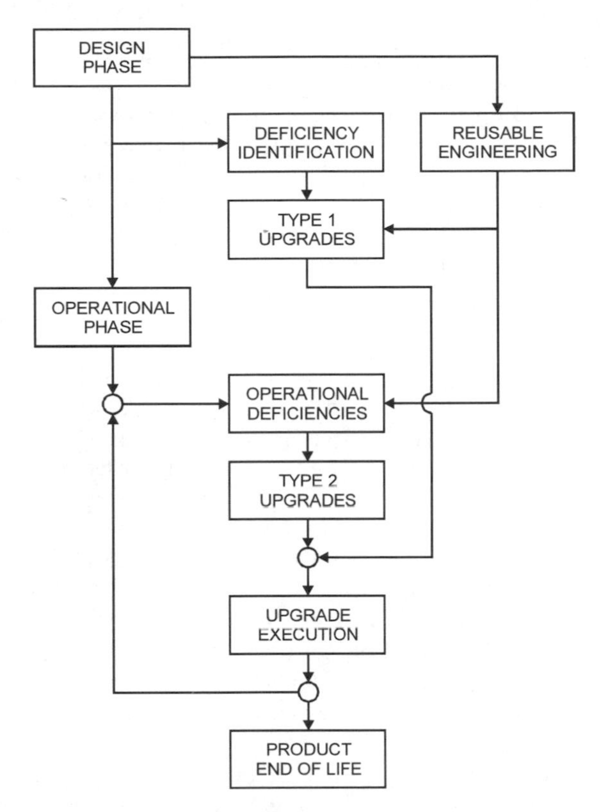

Figure 9.4 AdaptEx process flow chart

In order to have successful implementation of the upgrades it is very important that some key stages be completed within the methodology. Within the design phase attention will be placed on getting the design setup so that it can be easily maintained and improved in the operational phase. This includes establishing a core for the product that will allow for future growth through the implementation of product upgrades.

A significant element of any upgrade is the actual connection or attachment to the existing design. This will be improved through a good understanding of the designs overall strengths and weaknesses. The upgrades will be specifically attached to stable areas within the design, such as the core, to minimize problems that will occur due to weak attachments. Table 9.1 summarizes the various design levels and key elements that are contained within each.

The key to the AdaptEx process is managing these various design levels. When an upgrade to the system is made design information will be passed from high levels to lower levels to help stabilize the system. Each upgrade that takes place will focus on adding stability to the overall product design by adding new design functionality or repairing deficiencies that previously existed. Identifying and understanding the weaknesses of the design (high levels), within the design phase, will allow for upgrades and management of the system to be performed.

Table 9.1 Design levels

Level	Description	Key Elements
Core	Design Platform	Key elements of the design function that require little change and adjustment throughout the life of the project. Upgrades will attach to the core whenever possible
Level 1	Modular Framework	Implemented in various configurations to apply design alternatives with minimum reengineering
Level 2	Modular Building Blocks	Smaller modules that attach to the framework to enhance the design
Level 3	Weaknesses / Deficiencies	Information new to the design and system that has not been completely proven and validated

It is important for a design to stabilize itself by stepping through the levels. When a design upgrade takes place the design becomes stronger by moving system data up the various levels towards the core. Therefore the AdaptEx process will try and eliminate as much high-level design as possible. Table 9.2 summarizes the various cases that can occur during the implementation of an upgrade.

This case structure will define the various scenarios developed within the AdaptEx process. Using these various cases, upgrades will be developed. When things go wrong or not as expected it is important to know the severity of the problem. For example is it a high level assumption that was incorrect and can easily be rectified or does the problem lie within the inner levels of the design and requires further work and investigation. Whenever possible, design modules and subsystems are based on the core or lowest level possible. This helps eliminate the exponential growth of errors and problems caused by relying on inaccurate design information to make decisions.

Table 9.2 Various upgrade cases

Action	**Type**	**Occurrence**
Level 1	New	New Level 1
Level 1 to Core	Drop	Level 1 to Sub-Core
Level 1 to Level 1	Merge	Level 1 Merger
Level 1 to Level 2	Degrade	Level 1 to Level 2
Level 2	New	New Level 2
Level 2 to Level 1	Drop	Level 2 to Level 1
Level 2 to Level 2	Merge	Level 2 Merger
Level 2 to Level 3	Degrade	Level 2 to Level 3
Level 3	New	New Level 3
Level 3 to Level 2	Drop	Level 3 to Level 2
Level 3 to Level 3	Merge	Level 3 Merger
Remove Level 3	Degrade	Level 3 Removal

9.5 Concluding Remarks

This paper revealed the AdaptEx process and the benefits it can have to complex large scale engineering system design. AdaptEx will focus on the strategic implementation of technological and environmental upgrades that will stabilize the overall design. During the implementation of the upgrades the design will be enhanced and the product life cycle will be extended.

When applying this methodology to CLSES such as oil refineries and nuclear reactors the potential benefits can include an increased return on capital investment, improved operational functionality and extended operational life.

While initially developed for CLSES, future work on the methodology may include the transition to smaller design models. Many of the benefits that apply to CLSES will also prove applicable to small scale design.

9.6 Acknowledgements

The authors wish to thank the Natural Sciences and Engineering Research Council of Canada and an industrial consortium for providing financial support for this research.

9.7 References

[1] Alting, L. and Legarth, J.B., 1995, "Life Cycle Engineering and Design," *Annals of CIRP,* Vol. 44/2, p. 569.

[2] Gu, P., 2002, "Adaptable Design Using Bus Systems," *Proc. of IMCC'2002*, Xiaman, China, October 10-12, 2002, Keynote Paper, 11 pages.

[3] Kimura, F., Kato, S., Hata, T., Masuda, T., 2001, "Product Modularization For Parts Reuse In Inverse Manufacturing," *Annals of CIRP* Vol. 50/1, p. 89.

[4] Krause, F.L., Kimura F., Kjellberg, T., Lu, S.C.Y., Alting, L., ElMaraghy, H.A., Eversheim, W., Iwata, K., Suh, N.P., Tripnis, V.A., Weck, M., Van Der Wolf, A.C.H., 1993, "Product Modelling," *Annals of CIRP*, Vol. 42/2, p. 695.

[5] Lange, S., Schmidt, H. and Seliger, G., 2000, "Product and Assembly Design for a Fibre Reinforced Plastic Track Wheel," *Annals of CIRP*, Vol. 49/1, p. 105.

[6] Sand, J.C., 1999, "HOME: House of Modular Enhancement," University of Saskatchewan, M.Sc. Thesis.

[7] Sand, J.C., Gu, P. and Watson, G., 2002, "HOME: House of Modular Enhancement for Product Modularization," *Concurrent Engineering, Research and Applications*, Vol. 10, No. 2, p. 153.

10

Product Genetic Engineering

Kezheng Huang, Hongwu Chen, Yandong Wang, Zhengjun Song, and Liangmin Lv

Abstract: Creativity and high efficiency are still the essential requirements for product design with widc impact on current design research and engineering practice. Design automation aims to increase the efficiency and quality of design work. Creativity is receiving more attention but with essentially little progress so far, especially in automatic design. The rapid and automatic growth of organisms and the great potential for production of new species that Genetic Engineering shows are the two main reasons that lead to our work. A new design environment - Product Growth Design platform (DARFAD) - has been developed, new concepts such as Product Genetic Engineering (PGE) are proposed, a theoretical PGA framework is discussed, and an example of product design is introduced.

Keywords: Computer aided design, Methodology, Product Genetic Engineering (PGE)

10.1 Introduction

With an increase in global competition, it is important to make the product design process as efficient and high in quality as possible. In current product design practice, it is useful to distinguish two basic types of design work. In creative design, new solutions or schemes have to be explored. In routine design, there is a relatively well-structured solution or top-level scheme for the product to be developed, and the design work can be divided into sub-tasks. In a broad sense, design is essentially creative and innovative work based on new requirements, knowledge and experience.

In practice and theoretical study, there havc bcen a lot of ideas and opinions about what product design is and how the design process is conducted. Though they are helpful to human designers in stimulating new thoughts and guiding design work, none has been completely accepted for the whole design process. Due to the complicated nature of the design process, there is still no effective design theory for mechanical product design automation in truly top-down style and with operability, though developments in the design theory and method over the past decades, such as Axiomatic Design [1] and TRIZ [2], have helped researchers and designers in design research and practice.

Creativity and high efficiency are still the essential requirements for product design with serious/significant impacts in current design research and engineering practice. Design automation aims to increase the efficiency and quality of design work, but mainly in its later stages, such as in modeling and analysis. Creativity is obtaining more attention but has attained little essential progress so far, especially in the field of automatic design.

Design Automation

The product design process in an interactive single-user CAD system is also the basis for a multi-user collaborative design system. Due to the complexity of design work, interaction between designer and CAD system is still vital for successful and efficient results.

Actually, the rapid and wide application of computer technology makes the design information explode and alternative schemes increase more rapidly during the design process. Without the effective use of computer systems, human designers will find it difficult to obtain the best design solutions. The current design automation research is mainly based on human decision-making and design schemes evaluation. But in each scheme there are still a large number of problems to be verified. To rely in all these synthesis work on a human designer is inadequate. To what extent we should develop artificial intelligence and use it in design systems remains to be addressed.

The automatic generation and synthesis of design alternatives are two of many aspects that can facilitate design work.

The new design automation theory needs to be studied. Decomposition and Reconstitution (D&R) is one of the most important principles for design automation [3], which is developed as an innovation principle for the introduction of creative potential into the design automation system. Based on this principle, the presented theoretical study has established some new design automation methods and principles, such as General Positioning Principle (GPP), 'Cell Growth' Design Principle (CGDP) [5] and the D&R Based Design Process Model [4], in which the traditional design practice and process are decomposed into small steps and reconstituted in a different way. Applications include modular fixture design [6], dedicated fixture [7], and others.

A design automation system, which reflects these factors needs to be developed. Applying the D&R principle to product design, a generic structural design approach (GSDA) was set forth [8], and new software tools [9] have been developed. The DARFAD system has become a new design automation platform, in which the D&R Principle, GPP, and CGDP have been fully utilised and imaginative thinking implemented with easy operability [10].

Product Genetic Engineering

Whilst there has been considerable research in developing computational models of design based on how humans design, there is increasing interest in those based on non-human processes. Genetic algorisms, genetic programming, and evolutionary design are methods of design in which a search is restricted in its application to routine or parametric designing, and the processes of search map well onto those of optimisation [11]. Some extensions have been made to these methods, drawn from analogies with the more recently developed areas of genetic engineering and developmental biology [12]. But this only reveals a simple and interesting phenomenon in design: It is how far away from the way an organism grows.

Genetic engineering also shows great potential for the production of new species. Gu, *et al.* [13, 14] proposed the concept of "product gene" for the inheritance and transfer of product knowledge. Feng, *et al.* [15] put forward the "product gene" concept for product conceptual design. However, they did not mention how a real product can be designed and realized.

The essential difference between an organism and a man-made artefact is that there is a growing mechanism for organism, but none exists for man-made artefacts design. In this work, genetic engineering knowledge is applied to design. New concepts of Product "Genome" and PGE are proposed, a PGE framework is discussed, and finally an example is presented.

10.2 Product Growth Design Platform

'Cell Growth' Design Principle

Comparing with organisms, we can consider a product as an organism and a part within a product as a cell. Biology shows that cell division is fundamental to the growing process by which a cell divides to form two daughter cells.

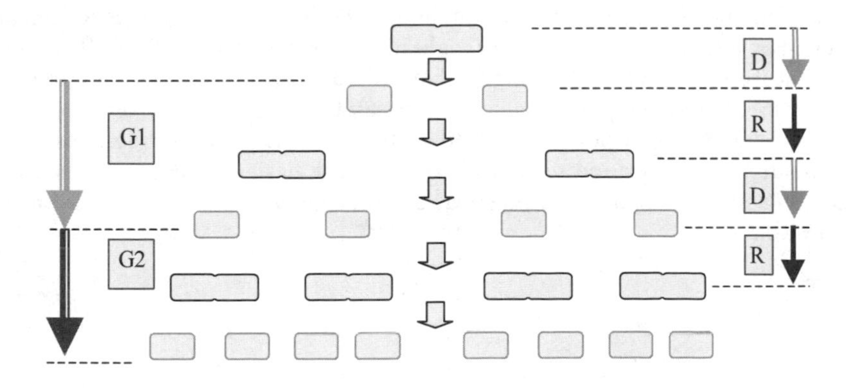

Figure 10.1 Cell division principle

In the design process of mechanical products, the key stage is the mapping process from functional requirements to product structure, in which a series of design activities are performed in sequence. If the design can progress like a growing process, its efficiency can be greatly improved. The Cell Division mechanism can be used as shown in Figure 10.1. To facilitate the structural design process, the D&R Principle [1] has been utilized to standardize the various design activities so that the mapping process can be easily made. Some design activities and their graphical representations are shown in Figure 10.2. The fundamental feature of all activities is organized as a "cell division and growing" process.

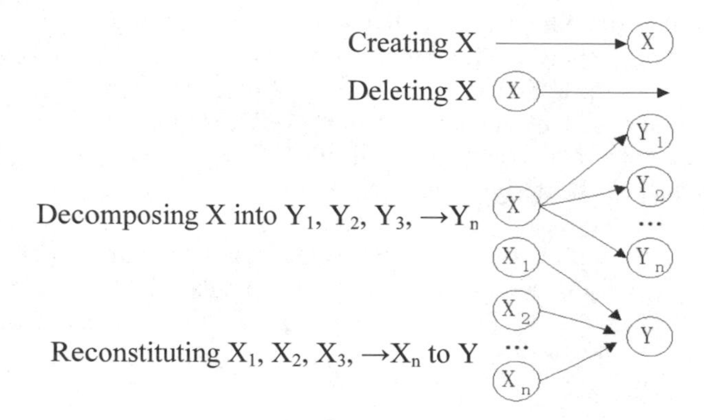

Where X, Y, X_i, Y_i are design objects

Figure 10.2 Partial design activities standardized in the DARFAD System

Growing Tree for A Multi-scheme Design Process

Applying the 'Cell Growth' design principle to product design, many product schemes can be obtained in a growing tree-like process with each leaf as a product scheme. This process integrates concurrent design and axiomatic design concepts in a unified and structure-oriented automatic design process for mechanical products. For instance, the concept of Qualitative Closing Space (QCS) [7], which is different from absolute positioning in that only the capability of moving in and out is required, was proposed as a theoretical basis for qualitative assembly condition. Due to lack of design knowledge, the dynamic growing tree is used to record multi-scheme data and their relationships. The design process is driven by the assemblability evaluation and proceeds as a tree growing process, as illustrated in Figure 10.3.

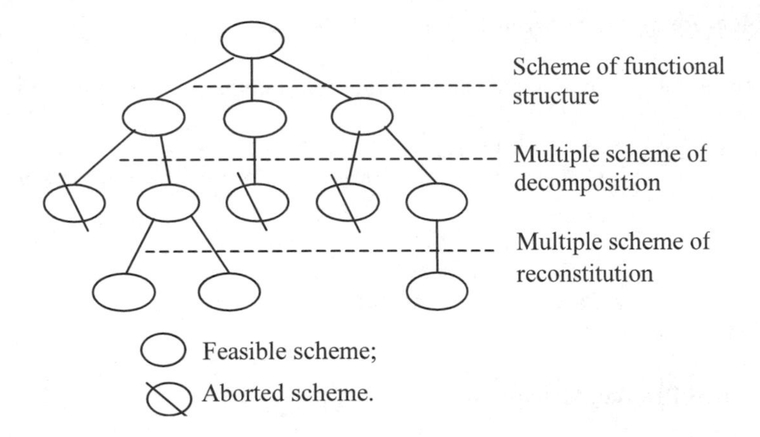

Figure 10.3 "Growing Tree" for multi-scheme design

10.3 Concept of "Product Genome"

Comparing with organism, we can take a product as an organism – a complex assembly of different kinds of cells that perform many different functions, - and a part of the product as a cell—the smallest structural unit of an organism that is capable of independent functioning. The rapid and automatic growth of organisms is a fascinating process, which lead to the idea of product gene.

Gene and its Mechanism

To get a scientific definition of Product 'Gene' (PG), firstly we need to investigate gene and the mechanism of gene function. Biologically, a gene is a segment of a cell's DNA. DNA is the blueprint of life containing codes for the proteins that make up an organism's specific characteristics including physical appearance, physiological functioning, *etc.*. That is, the segments of DNA that have been associated with specific features or functions of an organism are called genes. A gene is a functional and structural unit of a DNA. Genes consist of structural genes, operational genes and regulator genes according to functional actions in the process of the transcription. The genome is the entire DNA "recipe" for an organism, which comprises a certain numbers of genes of the organism.

According to the Central Dogma in molecular biology, two steps of gene expression are essentially the same in all organisms. The term gene is usually taken to represent the genetic information transcribed into a single RNA molecule, which is in turn translated into a single protein.

Analogical Relations between Organism & Product

Based on investigation and analysis, the following corresponding relations can be proposed:

Gene in biology → Product 'Gene';
RNA → Conceptual Architecture (Visual Concept based on Functional Surface);
Protein → Solid Product Element,
Cell Coat → Function Surface,
Cell → Part or component, and
Organism → Product.

Definition of Product Genome

After a systematic comparison between products and organisms, the hypothesis is proposed that there are similar genes in products, which should have:

- an Automatic growth mechanism;
- Inheritance – genetically transmit the product characteristics from parent parts to offspring parts;
- Self organisation - communicate and inform each other;
- Adaptability – allow different environmental constraints and user needs to be satisfied.

Here, Product "Genome" is defined as an integrated set of information which specifies a product's structures and functions and its mechanisms to "grow" automatically, and can under suitable conditions generate specific structures of a product which can accomplish its functions in appropriate environments.

How does the sequence of a strand of DNA correspond to the amino acid sequence of a protein? This concept is explained by the Central Dogma of molecular biology that is shown in Figure 10.4.

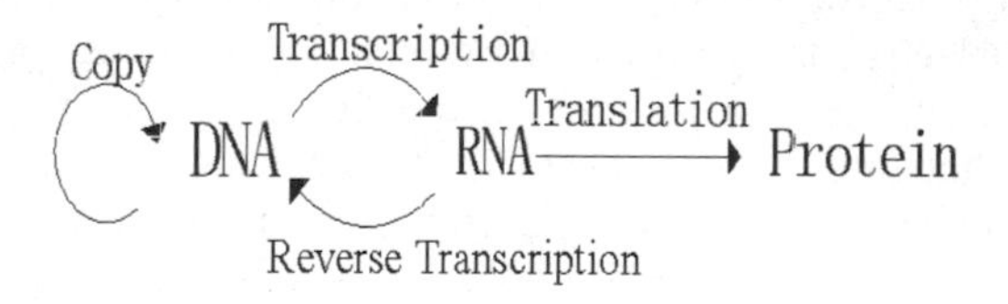

Figure 10.4 Central dogma

Similarly, we hypothesize that there exists a similar Central Dogma in man-made product as shown in Figure 10.5. The product genome consists of three kinds of genes, Functional Genes (FGs), Control Genes (CGs) and Structural Genes (SGs). FGs represent all kinds of functional units which draw parallels to the variation of requirements. Different FGs can form a specific genome that can be transcribed into the functional prototype of a product. Definitely, the requirement sub-function corresponds to certain FGs, which are turned into specific functional surfaces by transcription. Moreover, the overall functions embody the relations of functional surfaces. Therefore, requirements, FGs and functional surfaces together form the product prototype.

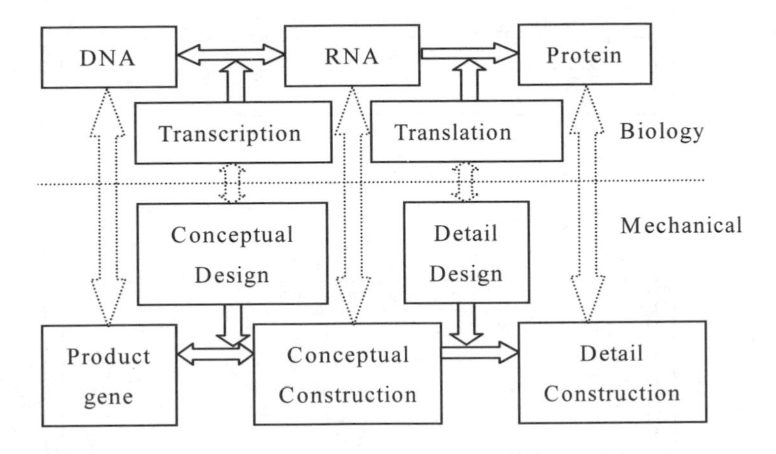

Figure 10.5 Product central dogma

SGs contain the detailed structural information of a product for use in its growth process, some of which is represented in modes, the generic stable features in components, such as executive part modes $SG_1=\{SG_{1i}, i=1,2...n_{s1}\}$, transmitting part modes $SG_2=\{SG_{2i}, i=1,2...n_{s2}\}$, and structure part modes $SG_3=\{SG_{3i}, i=1,2...n_{s3}\}$. Together with particular information for a specific product, some SG_S can constitute any component of any complexity, which in turn can form complicated artefact structures such as transmission chains.

CG_S play the roles of management and control in product design. It starts with situation evaluation or status sensing, continues with selection making of growing directions, control of the growing mechanism and selection of SG_S. As a result, the integrated sequence of SG_S manoeuvre is determined in transcription. Hence, one way of expressing the CG_S is by writing explicit series of SG_S codes. For example, a control gene CG_1 for a transmission chain can be described as follows:

$$CG_1 = \{SG_{2s1}, SG_{2s2}, \ldots, SG_{2si}, \ldots, SG_{2sm} | s_1, \ldots, s_i, \ldots s_m \in \{1,2, \ldots, n_{s2}\}\}$$

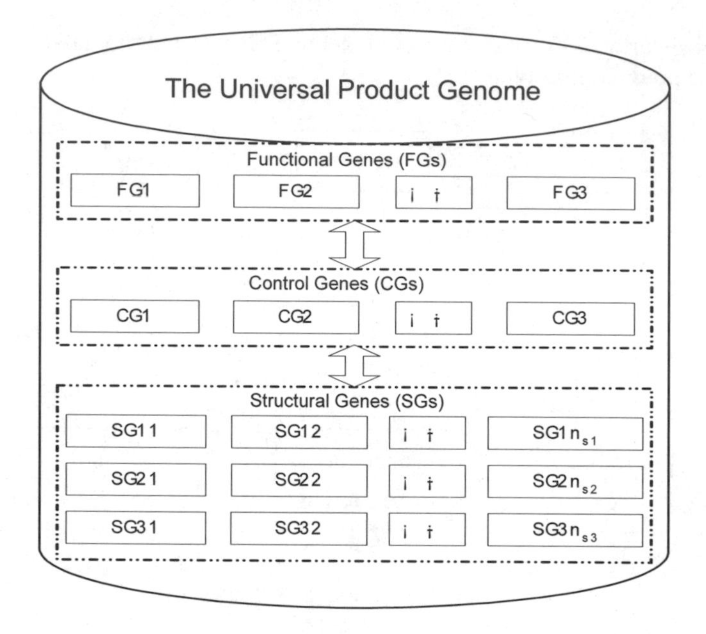

Figure 10.6 Product Genome

10.4 Product Genetic Engineering

Genetic Engineering

Genetic Engineering is the process of insertion of one or more genes from one organism into the DNA of a different organism. It can be thought of as a cut-and-paste process in which a specific gene is cut from a donor organism and pasted into the genetic material of another organism. It is the heritable, directed alteration of an organism. The mainstay of genetic manipulation is the ability to isolate a single DNA sequence from the genome. This can be considered as a series of four steps of a gene cloning experiment: Generation of DNA fragments, joining to a vector or carrier molecule, introduction into a host cell for amplification, and selection of a required sequence.

Product Genetic Engineering

PGE is taken here as a process that consists of the following steps: i) isolation and extracting of PGs from existing products, ii) formation of PGs database and rules on how PGs recombination meets user requirements, iii) evaluation of the new genome by recombination, iv) expression of PGs under certain environments, and

v) evaluation of the new product. It aims to create the new product in an innovative way.

The Main modules in PGE will include the following:

- Extracting and isolating PGs from existing products, as well as the corresponding requirements and environmental constraints;
- Establishing a PGs database, which consists of all the PGs from specific products, and a management system preparing for PGs recombination;
- Recombining PGs from the database under certain rules, requirements and environmental constraints;
- Evaluating the new genome obtained by recombination.

The design process model based on PGE is illustrated in Figure 10.7.

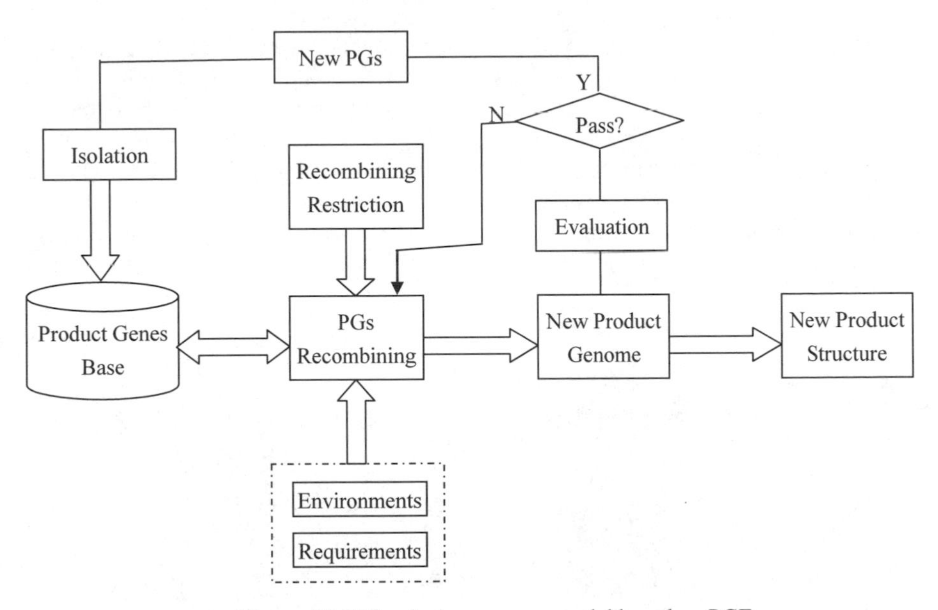

Figure 10.7 The design process model based on PGE

10.5 PGE-DARFAD System Framework

Finally, the theoretical framework for PGE is proposed as reflected in the PGE-DARFAD system, the key components of which are Product Requirements Acquisition, PGs Isolation, PGs Recombining, Genome Evaluation, Growth Design, and PGs database. The architecture of the PGE-DARFAD system is shown in Figure 10.8.

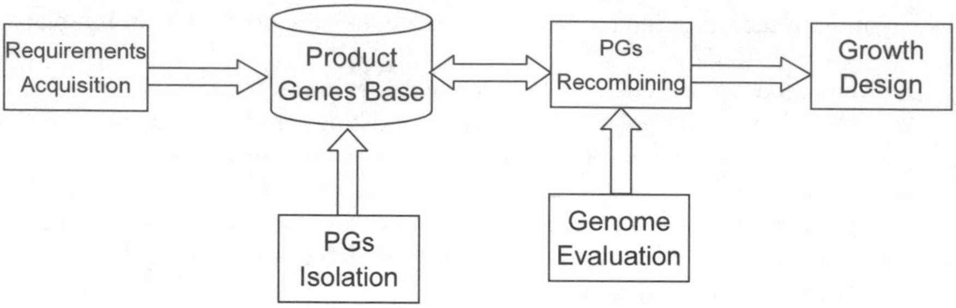

Requirements Acquisition
Product Genes Base
PGs Recombining
Growth Design
PGs Isolation
Genome Evaluation

Figure 10.8 PGE-DARFAD system architecture

Using the PGE-DARFAD System, a mechanical product 'GUNZI fixture' is designed. The growing design process is shown in Figure 10.9. Figure 10.9(a) shows the visual shape of the product prototype generated according to design requirements expressed in a 3D workpiece drawing. Figure 10.9(b) is the result of the first product growth step generating a supporting body with 2 rolling wheels. Figures 10.9(c) to 10.9(g) display a series of growth steps that produce an additional component each time. Figure 10.9(h) is the final 3D model for the fixture after translating the "genome" shown in Figure 10.9(g) to "proteins" in the form of solid structure.

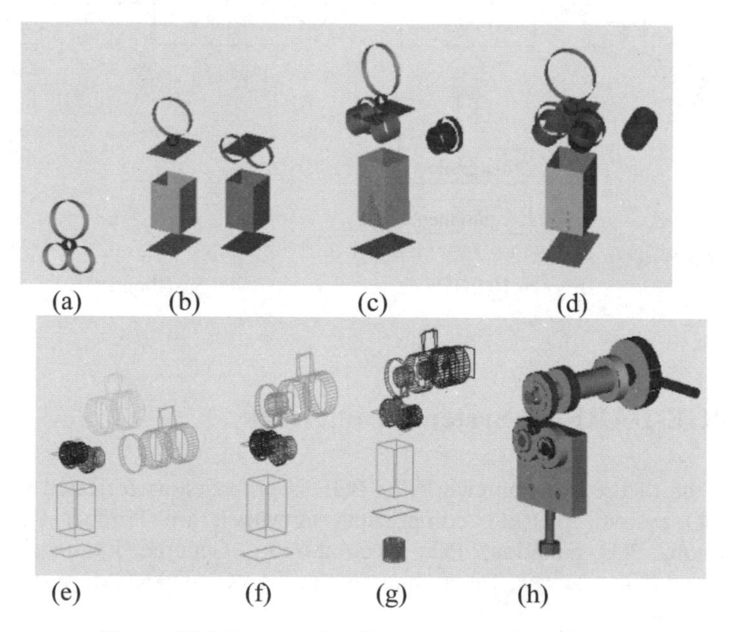

Figure 10.9 An example of product growth design

10.6 Conclusions

A Product Growth Design platform (DARFAD) is briefly introduced and new concepts in products engineering, are proposed. Through theoretical study and development efforts interesting and useful results have been obtained and conclusions can be drawn as follows:

(1) PGE is a significant extension to the current non-human process of design and is valuable for a new generation of CAD systems with both high efficiency and innovative potential;

(2) Growth Design is the bottleneck of PGE since there are natural growth mechanisms in organisms but none exist for products; PGE can utilize genetic engineering principles only at higher functional levels;

(3) An Example of mechanical product design shows that PGE is a general purpose process and that growth design is very effective in representing different kinds of product structures;

Although current efforts have provided a solid foundation for PGE, much more research needs to be done before it can be widely used in practice. As PGE is studied and developed further, more rules and principles will be established until one day new products can be designed by manipulating product genes, and final design results can be obtained automatically through "growing" processes in PGE environments according to their "Genomes".

10.7 References

[1] Suh, N.P., 1990, *The Principle of Design*, Oxford University Press.

[2] Altshuller, G., 1996, "And Suddenly the Inventor Appeared," *Technical Innovation Center*, Worcester, MA.

[3] Huang, K.Z., Ai, X., Zhang, C., 1997, "Decomposition And Reconstitution Principle For Complex Surfaces And Its Applications," *Science in China (Series E)*, Vol. 40(1), pp. 89-96.

[4] Huang, K.Z, *et al.*, 1997, "Design Process Model For Mechanical Product Based On D&R Principle," In: Sun Guozheng, Ben Qingyuan(Eds.) *Modern Design Sci. Facing 21st Century,* Wuhan, People's Jiaotong Press, pp.10-14.

[5] Cao, S., 2002, "Mechanics Systhesis Study and its Application for Growth Design of Product Conceptual Structure," *Dissertation*, Jinan Shandong University, 2002(12).

[6] Huang K., Y., Rong, 1999, "Decomposition and Reconstitution based Fixture Design," *Int. Mechanical Engineering Congress and Exposition*, Nov., 1999, Nashville, Tennessee, USA.

[7] Yang, Z., Huang, K.Z., 2003, "Fixture Planning Technique Study," *CIMS*, 2003(4).

[8] Huang, K.Z., *et al.*, 1998, "Generic Structural Design by Assemblability for Mechanical Product," *Proc. of 14th Int. Conf. On CAPE*, Sept. 8-10, 1998, Tokyo, Japan.

[9] Huang, K.Z., Li, X., 2001, "Co-DARFAD - The Collaborative Mechanical Product Design System," *The 6th Intl. Conference of Computer Supported Cooperative Work in Design (CSCWD)*, July 2001, London, Canada.

[10] Cao, S., Huang, K.Z., 2003, "Imaginative Design Technique and Product Model Based on Functional Surface," *Modern Manuf. Engineering*, 2003(9).

[11] Gero, J. S. and Shi, X-G., 1999, "Design Development Based On An Analogy With Developmental Biology," In: J. Gu and Z. Wei (eds), *CAADRIA'99*, Shanghai Scientific and Technological Literature Publishing House, Shanghai, China, pp. 253-264.

[12] Gero, J. S., 1999, "Extensions To Evolutionary Systems In Design From Genetic Engineering And Developmental Biology," *Proc. 1999 Congress on Evol. Computation – CEC'99*, IEEE, Piscataway, New Jersey, pp. 474-479.

[13] Gu, X., Tan, J., Qi, G., 1997, "Genetic Model for Mechanical Product Information," *Chinese Mechanical Engineering*, 1997(2), pp. 77-79.

[14] Gu, X., Qi, G., 1998, "Genetic Model in Process Information," *Chinese Mechanical Engineering*, 1998(11), pp. 80-84.

[15] Feng, P., Chen, Y., Zhang, S., Pan, S., 2002, "Conceptual Design Based on Product Gene," *J. of Chinese Mech. Engineering*, 2002, Vol. 38(1), pp. 1-6.

11

Gene Engineering-based Innovation of Manufactured Products

Ke-Zhang Chen, Xin-An Feng, and Xiao-Chuan Chen

Abstract: With the similarity between the evolution of living beings and the development of manufactured products, gene-engineering techniques have been applied to develop a systematic design theory and methodology for product innovation. This innovation method is different from the conventional one as it innovates products via artificial differentiation of the virtual product chromosomes. It provides a logically structured process, which can reduce blindness to innovation. Since products have no physical chromosomes, their virtual chromosomes must be reverse-deduced according to their function requirements so that the new innovation method can be applied. This paper introduces the gene engineering based innovation method, discloses the contents and data structure of virtual product chromosomes and applies database techniques to edit and store product chromosomes.

Keywords: Gene engineering, genetic engineering, product innovation, design theory and methodology, virtual chromosome

11.1 Introduction

Products are always developed based on existing or similar products from a lower level to a higher level. The development process of products is very similar to the evolution process of living beings in nature. The genetic information of living beings is stored in the chromosomes contained in their cells and the evolution of living beings is based on the variation of their chromosomes [1]. Since the evolution of living beings is very slow, genetic or gene engineering [1-4] has been developed to reform consciously the chromosomes for living beings to accentuate good characteristics or to clone living beings according to their chromosomes. In a similar manner, manufactured products also possess genetic information, and their innovation can also be actively implemented using a similar reforming method. Gene engineering techniques have thus been applied to develop a systematic design theory and methodology for product innovation [5, 6]. This design method is different from the conventional one as it innovates products via artificial differentiation of virtual product chromosomes. It provides a logically structured procedure, which can reduce blindness to innovation and can even clone a product with the aid of virtual manufacturing technology.

Manufactured products, however, have no physical chromosomes. Their virtual or analogous chromosomes need to be artificially created first so that gene-engineering techniques can be applied. This paper introduces the gene engineering based innovation method, discloses the contents and data structure of virtual product chromosomes and applies database techniques to edit and store product chromosomes.

11.2 Gene Engineering-based Design Method for Product Innovation

The design method for product innovation has been developed using genetic-engineering techniques [5, 6]. Its workflow is shown in Figure 11.1 and illustrated through the following steps:

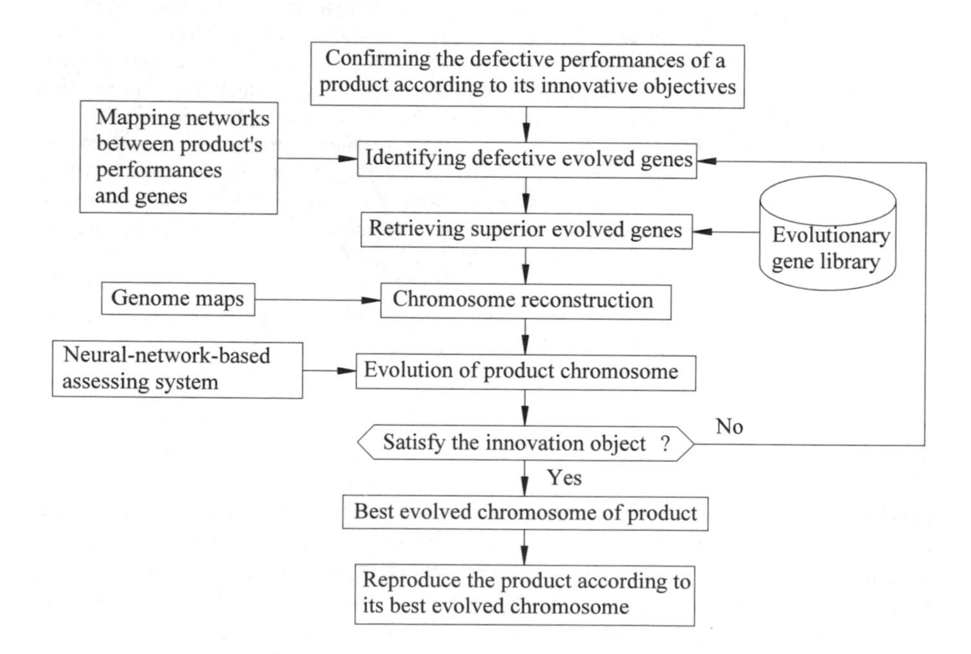

Figure 11.1 Workflow of the genetic-engineering-based innovation method

a) Confirm the defective performances of a product according to its innovative objectives. The approach can go through (1) to collect all the necessary information using the Innovative Situation Questionnaire [7], (2) to search a primary problem and break it down into many smaller problems, which are structured as contradictions, using the Problem Formulation [7], and (3) to confirm defective performances (*i.e.*, resolving one of these secondary problems will resolve the primary problem without coupling) using the Axiomatic Design principles [8].
b) Identify defectively evolved genes. The defectively evolved genes that affect the defective performances can then be identified from the mapping networks [6] between the product's performances and genes according to the defective performances determined. Their locations in the chromosome can be recognized from the genome maps [6] of the product.
c) Retrieve superior evolved genes from an evolutionary gene library [6]. The superior evolved genes are used to replace corresponding defectively evolved genes to improve the performance of a product and can be derived from successful products among the same and similar types of products in the world using the latest achievement in science and technology. According to the innovative objective, the code names and detailed information of the superior evolved genes that can resolve the contradictions can be retrieved from the evolutionary gene library.
d) Chromosome reconstruction. The reconstruction procedure will be first to endow the defectively evolved genes with zero or much lower probability for being selected as genetic information and then graft the superior evolved genes to the locations of the corresponding defectively evolved genes and endow them with higher probabilities. Sometimes, there are several superior evolved genes which can possibly be used to replace the same defectively evolved gene. These superior evolved genes should all be grafted in the location of corresponding defectively evolved genes and be endowed with different higher probabilities according to their effects on realizing the innovation objective. The sum of the probabilities for all the evolutionary information and genetic information in the node should be equal to 100%.
e) Evolution of product chromosome. The nodes in the chromosome of a product ($G_1, G_2, \cdots, G_n$), where the genetic information needs to be selected from the evolutionary information in the same nodes for product innovation, are arranged into a string as a special type of chromosomes (*i.e.* an algorithmic chromosome) specially for using Genetic Algorithms (GAs) [9,10]. Under each node (*i.e.* an algorithmic gene), all the pieces of evolutionary information with their endowed probabilities for being selected as genetic information are listed. The optimization process using GAs can be regarded as the evolution process of a product chromosome. Its survival-of-the-fittest mechanism is implemented by means of a neural-network-based assessment system of product performances [11].
f) Innovate the product according to its evolved virtual chromosome. According to the algorithmic chromosome, a best-evolved chromosome of the product can be obtained, based on which the new product can be redesigned and manufactured. If the best-evolved chromosome of a product does not meet the

innovative objectives satisfactorily, return to Step (b) to repeat the design procedure until the new product can satisfy the innovative objectives.

11.3 Contents and Data Structure of Virtual Chromosomes of Manufactured Products

From the second step, the design method involves the genes/chromosomes of manufactured products. Since manufactured products have no physical chromosomes, their virtual or analogous chromosomes need to be disclosed. Their contents should be determined according to their functions. Two functions are essential; *i.e.*, the products should be able to be created or cloned according to their virtual chromosomes, and secondly, they should be able to be evolved by the differentiation of their virtual chromosomes. Therefore, the contents of a virtual chromosome of a product include both genetic information and evolutionary information.

The former is needed to clone the product and includes all the information for reproducing the product, such as the lists of parts and units, the shapes, dimensions, material technical specification and manufacturing process of each part, and the topological relationships and assembly relationships among parts and units. It is much more than the information contained in the Bill of Materials (BOM). The latter is all the information that has ever been used as the genetic information in the chromosomes of previous generations from the very beginning to the present or grafted artificially from advanced products, and can make it possible to differentiate the product's chromosome for propagating an offspring with performance that is different and may be better than its parents'. For example, Steel is used for a part's material now; but Copper, Aluminum, and Plastics have been used for this kind of part in the past already. Thus, Steel is the genetic information. Copper, Aluminum, Plastics and the newly grafted materials, for instance, Titanium, whose properties are better for this kind of part than those of Steel, are the evolutionary information. To evolve or innovate a product, the virtual chromosome of the product should also contain all the evolutionary information from the very beginning, like the chromosomes of living beings in nature. According to the above analyses, it follows that its data structure can be in the form of multi dimensional networks, as shown in Figure 11.2. Its characteristics can be summarized as follows:

a) Its data structure is basically hierarchical

The information or data in a parent node of a higher layer is detailed by the information or data in its child nodes of a lower layer until all the necessary details have been given in the last node of the lowest layer. The last node forms a terminal node of the hierarchical tree. There are usually more than ten layers in a product's chromosome. The relationship between the parent node and child nodes is a "one to more" relationship and belongs to the CONSIST type.

b) Its data structure is multi dimensional

Each node in the networks contains many pieces of evolutionary information, whether used historically or grafted artificially from advanced products. However, only one of them can be used as genetic information for reproducing or cloning the product. The evolution of products is implemented by replacing some genetic information with one of the evolutionary information in the same node. The relationship among them belongs to the SELECT type.

c) *There are inter-relationships among data or information*

Much evolutionary information displays dependences on other evolutionary information in the same layer of the same branch or in the same or different layer of different branches, which make the data structure much more complicated and cannot be illustrated completely in Figure 11.1. For example, there are three pieces of evolutionary information in the node of "heat treatment methods", *i.e.*, Heat treatment method A, B, and C. The selection of a heat treatment method is dependent on the material selected in the node of "part materials". The relationship among them belongs to the DEPENDENCE type.

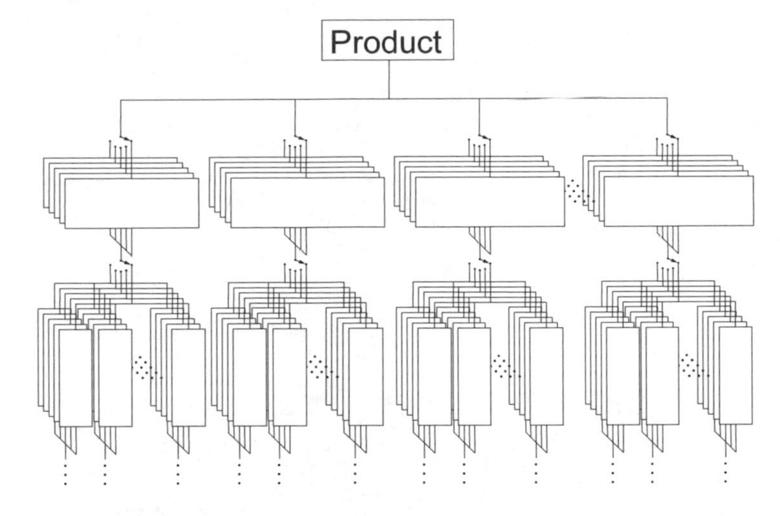

Figure 11.2 Schematic diagram of data structure of a product's virtual chromosome

The chromosome of a product contains so much information and is so complicated in data structure that it has to be stored and edited by database software. Some genetic and evolution information in the chromosome of a product can be acquired from the design and manufacturing documentation of products, related design handbooks, technical standards, technical specifications, technical regulations, technical documents, or the designers' brains, while others have not been collected and sorted out and need to be explored. As a result, the content and data structure of virtual chromosomes of products can be reverse-deduced, based on careful analyses of the evolution course of a product's design and manufacturing.

11.4 Database Structure for the Virtual Chromosomes of Manufactured Products

Main Framework of Databases

Based on the content and data structure of genetic information in a virtual chromosome, introduced in the last section, the main framework of databases for the virtual chromosomes of products is designed using IDEFIX notation [12] and shown in Figure 11.3. Commercial tools are readily available for constructing IDEFIX diagrams and generating database structure.

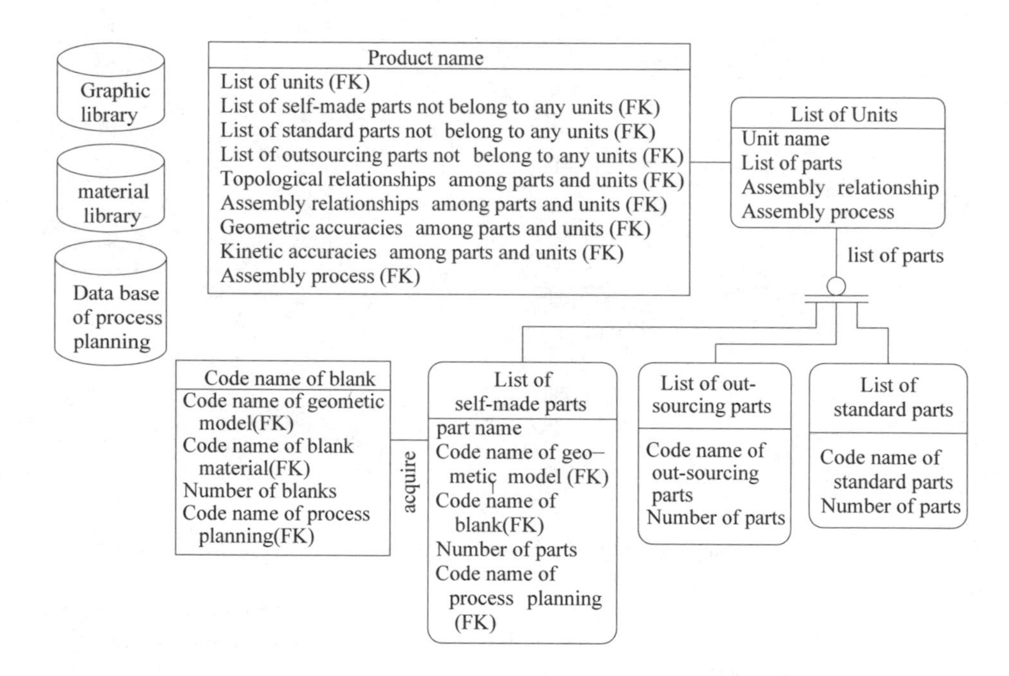

Figure 11.3 Main framework of database for the chromosomes of products

There are six entities in this main framework. The independent entity for products covers one primary key attribute (*i.e.* product name) and nine remaining attributes that are described as follows:

a) List of units or sub-assemblies: it includes the codes and names of all the units or sub-assemblies in a product.
b) List of self-made parts: it covers the codes and names of all the self-made parts that belong to the units and do not belong to any units.
c) List of standard parts: it includes the codes and names of all the standard parts that belong to the units and do not belong to any units.
d) List of out-sourcing parts: it covers the codes and names of all the bought-out parts that belong to the units and do not belong to any units.

e) Topological relationships among parts and/or units. The geometric model of every part or unit has its local coordinate system, the origin of which is the inserting point of the part or unit in the global coordinate system of the product. The topological relationships between two parts and/or units are the distance and angle between their local coordinate systems in the global coordinate system. The format of information about the topological relationship is:
(Part or Unit A, Part or Unit B, relative distance, relative orientation)
where, "relative distance" is a vector (Δx, Δy, Δz) for Cartesian coordinate system or ($\Delta \rho$, $\Delta \theta$, Δz) for cylindrical coordinate system, and "relative orientation" is a spatial angle and has three components ($\Delta \alpha$, $\Delta \beta$, $\Delta \gamma$). To reduce information in a database, only the topological relationships between two parts that contact each other will be stored in the database. With the information, the topological relationships between two parts that do not contact each other can also be calculated if needed.
f) Assembly relationships among parts and/or units. For the parts and/or units that contact each other, their assembly relationship must be defined. The format of information about it is:
(Surface Fa in Part or Unit A, Surface Fb in Part or Unit B, assembly type)
where "assembly type" has the following types: Non-dismountable fastening assembly (such as welded joints), Dismountable fastening assembly (such as bolted joints), Movable assembly, and Non-assembly.
g) Geometric and/or dimensional accuracy of assembly. After assembly, some geometric characteristics and/or dimensions should meet the required accuracies. The format of information about the topological relationship is:
(Surface Fa in Part or Unit A, Surface Fb in Part or Unit B, geometric characteristic or dimension, permissible errors or tolerances).
h) Kinetic accuracies between parts and/or units. After assembly, the relative motion between some parts and/or sub-assemblies also should satisfy the required accuracies. The format of information about the topological relationship is:
(Surface Fa in Part or Unit A, Surface Fb in Part or Unit B, kinetic characteristic, permissible errors or tolerances).
i) Assembly process. It includes all the working procedures for assembly, the inspections needed for each working procedure, *etc.*.

The independent entity for products is elaborated by one entity (list of unit) that is further elaborated by three entities: lists of standard parts, self-made parts, and purchased parts. The information about blanks or semi-finished parts for self-made parts can be acquired from the entities for blanks.

Database Structure for Evolutionary Information

As previously mentioned, every node in a chromosome has much evolutionary information, which must be stored in the same database. Based on the requirement, the database for each entity is further developed. Taking "product" entity as an example, its database structure is shown in Figure 11.4.

The value in brackets (such as p_{11}, p_{12}, or p_{13}) after each piece of evolutionary information represents its probability for being selected as genetic information and is between 0 and 1. But the sum of the probabilities for all the evolutionary information (including genetic information) in one entity should be equal to 100%.

Description of Inter-relationships among Data in the Database

A relational database can easily accommodate the CONSIST and SELECT types of inter-relationships among data by keywords. But the DEPENDENT type of inter-relationship among data in different layers of different branches needs to be built by using a special database file, as shown in Table 11.1, where D_i and D_j are the codes of "cause" information and "result" information in the product chromosome, respectively, and the relationship description can be of the following three types:

a) Logical expression

The DEPENDENT relationship can be described by a logical expression. If there is one "cause" and one "result" (*i.e.* $n = m = 1$), it is a *one-to-one* selection relationship. For instance, the logical expression, $C = A$, indicates that Information C must be selected if Information A is selected. When there are several "causes" and one "result", *i.e.* $n > 1$ and $m = 1$, it is a *more-to-one* selection relationship. For example, the logical expression:

$$\boxed{C = A \cap \overline{B}} \tag{11.1}$$

represents that Information C must be selected if Information A is selected and Information B is not selected.

b) Functional expression

This DEPENDENT relationship ($n \geq 1$ and $m = 1$) can be described by a functional expression and is usually used to define the dimension and precision relationship among parts and/or units. Taking a ball bearing as an example, the relationship among the diameter of a raceway on the outer ring (D_g), the diameter of a raceway on the inner ring (d_g), and the diameter of balls (d_0) can be presented as:

$$\boxed{D_g = d_g + 2d_0.} \tag{11.2}$$

c) Keyword expression

The DEPENDENT relationship ($n = 1$ and $m \geq 1$) can be described by a keyword expression. The "cause" information is a keyword. The "result" information can be searched according to the keyword. For instance, the catalog number of a ball bearing is the "cause" information and can be taken as a keyword. Based on this keyword, all the specifications of the ball bearing (such as outer diameter of outer ring, inner diameter of inner ring, and assembly width for ball bearing), *i.e.* "result" information, can be searched.

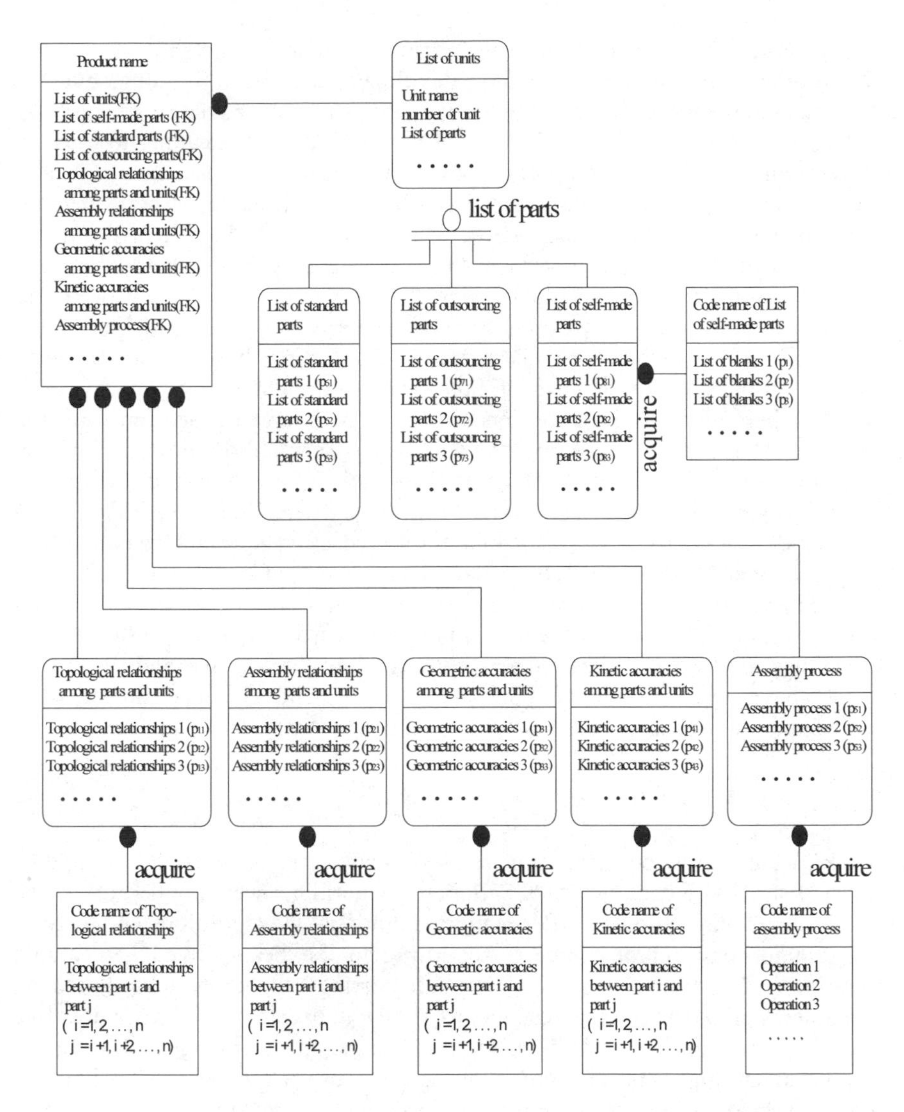

Figure 11.4 Database structure for evolutionary information of "product name" entity

Table 11.1 Form of database file

Codes of "cause"	Codes of "result"	Relationship description
D_i ($i = 1, 2, \ldots, n$)	D_j ($j = 1, 2, \ldots, m$)	

11.5 An Example

Since the length of a conference paper is limited, we take a simple type of mechanical product, Ball Bearings, as an example to trace back briefly to their innovation using the developed gene engineering based design method.

According to the innovation method developed, the innovative objectives of ball bearings have to be confirmed first. Its innovative objectives were: carrying both larger radical and larger bi-directional thrust loads (A), adapting to varying working loads, *i.e.* varying proportions of radical load to thrust loads (B), and lower cost (C). According to the innovative objectives, its defective performances can be obtained using the method introduced in Section 11.2. Then, the genes which need to be reformed can be identified from the mapping networks between the ball bearings' performances and genes as follows: (1) the method of filling balls, (2) the radius of its raceway's groove in comparison with the balls' radii, (3) the geometry of the raceway, (4) the method of adjusting the clearance between raceways and balls, (5) the number of ball rows, (6) the number of raceways, and (7) the numbers of outer rings and inner rings. These seven genes can form the main stem of an algorithm chromosome for using Genetic Algorithms. For each algorithmic gene, all the pieces of its evolutionary information and the grafted superior evolved genes retrieved from its evolutionary gene library were then listed with their endowed probabilities for being selected as genetic information. The probabilities were endowed based on the rules introduced in Section 11.2. The assessment system of the ball bearings' performances was designed using artificial neural networks. The inputs of the neural network based assessment system of the ball bearings' performances are the above seven algorithmic genes and its outputs are the above three innovative objectives (A, B, and C). The fitness of a chromosome is a function of these outputs, and can be expressed as follows:

$$\boxed{F = k_1 A + k_2 B + k_3 C} \tag{11.3}$$

where k_1, k_2, and k_3 are constants and are used to adjust the weights of A, B, and C. Their sum should be equal to 100%. Genetic Algorithms were applied to evolve the algorithmic chromosome. After many generations, the algorithms converge to the best algorithmic chromosome. According to the best-evolved algorithmic chromosome, the chromosome of a new ball bearing can be obtained, based on which a new ball bearing was redesigned as shown in Figure 11.5. This innovative ball bearing has one row of balls, a retaining cage, an outer ring, and a pair of separable inner rings. Since there is no need to use a second row of balls or another ball bearing to carry the thrust load in another direction, its cost is lower. The cross sections of the grooves on both inner and outer rings have two symmetrical arcs, the centers of which are located on two diagonal contact lines, respectively as shown in Figure 11.5. Moreover, the radii of the arcs are larger than the radii of balls. The clearances between raceway and balls can be adjusted accurately by grinding the inner face in one of the two inner rings, so that the contact angle between the balls and the raceway will not be changed much when the load varies, and the bearing can thus adapt to a varying working load. Since the inner rings can be separated while the balls are filled, this ball bearing can contain more balls so that it can carry not only larger radical loads but also larger bi-directional thrust loads. Therefore, this ball bearing can meet its innovative objectives satisfactorily.

Figure 11.5 Innovated ball bearing

11.6 Conclusions

With the similarity between the evolution of living beings and the development of manufactured products, gene-engineering techniques have been applied to develop a systematic design theory and methodology for product innovation. This innovation method is different from the conventional one as it innovates products via artificial differentiation of virtual product chromosomes. It provides a logically structured process, which can reduce blindness to innovation. Since products have no physical chromosomes, their virtual chromosomes must be reverse-deduced according to their function requirements so that the new innovation method can be applied. Two functions are essential; first, the products should be able to be reproduced or cloned according to their virtual chromosomes, and second, they should be able to be evolved by the differentiation of their virtual chromosomes. Therefore, the contents of a product's virtual chromosome include both genetic information and evolutionary information. Its data structure is basically hierarchical and multi dimensional, and there are three types of inter-relationships among data or information. Since the product's chromosome contains so much information and is so complicated in data structure, it has to be stored and edited by database software. The research on the reverse deduction of the virtual chromosomes of other products can be implemented based on the results.

11.7 Acknowledgements

The reported research is supported by University Research Committee Grants (CRCG) in the University of Hong Kong. The financial contribution is gratefully acknowledged.

11.8 References

[1] Nicholl, D.S.T., 1994, *An Introduction To Genetic Engineering*, Cambridge University Press, Cambridge, UK.

[2] Ho, M.W., 1998, *Genetic Engineering-Dream or Nightmare,* Gateway Books, Bath, UK.

[3] Winter, P.C., Hichey G.I., and Fletcher, H.L., 1998, *Instant Notes in Genetics*, BIOS Scientific Publishers Limited.

[4] Wright, P.K., 2001, *21st Century Manufacturing,* Prentice-Hall Inc., New Jersey.

[5] Chen, K.Z., Feng, X.A., 2002, "Exploring A Genetics-Based Design Theory and Methodology For Innovating Products," *Proceedings of the 6th Int. Conf. on Engineering Design and Automation,* 2002, Maui, Hawaii, USA, pp. 278-283.

[6] Chen, K.Z. and Feng, X.A., 2003, "A Framework of the Genetic-Engineering-Based Design Theory And Methodology For Product Innovation," *Proc. of 14th Int. Conf. on Engineering Design*, August, 2003, Stockholm, Sweden, No.1093.

[7] Terninko, J., Zusman, A., and Zlotin, B., 1998, *Systematic Innovation: An Introduction to TRIZ*, CRC Press LLC, New York.

[8] Suh, N.P., 1990, *The Principle of Design*, Oxford University Press, New York.

[9] Gen, M. and Cheng, R., 1997, *Genetic Algorithms & Engineering Design*, John Wiley & Sons, Inc., New York.

[10] Chen, K.Z., Feng, X.A., 2003, "Computer-Aided Design Method for the Components Made of Heterogeneous Materials," *Computer-Aided Design*, Vol. 35, pp.453-466.

[11] Graupe, D., 1997, *Principles of Artificial Neural Networks*, World Scientific Publishing Co. Pte. Ltd., Singapore.

[12] Blaha, M.R., 2001, *A Manager's Guide To Database Technology: Building and Purchasing Better Applications,* Prentice Hall, USA.

12

Use of Constraint Programming for Design

Bernard Yannou, and Ghassen Harmel

Abstract: Three families of methods coexist for managing (*i.e.* representing and propagating) uncertainty of product data during the preliminary design, namely: fuzzy methods, probabilistic methods and Constraint Programming (CP) methods. CP methods over reals are, up to now, the less frequently used approaches, but they are worth further study for use in design engineering thanks to a number of good properties and recent significant advances. They may be roughly considered as a collection of methods that are sophisticated evolutions of interval analysis. The objective of this paper is to assess four of these major methods; namely the {Hull, Box, 3B-weak, 3B}-consistency methods in the context of the preliminary design of mechanical products where large variable domains are considered and a representation of the remaining consistent design space turns out to be of practical interest to support the designers' understanding and decision making. A measure for comparing the level of consistency of the methods is then proposed in the context of engineering design. It consists of a pairwise comparison of the overlapping part of the remaining design spaces for a given splitting grain size. Numerical results are established for an example of a combustion chamber design with 6 variables and 12 constraints. Next, a sensitivity analysis of the consistency of the previous methods is performed in regards to a variable splitting grain size. Experiments have been performed on a research platform including up-to-date NCSP methods. The paper concludes that NCSP methods are not easy-to-use and that the designer must be aware of a number of concepts so as to select the best choices in the resolution strategies.

Keywords: uncertainty, preliminary design, consistency measure, constraint programming, design space

12.1 Introduction

A previous article [21] highlighted that the management (representation and propagation) of uncertainty was of the utmost importance in the preliminary design stages of product design. Indeed, the uncertainty reduction paradigm turns out to be much more virtuous for concurrent engineering in comparison to the try-and-test deterministic optimization paradigm [20]. Unfortunately, few uncertainty management systems exist for they face a number of severe limitations such as: size of the problems, computation times, and lack of consistency between the "uncertain" representations of the design variables. We presented [21] the outlines of the three families of methods for managing uncertainty, namely: fuzzy methods [1], probabilistic methods [16] and Constraint Programming (CP) methods. We advocated that CP techniques over reals outperform the two other families of methods for use in preliminary design or, at least, that they are worth further study because of the number of good properties and recent significant advances. We also provided a state-of-the-art review of CP techniques for solving design problems (see for example [9, 11]).

In the present paper, we want to tackle the practical implementation of a design problem with CP (over reals). The Modelling and solving stages of a design problem with CP techniques are peculiar since, first, a design problem is characterized by significant uncertainties on variable values (large variable domains) at the beginning of a dimensioning process and, second, the designers want to benefit from a precise and consistent representation of the remaining design space at any moment. In addition, the constraints linking design variables are often of a heavily intricate polynomial form. These considerations entail a particular strategy for the choice and tuning of a given CP resolution technique, among the Hull-, Box-, Weak-3B-, and 3B- consistency techniques [5, 6]. The present paper aims at advancing the comprehension of the drivers that influence the quality of the result for preliminary design.

In Section 12.2, the principles of CP techniques over reals are presented. In CP, variables are modelled as intervals of allowable values and CP techniques are sophisticated evolutions of interval analysis. Section 12.3 presents the "branch and prune" solving process of CP that results in a representation of the design space as a consistent Cartesian product of intervals. Four main consistency techniques for uncertainty reduction are briefly presented in Section 12.4. Some quality measures of the results are proposed in Section 12.5; they concern the consistency of the design space, which is assessed in a relative manner between techniques. Practical results are provided for the case study of the design of a combustion chamber as defined in [18]. Finally, the influence of the splitting grain size is studied over the design space quality in Section 12.6. Some practical advice for using CP in a preliminary design context is offered based on the experiments in Section 12.7.

12.2 The Principles of CP

Two major families of CP techniques exist: CP techniques on discrete domains and CP techniques over continuous domains. Moreover, more and more research is carried out on compound techniques.

CP techniques on discrete domains (*e.g.*, integer domains) have been developed for 3 decades [12, 19]. The arc-consistency technique is an efficient uncertainty *reduction* or *filtering technique*[1] for a number of high combinatorial problems such as scheduling [4] or space layout planning [10]. Some attempts have been made to use this efficient technique in design engineering for discretizing the domains. However, it has been found that some solutions lying between two discrete values may be lost in the process whereas the founding, paradigm of CP is paradoxically not to forget any possible solution.

CP techniques over continuous domains (or reals) are based on modelling of a variable domain by an interval whose bounds are known with a given accuracy. The foundation works on *interval arithmetics* are not recent (see [13]). These methods basically consist of replacing any variable occurrence in a mathematical constraint (*e.g.*, polynomial) by its current interval domain so as to shrink it at best and to eliminate the infeasible parts of the design space. Equation 12.1 shows the basic relations that must be fulfilled between thc bounds of intervals linked by the four elementary arithmetic operations.

$$\boxed{\begin{array}{l} [a,b]+[c,d]=[(a+c),(b+d)] \\ [a,b]-[c,d]=[(a-d),(b-c)] \\ [a,b]\times[c,d]=[\min(ac,ad,bc,bd),\max(ac,ad,bc,bd)] \\ [a,b]/[c,d]=[\min(a/c,a/d,b/c,b/d),\max(a/c,a/d,b/c,b/d)] \\ \qquad\qquad\qquad si \quad 0\notin[c,d] \quad \sin on \quad =[-\infty,+\infty] \end{array}} \tag{12.1}$$

For instance, let us take the constraint $x = y + z$. Starting from *initial variable domains* for the three variables, one must shrink the domains in a manner that is independent from the way the constraint is written. *Constraint inversion techniques* are then used (Equation 12.2.) to propagate domain reductions in all directions and to engage a reduction process until no more significant reduction can be achieved, *i.e.* until stabilization or quasi-stabilization (in certain cases infinite loops) may occur.

$$\boxed{\begin{cases} x_{\min} \leftarrow \max(x_{\min},(y_{\min}+z_{\min})) \\ x_{\max} \leftarrow \min(x_{\max},(y_{\max}+z_{\max})) \\ y_{\min} \leftarrow \max(y_{\min},(x_{\min}-z_{\max})) \end{cases} \quad \begin{cases} y_{\max} \leftarrow \min(y_{\max},(x_{\max}-z_{\min})) \\ z_{\min} \leftarrow \max(z_{\min},(x_{\min}-y_{\max})) \\ z_{\max} \leftarrow \min(z_{\max},(x_{\max}-y_{\min})) \end{cases}} \tag{12.2}$$

---

[1] A *filtering* technique attempts to rule most of inconsistent values out of the variable domains.

Constraint Programming methods have made this process more sophisticated by:

- allowing a dynamic updating of domains as soon as new variables and constraints are added in the constrained system (a facility well adapted to incremental design actions),
- improving the efficiency of interval reductions, that is to say the *consistency* of the resulting domains and, therefore, the minimal and pertinent representation of the remaining design space.

It has been made possible to adopt modelling of the constrained system in the form of a graph whose nodes represent variables and domains, and whose edges represent binary or n-ary constraints. *Constraint Satisfaction Problems* (*CSPs*) and *Numerical CSPs* (*NCSPs*) may be encountered in the special case of continuous domains. Some significant advances in *NCSPs* have been made in the last decade (see for example [2, 8, 14, 17]), resulting in a better understanding of practical industrial problems.

Why do some *reduction* or *filtering* techniques turn out to be more or less efficient and their result more or less *consistent*? The inconsistency sources can be explained simply:

- The *dependency problem* of interval arithmetics (see [6]) is generated by the fact that a variable occurrence is brutally replaced by its current domain during the solving process. Subsequently, the multiple occurrences of a given variable within a given constraint, and even between different constraints, are de-correlated. This de-correlation results in relaxed constraints and then in larger domains. This is why it is often necessary to reformulate constraints in decreasing the numbers of the same variable occurrences by appropriate factorization strategies (see [3]). For example, Equation 12.3 presents two mathematically equivalent forms of a function of x: $4x-x^2=4-(x-2)^2$. However, the two forms lead to different interval reductions. The factorized form leads to the best (even optimal) uncertainty reduction. In the same manner, it is proved that the domain of $x\times(y+z)$ is always included in, or is equal to, the domain of the following developed form $x\times y+x\times z$.

$$x\in[0,1]\Rightarrow\begin{cases}4x-x^2\in[-1,4]\\4-(x-2)^2\in[0,3]\end{cases} \tag{12.3}$$

- The scope (locality) refers to where the reduction/filtering/consistency technique operates. *Local consistency techniques* like the *hull-consistency* and *box-consistency* techniques, process constraints one by one. Then, final domains are computed from the local computations until stabilization. This is not as efficient as it could be if all the constraints were taken into account together. More globally, consistent techniques like weak-3B-consistency and 3B-consistency techniques partially solve the problem to the detriment of the computation time (see [8]).

12.3 The Solving Process of CP

A practical case study of a combustion chamber design originally introduced by Wagner and Papalambros [18] is used. It is a non-trivial constrained problem (see Figure 12.1) involving 6 variables: cylinder bore (B), compression ratio (C_r), exhaust valve diameter (D_e), intake valve diameter (D_i), and revolutions per minute at peak power (W). One of the target performances is to maximize the brake power per unit engine displacement (F).

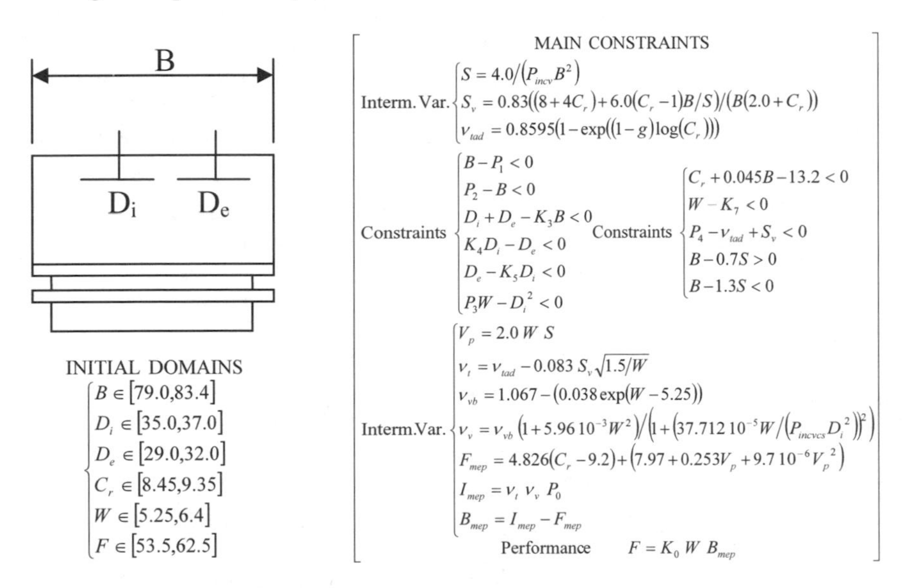

Figure 12.1 Parameterization of the combustion chamber and main constraints

For the CP calculations, an NCSP platform named RealPaver (see [5]), developed by the IRIN computer science department of the Nantes university (France), was used. The solving process of a constrained problem over a continuous domain follows a well–known "*branch and prune*" algorithm composed of 3 stages:

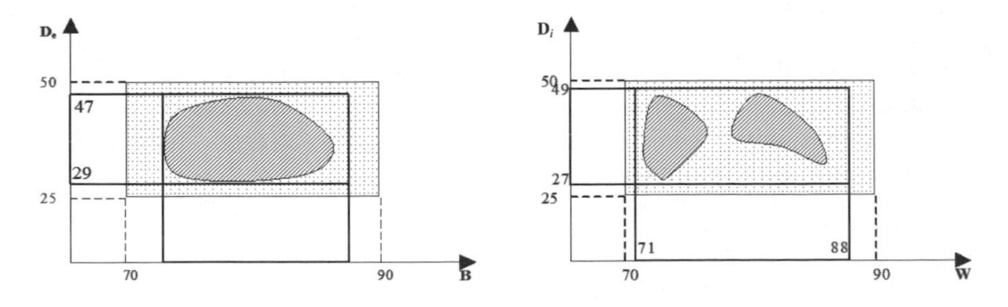

Figure 12.2 Initial domain reductions/filtering

Initial reduction (or filtering): The choice of a given filtering technique and the consideration of the constraints and initial domains lead to primary significant domain reductions. Figure 12.2 shows, in the combustion chamber problem, that W domain has been shrunk from [70, 90] to [71, 88][2]. But the Cartesian product of the shrunk/filtered domains does not represent the *design space* (hatched areas in Figure 12.2), *i.e.* the set of all thc valid design points. Indeed, this Cartesian product exceeds the design space since some points in the n-dimensional box of this Cartesian product (here $n=6$) do not belong to the design space. The method is not *consistent*, meaning that no in-domain solution is ensured to be actually valid, even if all PC methods ensure *completeness*. This means that any actual design solution is ensured not to be forgotten. In fact, no existing method is able to perfectly represent this design space (hatched areas). Moreover, there is no presumption that the ultimate reduction[3] around the design space has been performed.

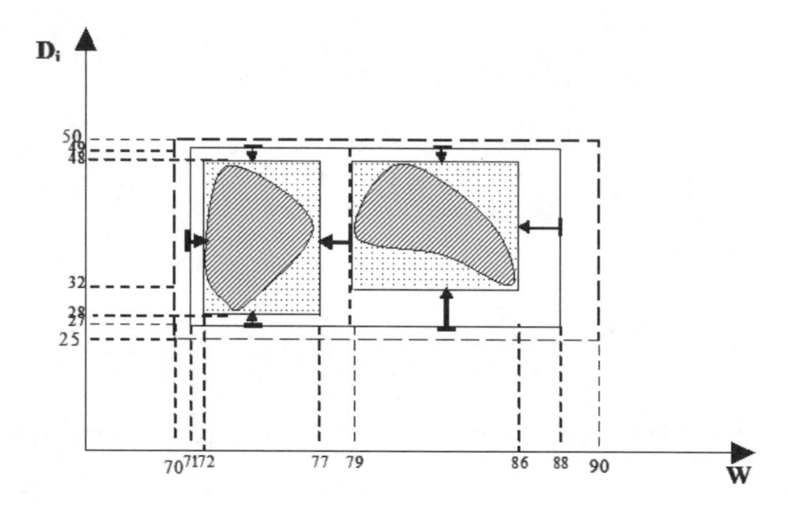

Figure 12.3 Bisection of W domain and reduction of the two constrained sub-problems

Alternation of domain bisections and domain reductions/filtering: In order to converge towards a more precise representation of the design space, one proceeds to the successive domain bisections for variables (W, B, D_i, D_e, C_r, F) in a predefined order or according to a given strategy for choosing the next variable domain to bisect[4]. After bisection has been achieved, once on each of the n variables, the process is repeated recursively until a given *level of splitting* is reached. Each bisection generates a *branching* in a binary search tree, leading to two new constrained sub-problems. For instance, the bisection on the first domain

[2] The numerical results have been approximated for clarity.

[3] When the tightest domain narrowing has been performed, the filtering technique is said to be *globally consistent*.

[4] A conventional splitting (or branching) strategy is proposed in RealPaver; it consists of choosing, for the next bisection, the remaining variable whose domain is the largest.

of the following domains ($[a,b]$, D_2, D_3) leads to the two constrained sub-problems ($[a,(a+b)/2]$, D_2, D_3) and ($[(a+b)/2,b]$, D_2, D_3). Each sub-problem is immediately filtered as well throughout the consistency technique so as to contract the variable domains (see Figure 12.3). Then the next bisections are made on domains recently shrunk. As soon as a domain is contracted to an empty set, it is certain that no design solution exists under this search branch. The more consistent the filtering technique, the earlier the detection of "no design space here", the more efficiently the research tree is pruned close to the root, avoiding useless explorations. A trade-off clearly appears between the filtering technique efficiency (consistency level) and its computation time. The ultimate level of splitting may be defined by an expected granularity or precision of domains; when all the domains have lowered their size under an expected precision without any empty domain, the resulting n-dimensional small box is considered as a piece of the potential design space. Once the splitting process is over, a *collection of small, disjointed n-dimensional boxes* approximates the potential design space (see Figure 12.4). As the number of the enumerated small boxes may be huge and unexpected when a splitting grain size is expected, an alternative mechanism is provided in *RealPaver* [5] for fixing a given number of boxes and ensuring that the whole design space is covered. This mechanism is used to compare the efficiency of the filtering techniques.

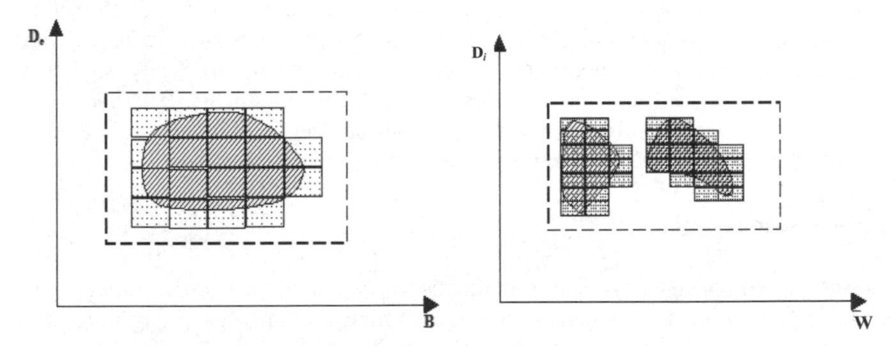

Figure 12.4 Representation of the potential design space after a process of domain bisections

Post-processing: representations of the design space. The graphical representation of this collection of n-dimensional small boxes is easy and convenient to get a good picture of the design space. Ultimately, the design space can be apprehended by its two or three-dimensional projections on couples or triplets of variables. Sam-Haroud [15] has proposed an economical representation of these n-dimensional boxes that allows quick 2D or 3D representations. We used a simpler representation tool (see Figure 12.5).

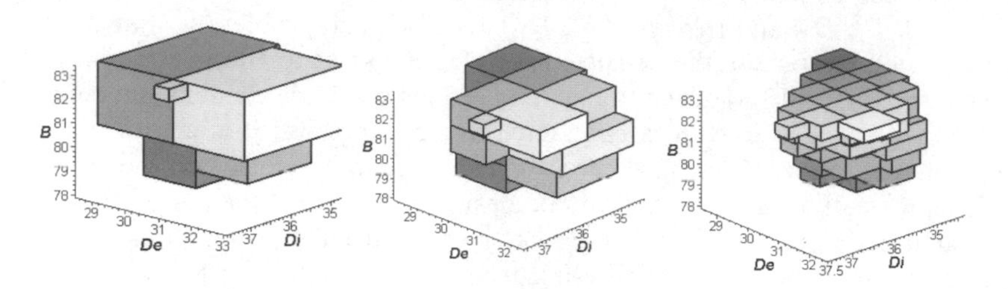

Figure 12.5 3D projections of the design space of the combustion chamber for different granularities of the enumerated boxes

12.4 The Consistency Techniques

Three major consistency techniques exist for continuous domains: *hull* (or *2B*), *box* and *3B*. The first two are called *local consistency techniques* and the last is said to be more global. All these techniques have variants that sometimes slightly differ in terms of results and sometimes get the same result but differ in terms of algorithms. Although our experimental platform, *RealPaver* [5], proposes such variants, we do not distinguish between them here since we did not observe any significant difference in the resulting potential design space, except for a *3B* variant called *weak-3B* that, consequently, needs further consideration. In this section, we briefly describe the major principles and properties of these techniques.

Hull-consistency

The most used consistency technique for discrete domains is *arc-consistency*. Arc-consistency is a local consistency technique which includes reasoning in terms of compatible values in domains only in the locality of a couple of variables. For a given couple of variables, a value is kept in the domain of a first variable if, and only if, a compatible value exists in the second domain relative to all constraints. In practice, user constraints are decomposed into ternary constraints through a number of intermediary variables. For instance, $C : x + y \times z = t$ might be decomposed into $C_{dec} = \{y \times z = \alpha, x + \alpha = t\}$ via the creation of the intermediary variable α. Then, the basic inference mechanism is the one presented in Section 12.2 under the name of constraint inversion techniques so as to propagate domain narrowings in all directions.

Hull-consistency is simply an adaptation of arc-consistency to continuous domains with paying attention to round-off errors at interval bounds.

Despite its simplicity, speed of execution and efficiency for simple constraints, the introduction of a number of intermediary variables significantly hinders the domain narrowing of the variables the designer wishes to focus on (*i.e.* the 6 variables for the combustion chamber). This problem is particularly enhanced by complex constraints having several occurrences of the same variable. This case is frequent in engineering as can be seen from the combustion chamber problem with v_v function of W (see Figure 12.1).

Box-consistency

This last issue is partly overcome by the *box-consistency* technique because user constraints do not require further decomposition. Here, in an elementary loop (see Equation 12.4), each variable is successively kept as a variable in the constraint set whereas all others are replaced by their domains, resulting in a new domain narrowing. Once all the domains have been successively narrowed, another loop may start until (quasi-) stabilization is accomplished. The dependency problem due to multiple variable occurrences is thus efficiently solved.

$$\left\{\begin{array}{l} C_1(x_1, D_2, ..., D_n) \\ C_2(x_1, D_2, ..., D_n) \\ ... \\ C_m(x_1, D_2, ..., D_n) \end{array}\right. \xrightarrow{\text{New } D_1} \left\{\begin{array}{l} C_1(D_1, x_2, ..., D_n) \\ C_2(D_1, x_2, ..., D_n) \\ ... \\ C_m(D_1, x_2, ..., D_n) \end{array}\right. \xrightarrow{\text{New } D_2} \left\{\begin{array}{l} ... \\ ... \\ ... \\ ... \end{array}\right. \rightarrow \left\{\begin{array}{l} C_1(D_1, D_2, ..., x_n) \\ C_2(D_1, D_2, ..., x_n) \\ ... \\ C_m(D_1, D_2, ..., x_n) \end{array}\right. \xrightarrow[\text{next loop}]{\text{New } D_n} \quad (12.4)$$

3B-consistency

Let us briefly note that 3B-consistency (see [7]) comes from the k-consistency techniques for discrete domains. It consists of obtaining consistent interval bounds in the sense of the 3-consistency, *i.e.* reasoning with triplets of variables instead of couples like for hull-consistency. The weak-3B is a relaxed form of 3B (see [5]). The combination of the presented techniques is not discussed here.

12.5 Consistency of Design Spaces

As we are especially interested in the result of the design space in design engineering, we already proposed in [22] to characterize the consistency of a filtering method or of its resulting design space by the size and location of the apparent design space. However, as the knowledge of the actual design space remains a utopia, we can only apprehend a relative consistency when comparing two apparent design spaces (noted DS_1 and DS_2), computed by two different filtering techniques. Finally, the three proposed indicators are the volumes of the common volume (between both design spaces) relative to each of the design space volume and the volume ratio itself. They are given by:

$$V_{\cap/1} = \frac{V_{DS_1 \cap DS_2}}{V_{DS_1}}, \quad V_{\cap/2} = \frac{V_{DS_1 \cap DS_2}}{V_{DS_2}}, \quad V_{1/2} = \frac{V_{DS_1}}{V_{DS_2}} \quad (12.5)$$

Since we expected that *3B* might be more consistent than *weak-3B*, itself is more consistent than *box*, that is more consistent than *hull*, we have preceded to relative consistency measures in that order. Table 12.1 shows the results by roughly representing in 2D the overlap of the design spaces. To make things comparable, the design spaces are all composed of 1000 boxes; of course the execution times *t* are not of the same order of magnitude.

The results stress the necessity to be conscious of the CP filtering techniques to assess result quality. Indeed, the size of the (potential) design space is drastically contracted from box to 3B by a ratio of 68%. *Box* dominates *hull* because *hull* is unable to contradict any potential solution provided by *box*. Moreover, Ds_{box} is completely imbedded in DS_{hull}. Let us recall that, due to the completeness property, all actual design solutions are necessarily within the calculated design spaces and, subsequently, within the intersected part of the two design spaces. Next, a design space contraction of 17.8% occurs when passing from *box* to *weak-3B* and another 20.9% when passing from *weak-3B* to *3B*. However, for both moves, each method brings on (or omits to filter) some inconsistent design solutions.

Experimental results confirm the theoretical expectations, providing at the same time some orders of magnitude. Indeed, the *dependency problem* and the *locality* of the filtering reasoning are the two major limitations of the CP techniques. These problems are further enhanced in the case of engineering design since large domains are considered[5] and multiple occurrences of a same variable exist over the set of constraints and mainly within some given constraints. This last case can explain the important gain between *hull* and *box*, since *box* partly overcomes this issue.

Finally, *weak-3B-consistency* appears to be a good trade-off in design engineering because the difference of volume is only 20% with 3B, but the technique is 100 times faster. This rougher design space can easily be compensated by a higher number of enumerated boxes as it is suggested in the next section.

Table 12.1 Relative comparisons of design space consistencies

Hull	**Hull vs box**	**Box**	**Box vs weak-3B**	**Weak-3B**	**Weak-3B vs 3B**	**3B**
V 91.3 t 90 ms	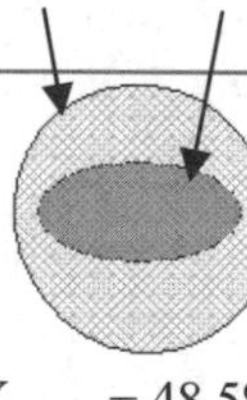 $V_{\cap/hull} = 48.5\%$ $V_{\cap/box} = 100\%$ $V_{box/hull} = 48.5\%$	V 44.3 t 660 ms	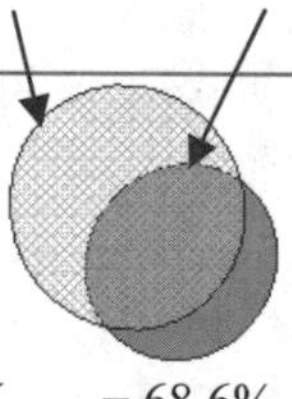 $V_{\cap/box} = 68.6\%$ $V_{\cap/w3b} = 83.5\%$ $V_{w3B/box} = 82.2\%$	V 36.4 t 9.7 s	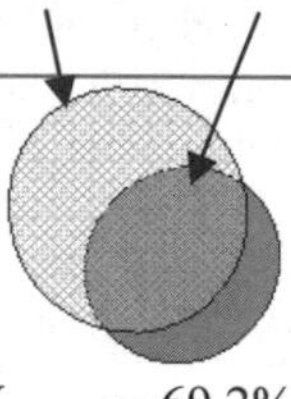 $V_{\cap/w3B} = 69.2\%$ $V_{\cap/3B} = 87.5\%$ $V_{3B/w3B} = 79.1\%$	V 28.8 t 1100 s

[5] Contrary to applications where domains represent physical dispersions, such as in tolerancing or imprecisions on initial conditions of a simulation.

12.6 Influence of Splitting Granularity

A finer granularity of the enumerated boxes can result in a more consistent (contracted) design space, as it appears in Figure 12.5 on a 3D projection of the design space of the combustion chamber. A more precise study of the influence of granularity has been performed for a number of boxes varying from 1000 to 10000 (see Figure 12.6). Classic asymptotic curves appear, but each limit is specific to a filtering technique, and an increase in the number of boxes does not always compensate a better native consistency. The influence of splitting is even greater with *weak-3B* (gain of 49% between 1000 and 10000) and *box* (gain: 41%) than with *hull* (gain: 27%).

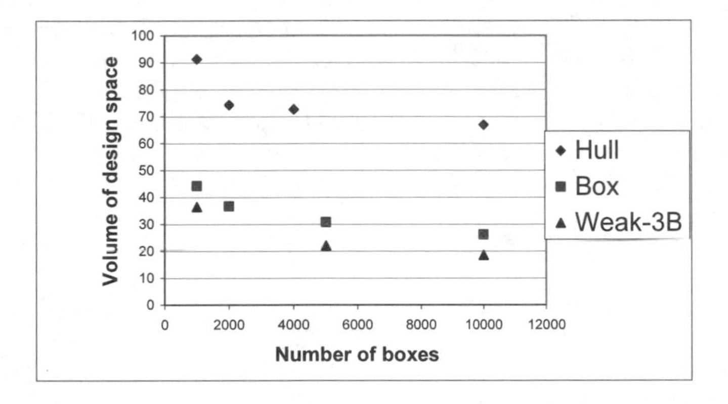

Figure 12.6 Influence of the splitting grain size on the volume of the design space

12.7 Conclusions

We believe that Constraint Programming over continuous domains is a valuable and promising technique to model and propagate uncertainties on variable values during the conceptual design stage (see [21]). It results in an encompassing design space that is easy to represent and that may bring pertinent information for personal or inter-personal negotiation [9]. But these techniques are not easy to tune (choice of filtering technique and splitting granularity) so as to obtain a sufficiently consistent representation of the design space. Moreover, we have mentioned that constrained problems in engineering were of a certain type. This work aims at providing some qualitative and quantitative information on CP techniques in design engineering. We tested four major techniques on an engineering case study of the combustion chamber, an example of non-trivial complexity that we believe to be representative. In conclusion, we believe that the *weak-3B-consistency* technique is a good trade-off because it is not a local consistency technique, and its computation time remains reasonable.

12.8 References

[1] Antonsson, E.K., Otto, K.N., 1995, "Imprecision in Engineering Design," *Journal of Mech. Design*, Vol. 117(B), pp. 25-32.

[2] Benhamou, F., McAllester, D., Van Hentenryck, P., 1994, "Clp(intervals) Revisited," In: *Logic Programming*.

[3] Ceberio, M., Granvilliers, L., 2000, "Solving Nonlinear Systems by Constraint Inversion and Interval Arithmetic," In: *AISC'2000: 5th International Conference on Artificial Intelligence and Symbolic Computation*, Madrid, Spain, pp. 127-141.

[4] Fox, M.S., Sycara, K., 1990, "Overview of CORTES: a constraint based approach to production planning, scheduling and control," In: *Fourth Int. Conf. of Expert Systems in Production and Operations Management*, 1990.

[5] Granvilliers, L., 2002, "RealPaver User's Manual. Version 0.2, http://www.sciences.univ-nantes.fr/info/perso/permanents/granvil/realpaver/main.html," *University of Nantes, Lab of Computer Science,* Nantes, France

[6] Granvilliers, L., Benhamou, F., Huens, E., 2001, "Constraint Propagation" (Chapter 5), In: *COCONUT Deliverable D1 - Algorithms for Solving Nonlinear Constrained and Optimization Problems, The Coconut Project*, pp. 113-149.

[7] Lhomme, O., 1993, "Consistency Techniques for Numeric CSPs." In: *IJCAI-93,* Chambéry, France, pp. 232-238.

[8] Lhomme, O., Gotlieb, A., Rueher, M., Taillibert, P., 1996, "Boosting The Interval Narrowing Algorithm," In: *ICLP*.

[9] Lottaz, C., Smith, I.F.C., Robert-Nicoud, Y., Faltings, B.V., 2000, "Constraint-Based Support for Negotiation In Collaborative Design," *Artificial Intelligence in Engineering,* Vol. 14, pp. 261-280.

[10] Medjdoub, B., Yannou, B., 2001, "Dynamic Space Ordering at a Topological Level In Space Planning," *Artificial Intelligence in Engineering*, Vol. 15, 1 January 2001, pp. 47-60.

[11] Merlet, J.-P., 2001, « Projet COPRIN : Contraintes, OPtimisation, Résolution par INtervalles, » Rapport, *INRIA* Sophia- Antipolis, 21 Septembre 2001.

[12] Montanari, U., 1974, "Networks Of Constraints: Fundamental Properties And Applications To Picture Processing," *Information Sciences*, Vol. 7, pp. 95-132.

[13] Moore R.E., 1979, "Methods and Applications of Interval Analysis," *SIAM Studies in Applied Mathematics*, Philadelphia, USA.

[14] Rueher, M., Solnon, C., 1997, *Concurrent Cooperating Solvers over Reals. Reliable Computing*, Vol. 3(3), pp. 325-333.

[15] Sam, J., 1995, « Constraint Consistency Techniques for Continuous Domains, » *Ph.D. Thesis 1423*, Ecole Polytechnique Fédérale de Lausanne, EPFL.

[16] Thurston, D.L., Liu, T., 1991, "Design Evaluation of Multiple Attributes Under Uncertainties," *International Journal of Systems Automation - Research and Application (SARA)*, Vol. 1, pp. 143-159.

[17] Van Hentenryck, P., Michel, L., Benhamou, F., 1998, "Newton: Constraint Programming over Nonlinear Constraints," *Science of Computer Programming,* Vol. 30(1-2), pp. 83-118.

[18] Wagner, T.C., Papalambros, P.Y., 1991, "Optimal Engine Design Using Nonlinear Programming and the Engine System Assessment Model," *Technical Report,* Ford Motor Co. Scientific Research Laboratories, Dearborn, Michigan and the Department of Mechanical Engineering at the University of Michigan, Ann Arbor, USA.

[19] Waltz, D., 1972, "Generating Semantic Descriptions From Drawings Of Scenes With Shadows," *Report*, MIT, MA, USA.

[20] Ward, A.C., Liker, J.K., Sobek, D.K., Cristiano J.J., 1994, "Set-based concurrent engineering and Toyota," In: *DETC/DTM*, pp. 79-90.

[21] Yannou, B., 2003, "Management of Uncertainty in Conceptual Design," In: *International CIRP Design Seminar,* May 12-14, 2003, Grenoble, France.

[22] Yannou, B., Simpson, T.W., Barton, R.R., 2003, "Towards A Conceptual Design Explorer Using Metamodeling Approaches And Constraint Programming," In: *DETC/DAC*, September 2-6, 2003, Chicago, Illinois, USA.

13

Model Infrastructures and Human Interaction in a Stereo Table Environment

Torsten Kjellberg, Christoffer Lindfors, Mattias Larsson, and Jonny Gustafsson

Abstract: The digital modelling of technical systems and physical phenomena and their environment forms the basis for design and development in the future. Based on models and visualisation technologies, the appearance and behaviour of a technical system can be studied, simulated and viewed for different purposes, in different disciplines, and at different detailing levels. It forms the base for human communication and interaction to meet the goals of all stakeholders in their study and development. Adding new technologies for man-model interaction and for the dynamic change of models and their viewing will further increase human possibilities for interaction and communication in the same co-located environment as well as in a virtual co-located environment. The paper will focus on a stereo interaction table environment as an important vehicle for integrated human–to–model and human–to–human interaction and some future possible developments.

Keywords: Modelling, Interactive system, Stereo viewing

13.1 Introduction

The digital modelling of technical systems, physical phenomena and their environment forms the basis for design and development in the future. Based on models and visualisation technologies, the appearance and behaviour of a technical system can be studied, simulated and viewed for different purposes, in different disciplines, and at different detailing levels. These technologies form the basis of human communication and interaction to meet goals in studies and development. Adding new technologies for man-model interaction and for the dynamic change of models and viewing will further increase human possibilities for interaction and communication in a co-located as well as a virtual co-located environment.

Man-model interaction technologies should form a natural way for humans to change and modify the models and the scene. Virtual clay modelling [1], haptic devices [2, 3] and supporting software are being further researched and developed to support engineering work. Technologies for Virtual Reality (VR) and human immersion into model space and for self-representation and association of a VR user in the environment can also be added. The CAVE has been popular for some time. One of its disadvantages is that only one person in the CAVE will have the right viewing perspective. Another disadvantage is the cost of investment, maintenance and operation.

In order to deal with these limitations, we propose a paradigm shift in bringing the model out to reality instead of having people going into the virtual world. We are projecting the digital model out of a table, called the stereo interaction table. People can sit around the table and, without special glasses, they can see the presented model in 3D on the table. They can discuss and interact and, at the same time, have eye contact and see each other's mimics and gestures. When talking about the presented model or VR model, we address the visualized model on the stereo interaction table (Figure 13.1).

In product development and realisation, human interaction and collaboration are of utmost importance. Communication, interaction and negotiations in interdisciplinary teams of marketing people, designers, manufacturing people, *etc.*. are the only way in which a successful product can be realized on the market.

Figure 13.1 Working at a stereo interaction table

Other applications for the stereo interaction table can be military command rooms, medical imaging and simulation, architectural and city planning, scientific visualization, chemistry and molecular analysis, games and a number of educational and information applications.

13.2 Principles of the Interaction Stereo Table

The stereo table uses a principle in which a number of 2D views of the 3D objects are presented in such a way that the display position of each view is limited and separated in space. The three-dimensional image is thus approximated as a number of two-dimensional images, each of which is sent to the correct angular segment of the viewing space. In this way, a viewer's right and left eye will see different images, which then should be the correct images for obtaining stereovision of the object. One of the most straightforward ways of achieving this is the *parallax stereogram,* conceived already in the 19th century [4].

In a *holographic stereogram* [5, 6] the optical effect separating the viewing positions is built into a hologram, but in this case the hologram also produces the actual images. In this work it, an Holographic Optical Element, HOE, was chosen as the component separating the two-dimensional images [7], while the images of the 2D views are produced in real-time by a number of digital projectors. This method is similar to the one described by Newswanger [8].

Hardware and Software Configuration

The computer projectors are mounted above the table projecting down on an HOE placed on the table surface. Each projector produces a view segment, which is reflected by the HOE. The projectors are driven by conventional PCs.

The software is developed using the Java 3D [9] programming package by Sun Microsystems. The Java3D API is an optional package belonging to a broader set of APIs called Java Media APIs. This makes it easy to integrate with the existing support for, *e.g.*, image processing, multimedia and networking. It also has built-in support for input devices and VR environments. The system is built in a client-server architecture where the server handles the communication to the underlying information structure with the geometric models and all necessary related information. Each client calculates its specific image with the right perspective and sends it to a digital projector. All transformations and modifications of the geometric model are handled by the server through input devices and are then sent to the clients to be presented. A user interface is presented on the server to make the control of the application easy.

13.3 Integration of Man Model with Man-to-Man Interaction

With the development of Virtual Reality and real three-dimensional visualization, there has been an increase in the use of 3D glasses and head mounted displays. There is also a tendency to use screens where all users are facing the same direction. Therefore, the social aspect of work, as we have known it, starts to decrease. In the use of VR technology today, wc tcnd to go into the VR environment rather than bring the virtual products and environments in to the real world. The social feeling of being present in a meeting is disappearing. In order to discuss a small detail, it may be necessary to remove the viewing glasses or head mounted equipment and verbally describe an issue rather than point and show, as would be done with a physical model. In a meeting we want to have the possibility to observe the expressed feelings and intentions of the participants [10]. This gives social quality to meetings and will mediate additional information through facial expressions, eye contact, focus, gestures and body posture. It will also add the strength of having more than one person in an environment focusing on examining a product or a part or its details. This will further increase the feeling of realism, control and engagement, and stimulate participants to be mentally present and active.

It is expected that a large auto stereoscopic display mounted as a tabletop can solve many of the problems in present VR systems addressed above. The interaction table shows a computer-generated three-dimensional image that can be viewed and interacted with by several people. The goals of the development of the interaction table are:

I. to have an autostereoscopic display, so there is no need for special glasses;
II. to allow all viewers to see the model simultaneously and with the correct perspective view;
III. to allow the objects shown on the display table to be moved and interacted with;
IV. to include haptic or tactil interaction which can be adapted to the display.

It is believed that a display fulfilling the possibilities of the stereo interaction table will be an important tool in all applications, which require a three-dimensional information display, and will enhance the possibilities for communication between people. Here we mainly talk about the design of manufacturing systems and products, but the interaction table can also be used in a number of other application areas as mentioned in the introduction.

13.4 Man-model Interaction in an Interaction Table Environment

There are a number of ways to interact with a VR model. One of the main possibilities is to just point at a specific feature or location in the model. That can be done by hand in the simplest way. However that is just an interaction with the viewing space for the co-located users and not with the actual presented model. The viewing space here is defined as the space where the model can be seen. To interact with the presented model, there is a need for an input device that can be interpreted by the system or a tracking system for the hand. That also is necessary in a virtual co-located environment where information must be sent through a network to be presented at a different location. With an input device it is possible to interact either in the viewing space or outside the viewing space by presenting a virtual cursor in the viewing space. If the interaction is done in the viewing space, the virtual model has to be presented above the table surface. In either case, the input device will be an extension to the human body and will put the device in relation to it. With haptic or tactil devices added to the VR environment, the control and feedback to the user are improved. Additional information to the user can also be displayed and interacted upon such as tolerances, functional requirements when shapes, features or surface structures are added.

System Functionality

The system tracks the position and movements of the pointing device. It is possible to interact with the whole scene or with a single object in the scene. This makes it possible to translate, rotate, zoom, mark-up or highlight an object or space. The system is also able to handle the starting and stopping of animations in the scene. Menus as well as information are not simple to visualise in a virtual environment. They tend to hide parts of the scene, and text information must be presented perpendicular to the user. Therefore, it is usually better to present menus and text information on a separate device next to the VR world. How the information is presented in the VR device is also critical. Colours and symbols may be used, but they tend to grow in numbers if the amount of information is large. Etiquettes or flags can be used to highlight specific application viewpoints or to detail the information to be displayed. Additional devices like Tablet PCs or personal digital assistants can also be used to interact and to present information.

How information is being presented in a scene is an important issue. The VR model is an information carrier, and it is important to consider how its underlying information model is represented. It is possible to have the interaction table on top of a CAD system or on top of other applications to interact with the presented information in three dimensions and have the underlying system as the information carrier.

13.5 Interaction Stereo Table in a Virtual Co-located Meeting

Even in a virtual co-located environment, the stereo interaction table has benefits. When linking additional stereo interaction tables, the natural eye contact between virtual co-located users is missing. Therefore, it would be necessary to integrate some kind of videoconference equipment to the system. This will partly add the social aspect of being present but will not solve all problems associated with virtual environments today. The users can, however, feel the social presence of the co-located users and the interaction with the presented model, which will be synchronized. Discussing the presented model on the table in front of them also adds eye and behaviour contacts. The same model is presented for all parties around the table, allowing simultaneous interaction among the participants who are facing one another.

13.6 Model-based Applications

Knowledge Representation in a VR Environment

Considering the potential benefits of VR models, it is important to take into account the context in which such models exist. VR models today mainly (re-) present objects in terms of shape and appearance. They should also present context–related information and knowledge. When ordinary geometric models are interpreted by the graphics processor, the code behind is of course based on know-how (from the programmer's viewpoint, who may not always be a professional engineer). Therefore, one could say that they represent knowledge to some extent. However, when we add interaction possibilities to the model, we take a step further because in such cases, it is no longer only a question of interaction through visual impressions. That means, the more complex the models are, in terms of interoperability, the more context–sensitive they get.

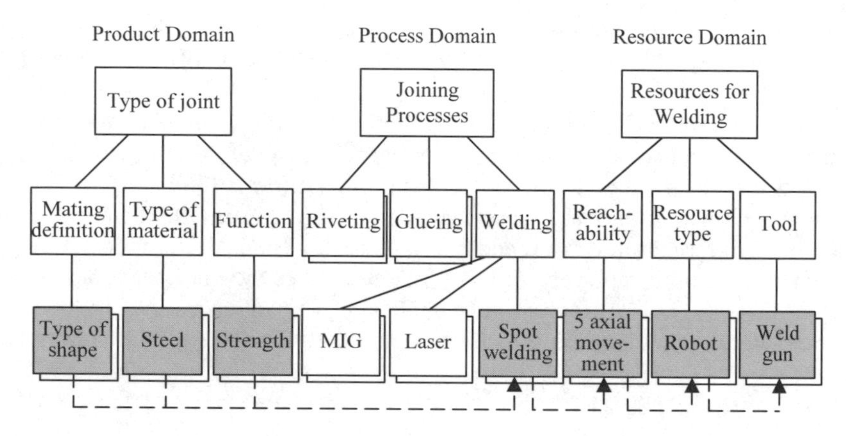

Figure 13.2 Schematic breakdown of the product, process and resource domains

Model Integration and Application Realization

Consider manufacturing and manufacturing system design as an affiliation case. When addressing manufacturing system design, by using the interactive stereo table approach, we have a tool that could potentially help shorten development times and produce a more optimal manufacturing system. However, in order to reach these goals, we must understand the process of manufacturing system design: What information is needed during the process of configuring a new manufacturing system? and, how to visualize and augment that reality? Our opinion is that by modelling manufacturing objects, their behaviour and intended usage, it will also be a lot easier to interact in a 3D environment.

The idea is to formulate domain knowledge by utilizing common product, process, and resource definitions, like geometric features and standardized processes (Figure 13.2). With a more or less systematic breakdown of manufacturing objects, derived from a global taxonomy [11] and built on logical reasoning, we get a framework for model interaction in which we can discuss using simple algorithms and heuristics such as if-then rules. This information will be of vital importance in order for the 3D application to interpret the model and thus support model interaction.

Consider Figure 13.3. In this scenario the project group is discussing the VR model in front of them, in this case an automated welding cell for automotive assembly. When interacting with the product model, in this case a model of a car, the system automatically visualizes how this affects the process, and resource models. Hence, this model allows us to verify concepts for a new as well as an existing manufacturing system.

Another reason for using the VR model in conjunction with the ontology–based framework was to simplify man-model interaction. If the system knows in what context the model resides, then it can be made more user friendly. Suppose we are planning the welding sequence. In that case we would like the system to highlight not only the surfaces that are to be joined but also the corresponding mating definitions. We would also like to know more about the robot and welding tools capability. In ordinary CAE tools this is usually impossible because the use of feature–based systems is not wide spread. But if we have an ontology model mapped to the product model, then the system can be programmed to reason out what surfaces are to be joined with a joining process and to visualize only the relevant information.

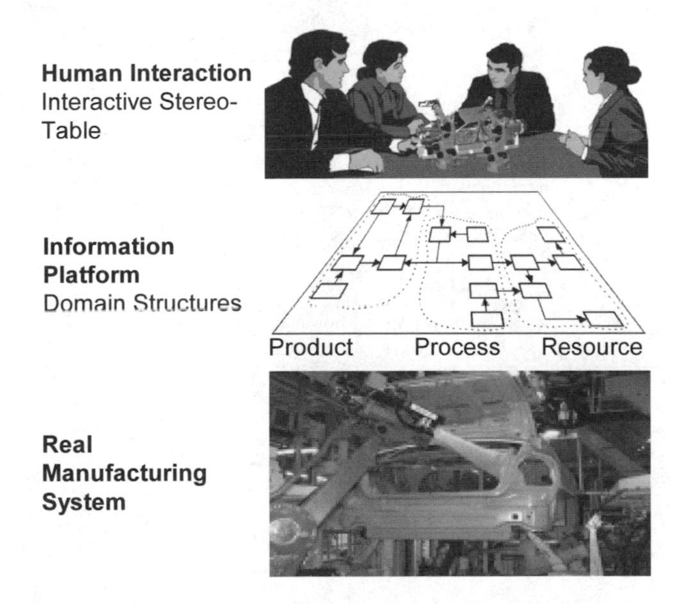

Figure 13.3 The information platform is a representation of the real manufacturing system, its requirements, capabilities, history, *etc.*. Interaction is enabled with the model or the real system through the stereo interaction table.

13.7 Conclusion

The stereo interaction table is an interface for information presentation and for user interaction. This information is generated from the data stored in the underlying data models and defined by an information model. It is therefore also vital to have a sufficient underlying information model that defines what type of information will be presented in a certain work task. This includes geometric information as well as structure and entity information. The presented model must be used in the right context, with a right purpose, viewpoint and detailing level.

The presented data is interpreted by the users as information and knowledge and gives the capability to generate new information (see Figure 13.4). In the case of the stereo interaction table, this is done by the interaction with the presented model through the generated information being fed back to the system and saved in the data structure model according to the information model. The system also has the capability to visualise different kinds of information from different domains in different ways. Information should then be linked between the domains and can be navigated through as seen in Figure 13.2. Different types of presentation techniques will be complementary in different situations to give a more complete information presentation. The stereo interaction table gives new dimensions to human–model and human interaction in development and decision-making.

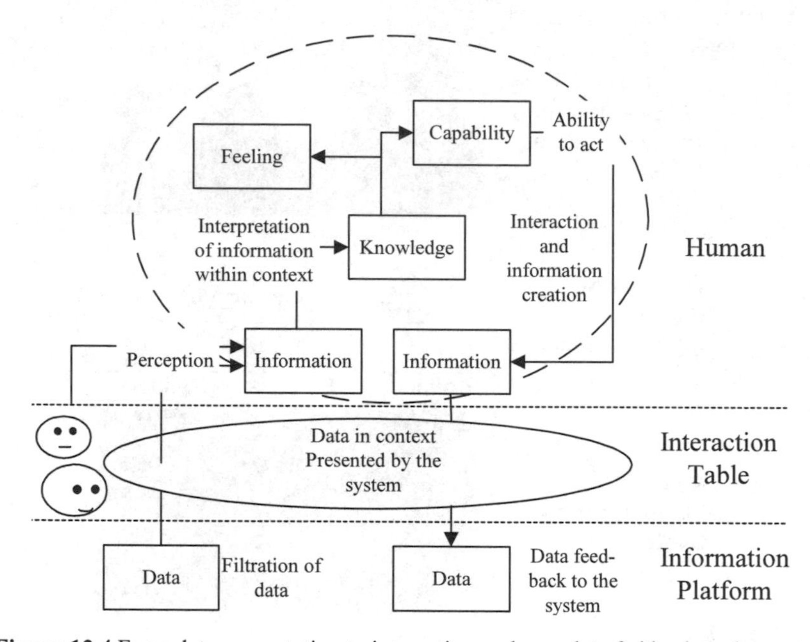

Figure 13.4 From data presentation to interaction and new data fed back to the system

13.8 Acknowledgments

We are grateful for the support we have received from VINNOVA – the Swedish Agency for Innovation Systems, by direct project support and through the WOXEN centre. We are also grateful for the interaction we have had with Professor Lars Mattsson, Industrial Metrology, KTH.

13.9 References

[1] Krause, F.-L., Luddemann, J., 1996, "Virtual Clay Modeling," *IFIP WG5.3 Workshop*, Airlie, May 19-23, 1996.

[2] Lu, S.C-Y., Shpitalni, M., Gadh, R., 1999, "Virtual and Augmented Reality Technologies for Product Realisation," *Annals of the CIRP*, Vol. 48/2, pp.471-495.

[3] Krause, F.-L., Neumann, J., 2002, "Haptic Interaction with Non-Rigid Materials for Assembly and Disassembly in Product Development," *Annals of the CIRP*, Vol. 50/1, pp. 81-84.

[4] Berthier, A., 1896, « Images st´er´eoscopiques de grand format, » *Cosmos* (591), pp. 227–233.

[5] Kasahara, T., Kimura, Y., R. Hioki, and S. Tanaka, 1969, "Stereo-radiography using holographic techniques," *Japan J. Appl. Phys.* 8, pp. 124–125.

[6] Bitetto, D. D., 1969, "Holographic panoramic stereograms synthesized from white light recordings," *Appl. Opt.* 8, pp. 1740–1741.

[7] Gustafsson, J., Lindfors, C., 2004, "Development of a 3D Interaction Table, Electronic Imaging," *IS&T/SPIE 16th Annual Symposium,* 2004.

[8] Newswanger C., 1987, "Real Time Autostereoscopic Display Using Holographic Di.Users." *United States Patent No. 4,799,739*, January 1989, Filed: August 10, 1987.

[9] Sowizral, H., Deering M., 1999, "The Java3D API and Virtual Reality," *IEEE Computer Graphics and Applications,* Vol. 19(May/June), pp. 12–15.

[10] Short, J., Ederyn, W., Bruce, C., 1976, "The Social Psychology of Telecommunications."

[11] Zimmermann, J.U., Haasis, S., van Houten, F.J.A.M., 2002, "ULEO – Universal Linking of Engineering Objects," *Annals of the CIRP*, Vol. 51/1, pp. 99-102.

14

Inventive Design Applied to Injection Molding

Thomas Eltzer, Denis Cavallucci, Nikolaï Khomenkho, Philippe Lutz, and Emmanuel Caillaud

Abstract: Increasing competition forces companies to put products on the market as soon as possible, hence the need for research in concurrent engineering. Invention is the second main issue: since today products must be cheaper and better than the competition's. This requires technological invention, which in turn necessitates research in creativity and problem solving theories. Our research interests are within these two academic domains: concurrent engineering processes and inventive solutions to technical problems. Starting from the specific situation of injection molding design, we identified the need to develop a new modeling approach for product and manufacturing molds that could link the powerful OTSM-TRIZ theory with concurrent engineering. We build our contribution on the parametric design model and cause-effect relationships; we propose guidelines to analyze and synthesize the resulting complex contradiction network in a single inventive redesign task. A plastic valve stem design is used for validation of the proposed approach.

Keywords: concurrent engineering, injection, design

14.1 Introduction

The field of injection molding technology does not escape the current need for fast design processes and high efficiency of final product. As competition is very fierce in this area, those two concerns are far more important here than in any other technical domain. Because of the nature of the injection molding process, those two issues are seldom dealt with well. Therefore, our research started with the goal of helping companies to speed up their design process and to develop powerful technical ideas through inventive design. In the following: Section 14.2 is dedicated to the description of current difficulties in injection molding design.

Section 14.3 presents the state-of-the-art of research in injection molding design, design process, as well as TRIZ, and points out the needed contribution. Section 14.4 presents our parametric problem modeling in design and its application, the use of which is shown in Section 14.5. Conclusions and perspectives are finally detailed in Section 14.6.

14.2 Injection Molding Design Issue

Injection molding technology is a widely applied manufacturing process which can rapidly produce finished plastic parts within a single process step. Plastic pellets are heated, and the resulting melt is introduced under high pressure in a metallic cavity which gives the required shape to the viscous material. After rigidifying, the part is pushed out of the cavity. Current mold functions (acting on the material) are "Distribute", "Shape", "Rigidify", and "Deliver". The part and mold design are crucial issues, and are very much related [11, 12]: a slight modification of the part can significantly reduce the tool complexity (for example, removing an undercut makes the use of simple mold possible); and a slight modification of the mold can increase part quality (for example, adjusting the gate dimension avoids jetting). Therefore, we can say that even if some mold design choices can be made independently from part design choices (and vice-versa), cooperation is strongly required in certain situations, as shown in Figure 14.1.

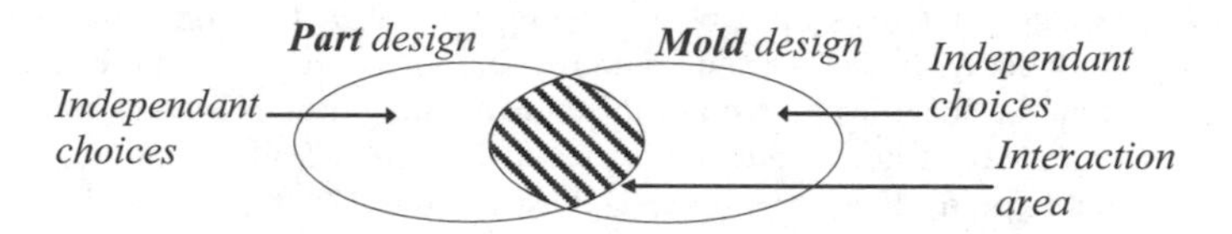

Figure 14.1 Independent and interacting design choices

Today, part and mould design are typically developed by two different companies [5]; mould design starts during the final stages of the part design process. The required cooperation described earlier creates many iterations in this classical design process between the part designer, who is making mold modification proposals according to the part requirements, and the mold designer, who is making part modification proposals according to the mold requirements, thus dramatically slowing down the process. The deadline for the product to be put on the market and the fuzziness of this "interaction area" (Figure 14.1) are the two main reasons why designers accept compromises when technical conflicts arise. The resulting design is hence accepted although it may have some inconveniences. Inventive approaches are, therefore, particularly required in this technological field.

Therefore, we can say that, on the one hand, concurrent engineering must be implemented in this special field to decrease development time, and, on the other, that inventive design approaches must be integrated as well in order to find real solutions to technical problems rather than merely compromises. The next section presents the state of the art of possibly useful research, and reveals what is to be achieved in order to answer this issue.

14.3 Integrate TRIZ in Concurrent Engineering

The TRIZ theory has been developed with the analysis of thousands of patents; technical problem formulating and solving tools have been built, and laws of technical system evolutions have been found [1,15]. We will be interested in the problem formulating part, the basic pattern of which is the contradiction summarized in Figure 14.2 with the OTSM-TRIZ (Russian acronym for General Theory of Advanced Thinking) approach [9]. "Parameter A", describing a certain "Element" should have a value V1 so that "Parameter 1" will have a satisfying value, but should have another value V2 so that "Parameter 2" will have a satisfying value. There is a physical contradiction in "Parameter A" (its value should be both V1 and V2) and a technical contradiction between "Parameter 1" and "Parameter 2" (they cannot both have a satisfying value).

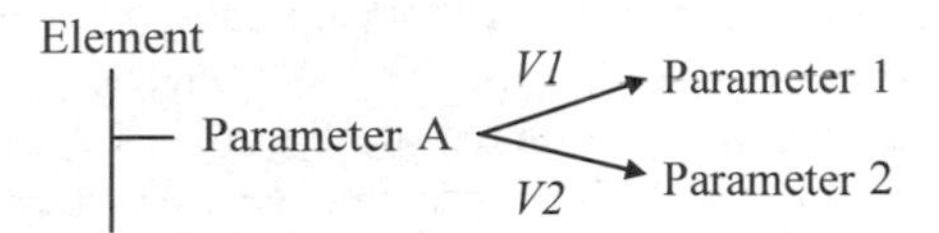

Figure 14.2 Contradiction representation

As the OTSM-TRIZ approach is dedicated to inventive problems, its use will help increase performance.

The design process has been presented in [2] as the evolution of a concept through four main steps: task clarification, conceptual, preliminary and detailed design. A resulting field of current research known as "concurrent engineering" [16,18] aims at helping designers perform all these design steps as simultaneously as possible, requiring cooperative design teams to take into account different points of view. The product and manufacturing process should be designed simultaneously [3], reducing needed time, and enhancing quality [10]. This simultaneous design is only possible with a clear representation of the links between the evolving part concept and the evolving mold concept.

Research in injection molding design usually does not make room for simultaneous part and mold design, studying either part design [4, 8, 11], or mold design [11, 12, 14]. Hence, concurrent engineering research still needs to be developed in this area.

As little research has been performed to integrate OTSM-TRIZ within concurrent engineering, the issue described in Section 14.2 cannot yet be easily solved. A new way to model the part and the manufacturing process, through the known design steps [2], using the OTSM-TRIZ principles [9] must therefore be developed. The model requirements are: to integrate OTSM-TRIZ in concurrent engineering, to focus inventive design and to store links between part and manufacturing processes.

14.4 A New Model for the Design Problem

We present here our contribution based on inventive redesign with product and tool parametric modeling [7, 13].

Parametric Model of the Design

Routine design is usually what engineers begin with; if the result is not satisfactory, they think in terms of inventive design. Therefore, routine design has been chosen as the representation from which we will shift to inventive design. Routine design can be seen as assigning values to a set of design parameters describing a generic product or tool. These parameters can be quantitative or not, and more or less fuzzy (for example: gate diameter, part position in mold, material entrance location). These parameters belong to the four stages of concept definition [2]. We consider therefore functionality, working principle, structure, and detailed dimensions as parameters of the concept. This parametric point of view has been chosen to fit the contradiction presented in Figure 14.2, and differs from the axiomatic model developed in [17] as precedence links between parameters are kept, and more than one entity is considered (the part and the mold).

These design parameters influence what we call "need parameters". Fixing a value to each of the former is done according to this influence and the desired value of the latter. For example, the value "low" is assigned to the detail level design parameter "Feature size" because it influences the need parameter "Amount of material" whose desired value is "low". As a consequence, the design parameters of any design stage can be linked when they influence the same need parameter (being then part of the interaction area presented in Figure 14.1). For example, the "Feature relative position" of the plastic part and the "Undercut release mechanism" of the mold are linked as they both act on the need parameter "Ejection deformation". Hence, assigning a value to each design parameter is done according to at least one need parameter it influences, and to the values of possibly linked design parameters. Routine design can now be seen as assigning values to a set of design parameters, in order to obtain the best ranking of a set of need parameters. Having presented routine design with part and mold design parameters as well as need parameters, explanation of how to answer the need is detailed in Section 14.3.

Parametric Model of the Problem

Invention is needed when the performance of routine-based design is not satisfactory. We first explain this problematic situation using the detailed Parametric Model of the Design.

The global need to "Reach a high global performance" is what the design process should answer. It can be decomposed into a few "local needs", each being a set of desired values of need parameters (introduced in the section on Parametric Model of the Design). “Local solutions”, the changing values of design parameters, answer those local needs one by one. Low performance can, therefore, be explained as follows: when local solutions require inconsistent values of the same design parameter, it is given a value that harms need parameters “neither too much, nor too few”. In such cases, the sum of local solutions is not the solution to the global need (see Figure 14.3).

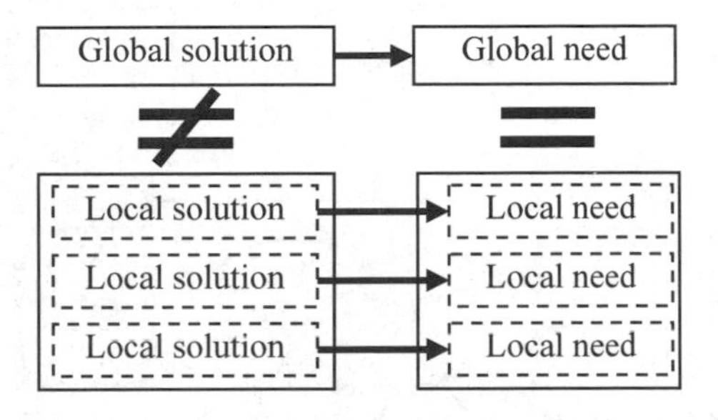

Figure 14.3 Global problem: the sum of local solutions does not answer the global need

The following definitions clarify the problem representation:

- The local solution is a real design action that changes the values of a set of design parameters (such as “mold feeding technology”, “number of cavities”, “feature thickness”, “location of material entrance on part”, and “cooling channels layout”). When a global problem arises, local solutions are “partial”;
- Local need is a level of real satisfaction represented as desired values of a set of need parameters such as “sink marks,” “cycle time,” “amount of plastic material,” and “mould life time”. They are influenced by design parameters;
- An intermediary parameter is influenced by either a design or another intermediary parameter and influences either another intermediary or a need parameter (for example: “cavity depth,” “skin viscosity,” and “mold core strength”);
- The route from a design parameter to a need parameter is the sequence of intermediary parameters between them.

Consequently, we formulate the global problem (Figure 14.4) as: design parameters (De. P.) influence need parameters (N. P.), directly or through intermediary ones (I. P.), within a complex network; inconsistencies between desired design parameter values create the global problem. The base pattern and an example of a network are given in Figure 14.4.

The first advantage in decomposing the link from design to need parameter is the clear representation of connections between design parameters. For example, in Figure 14.4, De. P.1 and De. P.2 are connected because they both influence IP.1. The second advantage is the description of ways to break this connection if required:

- Break the effect of De. P.1 on I. P.1, or the effect of De. P.2 on I. P.1;
- Break the effect of I. P.1 on N. P.2, and create De. P.7 to influence N. P.2.

The model shown in Figure 14.4 can be used to describe the reason of low performance design from the parametric model standpoint. In Section 14.5, its use for OTSM-TRIZ based concurrent engineering (see Section 14.3) is presented.

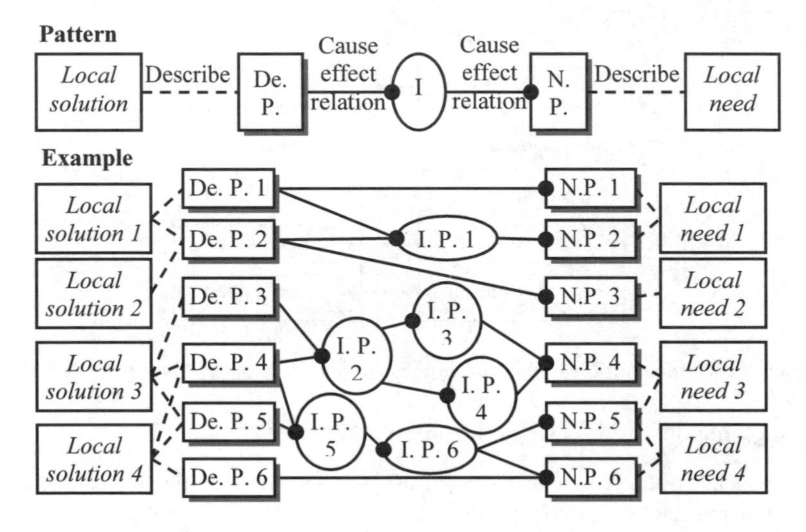

Figure 14.4 Pattern and example of complete network

Application of the Model

We present here how the model shown in Figure 14.4 fits the requirement listed at the end of Section 14.3.

Integrate OTSM-TRIZ in Concurrent Engineering

The representation proposed in Figure 14.4 enables us to find contradictions present in the design process. They are the set of different values each design parameter should be assigned in order to enhance need parameters. All along the process, designers can choose whether or not to solve them with TRIZ.

Focus Inventive Design

The representation proposed in Figure 14.4 enables us to identify the design parameter called "root parameter," which influences the greatest number of need parameters. As its value is a key issue in the global need described in the section on Parametric Model of the Problem, the redesign task has to focus on it. It bears a poly-contradiction; it should have many inconsistent values to achieve many local needs. As it describes a certain functional physical element, pointing out routes to be kept and routes to be broken (to solve this poly-contradiction) facilitates the description of the inventive functional mean to be developed.

Store Links between Part and Manufacturing Processes
The representation proposed in Figure 14.4 enables us to store the need parameters linking part and mold design parameters. A need parameter links two design parameters if they both influence it. It eases the introduction of concurrent engineering and clarifies the interaction area shown in Figure 14.1.

14.5 Validation in Injection Molding

We show in this section how our contribution is applied to a case study taken from the practicing engineer handbook [6]. The part, half presented in Figure 14.4, is a valve stem used to adjust water flow rate. The thread is released by rotating the part thanks to zone B, and the undercut groove by the slidings. The mold is made up of three plates, and a long core-shaped zone A. The global performance is not satisfactory due to mold complexity, scrap, absence of core cooling increasing cycle time and defects. Following the steps outlined in Section 14.4, the required invention and the effects of this case study on later concurrent engineering are explained below.

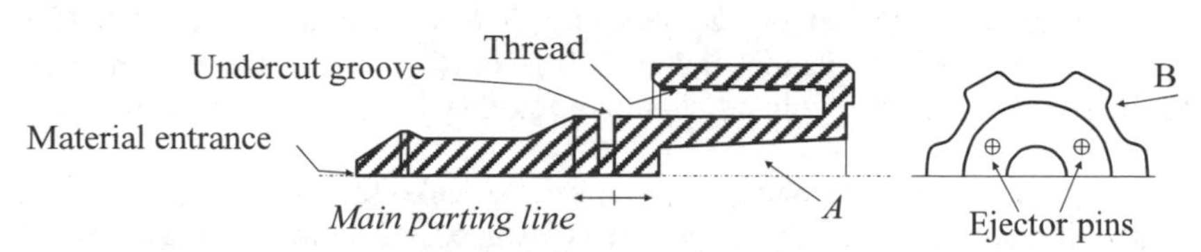

Figure 14.5 Plastic part

Parametric Model of the Design

Analyzing some designs shown in [6], we listed the routine design parameters of part and mold. The design steps proposed in [2] (Task clarifying, conceptual, preliminary and detailed design) are used to classify the parameters. Their values have been identified for the case study, and some of them are shown in Figure 14.6 and Figure 14.8. They are the basis of the problem model.

Parametric Model of the Problem

Listing the need and analysing the advantages and disadvantages of the part and mould design, as well as of the routine alternative solutions, identifies the parameters. They have been grouped into five local needs. Current design parameters, whose values can be changed in order to answer those local needs, have been grouped into five corresponding typical local solutions. The complete network, presenting design parameters linked by need parameters and intermediary parameters, is shown in Figure 14.9. This provides a precise evaluation of the global performance of the case study and helps to point out the inventive redesign task to increase it.

Application of the Model

We present how the complete network built in the previous section is to be used.

Integrate OTSM-TRIZ in Concurrent Engineering

Contradictions (see Figure 14.2) exist between need parameters (technical) and design parameters (physical). They are found by analyzing the network built in the section on Parametric Model of the Problem. We partly show them in Figure 14.5 by presenting the values some design parameters should be assigned to satisfy some need parameters. For example, the "core cooling channel" should not exist at all to have good "mold manufacturability," but, in order to avoid "shape changes after 24h," it should be exactly consistent with the core cavity shape. The presented contradictions have to be dealt with during the concurrent engineering process.

Focus Inventive Design

The root parameter influencing the greatest number of need parameters is "core cooling channel layout". This parameter has the greatest effect on the global performance, and the corresponding functional mean has to be changed to raise it to a satisfactory level. Related routes, taken from the complete network built in the section on Parametric Model of the Problem, are shown in Figure 14.6 (for example, the core cooling channel layout influences cavity surface temperature, which influences specific volume homogeneity, which, in turn, influences warpage).

The function of the core cooling channels is to make the melted plastic more regard after it has been correctly shaped. It is realized by running water in holed metal. The inventive functional mean has the following description:

- Do not change core internal shape (N. P.1 and 2);
- Avoid any channel discontinuities (N. P.3);
- Do not increase skin viscosity and thickness before end of filling (N. P.4,7,8);
- Give thermal rigidity when ejecting (N. P.5,6);
- Do not reduce bulk temperature before merging (N. P.7);
- Increase specific volume homogeneity (N. P.9).

Classical TRIZ tools can be used to further develop this description.

	Cycle time	Mold complex.	Mold robust.	Mold life time	Mold manufac.	Water leakage	Amount of scrap	Ejection deform.	Shape changes-24h
MOLD Conceptual level									
Cavity numbers	Max	Min.							
Mold struct.	Two plates	Two plates	Two plates	Two plates					
Feeding techno.		Cold runners					Hot runners		
Under-cut release mech.		No one	No one	No one		No one		At least two	
MOLD Preliminary level									
Core cooling channel layout				No one	No one	Around slide		Close under core surface	Close under core surface
Fixed cooling channel layout						Around slide		Close under fixed surface	Close under fixed surface
Part position in mold		Min. cavity depth		No under -cut				No under-cut	
MOLD Detailed level									
Gate diameter								Small	
PART Preliminary level									
Special feature								No one	
Feature relative position				Large distance between them	Large distance between them			Give rigidity	
PART Detailed level									
Feature size		Short						Thick	

Figure 14.6 Contradiction table

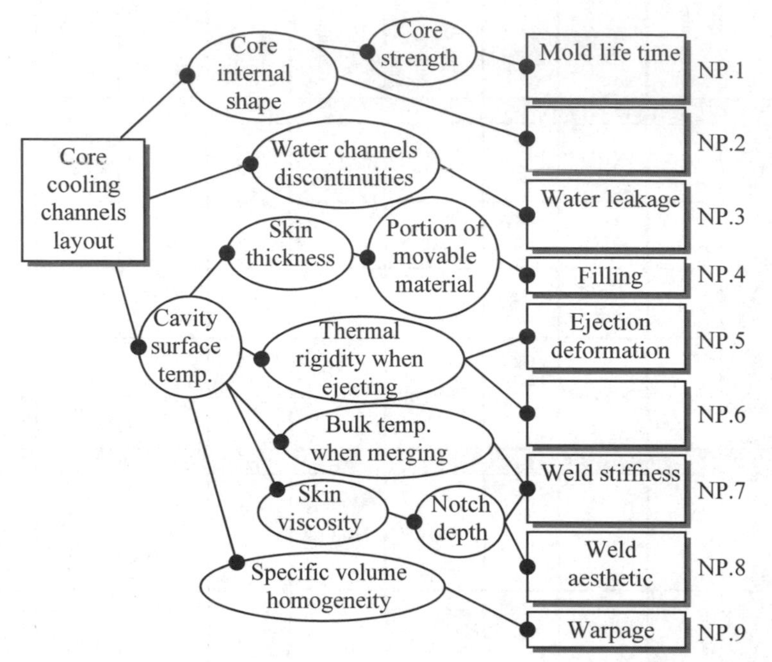

Figure 14.7 Need parameters linked to the most influencing design parameter

Store Links between Part and Manufacturing Process

This case study has identified a first vision of links between part and mold. They are shown in a chart where need parameters connect part and mold design parameters.

For example, this chart can be used to know beforehand that if a "special feature" (like a thread) is added to the part, the "part position in mold" must be verified (because of their common effect on "ejection deformation"). The values of those parameters should be determined simultaneously.

	PART P. level			PART D. level
	Special feature	Feature relative positions	Material entrance location	Feature size
MOLD C. level				
Number of cavities				Mold complex.
Mold structure		Mold life time	Runner fillability	Mold complex.
Feeding technology			Filling	Mold complex. Filling
Undercut release mechanism	Ejection def.	Mold life time Ejection def.		Mold complex. Ejection def.
MOLD P. level				
Core cooling channels layout	Ejection def.	Mold life time & manufa. Ejection def. Weld stiffness & esthetic Warpage	Weld stiffness & esthetic Warpage Filling	Ejection def. Warpage Filling
Fixed cooling channels layout	Ejection def.	Ejection def. Weld stiffness & esthetic Warpage	Weld stiffness & esthetic Warpage Filling	Ejection def. Warpage Filling
Part position in mold	Ejection def.	Mold life time Ejection def.	Runner fillability	Mold complex. Ejection def.
MOLD D. level				
Gate diameter			Jetting & Filling Sink marks	Jetting & Filling Sink marks

Figure 14.8 Couples of parameters linked by common need parameters

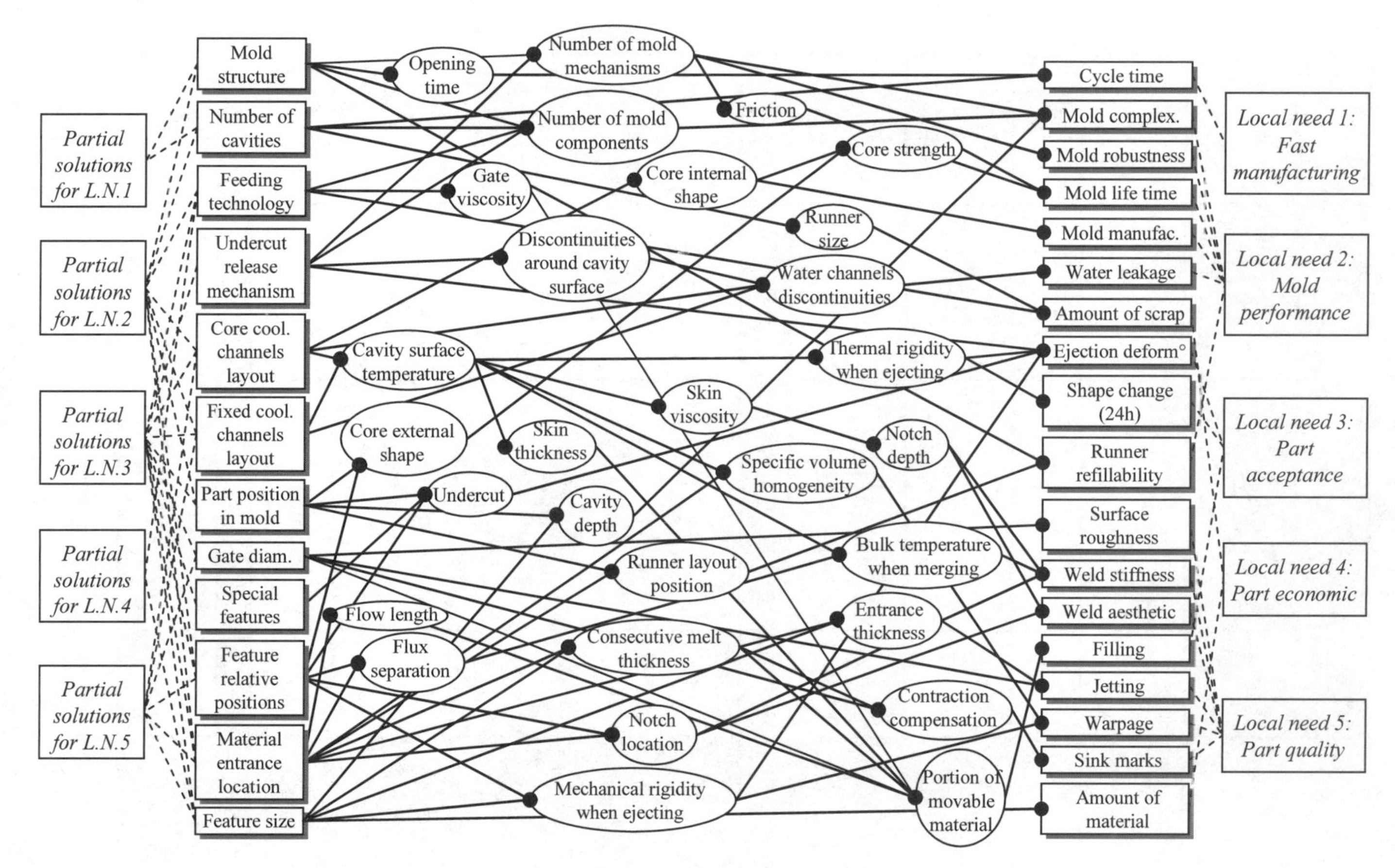

Figure 14.9 Complete network based on the specific case study

14.6 Conclusion and Perspectives

We have shown in this paper a new model of part and tool in injection molding, based on four entities: part design, mold design, intermediary and need parameters. We have given guidelines to apply the model in order to converge the unsatisfying routine-based design into a single inventive redesign task, and to store data for later OTSM-TRIZ based concurrent engineering.

Even if our results are already applicable as shown in a case study, further research has to be performed in the following directions:

- Test other rules to identify the root parameter, and to find the most effective way to increase performance (Focus Inventive Design section);
- Clarify the frontier between generic and specific contradictions shown in the Integrate OTSM-TRIZ in concurrent engineering section, to focus the general applicability of our model.
- Adapt classical TRIZ tools to our model, in order to integrate problem solving approaches rather than only formulating them.

14.7 References

[1] Altschuller, 1973, "The Invention Algorithm," *Technical Invention Center*.

[2] Beitz, W., Pahl, G., 1996, *Engineering Design: A Systematic Approach*, Springer.

[3] Chakravarty, A.K., 2001, "Overlapping Design And Build Cycles In Product Development," *European J. of Operational Res.*, Vol. 134, pp. 392-424.

[4] Chen, Y.M., Liu, J.J., 1999, "Cost-Effective Design for Injection Molding," *Robotics And Computer Integrated Manufacturing,* Vol. 15, pp. 1-21.

[5] Chung, J., Lee, K., 2002, "A Framework Of Collaborative Design Environment For Injection Molding," *Computers in Industry,* Vol. 47, pp. 319-337.

[6] Gastrow, 1983, *Injection Molds: 102 Proven Designs,* Hanser.

[7] Giachetti, R., E., Young, R.E., Roggatz, A., Eversheim, W., Perrone, G., 1997, "A Methodology for the Reduction of Imprecision in the Engineering Process," *European Journal of Operational Research*, Vol. 100, pp. 277-292.

[8] Hui, K.C., 1997, "Geometric Aspects of the Mouldability of Parts," *Computer-Aided Design,* Vol. 29, pp. 197-208.

[9] Khomenkho, N., 1997-2001, *Materials for Seminars*, Jonathan Livingston Project.

[10] Koufteros, X., Vonderembse, M., Doll, W., 2001, "Concurrent Engineering and its Consequences," *J. of Operations Management*, Vol. 19, pp. 97-115.

[11] Malloy, R.A., 1997, *Plastic Part Design for Injection Molding,* HANSER/SPE.

[12] Menges, Mohren, 1993, *How to Make Injection Molds,* HANSER / SPE.

[13] Myung, S., Han, S., 2001, "Knowledge-Based Parametric Design of Mechanical Products Based on Configuration Design Method," *Expert Systems with Applications*, Vol. 21, pp. 99-107.

[14] Putnik, G.D., DeLima, M.A.S., 2002, "Concurrent Engineering Based Mold Development," In: *ANTEC*.
[15] Salamatov, Y.P., 1991, *Chance to Adventure - System of Laws of Technical Systems Evolution,* Vol. 5, (A.B. Selutskii).
[16] Sohlenius, G., 1992, "Concurrent Engineering," *Annals of the CIRP*, Vol. 41, pp. 645-655.
[17] Suh, N.P., 1990, "The Principles of Design," *Oxford Series on Advanced Manufacturing*.
[18] Young, A.R., Allen, N., 1996, "Concurrent engineering and product specification," *J. of Materials Processing Technology*, Vol. 61, pp. 181-186.

Part IV

Design Frameworks

15

Supporting Problem Expression within a Co-evolutionary Design Framework

Pierre Lonchampt, Guy Prudhomme, and Daniel Brissaud

Abstract: From a generic point of view, the engineering design process can be considered as the transformation of needs into a complete product definition. Besides the needs, designers have to take into account some constraints. In this sense, the design process can be seen as a problem solving process, with some specific properties. Several models of the design process exist that consider this problem and its expression differently. We will question the consideration of the design problem in the classical approaches, and identify the relevance of co-evolutionary models to describe the design process, including its cognitive aspects. An activity-based co-evolutionary model of the design process is thus proposed, which defines and situates the objects implicated in evaluation and problem expression. A well-established design corpus is used to evaluate the relevance of the proposed approach to fit a real design process. The objective of the work described in this paper is to use this model as a basis to investigate the support of problem expression, and the activities that refer to it, in a concurrent engineering context.

Keywords: Integrated design, Evaluation process, Co-evolution

15.1 The Design Problem

The design process can be seen as a problem solving process. Indeed, its achievement corresponds to the shift from a problematic situation, in which the needs are considered unsatisfied, to an objective situation in which they are. Considering a design context, the problem has some particular properties.

This problem is open-ended, as its solving does not consist of finding the only solution, but in finding a satisfactory one (or several ones). In this sense, the number of satisfactory solutions, if any exists, cannot be known initially. The proposed solutions to a design problem are not true or false, but more or less acceptable. Design problems are considered ill defined, because initially, designers have only an incomplete and imprecise mental representation of the design goals or specifications [1]. After all, the design problem is something complex. Indeed, when judging an acceptable solution, this judgment implicates several different, non-comparable and non-independent aspects [2].

The work presented in this paper deals with the design process in a modern industrial context. This situation implies several stakeholders in the design process. The topic treated here addresses, on the one hand, the problem expression in such a context, and, on the other hand, the evaluation needed to judge whether or not proposed solutions are satisfactory.

15.2 The Design Problem as Described in Classical Models of the Design Process

The Systematic Approach

One of the most well-established models of the design process is the systematic approach [3]. It considers the design process as a set of successive stages that correspond to the achievement of associated tasks by concerned stakeholders. The first stage, called product planning and clarification of the task, consists of analyzing, expressing and decomposing the design problem. The following stages then deal with the solution definition. They aim at solving the expressed problem, according to a generic progression, from the most abstract and global aspects to the most detailed and physical ones, and involve evaluations and choices at the end of each stage.

Concurrent Engineering

During the last years, the increasing complexity of products, market competition and pressures on quality, cost and lead-time have resulted in an evolution of the industrial organization. The design tasks that were performed successively are now treated in parallel [4]. Consequently, the relation between the different design stakeholders, which was previously a one-way contractual prescription, is nowadays a cooperative and interactive link. Some works take into account this evolution by describing, besides the parallelism of tasks, the parallelism of several domains, spaces or worlds. These domains relate to the different possible points of view, with different abstraction levels, on the designed solution or on the expressed problem [5, 6]. Those domains are constituted of several objects whose emergence drives the domains' evolution.

Their Limits

The two approaches presented above fail, in our opinion, in taking into account all the design process characteristics. Indeed, the systematic approach pre-supposes that the design problem can be initially known, expressed and decomposed into independent sub-problems, which will be resolved during the achievement of independent successive stages. The unavoidable iterations, which occur during real design processes, are proof of the non-relevance of this assumption. Moreover, the initial problem expression tends to limit the solutions to those that fit the chosen decomposition. Those two points are incompatible with, on the one hand, the design problem properties mentioned in Section 15.2, and, on the other hand, with some cognitive aspects of the design activity. Indeed, works in cognitive psychology tend to prove that designers, rather than following a top-down and initially established planning and problem decomposition, plan their tasks according to opportunistic iterations [7]. Domain-based approaches offer an alternative to the classical sequential approach by decoupling the two axes of points of view and detail. Nevertheless, according to the authors, they are still in disagreement with the cognitive aspects of the design process. Existing models consider the different domains or worlds taken into consideration as successive points of view on the product, while cognitive science has revealed their simultaneity [8]. Moreover, the problem is expressed within some domain(s) that are the first to be defined, and thus it creates a one-way prescription. This bias was bypassed by considering relations between domains as zigzags in more recent publications on domain-based design process models [9].

15.3 The Co-evolutionary Approach

Existing Background

Domain-based approaches mention the coexistence of different domains or spaces, corresponding to different points of view on the designed product, within the design process. Those domains are progressively defined and detailed throughout the achievement of the design process.

Other approaches, based on a similar basis, consider the explicit coexistence and co-evolution of two domains, the one of problem expression and the other of solution definition. Thus, according to Simon [1], the design process is composed of problem solving activities, *i.e.* proposals and definitions of solutions according to the expressed problem, but these activities intrinsically alternate with problem setting and framing ones, in regard to the defined solution. Maher [10] and Cross [11] proposed a so-called co-evolutionary model that describes the design process as parallel evolutions of both problem-space and solution-space dimensions, which are linked by focus and fitness activities. Brissaud [12] adopted a model based on the same spaces, but in which the shifts between the two spaces are associated with alternative conjectures proposals and criteria emergences.

Interests

These models, seen as representation modes of the design process, completely decouple the three dimensions of led activities, of planning and of adopted points of view. In this sense, they offer an opportunity to situate design activities according to thc domains concerned rather than to an a priori schedule. Thus it correlates the opportunistic aspect of decomposition and planning, together with the noticed simultaneity of the points of view naturally adopted by design stakeholders. Moreover, explicitly treating an evolutionary problem expression fits totally the openness and ill-definition properties of the design problem.

The co-evolutionary models offer a basis to investigate the problem expression seen as a dynamic aspect of the design process. Indeed, this problem expression can evolve during the whole design process, according both to intrinsic shifts between two successive problem expression states and to interactions with the solution definition state.

15.4 Our Approach

From the generic models described above, we are developing a co-evolutionary model that aims at offering a basis to problem expression support in a concurrent engineering context. To achieve such a task, it is advisable to detail the existing representations. In our opinion this detailed work has to focus first on the collective aspect of concurrent engineering, then on both activities and objects. Indeed, to support problem expression requires supporting the activities whose implementation implicates problem expression elements, as well as identifying precisely these elements and their relationships.

The Co-evolutionary Model in a Concurrent Engineering Context

We have mentioned in Section 15.3 the nowadays-increasing implication of several stakeholders in the design process. These stakeholders are experts in different steps of the whole product life cycle (from the cradle to the grave). Their needs have to be taken in account during design. While many works aim at integrating the stakeholders, their objectives and skills in the solution definition process, the problem expression in such a context is quite evaded in the literature. The approach presented in this paper aims at treating this issue. In this sense, the problem expression and solution definition are defined as shared and common within the design team. We assert here the relevance of the co-evolutionary model to integrate stakeholders by a shared, common and evolutionary problem expression.

The Co-evolutionary Design Process Activities

It is possible to define generic activities according to the chosen basis for our approach. Indeed the model adopted describes the design process as the co-evolution of two domains. Considering an activity as an elementary process that allows to shift from one situation to another where either the solution definition or the problem expression, or the shared knowledge about them has changed, the adopted representation distinguishes four activities (Table 15.1).

Table 15.1 The four design activities

Activity	Description
Conjecture (C)	This activity is the one led by a design stakeholder proposing a new solution, or a new element to an already considered solution, supposed to solve the problem expressed.
Definition (D)	This activity is the one consisting in defining, setting, explaining and communicating a proposed solution (or proposed elements of a solution) among design stakeholders.
Evaluation (E)	This activity is the one consisting in judging a proposed solution in regard to the expressed problem.
Reformulation (R)	This activity consists in setting a new problem expression, or modifying the existing one. The first initial problem expression is considered as a particular reformulation.

Those four activities occur all along the design process, alternatively but apart from any pre-established scheme [13], in that we retrieve the classical triplet {specify generate evaluate} but independently from any order. Nevertheless, the chosen representation allows us to distinguish, inside the generated activity, between the conjecture and the definition. In our opinion, and due to the collective context considered, the conjecture refers to the individual, imaginative and creative act, which results in a new idea, while the definition denotes the communication acts that occur within the team to share this idea. Moreover, the distinction between reformulation and specification activities is considered as relevant due to the importance of the evolutional aspect of the expressed problem.

Objects Implicated in Activities Implementation

Together with the proposal of a four-activity model of the design process, the work presented here aims to identify the objects implicated in each of the activities. Those objects are defined by the concept of class, which refers to the set of generic properties shared by the objects it holds. In this sense, an activity implementation results in the instantiation of an object from the associated class.

Conjectures and Definitions

Those two activities are not within the scope of this paper. On the one hand, conjectures are associated with the individual, so they are naturally non-verbalized acts that do not implicate intrinsically shared objects. On the other hand, definitions are well treated in design research literature. The reader is for example invited to refer to [5], in which definitions are defined as the setting of solution characteristics. In this paper we will consider solution definitions as described by objects of the generic class ***solution***.

Reformulations

In our opinion, a design problem is composed of objects from at least two fundamentally different classes. Designers define a product with a purpose to fulfil expressed customer needs. They are the justification, the 'raison d'être' of the product designed and consequently of the design process. On the other hand, the designer's creativity has to be expressed within a limited space, according to several constraints, impossibilities, non-negotiable past choices or norms. This primordial distinction is kept in our approach. In this sense we consider the classes ***needs*** and ***constraints***, whose instantiations correspond to an implementation of the reformulation activity. Moreover, needs as constraints can be quantified by the instantiations of the class ***appreciation criteria.*** Those objects are a measure of the level specified for the different needs and constraints considered. In this sense the chosen definition for this concept reaches the one proposed in functional analysis formalism [14].

Evaluations

Evaluation, as an individual cognitive process, goes together with the emergence of an evaluation criterion [12]. This emergence is due to the (mental) meeting between the evaluated solution and an evaluation reference, and results in the expression of the evaluating author's opinion and judgment through the connotation (positive or negative) that is applied to this criterion. In the chosen approach the evaluation is considered as implicating the instantiation of an object from the class ***evaluation criteria***.

Different Reformulations

The evaluation reference is, by nature, composed of elements that belong exclusively to the evaluating stakeholder. In this sense the concept of evaluation reference denotes something that is not shareable. Nevertheless, according to the chosen approach, in a collective design context, evaluation refers to the problem expressed. The co-evolutionary paradigm adopted does not prescribe any order within the triplet {specify generate evaluate}. In our opinion if an evaluation requires the existence of a proposed solution, the ill-definition and openness problem properties can find expression in the possible implicit simultaneity between evaluation and problem reformulation. Indeed, some evaluations, *i.e.* emergences of evaluation criteria, can refer to problem elements that have not been expressed yet. Those problem elements belong to the evaluation reference called up, and can be identified, formalized, expressed and shared following the evaluation. There is a distinction, within the problem objects instantiated between those that appeared in association with a pure specification activity, *i.e.* a reformulation that was led independently from any solution and those that were implicated or created by the achievement of a solution evaluation [15].

The Relations between Objects
Apart from the intrinsic properties owned by the proposed objects, it is possible to define the properties of the relations that link them together. Indeed, first the design problem is defined as complex. It means, among others, that the elements that compose it are not independent, and can be decomposed into a hierarchical structure [5, 9]. Moreover, they can correlate positively or negatively. Secondly, as evaluation consists of judging a solution in regard to the expressed problem, its achievement is associated with the building of a relation between a solution definition object and a problem expression object, needs or constraints, through evaluation criteria. The proposed relations associated with problem expression and evaluation objects are thus illustrated in Figure 15.1.

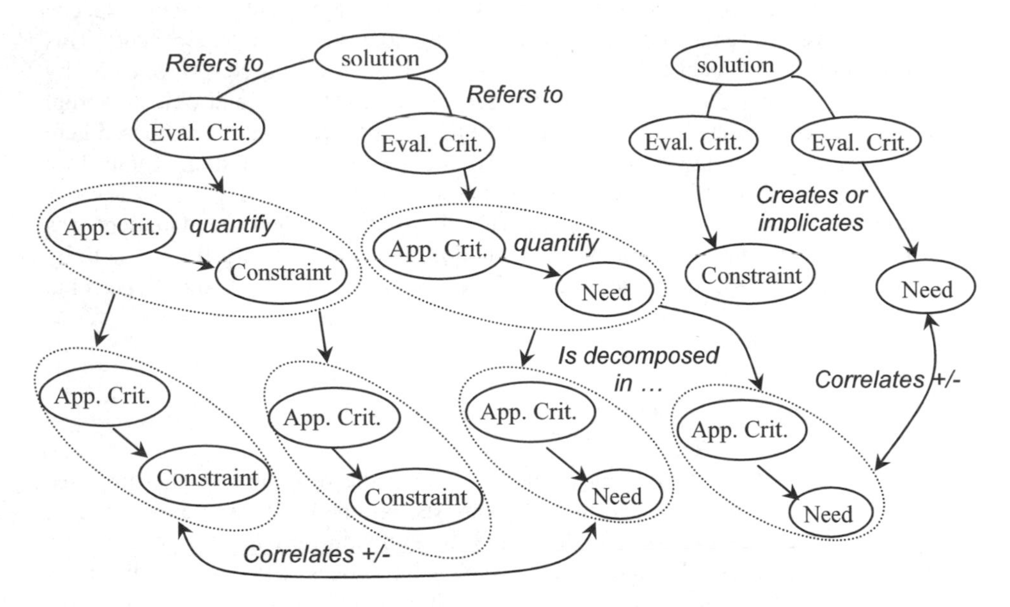

Figure 15.1 The relations between proposed objects

15.5 Example

The Design Corpus

To illustrate our approach, we have recorded the contents of a well-established design corpus using the proposed model. The design corpus used results from the Delft Design Protocol [16], and is referenced as DPW94.1.14.5. This issue implicated a team of three designers of various skills, who were asked to produce, within a limited period of time, the set of documents and annotated sketches that describe a chosen solution concept to a given assignment. The task consisted of defining a carrying/fastening device that aims at carrying a given backpack on mountain bikes, focusing on some precise aspects such as ease of use and price. The corpus is composed of an audio/video recording of the whole process (with four simultaneous frames), together with a written transcription of the dialogues and pictures of intermediate objects produced.

Recording

The written transcription was the only part of the design corpus to be used here. It consists of a spreadsheet containing, in each row, a time moment and length indication, the name of the designer speaking, and the content of his/her speech (and the possible remarks of the researcher who supervised the experiment). This transcription was recorded according to the model proposed above, *i.e.* by associating to each statement one of the four generic activities and object(s) from the implicated classes. Conjecture and definition activities are not considered here as the focus is on the aspects related to the problem expression. Table 15.2 illustrates the instantiation for several extracts of the design corpus. The six columns of Table 15.2 contain respectively the moment when the statement was pronounced, its author's initial (I, J or K), the statement itself, the activity identified (E or R), the associated object(s) (N, C, AC, EC) and a personal interpretation.

15.6 Results

The recording of the corpus allows us to validate the relevance of the proposed model. Indeed, it confirms the main hypothesis assumed, *i.e.* the co-evolving of problem expression and solution definition, by showing how the activities led by the designers alternate throughout the whole design process. At this time, it refutes the existence of a natural order within the triplet {specify generate evaluate}. This recording also reveals the ability of the activity-based model proposed to describe a design process without ambiguity.

A critical point appears using the proposed model related to object classes: we can notice the difficulty to distinguish between implicated objects in solution evaluations, that is to say, between constraints and needs. In our opinion this difficulty is due to the ambiguity that intrinsically exists in the concept of "internal needs". This concept denotes needs that result from the choice of a solution, and refers more to a limit to the freedom of designers than to the goal of the product. Nevertheless, the adopted scheme explains the causal relation that links the needs or constraints, via the evaluation criteria, to the conjectures that implicate them.

Table 15.2 Examples of activities and objects recorded

Moment	.	Statement		Object(s) identified	Comment
00:17:07	J	that's the target group does it say how many they'd sell per year	R	C: fit with estimated annual volume	J reformulates a given constraint
00:17:27	J	fifty thousand fifty thousand units a year products	R	AC: 50 000 / year	And quantifies it
00:17:30	K	that's certainly in the range of ( ) injection mould stuff	E	EC: injection moulding fits with the target.	K evaluates the solution (injection moulding) considered in reference to a previously quoted constraint.
00:32:24	K	if em this rack was used for something else like you take your backpack off and then this rack you can still put stuff on it but …	R	N: to hold something else than the backpack	K proposes new needs
00:32:28	K	…be if you could flip it out and it becomes a bike lock	R	N: to become a bike lock	to fulfil.
00:35:32	J	we have two joining problems we have the frame to the bike and then we have the pack to the frame	R	N: to join frame to bike N: to join pack to frame	J decomposes a need
01:47:10	K	bungees would actually work great	E	EC: bungees "work" well	K evaluates positively a solution without formulating the evaluation criterion
01:47:12	K	'cos then you don't have to cinch 'em	R	N: not to need to cinch	And then formulates the need to which it refers

15.7 Future Work

We have proposed a co-evolutive model of the design process based on activities. The recording of a corpus was then used to validate the relevance of this model to describe a design process, including the dynamics of its progress. We also identified generic object classes associated with those activities, whose ability to include aspects raised during a real design process was validated and discussed. This descriptive work is the first part of a more comprehensive study that aims at offering support to evaluation and problem expression within the design process. This model can be used to identify which classes the objects proposed in existing design tools (as functional analysis [14] or QFD [17]) belong to. Thus this allows investigating an integrated and more appropriate use of these tools to support evaluation and problem expression within a design team [17], in accordance with the real design process dynamics.

15.8 References

[1] Simon, H., 1981, *The Sciences of the Artificial*, MIT Press, Cambridge, MA.
[2] Ishii, K., 1990, "Role of Computers in Concurrent Engineering", *ASME Computers in Engineering*, Vol.1, pp. 217-224.
[3] Pahl, G. and Beitz, W., 1996, *Engineering Design: A Systematic Approach*, Springer Verlag, 2nd edition, London.
[4] Solhenius, G., 1992, "Concurrent Engineering," *Annals of CIRP*, Vol. 41.
[5] Suh, N.P, 1990, *The Principles of Design*, Oxford Univ. Press, New York.
[6] Mortensen, N.H. and Andreasen, M.M., 1999, "Contribution to a Theory of Detailed Design," *10° Symposium, Fertigungsgerechtes Konstruieren*, V. 99-11, U. of Denmark, Schnaittach.
[7] Bonnardel, N. and Sumner, T., 1996, "Supporting Evaluation in Design," *Acta Psychologica,* Vol. 91, pp 221-244.
[8] Darses, F., Falzon, P. and Béguin, P. 1996, "Collective Design Processes," *Proc. of COOP'96, 2nd Int. Conf. on the Design of Cooperative Sys. INRIA*, Sophia-Antipolis.
[9] Suh, N.P, 2001, *Axiomatic Design: Advances and Applications*, Oxford University Press, New York, USA.
[10] Maher, M.L., Poon, J. and Boulanger, S., 1996, "Formalising Design Exploration as Co-Evolution: A Combined Gene Approach", In: J.S. Gero and F. Sudweeks (eds), *Adv. in Formal Design Methods for CAD*, Chapman and Hall, London, UK.
[11] Cross, N. and Dorst, K., 2001, "Creativity in the Design Process: Co-evolution of Problem-Solution," *Design Studies*, Vol. 22, No. 5, pp. 425-437.
[12] Brissaud D., Garro O. and Poveda O., 2003, "Design Process Rationale Capture and Support by Abstraction of Criteria," *Research In Engineering Design*, Vol. 14-3, pp. 162 – 172.
[13] Lonchampt, P., Prudhomme, G. and Brissaud, D., 2003, "Assisting Designers in Evaluating Proposed Solutions throughout the Design Process," *Proceedings of ICED03*, Stockholm, Sweden.
[14] Zwolinski, P. and Prudhomme, G., 2003, "A Detailed Comparative Analysis of Two Mechanical Products with a Value Analysis Approach," *Proceedings of ICED03,* Stockholm, Sweden.
[15] Lindholm, D., Tate, D. and Harutunian, V., 1999, "Consequences of Design Decisions in Axiomatic Design," *Transactions of the Society for Design and Process Science (SDPS),* Vol. 3 N°4, pp. 1-12.
[16] Cross, N., Christiaans, H. and Dorst, K., 1997, *Analysing Design Activity,* John Wiley & Sons, Chichester.
[17] Schueller, A. and Basson, A.H., 2003, "Case Study on Low Cost Distributed Conceptual Design Support for Small Teams," *Proc. of the CIRP Design Seminar'03*, Grenoble, France.

16

A Four-stage Approximation Strategy for the Exploration of a Mechanical Concept

Bernard Yannou, Abdelbasset Hamdi, and Eric Landel

Abstract: The assessment of mechanical performances, in the automotive engineering domain, is mainly at present time, the result of late finite element analysis processes (FEA) which remain computationally expensive, limiting their use to the analysis of a limited number of design alternatives. But, in the conceptual design stage, the quality depends on the comprehension and on the exploration capabilities of the design space. This paper describes a strategy for building and more systematically exploring mechanical conceptual models, in the case of non-trivial expected mechanical performances. This strategy consists of a series of consistent stages: simplification of the parameterized structural model, choice of a subset of determining design parameters, computation of a limited number of approximate models of performances (metamodels obtained after a design of experiments and a model fitting) and a concept exploitation stage (deterministic exploration, optimization, non-deterministic exploration). This strategy has been successfully applied to assess vibro-acoustic performances of an automotive sub-frame. In this example, we show that designers have obtained useful information from the graphical and the numerical exploitation of this conceptual model. Moreover, this is now possible to take vibro-acoustic performances into account since the determining stage of envelope volumes allocation for sub-systems, a stage that is necessary in concurrent engineering for the automotive architecture deployment. Before, acousticians were not even able to negotiate with architects for a given volume allocation in regards to the possible consequences on the performances of which they were in charge.

Keywords: Conceptual design, conceptual modeling, metamodels, design space

16.1 Introduction

The concurrent engineering of an automotive development project requires paralleling the design tasks dealing with the sub-systems or organs. Consequently, a determining decision that could appear somewhat arbitrary has to be made during a preliminary design stage. It concerns the *envelope volumes allocation stage* for the vehicle's organs and components, which defines approximate geometrical areas outside which the designers are not allowed to propose shapes. It is not rare that some volume allocations be performed without any actual pre-existent assessment of the leading mechanical *performances*. The reason for that is the impossibility to assess the potential ranges of these performances, given, at the beginning of the process, the possible variations of the *design variable* values of a given organ or component design concept. This is why the fulfilment of some functional performances such as noise performances in the passenger cell, or the contribution of an organ to the noise in the passenger cell, may be penalized by the choice of a concept and of the envelope volumes for sub-systems or organs that will only be slightly modifiable afterward. Traditionally, the assessment of vibro-acoustic performances of an organ requires a meshing of a CAD model, to take boundary conditions into account and to perform a finite elements computation (FE). However, it may take longer than one month to build a reliable and accurate analysis model for a car body. But, during a preliminary stage of design, qualitative, approximate or tendencies results are more tractable than quantitatively accurate ones from a carefully prepared detailed FE model.

In this article, we propose a strategy to more systematically build a conceptual model that allows considering vibro-acoustic performances at the stage of envelope volumes allocation (see also [10]). This strategy has been applied to the vibro-acoustic preliminary design of a car sub-frame. Section 16.2 introduces the frameworks of the car sub-frame case study and of the four-stage overall strategy. The preparation of the conceptual model is presented in Section 16.3 and the design problem setting in Section 16.4. The construction of an approximate model of design criteria – named metamodel – is established in Section 16.5. Three types of exploitation of the conceptual model are described in Section 16.6 before concluding.

16.2 The Overall Strategy

Variables Notation

For reasons of clarity and simplicity, we already introduced a brief vocabulary about the variables (see [9]) and constraints (see [11]) involved in an elementary design analysis loop (see Figure 16.1). Basically, a design concept may be structurally defined by a number of *Design Variables* (DVs) which might be sufficient to assess the current *Performance Variables* (PVs) of the studied concept. These latter PVs are compared to the expected performances, namely the

Functional Performances (FPs). Sometimes, a performance PV may be expressed through a parametric function of some DVs or after a fast analysis like the *mass* assessment from a CAD model. But, most of the time, some hard mechanical performances are assessed after an heavy pre-processing (*e.g.* meshing) and analysis or simulation (*e.g.* finite elements calculations) as for acoustic and vibration performances. The objective of the present paper is to facilitate the assessment of such hard mechanical performances so as to be performed earlier in the design process and faster to allow an extended exploration of the conceptual possibilities.

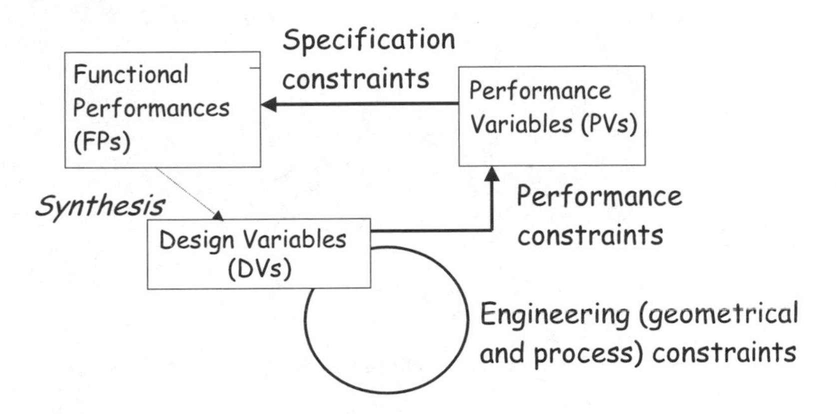

Figure 16.1 Variables and constraints in an elementary design loop

The Sub-frame Case Study

A sub-frame is a preponderant link between different sub-systems located in the front part of a vehicle. It is a platform upon which the car body lies, and itself lying on suspension triangles and supporting the steering box, the stabilizer bar and some elements of the water cooling system. Thus, the sub-frame transmits solid vibrations coming from the engine as well as from the tires towards the passenger compartment. The sub-frame vibro-acoustic performances are its attenuation mismatch efficiency and its dynamic structural integrity at its interfaces with the car body.

After the choice of a global car concept, *architectural* or *location constraints* are defined for the organs like the sub-frame (see Figure 16.2) in representing the approximate locations of the engine, the steering volume of the front wheels, some parts of the car body and of some additional components. Once a concept chosen for the sub-frame, the architectural constraints are transformed into a maximal volume allotted to the sub-frame, namely the envelope volume (see Figure 16.3). The objective of the present paper is to provide acousticians with the ability to negotiate this envelope volume with architects so as to permit the acoustic PVs to match the expected FPs, given that this is not possible so far.

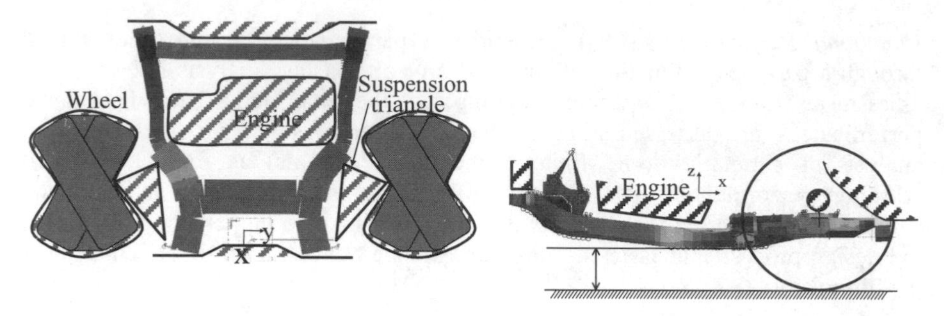

Figure 16.2 Architectural constraints on the sub-frame

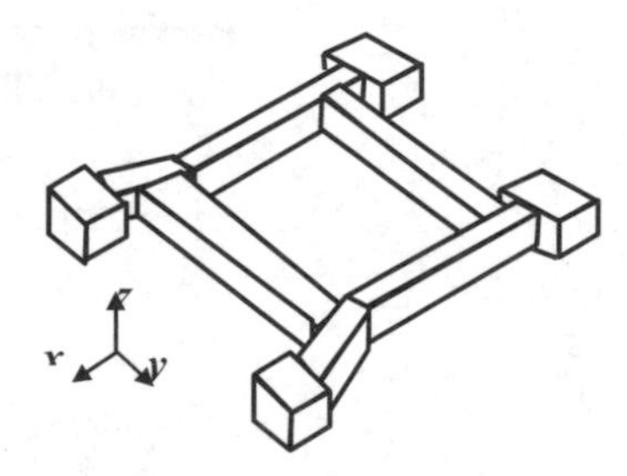

Figure 16.3 The envelope volume allotted to the sub-frame

The Four-stage Overall Strategy

We propose a strategy that would result in a conceptual model of mechanical performances in four stages (see Figure 16.4):

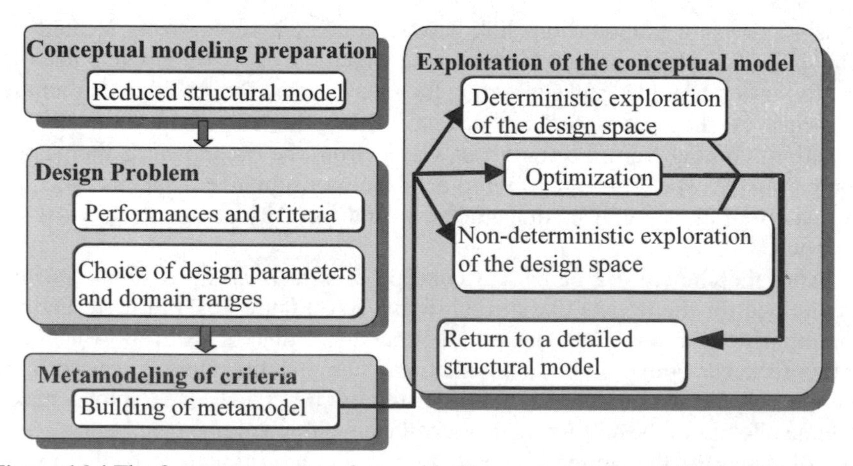

Figure 16.4 The four-stage strategy for resulting in a conceptual model of performances

1. *Preparation of the conceptual model.* A rough structural model must be defined from a set of simple elements such as beams, plates, isolated/concentrated masses and joints. For that, one can possibly start from a pre-existing detailed CAD model but this is not a necessity. The rough structural model must be defined as a parametric CAD model.
2. *Design problem setting.* This stage firstly consists of defining performances, constraints and criteria for assessing the pertinency of the sub-frame. Next, one has to choose the most influential design variables DVs of the rough structural model on the design criteria. The allowable ranges of the influential DVs must also be defined.
3. *Metamodeling of criteria.* This stage corresponds to the building of an approximate mathematical model for straightforwardly assessing the criteria from the influential DVs. An appropriate *design of experiments* must have preliminarily been carried out.
4. *Exploitation of the conceptual model.* We consider three types of exploitation of the fast assessment possibilities of these design criteria:
    a. *Deterministic exploration of the design space*
    b. *Optimization*
    c. *Non-deterministic exploration of the design space*

The exploitation of the conceptual model provides more qualitative than quantitative information on the influence of some DVs on the rough structural model. The designers must proceed to a last stage of interpretation of this knowledge about the rough model onto given actions on the detailed structural model.

16.3 Preparation of the Conceptual Model

The construction of a rough structural model lies onto the representation of a real structural model by simple elements: hollow beams, plates, punctual masses and perfect joints. This is yet a conventional procedure in a car design process [1-3 ; 8]. The validity of such rough structural models has been proved for low vibration frequency ranges (0 to 100 Hz) [8].

This stage is delicate because the sub-frame must be divided into beam sections, each of them having a thickness resembling the one corresponding to the actual detailed model and having a cross-section size roughly in accordance with the stiffness of the cross-section in the actual detailed model. In addition, the beams are connected by generalized joints whose stiffness and damping characteristics must be optimized so that the overall vibration behavior of the rough model fits at best the detailed models if existing.

For the sub-frame case study, the number of finite elements is 7762 in the detailed structural model (see Figure 16.5) whereas 224 simple elements represent the rough model (see Figure 16.6). 12 influential DVs have been chosen in the rough model; they correspond to 11 thicknesses of beam sections e_1 to e_{11} and an elongation ratio *alpha* for when allowable value domains have been defined (see Figure 16.6).

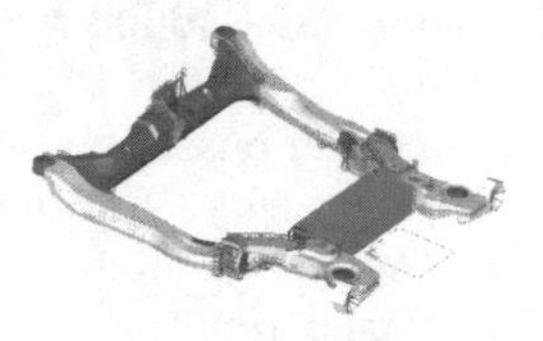

Figure 16.5 The detailed structural model of the sub-frame

The free variables of the rough model as well as the stiffness of the generalized joints and the value of some punctual masses have been optimally determined such that the rough model represents at best the vibration behavior. It has been made possible because a detailed model of the sub-frame, corresponding to the mid-point of the allowable DV's domains, was available. This optimization revealed that a highly accurate match exists for the first four frequencies and modes of both the rough and detailed structures (see Figure 16.7 for the first vibration mode).

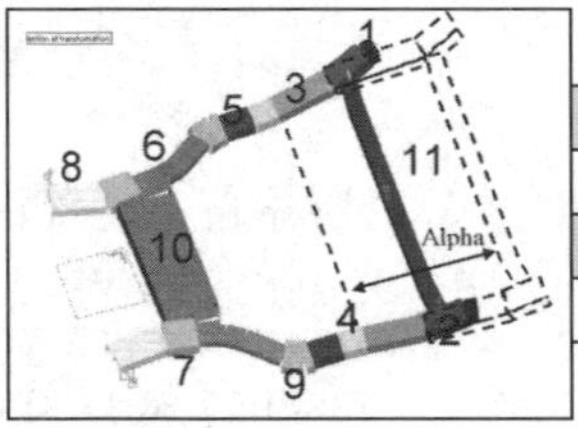

e_1	e_2	e_3	e_4	e_5	e_6	e_7
[1..3]	[1..3]	[1..3]	[1..3]	[1..3]	[5.6..9]	[1..3]
e_8	e_9	e_{10}	e_{11}		alpha	
[1..3]	[1..3]	[1..3]	[1.5..4.5]		[1..1.25]	

Figure 16.6 The rough structural model of the sub-frame, the 12 influential design variables DVs and their allowable domains

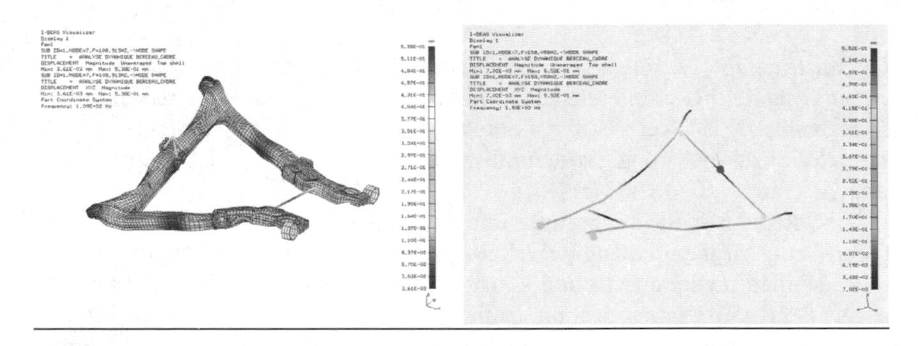

Figure 16.7 The first vibration mode for the detailed (resp. rough) structural model

16.4 Design Problem Setting

In a first attempt, five performances variables (PVs) have been defined to help in the negotiation phase of envelope volumes between acousticians and architects for the sub-frame. A performance is often compared to its expectation FP, resulting in design criteria that are easier to assess. For the sub-frame, the different performances and criteria are:

- a mass performance m, that one would like to minimize.
- an obstruction criteria obs, equal to 0 if the current rough sub-frame (in Figure 16.6) stands inside the envelope volume (in Figure 16.3) and equal to the exceeding volume otherwise,
- the fundamental frequency f_1 (corresponding to the first mode) that one would like to maximize, or at least to be greater than a minimal value f_{min},
- two typical acoustic criteria B_1 and B_2.

The m and obs criteria are traditionally obtained by CAD calculations that can be characterized to be of respectively low and medium complexity, whereas the f_1, B_1 and B_2 criteria are traditionally obtained after a heavy meshing and FE calculation process. But for the 5 criteria, a unique metamodeling process (construction of an approximate model) has been performed.

The B_1 and B_2 acoustic criteria are based on an aggregated dynamic performance of the structure $RMS(f)$ that is considered sufficiently representative of the contribution to the noise spectrum in the passenger compartment. Indeed, let us denote $\mathbf{H}_{ij}(f)$ an elementary transfer function, named FRF for *Frequency Response Function* expressing the displacement at joint j from an elementary force excitation in joint i in a frequency range of interest. Our aggregated criterion $RMS(f)$ is the root mean square of the sum of the squares of these FRFs and is expressed by the following formula:

$$RMS(f) = \sqrt{\frac{1}{M \times N} \sum_{i=1}^{N} \sum_{j=1}^{M} \left|\mathbf{H}_{ij}(f)\right|^2} \tag{16.1}$$

Finally, both B_1 and B_2 criteria come from the comparison of the aggregated frequency spectrum $RMS(f)$ with a given maximal allowable profile spectrum (see Figure 16.8). For reasons of confidentiality, this profile spectrum is represented flat. B_1 criterion is the exceeding surface area of $RMS(f)$ beyond the profile spectrum; B_1 must be equal to 0 in the best case or be minimized. B_2 criterion is the surface area between both profiles in the section where $RMS(f)$ is below the profile spectrum; B_2 must be maximized.

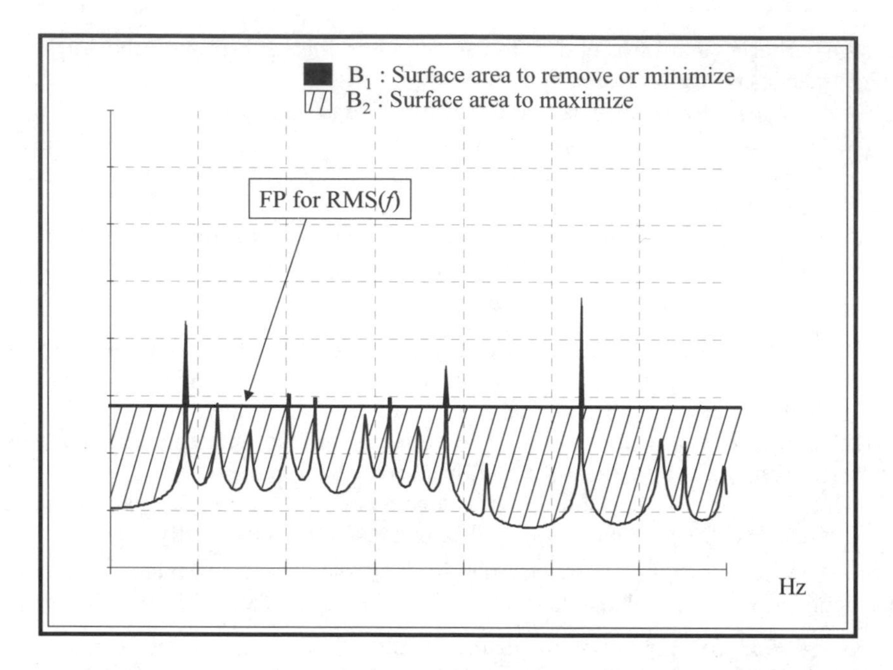

Figure 16.8 The two acoustic criteria obtained from a comparison of an aggregated vibration profile with a maximal allowable profile spectrum

16.5 Metamodeling of Criteria

Metamodeling techniques (see [7] for a recent state-of-the-art) are a collection of techniques that permit an approximate but instantaneous assessment $\hat{f}$ of the performances PVs, function of the set of influential DVs. Let us denote $P\hat{V}s$ this approximate vector of performances, one wants to establish: $P\hat{V}s = \hat{f}(DVs)$. This function $\hat{f}$ must be determined from a very limited number of finite elements trials (DVs_i, PVs_i) on the rough structural model to obtain a sufficiently good image of the actual function *f*. Consequently, the value sets of DVs must be wisely chosen within the allowable domains through a *design of experiments* (see [4]). The determination of the approximate mathematical function $\hat{f}$, namely the *metamodel*, requires:

- the choice of a model type among *Response Surface Models* (*i.e.* polynomial), *Kriging* models, *Radial Basis Functions* models or *Artificial Neural Networks*,
- a fitting of the parameters of the mathematical model by an optimization procedure so as to respect at best the 64 relations: $PVs_i = \hat{f}(DVs_i)$.

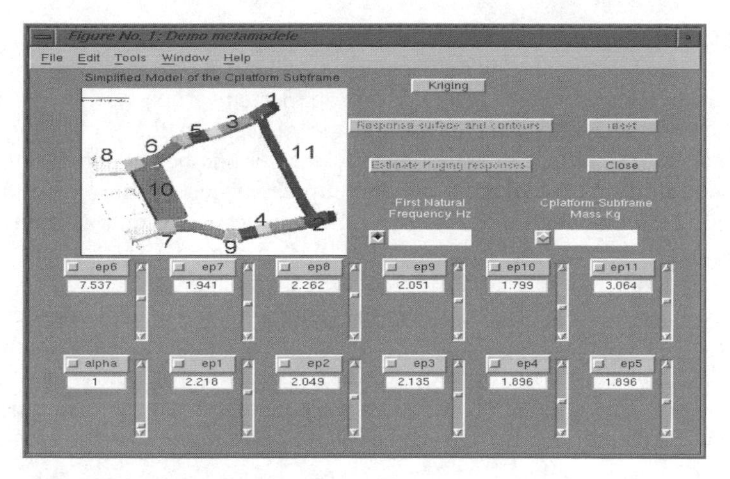

Figure 16.9 A graphical interface for instantaneously assessing performances f_1 and m from a given DV value vector through the kriging metamodel

For our case study of sub-frame, FE calculations were carried out with MSC/NASTRAN. The design of experiments (DOE) has been generated with the software iSGHIT[1]; this is an orthogonal DOE of 64 trials with 12 factors, each of them with 4 levels. The simulations lasted 90 minutes on an SGI/Octane workstation of 1 GHz. The metamodel type is a *kriging* model (see [5; 6]). The fitting of the metamodel parameters has been carried out by a specific procedure on Matlab. The accuracy of the metamodel is judged satisfactory: the root mean square of the errors in the 64 trials is about 0.2 kg for mass m varying between 18 and 24 kg and about 2 Hz for the fundamental frequency f_1 varying in a range of 30 Hz around f_{min}. Only these two performance outcomes are discussed later. A graphical interface (see Figure 16.9), developed on Matlab, allows an instantaneous assessment of performances f_1 and m from a given DV value vector through the kriging metamodel.

16.6 Exploitation of the Model

Deterministic Exploration of the Design Space

The first use of our conceptual modelling process is to provide acousticians and architects with an exploration tool of the sub-frame design space, a tool that has proved to be useful in the envelope volume negotiation phase and even later. The performance metamodel is then used to graphically build in 3D surfaces, which represent the assessed performance values DVs (f_1 et m) function of a couple of DV values (by example *alpha* and e_{10} in Figure 16.10 and e_6 and e_{10} in Figure

[1] iSHIGHT is a product of Engineous Software, NC, USA. http://www.engineous.com

16.11). For this surface construction, 900 performance assessments are performed by the metamodel in only 30 seconds. This undeniable fast performance in the performance assessment is precisely the expected property for an efficient exploration of the potential of the sub-frame concept in the preliminary design stage. This tool has allowed the designers to draw a number of conclusions on the sub-frame behavior, potentialities and comprehension. Notably, the designers realized that the mass and the fundamental frequency were quite sensitive to the elongation ratio *alpha* (see Figure 16.10) and to the two thicknesses of beam sections e_6 and e_{10} (see Figure 16.11). Indeed, the elongation of the lateral beams (#6 and #10) by a factor of 10%, while the other DVs being kept constant at their initial value, results in a decrease of f_1 of about 15 Hz (its initial value being around 100 Hz). A graphical study also showed that an important reduction of weight was possible in decreasing the thickness e_6 without significant deterioration of f_1. This has been confirmed by a further investigation with a detailed structural model.

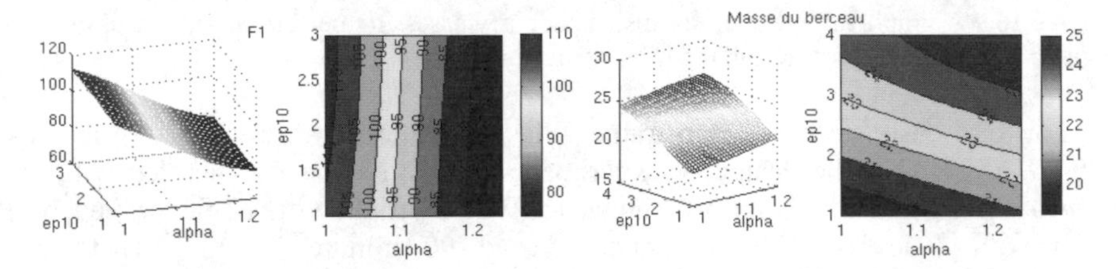

Figure 16.10 Influence of the alpha and e_{10} DVs on f_1 (Hz) and m (kg)

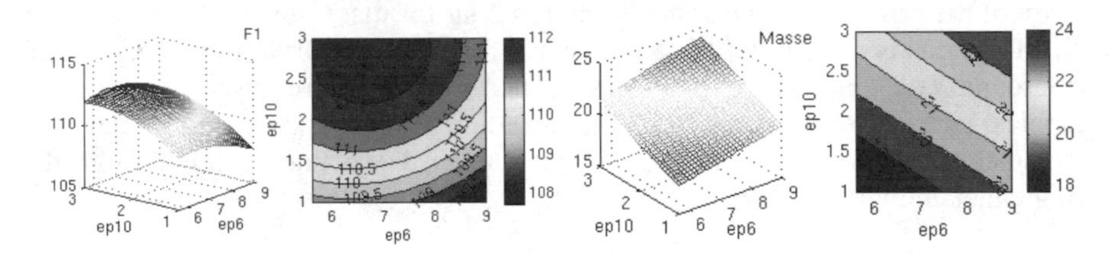

Figure 16.11 Influence of the e_6 and e_{10} DVs on f_1 (Hz) and m (kg)

Optimization

The metamodel may also be used within an optimization loop. Then, the FPs can be interpreted as optimization constraints or in elementary contributors to the objective function to minimize.

For the sub-frame case study, the objective function was to minimize the mass *m* under the constraint $f_1 > f_{min}$. The optimization algorithm has been the conjugate gradient. The optimization outcome has been quite surprising (see Table 16.1) in reducing by 10%, *i.e.* 2 kg, the sub-frame mass compared with the initial dimensioning of the designers (20 kg) without any substantial modification of the fundamental frequency f_1.

Table 16.1 Optimal dimensioning of the sub-frame

DVs	e_1	e_2	e_3	e_4	e_5	e_6	e_7	e_8	e_{10}	e_{11}	alpha	m *(kg)*	f_1 *(Hz)*
initial	2.2	2.	2.	1.8	1.8	7.	2.	2.2	1.8	3.	1.	20	$f_{initial}$
optimal	1.3	1.	1.	1.2	1.1	5.6	1.3	1	1	1.5	1.07	18	$f_{initial}$

Non-deterministic Exploration of the Design Space

So far, the performance metamodel has been used for deterministic or *crisp* assessments of a performance vector for a particular valued DV vector[2]. All the benefits one could have in terms of concurrent engineering are in representing the uncertainty in the variable values at any moment and in managing the consistency between these variable uncertainties [9]. Techniques of *constraint programming* have been particularly praised. All variable domains are then represented as intervals of possible values which are tightened as much as possible so as to rule out values that are ensured not to figure in any deterministic solution. The uncertainty management process can be globally summed up (Figure 16.12) with multi-directional interval reduction propagations. This figure is to be compared with the conventional causal bottom-up deterministic assessment process of Figure 16.1. Yannou *et al* recently proposed in [11, 12] a strategy for optimally coupling a metamodeling of mechanical performances and a constraint programming resolution procedure, which is currently being applied to the sub-frame case study.

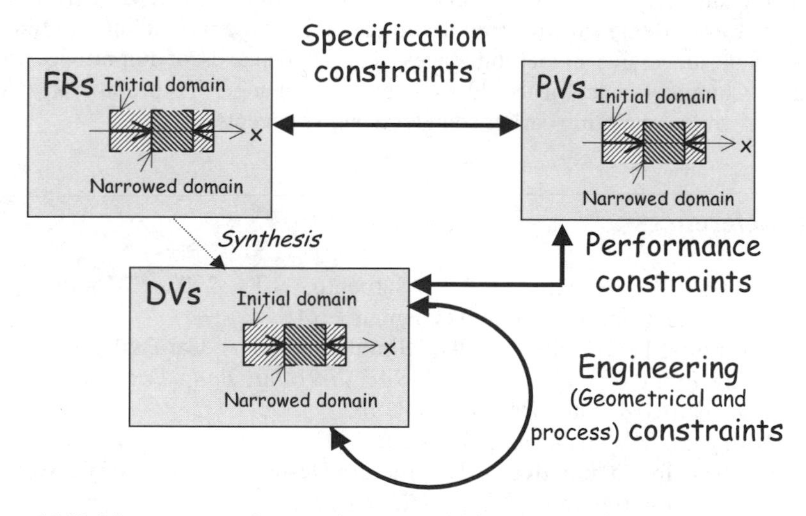

Figure 16.12 Management of uncertainty in variable values within an elementary design loop

[2] One also speaks about a *crisp* or *instantiated* DV vector.

16.7 Conclusions

We have presented a four-stage strategy for the construction of a conceptual model of mechanical performances. A rough structural model must be established which, in turn, helps in resulting in an approximate model (metamodel) of the relation between influential design variables and performance variables or design criteria. Next, different ways for exploiting the metamodel have been presented.

This strategy is adapted to the issues of fast performance assessments and graphical exploration for a better concept comprehension in a preliminary design stage. Moreover, sensitivity analyses can be extracted from the graphical outcomes and a first evaluation of an optimal dimensioning is made possible. This first evaluation is important to better coordinate the designer teams so as to avoid costly design backtracks and to quickly focus on the dimensioning areas to explore in a detailed manner. We believe we can bring a satisfactory answer to the determination envelope volume negotiation stage in a car design process. Indeed, this conceptual modelling strategy has provided fast assessments of performances such as the mass and the fundamental frequency and of criteria like respecting architectural and noise transmission constraints, given an envelope volume and a parameterised sub-frame concept.

In short term, one would like to better apprehend the overall approximation made all along the conceptual modelling process. In a medium term, one wants to be able to rerun such a process of conceptual modelling (as described in section 16.3) on different car subsystems and for different mechanical expected performances and thus our research has to encompass the production of guidelines for the four successive modelling stages. The challenge is of importance for the Renault Company since it could considerably change its preliminary design processes while generating earlier sounder design concepts.

16.8 References

[1] Brown, J.C., Robertson, A.J., Serpento, S.T., 2002, "Motors Vehicle Structures: Concepts and Fundamentals," *SAE.*

[2] Eriksson, M., Bylund, N., 2001, "Simulation Driven Car Body Development Using Property Based Models," *SAE 2001 Int. Body Eng. Conf. - IBEC*, October 16-18, 2001, Detroit, MI, USA.

[3] Nishigaki, H., Nishiwaki, S., Amago, T., Kikuci, N., 2000, "First Order Analysis for Automative Body Structure Design," *ASME/DETC,* September 10-13, 2000, Baltimore, MA, USA.

[4] Pillet, M., 1997, "Les plans d'expériences par la méthode de Taguchi," *Les éditions d'organisation*, Paris, France.

[5] Sacks, J., Welch, W.J., Mitchell, T.J., Wynn, H.P., 1989, "Design and Analysis of Computer Experiments," *Statistical Sci.*, N°24(4), pp. 453-473.

[6] Simpson, T.W., Mauery, T.M., Korte, J.J., Mistree, F., 2001 "Kriging Metamodels for Global Approximation in Simulation-Based Multidisciplinary Design Optimization," *AIAA J.,* N°39(12), pp. 2233-2241.

[7] Simpson, T.W., Peplinski, J.D., Koch P.N. and Allen, J.K., 2001, "Metamodels for Computer-based Engineering Design: Survey and Recommendations," *Engineering with Computers,* N°17, pp. 129-150.

[8] Sung, S.H., Nefske, D.J., 2001, "Assessment of a Vehicle Concept Finite-Element Model for Predicting Structural Vibration," *SAE Noise and Vibration Conf. & Expo.*, April 30 - May 3, 2001, Traverse City, MI, USA.

[9] Yannou, B., 2003, "Management of Uncertainty in Conceptual Design," *International CIRP Design Seminar*, May 12-14, 2003, Grenoble, France.

[10] Yannou B., Hamdi, A., Landel, E., 2003, "Une stratégie de modélisation conceptuelle pour la prise en compte de performances vibro-acoustiques en préconception d'un berceau automobile," *Mécanique et Industries,* N°4, pp. 365-376.

[11] Yannou, B., Simpson, T.W., Barton, R.R., 2002, "NCSP in Design Engineering: Capturing Performance Constraints through Metamodeling Approaches," *COCOS'02: 1st Int. Wkshp. on Global Constrained Optn. and Constraint Satisfaction*, 2-4 October, Sophia Antipolis, France.

[12] Yannou, B., Simpson, T.W., Barton, R.R., 2003, "Towards a Conceptual Design Explorer Using Metamodeling Approaches and Constraint Programming", *ASME/DETC/DAC Design Automation Conf.*, 2-6 September, Chicago, IL, USA.

17

A Framework of Product Styling Platform Using Case-based Styling Indexing

Dr. Richard Y. K. Fung

Abstract: Incorporating manufacturing flexibility into product styling is the next challenge of mass customisation. Fashionable or preferable style is widely accepted to visually enhance a product and satisfy the demands of today's consumers. Considerable work remains to be done to integrate product styling (the process used to enhance visual aesthetics of a product) with manufacturing flexibility, *i.e.* platform approach. The initial findings from an exploratory study that consists of interviews and a research on "*iMac* look" style are reported. It examines the relationships between product style/styling and fashion trends on consumer's preferences. The findings indicate that style can be manipulated by a proposed set of complex attributes. A particular style plus its application method(s) function together as an intangible module to refresh ordinary products. A framework of case-based indexing device is developed to support the above product styling platform approach. This study has opened up a wealth of interest towards the understanding and applying of the visual aesthetic aspects in meeting product styling challenges in the dynamic marketplace.

Keywords: Mass customisation, product styling, product attributes, case-based indexing

17.1 Introduction

Mass customisation is the paradigm-breaking manufacturing reality that attempts to summarize recent trends towards manufacturing flexibility, and which aims at responsively offering *individual* consumer satisfaction. The common topic found in literature of mass customisation is how to achieve the manufacturing efficiencies and how to meet the technological challenge of mass customisation. There are comparatively few studies of mass customisation on how to offer styled products to satisfy today's individual consumer demands. The process to style a product is popularly called product styling. In fact there is almost no literature looking into manufacturing flexibility with the process of product styling. This turns out to be a research niche because to style the exterior of a product becomes the final criterion for developing successful products, especially when the functionality and quality of most products have been improved to be identically the same [1]. It is argued that incorporating manufacturing flexibility into product styling is the next challenge of mass customisation if visual aesthetic aspects become the final criterion for developing successful products.

1. What is the reusability and commonality of a product style?
2. How can product styling be integrated with the platform approach?

A model of consumer preferences to product styling is proposed to answer the first question. With the support of a case study, a framework of product styling platform approach, *iMac*, has been developed to answer the last question.

17.2 Meyer and Lehnerd's Concepts of Product Platform

The platform approach is a good start for incorporating manufacturing flexibility into product styling, since it embeds reusability / commonality and leads to integrated product development. Meyer and Lehnerd's platform approach [13] is essential, which defines product platform as sets of subsystems and interfaces, and these subsystems together construct a common architecture spinning across multiple products. In short, there are two key processes involved: The first is product modularization that decomposes a product into modules. A module can be defined as a grouping of physical or conceptual components including mostly tangible attributes [10]. The second is re-assembling modules into derivative products. The next section presents the proposed style modules of a product style.

17.3 A Set of Complex Attributes to Represent Product Style

To understand styles is mostly through art theories [12]. Lloyd-Jones defines style as the collection of characteristically interrelated symbols and forms. It is echoed by Crozier's [2] findings of core entities in design: meaning and form. These findings give an idea of what a product style is and imply how it can be represented by both physical and symbolic (or meaningful) attributes. To understand physical attributes is straightforward: Physical form features of a product style. However, to understand symbolic attributes is not easy, because a symbol is never precisely defined or fully explained, as Jung claims [11]. By borrowing Jung's ideas [11], symbolic attributes can be defined as the further and additional meanings of a product style. In a marketplace, such additional meanings are accumulated from product messages and design images found in the marketplace. Such information is accessible and can partially reveal the symbolic and additional meanings of a product style. It is proposed to call these informative attributes of a product style. Hence, the physical attributes and the proposed informative attributes together can tangibly represent a product style and vice versa. It is proposed that the collection of these two attributes be named a set of complex attributes.

17.4 Research Hypotheses

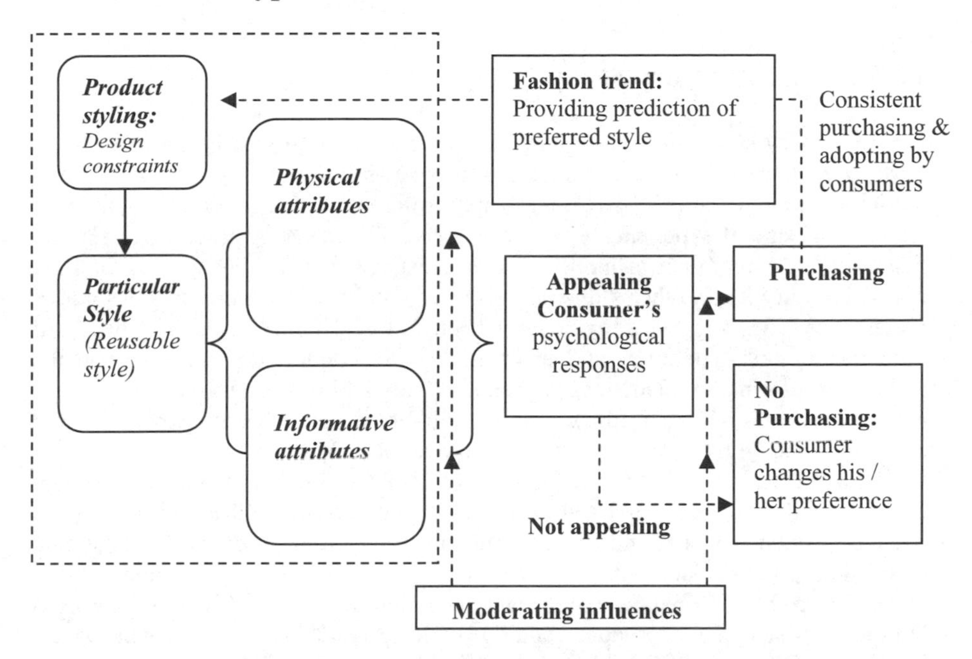

Figure 17.1 A model of consumer's preferences to "reusable" product style

Figure 17.1 shows that physical attributes consist of characteristically interrelated symbols or forms, while informative attributes consist of mutually agreed product messages. It illustrates the cycle of reusing a product style in product styling. This is elaborated from Bloch's model [1] by considering that consistent purchasing can help continuous preferences. The model agrees with the market phenomena that fashion trend can predict the consumers' positive response to the similar stylistic product form [7]. It also accepts the common belief in fashion / innovation diffusion literature that the majority of consumers' major psychological responses have been led by a fashion trend, unless new fashion trend emerges. This means it adopts Sproles' idea[1] – fashion trend – and applies it to consumer electronic products to predict the preferred styles. As a whole, this paper hypothesizes:

[1] Fashion trend is a common marketing phenomenon while a particular fashionable design and its images become preferred style to apply to products (textile and clothing), as Sproles [15, 16] claimed.

H1: In consumer electronic industry, product styling can also reuse a particular style of the fashion trends to differentiate derived products.

H2: The commonalties of the styles of a consumer electronic product can be distinguished and managed by the set of physical and informative attributes.

17.5 The Exploratory Study

An exploratory study of product style was conducted to provide empirical support for the hypotheses H1 and H2. It includes interviews and a case-based research of a particular "product style". Five Hong Kong professional design experts with more than seven years of experience were interviewed. The interviews supported H1 and H2. They all agreed that fashion trends do exist not only in fashion and clothing industries, but also in other industries, and that can be used to predict consumers' preferences. They suggested that tracing the style trends could be done by means of tracking the design features and investigating the connotation embedded in the marketing information. This means product style can be distinguished by a set of physical and informative attributes. As a whole, the design experts' experience and opinions agree with the proposed working model.

A case study of the style of "*iMac* feel" was conducted. Visual and textual contents analyses were employed as analysis methodology and to count and sort out the distinctive visual and textual contents of different products from the data sets. The capture of data took place over a period of five years (December 1998 – November 2003). The data were retrieved from the available archives ranging from the publications of IT products and information found in general magazines, including PCXPress, PCWorld, e-Zone, PC weekly, PC market and other related websites. The case findings revealed that a distinctive fashion trend of "*iMac*" as appealing product style has emerged within a short time since 1998. *iMac* was a top-Selling PC in the United States in 1998. The boom in the sale of *iMacs* attracted many followers to copy its form features. Figure 17.2 shows the results of a survey on new consumer electronic products having an "*iMac* style" over a period of five years delineating the complete life cycle of *iMac*.

Complex Attributes of "*iMac*" Style

Physical (Phy) Attributes of iMac Style

The composition of the *iMac* style is stated as the jelly-bean-shaped translucent "Bondi Blue" plastics with white colour and all-in-one casing. It is the definition of a particular style (*iMac* style) to distinguish which are its followers through pair-wise comparison[2].

---

[2] The procedures of pair-wise comparison can be conducted as follows:

- The composition statement is set as the standard of the *iMac* style.
- Use this standard to compare with the new products appearing in the selected magazines in pairs.

Informative (Info) Attributes of iMac Style

Since 1999, many marketing messages have emerged to describe products with *iMac* style. They started with describing the physical features of *iMac* Computer style as a stand-alone label for *iMac* style symbolising an "*i*" as a symbol of Internet. Table 17.1 shows the cumulative messages adding to the style of *iMac,* which finally symbolise the Internet, brands of Apple Computer, and *trendy* (trendy as a description only validated in 2001-2002) digital products.

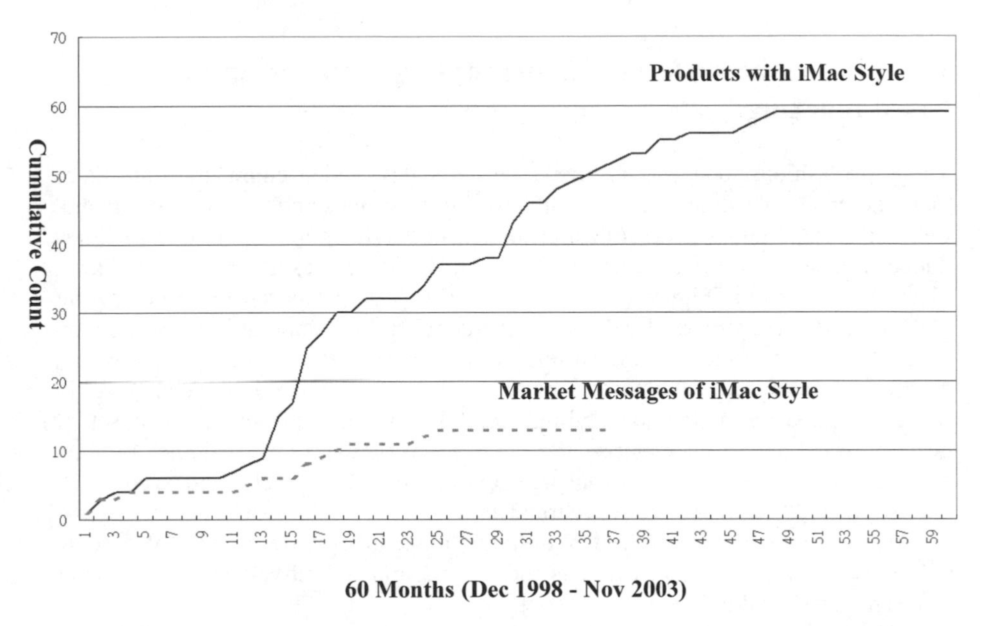

Figure 17.2 The number of products with iMac style: The life cycle of iMac style

Table 17.1 Product messages of "iMac look" from 1999 to Dec 2001

Time	1999_ 11	1999_ 12	2000_ 3	2000_ 4	2000_ 7	2000_ 8	2001_ 11
Messages	iMac trend	iMac concept	iMac, iMac feel	iMac feel, i-base	i- colour	iMac colours	i-SCSI

17.6 Summary and Discussions

The case of *iMac* illustrates that fashion trends do exist in consumer electronic products, and a particular product style can be reusable if it becomes a fashion. It

- The pairs include the style of a new product and the standard. If they match with each other, the new product is deemed to have an *iMac* style.

means that the experience from the fashion and clothing industries could be applied to consumer electronic products. The case further reveals that a particular style (including the set of complex attributes) plus the application of how to reuse that style could together formulate the specific product styling process (styling) to apply ordinary products, and offers a method for how to modularise a particular product style. Hence, a particular "styling" can function as an alternative intangible component/ module to "refresh" existing products.

17.7 The Frameworks of Product Styling Platform and Case-based Indexing

There are fashions today, as Featherstone [3] states, which claim there are more than one preferable style and thus many "*stylings*" at the same time. As Meyer and Lehnerd's [13] suggest, the collection of learnt "stylings" as intangible modules that can create a set of products with derivative visual aesthetics can act like a platform, as shown in Figure 17.3. Therefore, this paper proposes a product styling platform that is constructed by the subsystems of "styling" modules into a common product platform to increase the variety of derivative products. An example would be the customisation of the colours of Nike shoes (http://nikeid.nike.com/). As in the case shown, a particular "styling" module can be represented by a set of descriptions: First to characterise the complex attributes of a particular product style like the application of visual and textual content analysis; and second, its application method(s) should be recorded as well.

As shown in Figure 17.3, if a designer aims to provide a series of secure solutions to style a product, such as the S1, S2 and S3 subsystems, three major issues have to be addressed:

1. What are the available appealing styles that can act as intangible modules for differentiating the product?
2. What are the design circumstances and the related technological or aesthetic constraints?
3. What are the application method(s) suitable for implementing the selected style?

The first two issues require codified information, while the application method(s) are primarily designers' intents. Despite the existence of uncertainties, design experts do learn how to create and apply new / fashionable product style as styling solutions using their experience under specific design circumstances. Although the styling solutions have been embedded in derived and styled products, many difficulties remain in the understanding of the styling solutions. Hence, to retrieve or decode the appropriate information from available archives or a particular style requires extensive design experience. To optimise the reward of a model, it is important to develop a systematic framework with practical applicability [4, 5, 6]. Therefore, a framework of case-based styling indexing is developed to support the above product styling platform approach in organizing the information of cases of product styling as shown in Figure 17.4.

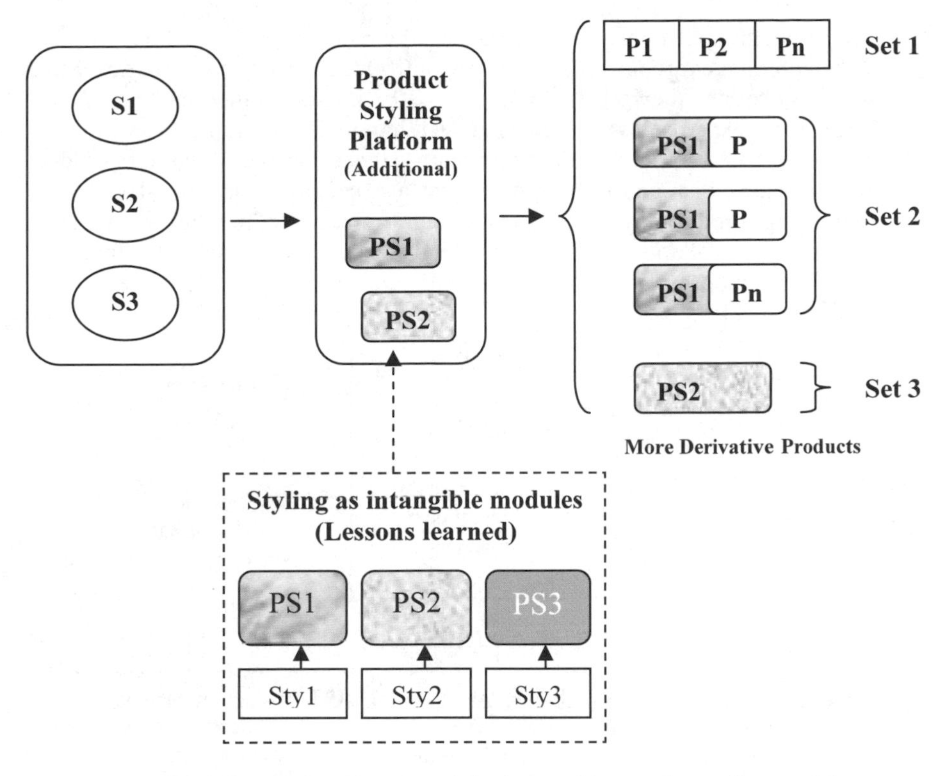

Figure 17.3 The creation of new modules by attaching product styling modules

The framework addresses the issues related to the storage and reuse of experience in product styling by using the index to create and retrieve case memory. In principle, the proposed case-based indexing approach employs a partial concept of case-based reasoning. In the interim, the approach serves as case-based aiding systems or retrieval-only systems as Ockerman and Mitchell defined [14]. The approach of case-based styling indexing is to construct the life cycle history of a style reflected by those products carrying that style, such as the case of *iMac* shows. Figure 17.4 shows the index structure of product styling, depicting the history through recording the complex attributes of that style, and illustrating the marketing phenomena of the products carrying that style:

1. The complex attributes can be grouped in two categories for indexing the products, 'lesson learned':
   a) Physical attributes of a styled product which embed the styling solutions relating to production constraints, and
   b) Informative attributes of a styled product which embed the information relating to meanings and intentions of the design.

These two types of attributes can be presented in written statements.

2. The marketing phenomena are recorded as the marketing information; Although images are the core of styling platform, the index can act as a simple word processing device, since it is mostly coded by written statements. Designers' expertise is needed to select and offer appropriate briefing to retrieve the relevant cases and information. Hence, further research is needed. All the same, this approach does suggest a systematic and formal means of expressing product styling information to relieve the burdens in a multi-disciplinary working environment. This research can help facilitate an intelligent information framework for managing intangible consumer requirements [8].

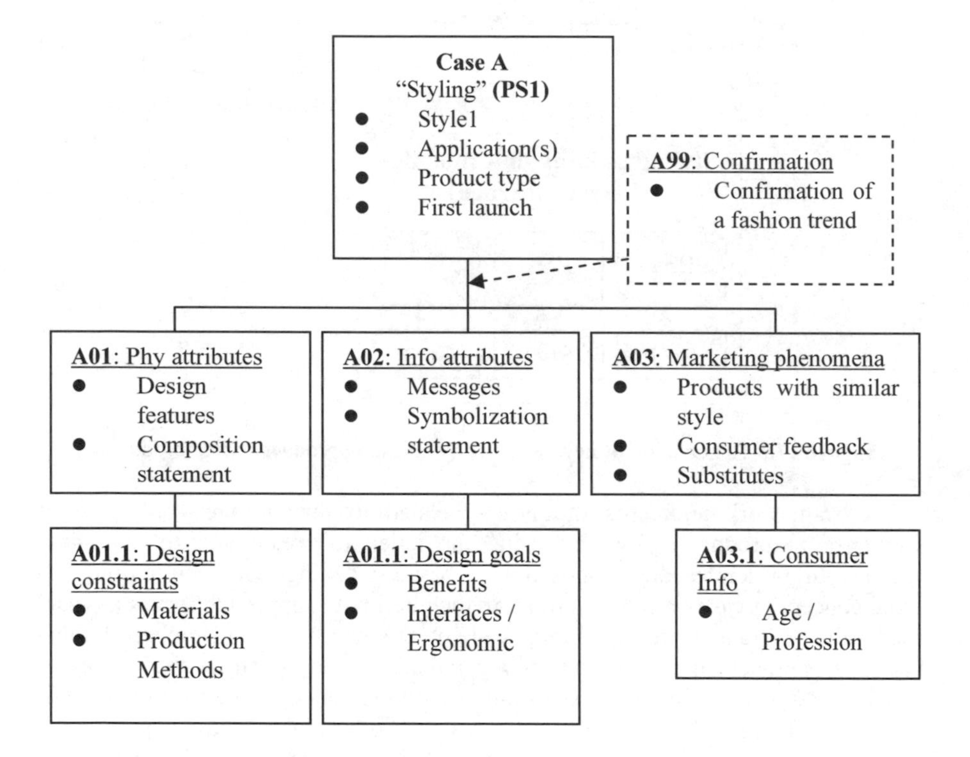

Figure 17.4 The framework of index structure of product styling in case memory

17.8 Conclusions

From the interviews of experienced designers, it became apparent that fashion trends do exist and can help predict consumer preferences. A case study of "*iMac* style" has been conducted to identify a set of complex product attributes. It showed that a particular product style can be reused if it has become a fashion. It is understood that product styling is an intangible module to bestow on ordinary products. A framework of product styling platform approach is developed to answer the question: how product style can be integrated into the platform approach. A framework of case-based indexing device is developed to support the above product styling platform approach. This study has opened up a wealth of interest towards the understanding and applying of the visual aesthetic aspects in meeting product styling challenges in the dynamic marketplace.

17.9 References

[1] Bloch, P.H., 1995, "Seeking the Ideal Form: Product Design and Consumer Response," *Journal of Marketing,* Vol. 59, July, pp. 16–29.

[2] Crozier, R., 1994, In: *Manufactured Pleasure: Psychological Responses to Design,* Manchester University Press, Manchester.

[3] Featherstone, M., 1994, In: *Consumer Culture and Postmodernism,* Sage Publications, London, pp. 83-84.

[4] Fung, R.Y.K., Popplewell, K., Xie, J., 1998, "An Intelligent Hybrid System for Consumer Requirements Analysis and Product Attribute Targets Determination," *Int. J. of Production Research*, Vol. 36, No. 1, pp. 13-34.

[5] Fung, R.Y.K., Tang, J., Tu, Y.L., Wang, D., 2002, "Product Design Resources Optimisation Using a Non-Linear Fuzzy Quality Function Deployment Model," *Int. J. of Production Res.,* Vol. 40, No. 3, pp. 585-599.

[6] Fung, R.Y.K., Tang, J., Tu, P.Y., Chen, Y., 2003, "Modelling of Quality Function Deployment Planning with Resource Allocation," *Research in Engineering Design*, Vol. 14, pp. 247-255.

[7] Fung, R.Y.K., Chong, S.P.Y., Wang Y., 2004, "A Framework of Product Styling Platform Approach: Styling as Intangible Modules," *Concurrent Engineering: Research and Applications,* Vol. 12, No. 2, pp. 89-103.

[8] Harding, J.A., Popplewell, K., Fung, R.Y.K., Omar, A.R., 2001, "An Intelligent Information Framework Relating Consumer Requirements and Product Characteristics," *Computers in Industry*, Vol. 44. No. 1, pp. 51-65.

[9] Jiao, J., Tseng, M.M., Duffy, V.G., Lin, F., 1998, "Product Family Modelling for Mass Customisation," *Computers and Industrial Engineering,* Vol. 35, Nos. 3-4, pp. 495-498.

[10] Jiao, J., Tseng, M.M., 2000, "Fundamentals of Product Family Architecture," *Integrated Manufacturing Systems,* Vol. 11/7, pp. 469-483.

[11] Jung, C.G., 1968, In: *Man and His Symbols,* Dell Publishing, New York, pp. 3-5.

[12] Lloyd-Jones, P., 1991, In: *Taste Today: The Role of Appreciation in Consumerism and Design,* Pergamon Press, New York.

[13] Meyer, M.H., Lehnerd, A.P., 1997, In: *The Power of Product Platforms: Building Value and Cost Leadership,* Free Press, New York.

[14] Ockerman, J.J., Mitchell, C.M., 1995, "Case-based Design Brower to Support Software Reuse: Theoretical Structure and Empirical Evaluation," *International Journal of Human-Computer Studies,* Vol. 51, pp.865-893.

[15] Sproles, G.B., 1981, "Analyzing Fashion Life Cycles – Principles and Perspectives," *Journal of Marketing,* Vol. 45, pp.116-124.

[16] Sproles, G.B., Burns, L.D., 1994, In: *Changing Appearances: Understanding Dress in Contemporary Society,* Fairchild Publications, New York.
[17] Tseng, M.M., Jiao, J., 1998, "Concurrent Design for Mass Customisation," *Business Process Management Journal,* Vol. 4, No. 1, pp. 10-24.

18

A Systematic Design Approach for Reconfigurable Manufacturing Systems

Ahmed M. Deif, and Waguih H. ElMaraghy

Abstract: The evolution of manufacturing systems is triggered by the dynamic customer environment of its time. The main characteristics of today's customers' environment are mass customization and responsiveness to market demand and thus the reconfigurable manufacturing system was suggested for such environment. This paper presents a systematic approach for the design of reconfigurable manufacturing systems and how to control that design process through developing an open mixed architecture for that purpose. The architecture prescribes the different design activities starting from capturing market demand to the system-level configuration and finally the component-level implementation, and also provides some performance measures that are used to control the design process. An example of a reconfigurable automatic PCB assembly line is used to illustrate an application of the developed architecture in real world manufacturing system design.

Keywords: Reconfigurable Manufacturing System, Architecture, Design Methodology

18.1 Introduction

Shorter product life-cycles, unpredictable demand, and customized products have forced manufacturing systems to operate more efficiently and effectively in order to adapt to changing requirements. Traditional manufacturing systems, such as job shops and flow lines, cannot handle such environments. Flexible manufacturing systems are suitable for such environment; however the high initial capital cost is considered a disadvantage. Reconfigurable manufacturing system (RMS) is a new class of manufacturing systems proposed recently, which aims at combining the high throughput of dedicated manufacturing lines (DML) with the flexibility of flexible manufacturing systems (FMS), Koren *et al.* [1]. This could be achieved through the fast scaling of capacity and functionality, in response to new circumstances, by rearrangement or change of its components, Mehrabi *et al.* [2]. The RMS components could be classified to physical components (machines, tools…etc) and logical components (programs, control, plans...etc), ElMaraghy [3]. The key feature of RMS is that its capacity and functionality are modular and not fixed, thus they can be integrated (added) to the system and removed to adapt for the market demand through what may be called capacity and functionality scalability. This reconfiguration characteristic enables the system to produce different products mix (volume and variety) at low cost. Examples of RMS can be found in Heisel and Meitzner [4] and Urbani *et al.* [5].

18.2 Manufacturing Systems Engineering

Existing manufacturing systems engineering frameworks can be classified into frameworks that address the manufacturing system selection process, others that approach manufacturing system design and a third category that deals with manufacturing system control. Among the frameworks that were proposed for the manufacturing systems selection are Hayes and Wheelwright [6], Black [7], Chryssolouris [8] and Miltenburg [9]. Numerous frameworks and models were developed to guide the design of manufacturing systems. Some of these frameworks approached the manufacturing system design from the layout perspective. Meller and Gau [10] provided a comprehensive literature review of these approaches. Other approaches addressed the design from a more systematic perspective. Examples of the manufacturing systems design approaches are the manufacturing systems design decomposition MSDD by Cochran *et al.* [11], the core manufacturing systems design process by Duda [12] and the improved manufacturing systems design methodology by Katzen [13]. Some of the manufacturing systems design frameworks were only dedicated to the lean production design as in Monden [14] and Sakakibara *et al.* [15]. The control of manufacturing systems' frameworks is much less in number. The "Graphe à Resultats et Activites Interlies" (GRAI) by Doumeingts *et al.* [16] and CIMOSA by Vernadat, [17] are examples of these frameworks. As for the reconfigurable manufacturing systems there is no framework developed to analyze the design process for these modern manufacturing systems from recognizing the customer needs through system configuration generation, selection and implementation. This paper presents an architecture that captures the full reconfiguration process in reconfigurable manufacturing systems.

18.3 Design of Reconfigurable Manufacturing Systems

Figure 18.1 shows the proposed architecture for the design of reconfigurable manufacturing systems. The architecture was developed through adopting the information system design methodology. The architecture is composed of two modules; the first module describes the design process of the reconfigurable manufacturing systems and the second module describes the control of the design process at each level. The control module is based on performance measurements that reflect the strategic objectives and constraints indicated by the high-level decision makers at each level. Such measures ensure the consistency of design activities with the strategic objectives of RMS. The architecture as shown is open as its information flow is accessible through any layer and it is mixed since it is composed of both hierarchal and partitioned layers. The architecture is made of three layers; the market-capture layer, the system-level reconfiguration layer and finally the component-level reconfiguration layer. In the following section each layer, with both its design and control aspects will be explained.

Market Capture Layer

This layer describes how the reconfigurable manufacturing system responds to different market demand profiles (deterministic or stochastic) and converts these customer needs into required capacity and functionality levels. Reconfigurable manufacturing systems are designed to have scalable capacity and functionality as explained earlier. The main objective of this layer is to capture the customer needs to generate the required capacity and functionality levels that will act as the design parameters or inputs to the system-level reconfiguration layer. This layer in the design process is the basis of the development of the required scalability policy that minimizes time and cost. The market capture layer is explained as an $IDEF_0$ model in Figure 18.2.

System-level Reconfiguration Layer

This layer is the heart of the reconfigurable manufacturing systems design process. The required capacity and functionality levels together with the process plans are taken as inputs to the system configurator that generates different system configurations. This process is controlled by the reconfiguration constrains (cost, space…*etc..*) which are system specifications. Generation of multiple configurations is basically enabled by the modular design of the manufacturing system components. The selection of the best feasible configuration among the generated ones is selected using predetermined performance measurements such as quality, throughput, complexity or other criteria. The system design process is completed by planning for the reconfiguration of the existing system to the new selected configuration. This plan includes physical or hard reconfiguration plan (like add or remove a machine or a tool), logical or soft reconfiguration plan (like reprogram a machine, re-route or reschedule the production flow) and finally human participation reconfiguration plan (like reallocate human recourses or reconfigure the job tasks). The planning for system reconfiguration is controlled by a smoothness index that measures the smoothness of this reconfiguration process. The system-level reconfiguration layer is explained as an $IDEF_0$ model in Figure 18.3.

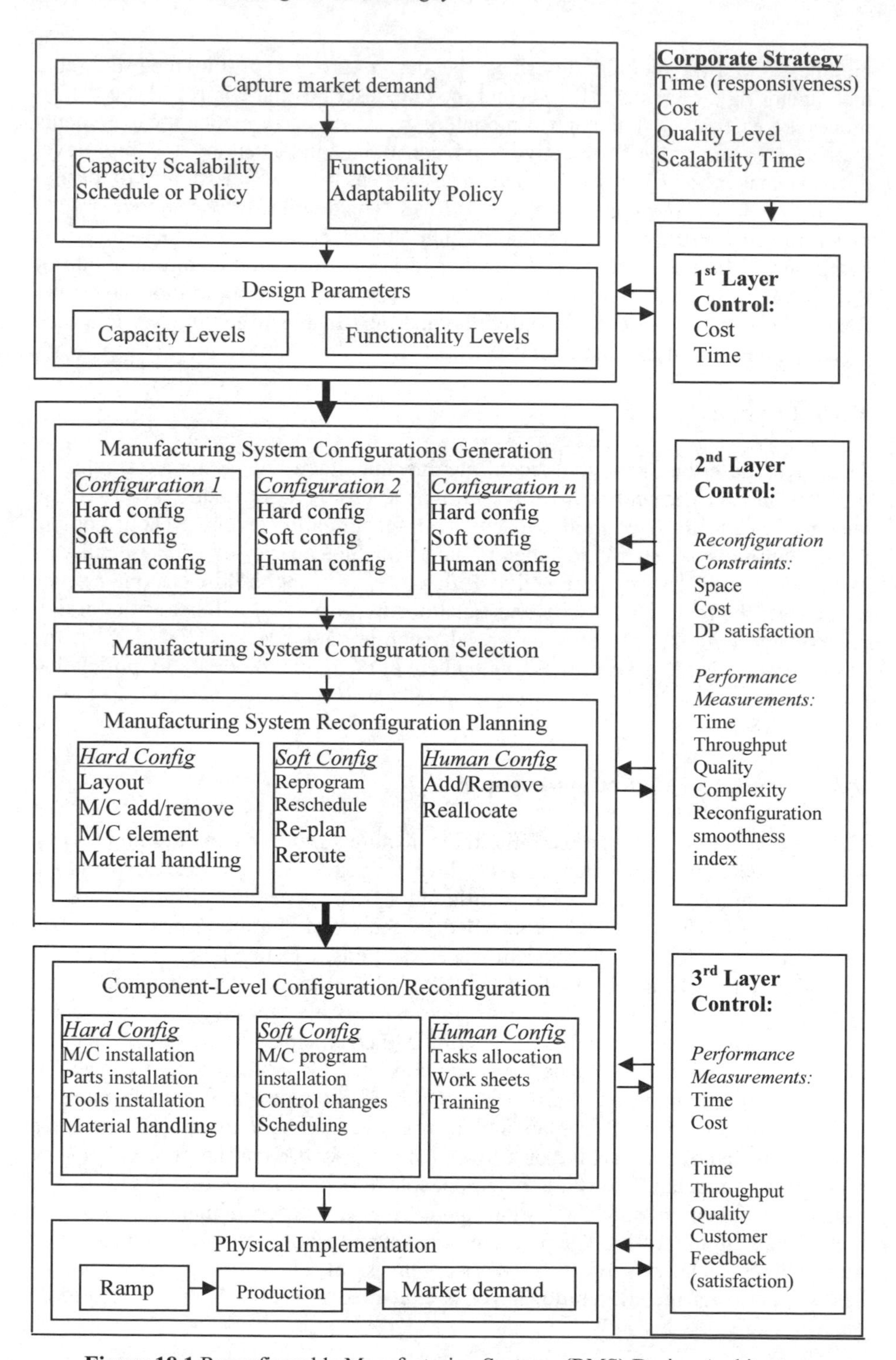

Figure 18.1 Reconfigurable Manufacturing Systems (RMS) Design Architecture

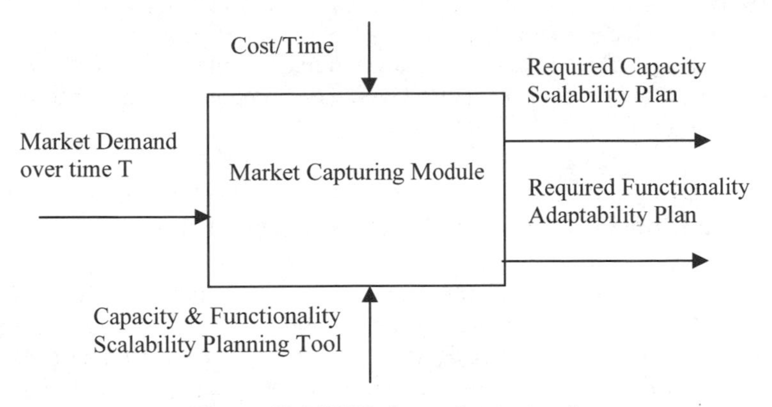

Figure 18.2 $IDEF_0$ for market capture layer

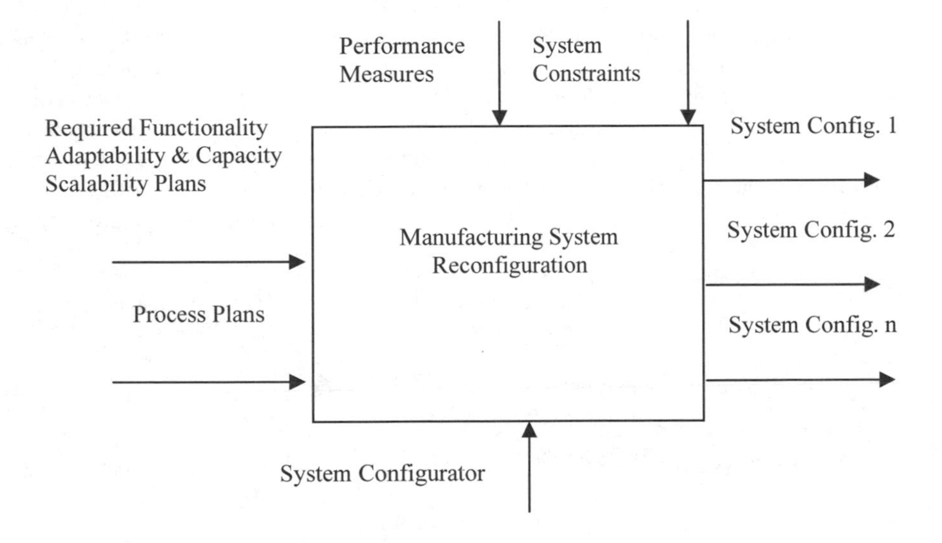

Figure 18.3 $IDEF_0$ for System-level reconfiguration layer

Component-level Reconfiguration Layer

This layer in the reconfigurable manufacturing systems design process deals with physical implementation of the selected generated configuration. The implementation affects the mentioned system components (physical, logical and human). Integrabilty and machine open control architecture are the major enabling technologies responsible for the successful of the real physical implementation of the reconfigurable manufacturing systems. This step is controlled by previously mentioned performance measurements especially quality, cost and time. After system reconfiguration, the system should ramp up in minimum time in order to achieve the responsiveness strategic advantage of the reconfigurable manufacturing systems. Finally, production takes place to manufacture the product mix required that was captured by the market capture layer. The component configuration layer is explained by $IDEF_0$ model in Figure 18.4.

18.4 Application to Automatic PCB Assembly Line

A traditional PCB automatic assembly line, shown in Figure 18.5, consists of a loader/unloader magazine, a printing machine, automatic pick and place machines, reflow oven and some inspection devices like the ICT (in-circuit tester). In a reconfigurable PCB automatic assembly line, these components are reconfigurable. For example, the automatic pick and place machines are designed to assemble different types of electronic components and IC chips by its modular design that can accommodate different types of cameras, according to the size of the components and chips and different types of nozzles to pick these components and chips. This is assisted by reconfigurable open control architecture of those machines. Also the printing machine is modularly designed to be reconfigured to act as screen-printing machine for the solder paste or as a glue dispenser according to the application by just adding the required modules and some control changes. The reconfiguration of the reflow oven is done through reprogramming the settings according to the type of the solder paste and product (logic or soft reconfiguration). The ICT machine is reconfigured by changing modules for the jig and testing probes.

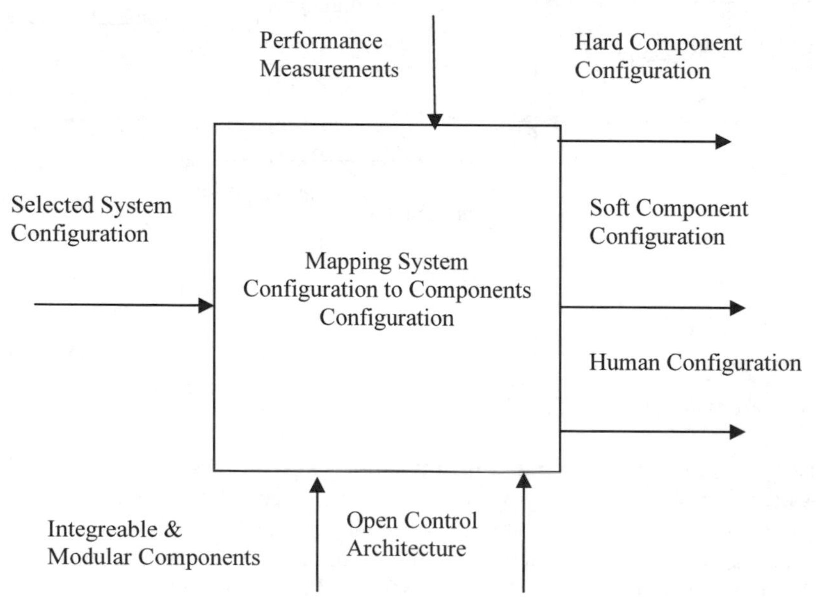

Figure 18.4 $IDEF_0$ for Component-level reconfiguration layer

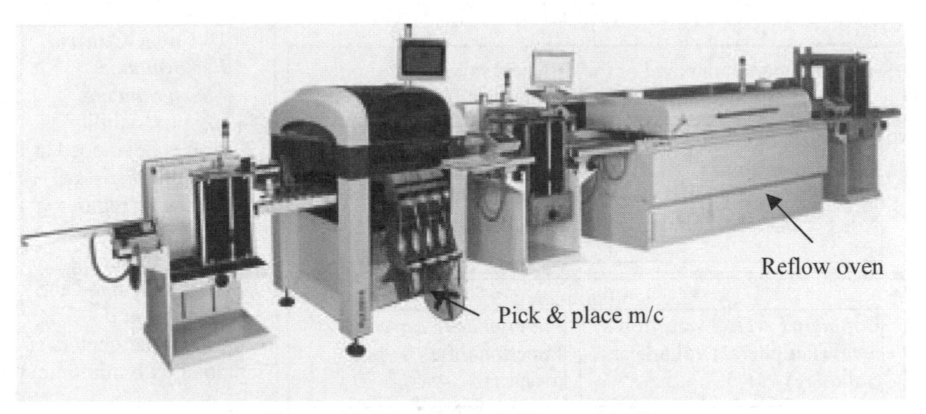

Figure 18.5 Automatic PCB asembly line

Applying the RMS Design Architecture to a Reconfigurable Computer Peripherals PCB Automatic Assembly Line

The proposed framework is applied to a computer peripherals automatic assembly line (main boards, VGA cards, sound cards, memory cards and fax modem cards). This type of market is characterized by being very turbulent due to the short life cycle of the products and the high need for mass customization. In this environment the need to apply the RMS technology is recognized. Example of system-level reconfiguration where the capacity was scaled up by adding two extra pick and place machines in series shown in Figures 18.6. As for machine level, examples of physical and logical configurations are listed in Table 18.1. The reconfiguration process of that line is illustrated through adopting the proposed architecture as shown in Figure 18.7. The architecture guides the system designer to the required reconfiguration process on the system-level as well as for the machine-level as discussed in the previous section.

Figure 18.6 System-level capacity scalability example

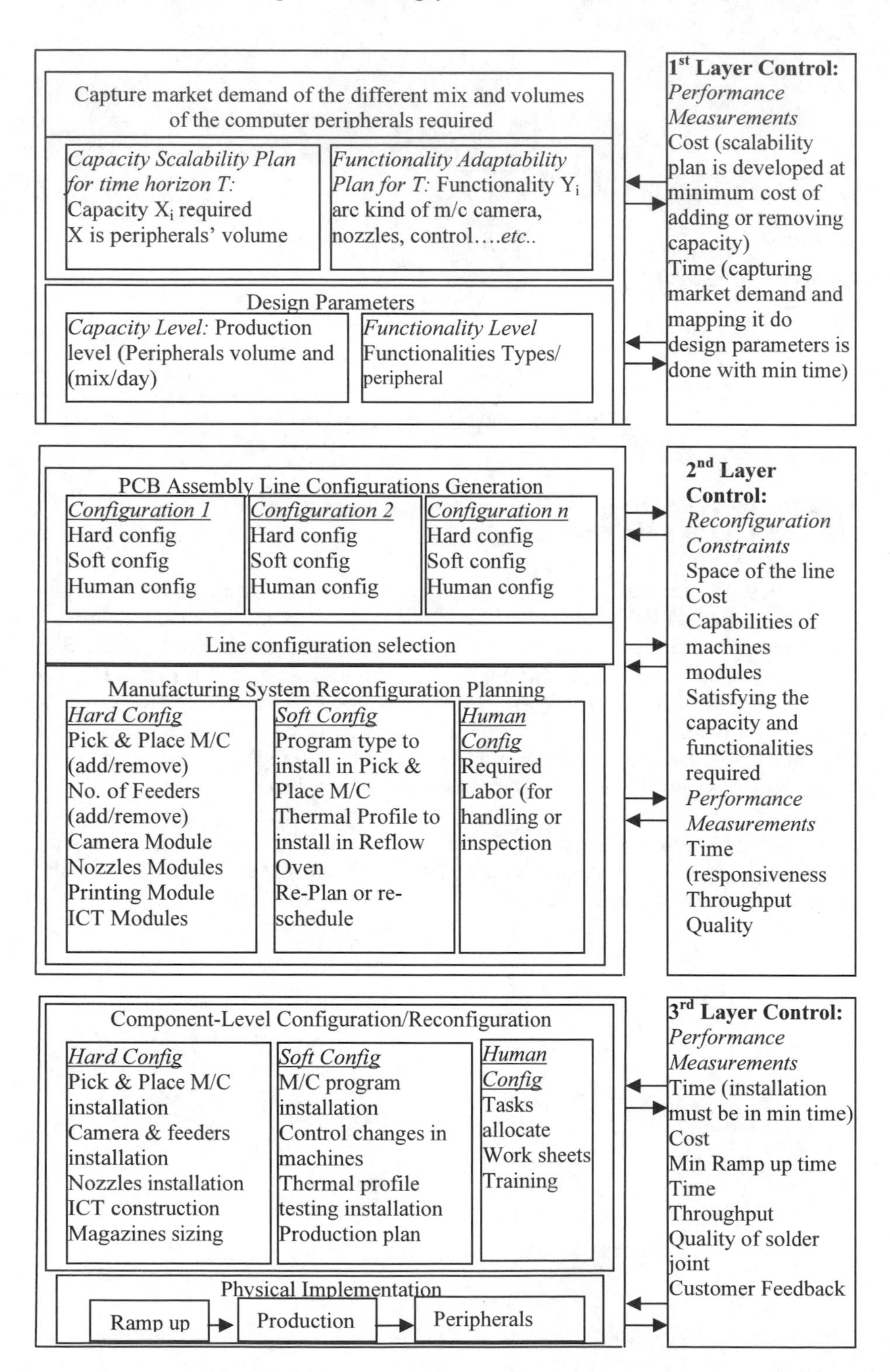

Figure 18.7 Design architecture applied to reconfigurable automatic PCB line

Table 18.1 Machine-level reconfiguration examples

Machine	Physical Reconfiguration	Logical Reconfiguration
Screen Printing	Different types of cameras (based on components' sizes) Different number of feeders (based on components' number) Different types of nozzles (based on components' sizes) Different types of PCB clamping	Different types of programs (based on the type of the product)
Pick and Place	Add/remove dispensing module (to have both types of printing options)	Different printing modes
Reflow Oven		Different thermal profiles (based on the type of solder paste and PCB layout)

18.5 Conclusions

The paper presented an open mixed architecture for the design of reconfigurable manufacturing systems and how to control this design process. The architecture describes the different design processes starting from capturing the market demand to generating and selecting the best configuration that satisfies this demand to the final physical implementation of that system configuration. The architecture showed how each design layer is controlled by different performance measurements that reflect the strategic objectives of the reconfigurable manufacturing system. The architecture could be considered a comprehensive explanation of the reconfiguration process in reconfigurable manufacturing systems and opens the door for researchers to visualize the different areas that need to be developed in such systems. The Application discussed in this paper presented a practical example of how such architecture can be applied in real reconfigurable industrial environment. Further work is needed to model each of the layers prescribed in order to combine the qualitative systematic approach with the quantitative one. Such combination has the potential to produce a generic design tool for reconfigurable manufacturing systems

18.6 References

[1] Koren, Y., Heisel U., Jovane, F., and Moriwaki, T., 1999, "Reconfigurable Manufacturing System," *Annals of CIRP*, Vol. 48/2.

[2] Mehrabi, M., Ulsoy, G., and Koren, Y., 2000, "Reconfigurable manufacturing systems: Key to future manufacturing," *Journal of Intelligent Manufacturing*, Vol. 11, pp. 403-419.
[3] ElMaraghy, H., 2003, "Fundamentals of Flexible Manufacturing Systems," *Lecture Notes, University of Windsor*.
[4] Heisel, U., and Meitzner, M., 2003, "Progress in Reconfigurable Manufacturing Systems," *CIRP 2nd International Conference on Reconfigurable Manufacturing Systems*, Ann Arbor, MI.
[5] Urbani, A., Molinari-Tosatti, L., Pedrazzoli, P., Fassi, I., and Boer, C.R., 2001, "Flexibility and Reconfigurability: An Analytical Approach and some Examples," *CIRP 1st International Conference on Reconfigurable Manufacturing Systems*, Ann Arbor, MI.
[6] Hayes, H., and Wheelwright, C., 1979, "Link Manufacturing Process and Product Lifecycles," *Harvard Business Review*, Jan/Feb, pp. 2-9.
[7] Black, J.T., 1991, *The Design of a Factory with a Future*, McGraw Hill.
[8] Chryssolouris, G., 1992, *Manufacturing Systems: Theory and Practice*, Springer-Verlag, New York.
[9] Miltenburg, J., 1995, *Manufacturing Strategy: How to Formulate and Implement a Winning Plan*, Productivity Press, Portland, Oregon.
[10] Meller, R.D., Gau, K.Y., 1996, "The Facility Layout Problem: Recent and Emerging Trends and Perspectives," *Journal of Manufacturing Systems*.
[11] Cochran, D., Arinez, J., Linck, J., 2002, "A Decomposition Approach for Manufacturing System Design," *Journal of Manufacturing Systems*, Vol. 20/6 pp. 371-389.
[12] Duda, J., 2000, "A Decomposition-Based Approach to Linking Strategy, Performance Measurement, and Manufacturing System Design," *Ph.D. Dissertation, Massachusetts Institute of Technology*.
[13] Katzen, J., 2003, "Concurrency Designing a Physical Production System and an information System in a Manufacturing Setting," *M.A.Sc. Thesis, Massachusetts Institute of Technology.*
[14] Monden, Y., 1998, *Toyota Production System - An Integrated Approach to Just-In-Time, Engineering & Management Press*, Third Edition.
[15] Sakakibara, S., Flynn, B.B., and Schroeder, R.G., 1993, "A Framework and Measurement Instrument for Just-In-Time Manufacturing," *Production and Operations Management Journal*, Vol. 2, No.3, pp.177-194.
[16] Doumeingts, G., Chen, D., Vallespir, B., Fenie, P., and Marcotte, F., 1993, "GIM (GRAI Integrated Methodology) and Its Evolutions - A Methodology to Design and Specify Advanced Manufacturing Systems," *Proceedings of the JSPE/IFIP TC5/WG5.3 Workshop on the Design of Information Infrastructure Systems for Manufacturing, DIISM '93*, 8-10 November 1993, Tokyo, Japan.
[17] Vernadat, F., 1993, "CIMOSA: Enterprise Modeling and Enterprise Integration Using a Process-Based Approach," Proceedings of the JSPE/IFIP TC5/WG5.3 Workshop on the Design of Information Infrastructure Systems for Manufacturing, DIISM '93, 8-10 November 1993, Tokyo, Japan.

19

Crosstalk: Collaborative Framework for Electro-mechanical Product Design

Michael Montero, Noe Vargas, Paul Wright, and Jami Shah

Abstract: Design of electronic-mechanical assemblies involves many disciplines, tasks, and a disparate set of CAD/CAE tools. The problem is to facilitate collaboration between them by increasing the interoperability between these tasks and their respective tools while maintaining the integrity of the designs, including constraint management. This paper will present the requirements envisioned for a ME-EE co-design environment called CrossTalk. It is hoped that the knowledge gained from this bottom approach will be used to design a framework for commercial ECAD-MCAD-CAE collaboration and communication that is specific to this cross-domain design but independent of proprietary file formats and software.

Keywords: collaborative/concurrent design, electronic-mechanical design

19.1 Introduction

Designers of consumer electronics (PCs, hand-helds, cell phones and games) are driven by short delivery-times and globally distributed supply chains. There is high pressure to reduce device size and improve aesthetics or ergonomics, which necessitates mechanical and electronic engineers to work closely to handle packaging design with thermal, RF, and electromagnetic considerations. One of the challenges of streamlining the design process is the lack of coupling between ECAD, MCAD, and CAE tools that cause delays in design and fabrication as well as quality problems. Examples of typical problems encountered in electronic-mechanical design are listed in Table 19.1. Another challenge entails the reuse of ECAD/MCAD models within CAE applications. Design engineers indicate that it is very desirable to combine unchanged portions of older versions of simulation models with new versions of changed models to save time. Unfortunately, design changes require regeneration of complete analysis models. CrossTalk will address these issues. By focusing on the specific domain of electronic-mechanical design, there is the potential to have a broad impact on the efficiency of the consumer electronics industry by developing pragmatic techniques for reducing product design time.

Table 19.1 Typical electronic-mechanical design problems

Problem	**Cause**
Connector off by 3mm; pin & hole interference	Lack of coordination between ECAD & MCAD
Design change caused electromagnetic problem	Lack of coordination between ECAD & E-CAE
Thermal source moved causing drift problem	Lack of coordination between MCAD & M-CAE
Physical pin does not match model port name	Misinterpretation of data by Fab house
Design change causes complete regeneration of analysis models	Inability to identify exact effect of change and to re-use portions of models from past iterations

19.2 Frameworks and Data Exchange

In reviewing framework architectures such as the following MIT/DICE [12], DOME [6], Shared Design Manager [14], X-DPR [1] and others we find two extreme approaches. Much of the academic work such as DICE, has developed integration environments around in-house software applications over which one has complete control. These frameworks typically work at a fine level of data granularity in shared databases. For example, the MIT/CMU DICE project used its own CAD system (GNOMES) rather than a commercial package [12]. The other extreme is found in large companies that use PDM systems to manage native format files produced by commercial CAD/CAE tools [CIMData]. PDM systems allow version and configuration management via product structures but at a very coarse level of granularity (file management rather than data). Given the pervasiveness of commercial CAD/CAE tools in industry, we need to accept the fact that most of the product data is primarily generated by these tools and resides in native files of these applications. However, to enhance the interoperability between related tasks and constraint management we need to devise a scheme in which data is shared at a more meaningful level of granularity. How to achieve this in the presence of commercial CAD/CAE tools is one of the challenges.

Although not widely embraced by industry, which continues to use non-standard formats (*Gerber, IDF, etc..*), AP210 has the potential for aiding static data transfer between ECAD and MCAD. Currently, Cadence is developing an output format, which adheres to the AP210 specification. Although the geometric information in AP210 is interoperable with AP203 which MCAD system can understand, there is no support for parametric level information needed to support constraints and couplings. PTC/Windchill has recently introduced the ATB Associative Topology Bus to support the latter. Designated parameters are exported from CAD models to Windchill PDM, which stores the linkages between them. However, this capability is a custom point-to-point interface between proprietary systems. Vendor independent solutions are desired. Problems in using AP210 need to be identified and solved. Another format, which has become a *de facto* standard, is the Intermediate Data Format (IDF). The format was initially developed by SDRC and MentorGraphics in 1992 and has grown to include many other electrical CAD vendors such as Bentley Systems, Cadence, CoCreate, Incases, Unigraphics, PADS, PTC, VeriBest, and Zuken-Redac. The IDF 4.0 data model is based on a hierarchy of assemblies, parts, and features [4]. MCAD systems rely on third-party software vendors to provide the IDF translators necessary for importing the ECAD models into the MCAD applications. Problems arise from versioning, unlinked ECAD geometry libraries (necessary for full PCB component renderings), and instability due to variation in IDF formats outputted have caused a slow adoption of this standard into the CAD community.

19.3 Domain-specific Design

There are some characteristics specific to electronic and mechanical design. Geometric layout designs of the electronic and mechanical components (Figures 19.1-19.3) are inter-twined. From cellular phones, PDAs, personal computers and laptops to laboratory and medical equipment, the design cycle often involves electrical engineers for the integrated-circuit design and mechanical engineers for the thermal, structural, and packaging design. Electrical engineers do the primary functional design by selecting standard electronic components, devices and special-purpose chip sets specific to the system. As the capabilities of electronics continue to increase while their size decreases, these products are dissipating more heat. Proper thermal management is therefore becoming essential. The packaging strategy depends on the size of the printed circuit board, and the cooling methodology depends on the heat dissipation by the electronics. Due to the limited design and development time, already-proven electronic cooling methods are often preferred as shown in Figure 19.4 [5].

During layout design a CAD model of the mechanical package is developed based on appropriate material, manufacturing and cooling method selections that, in turn, are based on the electrical design. This is followed by thermal and structural analysis. Mechanical engineers are also responsible for determining how different pieces of the housing will assemble, snap, or lock together. Design phases "overlap" or "iterate" during the evolution of a new consumer product requiring repeated data transfer and regeneration of analysis models in CAE packages.

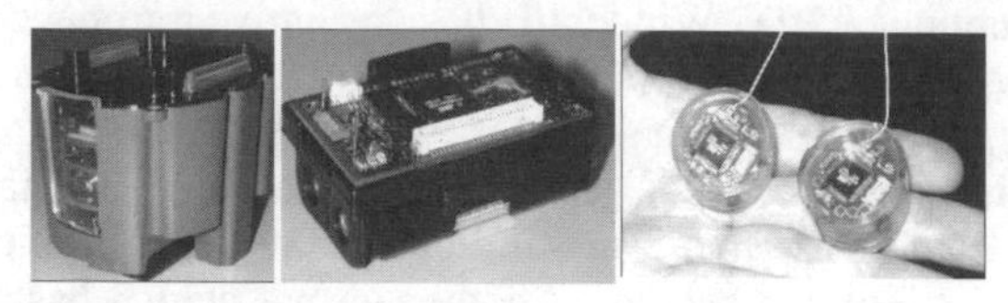

Figure 19.1 Evolution of wireless, sensor platforms at Berkeley

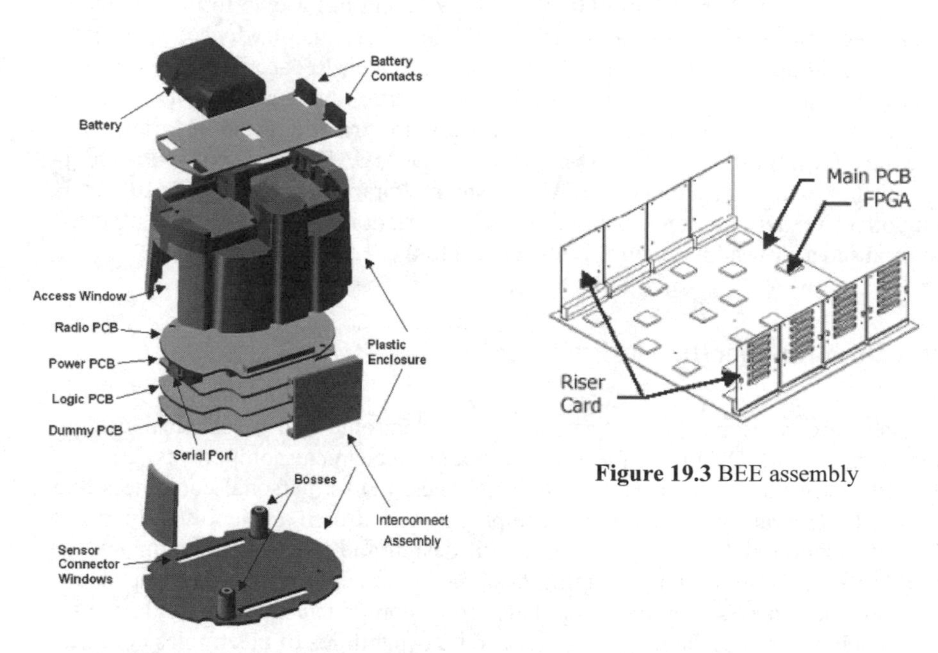

Figure 19.2 PicoRadio assembly

Figure 19.3 BEE assembly

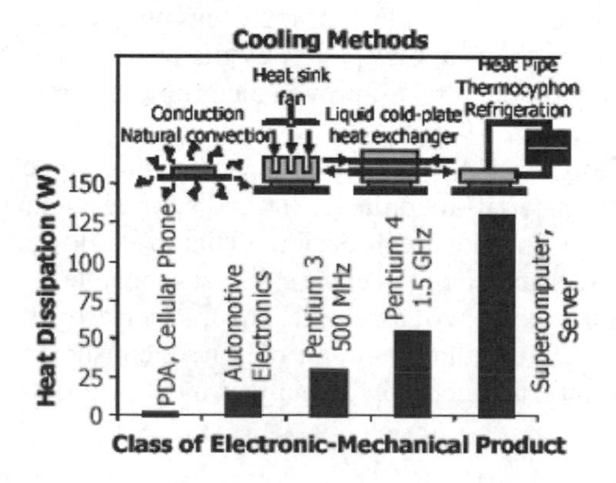

Figure 19.4 Cooling methods

19.4 Framework Development

The first task is to get an in-depth understanding of the requirements for CrossTalk (Task 1). This consists of 3 subtasks, beginning with studying the "as is" design process for this particular class of electronic-mechanical products to gain an understanding of the inter-dependencies between the design tasks. Also, to study each individual application in detail, the models created, their native formats, data used and design reasoning in each. Finally, to study the data flow between tasks and identify mismatches in abstraction levels, semantics and incompatible file formats. Knowledge gained from these three subtasks will be used to design a collaboration framework from commercial ECAD-MCAD-CAE, specific to this domain but independent of proprietary file formats or software (Task 3). Associated model entities within each application will be identified and formalized as MCAD, ECAD features. Constraints across application models will be represented by structures called "cross-couplers" (Task 2), which will be explained in the following sections. Features and cross-couplers will provide the mechanism to manage constraints and archive design knowledge in feature and coupler libraries. We will determine logical partitioning of monolithic CAX files into units based both on product structure and cross-couplers. These partitioned shared objects will be stored in a shared database. A constraint engine is proposed for constraint validation and propagation (Task 4). A proof of concept testbed will be implemented to conduct case studies at the Berkeley Wireless Research Center (BWRC) (Task 5). These tasks are discussed further in the following sub-sections.

Determination of Collaboration Requirements

Figure 19.5 shows a sequence of tasks typical of electronic design. The design may follow an "inside-out" or "outside-in" process. An example of "inside-out" is the BEE device in Figure 19.3 where the electronic design proceeds largely free of packaging constraints and the enclosure is designed afterwards. By contrast, an example of "outside-in" is shown in Figure 19.6 - a marine intercom project for DARPA. The goal was to generate a new wireless intercom system for use in marine tanks. It consisted of two earpieces and a handset. The earpieces contained a digital board, a power board, and a radio adaptor board connected to a radio. To accommodate the flat-sided oval shape of the existing earpieces without wasting printed circuit board (PCB) area, the PCBs were also shaped as flat-sided ovals. Clearly, in designs such as this, the PCB geometry is dictated by the mechanical outside design, in this case the helmet assembly.

ECAD packages, such as CADENCE, MentorGraphics, *etc..* are collections of integrated modules to support major electronic design functions: IC Schematic, Board Layout, and PCB Assembly physical layout. ECAD modules contain libraries of functions, components and layout rules to aid the designer. Mechanical Packaging is carried out in MCAD systems (AutoCAD, SolidWorks, *etc..*) and involves both component design and assembly design modules. Mechanical design may be carried out at the 2D level (AutoCAD) or 3D level (SolidWorks, Pro E, *etc..*). For the former, a DXF file of 2D geometry of the PCB layout from ECAD is sufficient, while an IDF file (2D+heights) may be used for 3D design. Three distinct manufacturing functions (PCB Fab, Housing manufacture and final assembly, including SMT components) are typically outsourced. Various non-standard file formats (Figure 19.5) are in use for communicating the specs to manufacturers: *Gerber* for board layout, NC files for holes, IDF and unstructured text files in PDF or .doc formats.

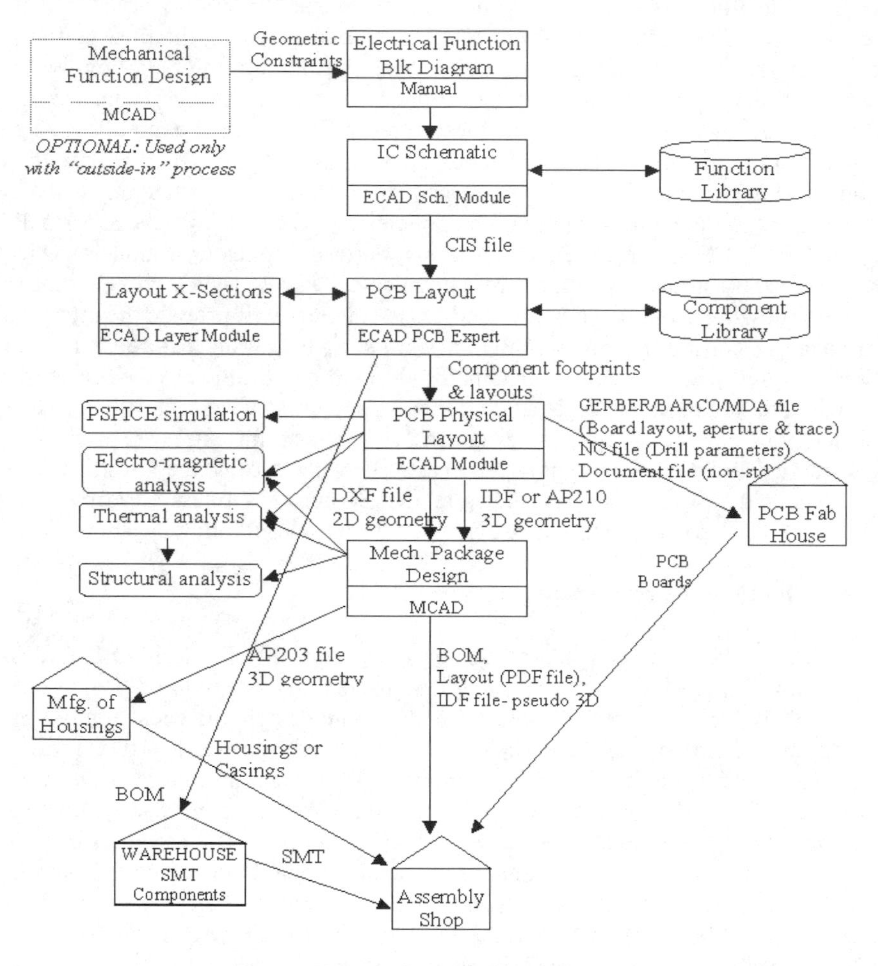

Figure 19.5 "As-is" typical design process

As shown in Figure 19.5 several types of engineering analyses are conducted at various stages. Analysis models are created in each package from data extracted from ECAD, MCAD or both.

The standardization effort for analysis date (STEP AP209) has been limited so far to static structural analysis models and geometry. Considerable time is lost in preparing analysis models from ECAD-MCAD data. Since design usually involves iterations using a "patch and refine" process based on analysis results, ECAD-MCAD model changes need to be transmitted to analysis applications repeatedly and models rebuilt.

Features and Cross-couplers

All major MCAD systems today support the creation of feature libraries and design by features in association with direct geometry creation. However, features are only being used as "construction macros" to speed up design rather than as *persistent* objects that strictly enforce constraints specified upon them in their original library definition; once a feature is instanced the user can modify it arbitrarily, in every possible way that the directly created geometry can be used. Consequently, features lose their meaning, *i.e.*, are *not persistent* in the model. Also, they are not represented explicitly in the final geometric model and may not even be implicit because of arbitrary modifications. Thus, downstream applications, such as CAE and CAPP cannot exploit this level of information. In order to understand the technical issues involved, one needs to examine the nature of MCAD models further. They have 3 major components: construction history, constraint graphs and evaluated boundary representation (BRep). When a design change is made, the modeler rolls back the model to the point of change in the history tree, solves the constraints at the 2D level, performs the 3D operation corresponding to the feature (sweep, loft) and rolls forward through the history tree to regenerate the design with modifications [10]. Features are nodes in the history tree; it does not matter whether the nodes came from pre-defined features in libraries or were created by direct geometric construction on the fly. Apart from their current role of speeding up design and re-use of shapes in MCAD, features have the unrealized potential to serve as key elements in data exchange between applications and for change management. But to exploit this latter possibility, some mechanism to achieve persistence in feature attributes is needed. This is another technical challenge that this project proposes to address.

Electronic devices are assemblies of many parts: PCBs, interconnects, surface mount components, product enclosures, *etc.* (Figure 19.2). They may be embedded in larger mechanical systems, such as the helmet of Figure 19.6.

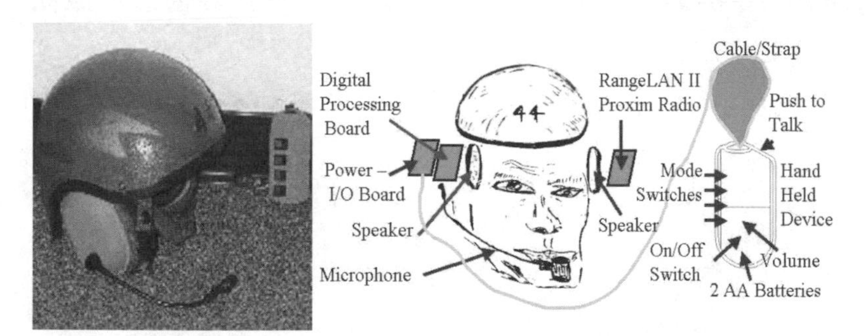

Figure 19.6 Example of "outside in" design (DARPA)

There are physical and functional relationships between the assembly components. The concept of features can also be applied to encode recurring, stereotypical functional and physical relationships. Assembly features, such as pin-in-hole, screw, press, heat shrink, glue, snap, etc need to be defined formally and stored in libraries. This has not been done in MCAD yet where designers must now use geometric constraints directly between each pair of geometric entities in mating features to position parts in assemblies.

Taking the concept of explicit encoding of stereotypical sets of entities and relations one step further, we propose the idea of "cross-couplers" as structures relating objects across different models (Figure 19.7). These could be used for maintaining relations between MCAD, ECAD, MCAE, ECAE models, aid constraint validation, change propagation and patching together portions of affected and unaffected pieces of models from previous iterations. It is anticipated that four different types of relationships will need to be supported: *topological, geometric, parametric and logical* (partly illustrated in Figure 19.8).

Framework Architecture and Data Sharing Issues

The elements of the framework include MCAD, ECAD, a representative MCAE application (thermal analysis) and a representative ECAE application (PSPICE simulation). The purpose of the framework will be to streamline workflow, link data repositories/files, and support change propagation between the applications. At the most basic level, data is exchanged between the applications in a batch mode using standard (AP209, 203, 210) or proprietary format files (Gerber, IDL, DXF).

This is the most common and least efficient situation found in small and medium size companies. In larger companies a PDM system may be used to manage the workflow, versions and configurations using product structure definitions. This may be considered a medium level of integration with a focus on file or document management. PDM setup is a major and costly effort, not easily amenable to changes in workflow or product structure. The deepest level of integration is achieved with all data residing in shared databases, federated or centralized. These have been demonstrated only in academic systems [11]. *While at a theoretical level this architecture sounds elegant, we contend that it is neither necessary nor efficient, nor is it feasible in real world environments involving disparate CAD/CAE tools from different vendors.*

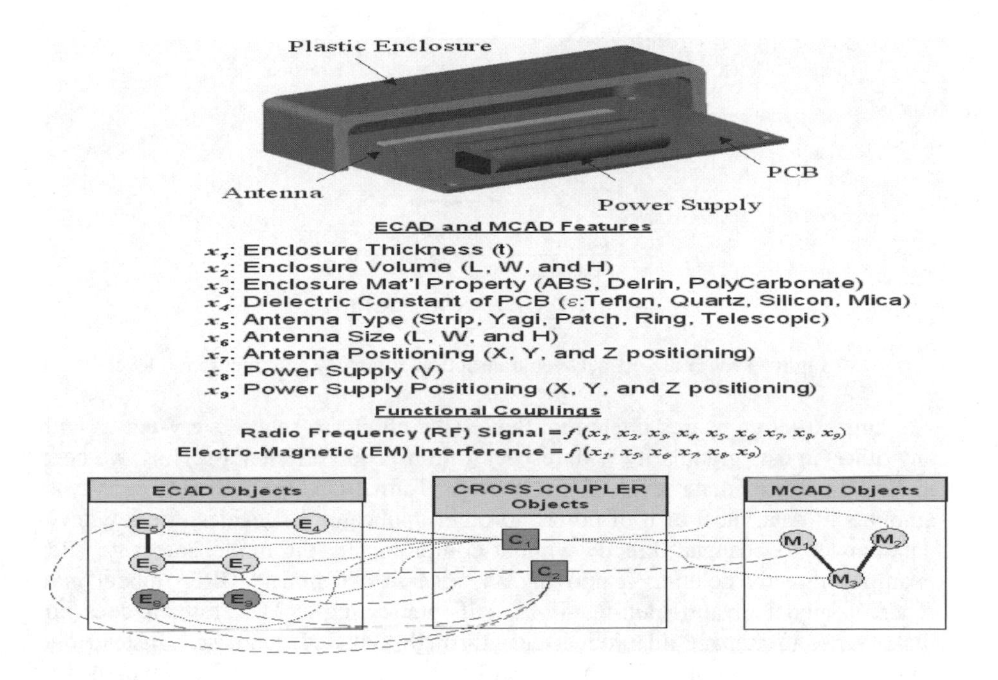

Plastic Enclosure
Antenna
PCB
Power Supply
ECAD and MCAD Features
x_1: Enclosure Thickness (t)
x_2: Enclosure Volume (L, W, and H)
x_3: Enclosure Mat'l Property (ABS, Delrin, PolyCarbonate)
x_4: Dielectric Constant of PCB (ε:Teflon, Quartz, Silicon, Mica)
x_5: Antenna Type (Strip, Yagi, Patch, Ring, Telescopic)
x_6: Antenna Size (L, W, and H)
x_7: Antenna Positioning (X, Y, and Z positioning)
x_8: Power Supply (V)
x_9: Power Supply Positioning (X, Y, and Z positioning)
Functional Couplings
Radio Frequency (RF) Signal = $f(x_1, x_2, x_3, x_4, x_5, x_6, x_7, x_8, x_9)$
Electro-Magnetic (EM) Interference = $f(x_1, x_5, x_6, x_7, x_8, x_9)$
ECAD Objects
CROSS-COUPLER Objects
MCAD Objects

Figure 19.7 Features and Cross Coupler examples

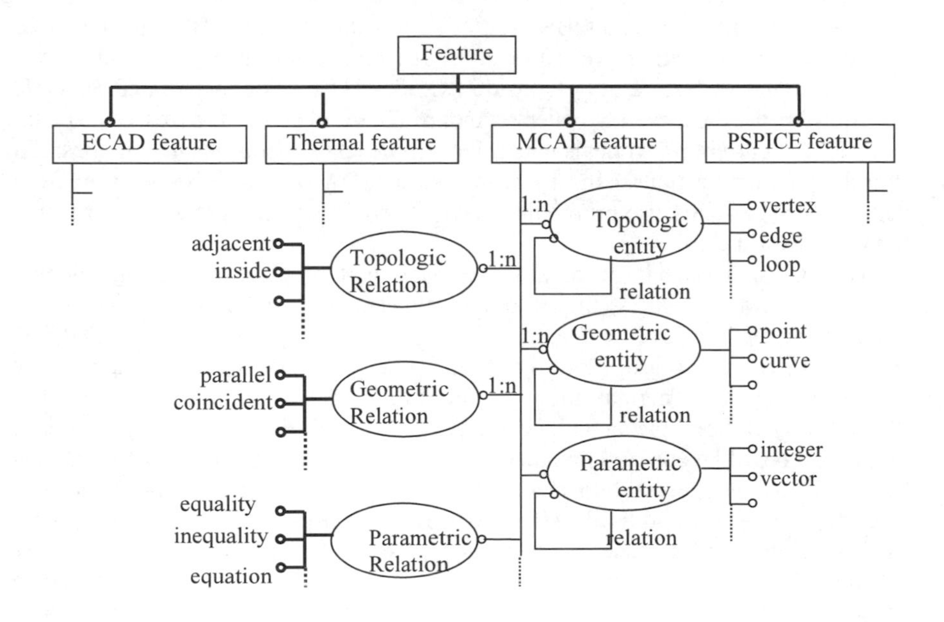

Feature
ECAD feature
Thermal feature
MCAD feature
PSPICE feature
adjacent
inside
Topologic Relation
1:n
Topologic entity
relation
vertex
edge
loop
parallel
coincident
Geometric Relation
Geometric entity
point
curve
equality
inequality
equation
Parametric Relation
Parametric entity
integer
vector

Figure 19.8 Feature definitions

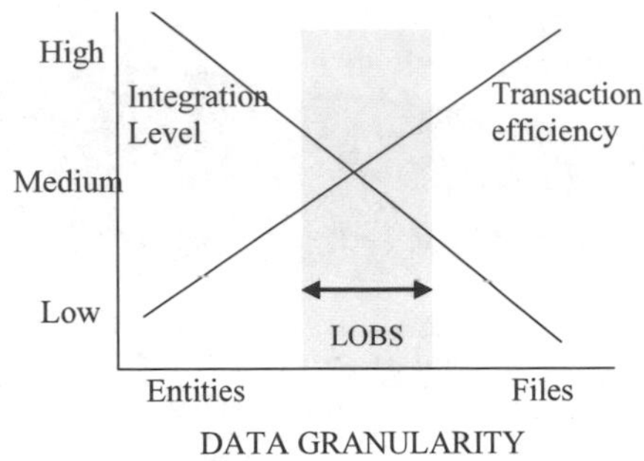

Figure 19.9 Trade-off between transaction efficiency and integration level

Figure 19.9 compares integration level (the ability to relate every entity with any other) to data granularity with respect to transaction efficiency. Thus, we need to find an intermediate level of granularity. Native CAD, CAE files carry vast amounts of data, most of it of no use to other applications. Besides, the low level data cannot be changed directly without compromising the model integrity. For example, directly deleting or moving an edge in a manifold BRep model may render a model invalid (non-manifold, self-intersecting). Therefore, it does not make sense to manage all product data through a shared database. Instead one should identify key objects/attributes that can be used to drive the low level data in application specific models. Creating an intermediate level of granularity between monolithic files and microscopic level data means that we must partition the files into sets of objects that are shared between applications. Let us call these "Large Objects of Shared Data" or LOBS. As far as possible, these LOBS must be based on data standards, such as AP210 or 209. It is expected that cross couplers will relate attributes of feature objects inside LOBS. STEP has concentrated on static file transfer; the database equivalent standard (SDAI) has not progressed very far [SDAI]. The current SDAI deals with microscopic level operations that are best left to individual applications. In any case, use of SDAI requires cooperation from vendors who currently use proprietary APIs. Therefore, this approach is not viable in the real world.

Thermal analysis has become an essential part of electronic design due to increasing power densities, higher reliability demands, and lower limits on die junction temperature. Commercial packages like FLOTHERM [www.flomerics.com] are used for fluid flow and heat transfer analysis by numerical methods. Thermal analysis builds its model from simplified geometry that is derived from the actual geometry. It also requires SMT data, material properties, board layout, and boundary conditions. Depending on the temperatures found, it may be necessary to make changes to the placement of the devices or even the cooling method used. After changes, the analysis is started from scratch. CrossTalk will aim at reuse of unchanged parts of the old models. This is only possible if we associate parts of CAE meshes and their associated properties (material, boundary conditions) with MCAD and ECAD features and procedures for extracting this information and dropping it into the CAE models. If this capability was available it will have uses beyond electronic design.

Constraint Management

CrossTalk will concern itself only with global constraints (inter-model), leaving intra-model constraints to be managed by the application to which they belong (Figures 19.7 and 19.8). For example, constraints applied to a 2D sketch in MCAD are considered "local" and managed within MCAD. On the other hand, PCB boundaries (ECAD) must stay within the geometric envelope from packaging design (MCAD) – this is the type of constraint that is considered global. These constraints may be topological, geometric, parametric (algebraic, differential) or logical. This representation provides a natural way to create cross-couplers between feature attributes. The primary issues for constraint management are representation, validation, solving and conflict resolution. Secondary issues are those related to specific implementation: system for constraint specification and exporting/importing constraints from/to CAD/CAE application tools.

Figure 19.10 shows a candidate design of an electronic-mechanical device, Intel's Personal Server. The arrows shown in the figure indicate geometric couplings between electronic design parameters, such as Reset Button and LED, with their respective mechanical domain counterparts such as Reset Button Access Hole and LED Window. Conflicts may arise when such design parameters are changed dimensionally in one domain without verifying the impact on the related design parameter in the other domain. A simple example would be the re-positioning of the LED on the layout by electrical engineering. If such information is not captured or communicated over to mechanical engineering, the LED window may not align properly with the LED. One might think that constraint management is a well-researched and mature topic and many techniques and solvers are available. However, none operate on ***all*** types of constraints and entities needed in this framework. The characteristics of constraints typical of electronic-mechanical device domain will be investigated from the point of view of solution strategy.

We propose to represent this mixed set of constraints and entities by bipartite graphs [9] where the nodes and arcs are drawn from a standard set of topological, geometric, parametric entities and the relations, respectively. Standard algorithms

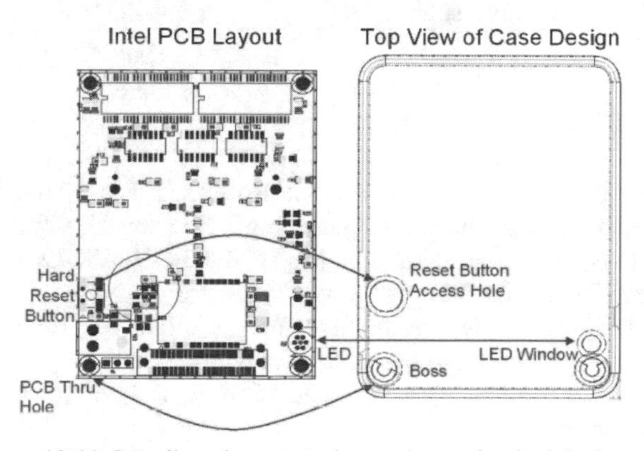

Figure 19.10 Couplings between electronic-mechanical design parameters

are also available for the decomposition of such graphs [8] into not only strongly connected components (SCC), but also into sub-problems that can be sent to different types of solvers. Graph based methods facilitate identification of over, under and improperly constrained conditions, even without solving. This will be used as the basis for constraint specification validation. Many of the symbolic solvers have difficulty with large sets of equations, particularly in the presence of transcendental functions. Additionally, specialized geometric solvers may be more efficient for maintaining geometric constraints, instead of general-purpose equation solvers. DCM3D from D-Cubed is the leading commercial solver now embedded in many MCAD systems [3]. Geometric solvers and solution selectors need to be combined with other types of solvers for CrossTalk.

19.5 Conclusion

In order to address the issue of design cycle compression for the growing market of wireless consumer electronics, multi-disciplinary product realization tools and frameworks are needed to create these complex products and systems. In addition, methods are needed to resolve the interfaces between these multi-disciplinary design and manufacturing teams. Customized, or customizable methods for shared design and constraint resolution is the approach CrossTalk takes in order to blend ECAD and MCAD design tools as well the discipline specific engineering analyses applications associated with them. CrossTalk will eventually abstract from the proprietary or customizable process to provide an open framework for various ECAD-MCAD-CAE applications.

19.6 Acknowledgements

We would like to thank Intel, especially Dr. Trevor Pering, Hewlett-Packard, and the Berkeley Wireless Research Center, especially Professors Jan Rabaey and Bob Broderson, for their continuing support.

19.7 References

[1] Choi, H.J., Panchal, J.H., Allen, J.K., Rosen, D., and Mistree, F., 2003, "Towards a standardized Engineering Framework for Distributed, Collaborative Product Realization", DETC 2003/CIE-48279, *Proceedings of ASME DETC*, Chicago, Illinois.

[2] CIM Data: "Collaborative Product Definition (cPDm) – growth, key players and market share," http://www.cimdata.com/articles/.

[3] 1999, "DCM: D-Cubed," *The 2D DCM Manual Version 3.7.0*, D-Cubed Ltd., Cambridge, England.

[4] Kehmeier, D.J., 1998, "Electrical/Mechanical Design Integration: An Introduction to IDF 4.0 and What it Can Do for You," *Intermediate Design Integration*.

[5] Lee, R.K., Montero, M.G., and Wright, P.K., 2003, "Design Methodology for the Thermal Packaging of Hybrid Electronic-Mechanical Products A Case Study on the Berkeley Emulation Engine (BEE)," DETC2003/DAC-48790, *Proceedings of ASME DETC*, 2003, Chicago, Illinois.

[6] Pahng, G.D.F., Bae, S., Wallace, D., 1998, "A Web-based Collaborative Design Modeling Environment", *WETICE,* pp. 161-167.

[7] 1994, SDAI: "STEP Part 22, Standard Data Access Interface", *ISO TC184/SC4 WG7 Document N350.*

[8] Sedgewick, R., 1998, *Algorithms in C++*, Addison-Wesley.

[9] Serrano, D., Gossard, D., 1987, "Constraint Management in Conceptual Design," Knowledge Based Expert Systems in Engineering: Planning and Design, D. Sriram and R.A. Adey (ed.'s), *Computational Mechanics*, Southampton.

[10] Shah, J.J., and Mantyla, M., 1995, *Parametric and feature-based CAD/CAM: Concepts, Techniques, and Applications*, Wiley, New York.

[11] Sheth, A., and Larson, J., 1990, "Federated Database System for Managing Distributed, Heterogeneous, and Autonomous Databases," *ACM Computing Surveys*, Vol. 22, No. 3, pp. 183-236.

[12] Sriram R., 2002, *Distributed and Integrated Collaborative Engineering Design*, Sarven Publishers, Glenwood, MD 21738.

[13] Thurman, T., and Smith, G., 1999, *Overview and Tutorial of STEP AP210, Standard for Electronic Assembly Interconnect and Packaging Design*, PDES, Inc.

[14] Urban, S., Dietrich, S.W., Saxena, A. and Sundermier, A., 2001, "Interconnection of Distributed Components: An Overview of Current Middleware Solutions," *Journal of Computing and Information Science and Engineering*, Vol. 1, No. 1, pp. 23-31.

20

Integrated Architecture of Geometric Models and Design Intentions

Kazuhiro Takeuchi, Akira Tsumaya, Hidefumi Wakamatsu, Keiichi Shirase, and Eiji Arai

Abstract: Recently, 3D CAD systems have been rapidly improving. However, the principal improvements have been focused on the geometric modeling and developing user-friendly operational improvements, while neglecting improvements to the treatment of the design information and the intention generated in the design process, especially to support the design process flow. In this process, it is important to transmit the design information and intention, considered by each design phase, to the downstream process. In this paper, we explain the framework to handle the design information including designers' intention, and we discuss the architecture that supports the design process flow. First, we describe the integrated model of the geometric model and the design information. We propose the framework to treat various kinds of design information and the intention and present a typical example. Next, we describe the methodology to extend the proposed integrated model and also the architecture to support the design process flow. We define what design information should be handled and explain how to extend the integrated model to support the design process flow. Finally, we describe an application example of design process flow. The result shows an immediate specification of the design information and intention at the downstream process. The system supports the designers creating and modifying the geometric model correctly to satisfy the designers' intention as the design progresses.

Keywords: Design process, Design Information, Design Intention, CAD, Geometric modeling

20.1 Background

In recent years, 3D CAD systems have been rapidly improving. However, the major improvement has only been made for geometric modeling and developing user-friendly operations. Therefore, treating the design information and intention is currently one of the very important problems for CAD systems. Recent studies on 3D CAD system describe the problems to be solved as shown below [1]:

1. Not integrated:
   The system cannot understand state of design process.
   The system cannot support initial stage of design.
2. Not Intelligent:
   The system cannot understand the designer's intension.
   The system does not have design knowledge.
   The system lacks designer's common sense.
3. Poor Human Computer Interaction (HCI):
   The system does not have efficient error detection ability.
   It is necessary to input all designer's information.

A number of studies for conceptual design discussed functional modeling and its intention in the conceptual design stage [2, 3, 4, 5, 6]. The research for synthesis of each functional design was discussed in [7]. Those studies propose how to implement functional intention and discuss about the conceptual design stage.

We introduced a basic framework for integrated geometric model, design information and intention [8]. In this paper, we discuss about the architecture of transmitting the design information and intention that is decided in the conceptual design stage. Applying this architecture to the design process flow, by introducing a progressive refinement of the design details in stages, coupled with integrated system support for this information, enables improved design workflow.

20.2 Objective

In many product designs, after the entire plan is decided, the design is advanced to detail in stages. In this paper, we named such a process to Break-Down design process. In the Break-Down process, it is very important to transmit the design information and intention from the upstream design stage to downstream design stage. Especially, if the design object is complex, it becomes difficult to understand the design intention mutually among designers. In this paper, we discuss about what kind of design information is necessary, and propose the architecture to support the Break-Down design process.

First, we assume the initial stage of design, where the layout of the entire product becomes roughly possible as shown in Figure 20.1.

At this stage, detailed shape is not decided. For example, there are only several lines, work planes, sheets, and simple solids. However, despite their simplicity, the geometries play an important role in setting design information at the initial stage and thus they become a key to determine the entire composition. Lines or sheets sub-assembly or show the some restrictions.

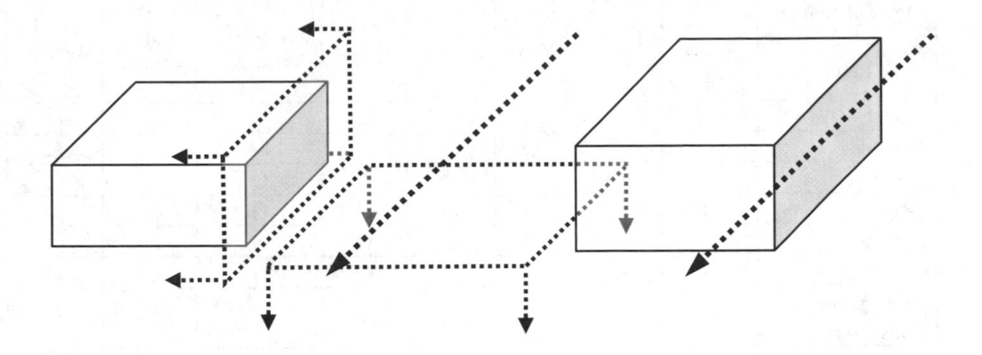

Figure 20.1 Example of the initial stage image

20.3 Basic Framework

To transmit design information and intention, the following requirement must be satisfied.

- Accurately transmit design information, even if any design change occurs in the design process.
- Create a mechanism capable of storing the various types of design information.
- Correlate CAD systems behavior to the implied meaning of design information.

The important items required to satisfy these requirements are described in the following sections.

Understanding Modification

3D CAD systems primarily deal with geometric models. During the process of adding and changing the geometric model, the design information and design intention have to be considered. It is important to consider the geometric model as an object to add design information and intention.

Basic operations in the 3D CAD systems are shape addition/modification and dimension setting/changing. Therefore, CAD system should accurately understand what changes were performed. Table 20.1 shows the targets of design information and intention, and types of changes.

Table 20.1 Target and type of change

Type of change / Target	Marge	Divide	Move	Rotate	Mirror	Geometry change	Value change	Remove
Model	---	---	O	O	O	---	---	O
Feature	---	---	O	O	O	---	---	O
Face	O	O	O	O	O	O	---	O
Edge	O	O	O	O	O	O	---	O
Surface	---	---	O	O	O	O	---	O
Curve	---	---	O	O	O	O	---	O
Arrangement Constraint	---	---	---	---	---	---	O	O
Dimension	---	---	---	---	---	---	O	O
Work	---	---	O	O	O	O	---	O

Handling Design Information Diversity

Table 20.2 shows various kinds of accuracy information, typical examples of design information. These examples show that there are two types of design information; one is attached to a single geometric element and the other details the relationship between geometric elements. For example, surface roughness and flatness are categorized to the former type, and the latter type contains parallelism, coaxiality, *etc.*. Both types of design information should be handled by CAD systems.

System Behavior by Intention and Situation

Peculiar behavior regarding the content of design information and intention should be able to be defined in the CAD systems. For example, the left-hand side of Figure 20.2 shows that the coaxiality is defined between axis-1 and axis-2. The Right-hand side shows that some operation is to be performed on only axis-2. The behavior of the coaxiality information should be different depending on the design intention. In some cases, the system should reject this changing operation. In other cases, the systems should execute the instruction and delete the related coaxiality information with cautionary feedback. Moreover, if the allowable range has been decided, such as the surface roughness *etc..*, the CAD system behavior should be different according to the value.

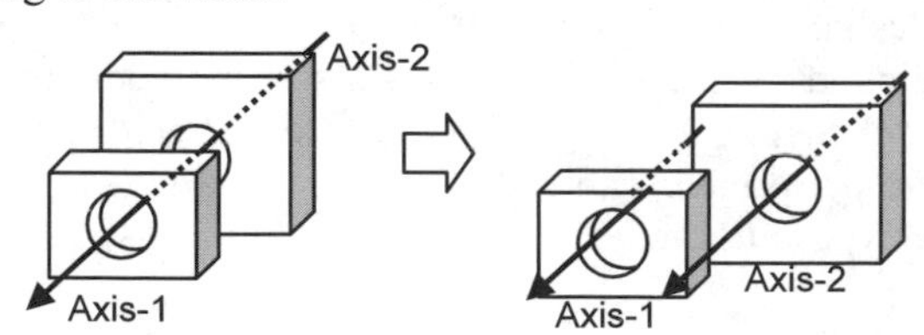

Figure 20.2 Example of geometry change

Behavior Definition for Single Design Information

The single design information is attached to a single element. The system behavior definition for single design information is proposed by evaluating information from the following three points of view:

1) Type of changing:
   These are basic information, such as movement, rotation, division.
2) Value of design information:
   This is a value that design information has, such as accuracy.
3) Characteristic value of element:
   These are mass properties and special vector of each element, such as area, and normal vector.

Table 20.2 Accuracy information

	Accuracy Information		Target Element	
Geometry deflection	Shape	Straightness,	Edge, Plane, Cylindrical.	Single
		Cylindricity, *etc.*.	Free form, surface/wire	Single, Relational
	Posture	Parallelism, *etc.*.	Edge, Plane, Centerline	Relational
	Positional	Location level	Edge, Plane, Cylindrical	
		Coaxiality *etc.*.	Center axis	
		Symmetry	Vertex, Edge, Center axis	
Surface roughness			Face	Single
Dimension Tolerance			Vertex, Edge, Center axis	Relational
Angle Tolerance			Edge, Plane, Center axis	

Figure 20.3 shows an example of behavior definition for the single design information, which conforms to XML format.

```
<behavior definition>
   <name>Volume limitation </name>
      <characteristic value of element editing method=”message output”>
                  <charac ele> volume </charac ele>
            <comparison ope ><!CDATA[<=]]></comparison ope>
                  <comparison val>100.0</comparison val>
            <message>alarm:The volume exceeded 100.</message>
   </characteristic value of element>
</behavior definition>
```

Figure 20.3 Example of behavior definition for single design information

Behavior Definition for Relational Design Information

Relational design information is a concept to treat the design information between targets. The behavior definition for relational design information proposed by evaluating information from another point in addition to the ones mentioned in 1), 2), and 3). Relational or characteristic value between elements shows the spatial relations between them. Figure 20.4 is a typical example of a calculation between characteristic values.

Another typical example of relational design information is the coaxiality or parallelism, which are accuracy related information. Figure 20.5 shows the relational design information of Face-A parallel to Face-B. Figure 20.6 shows an example behavior definition for the relational design information.

Line(Start P1, Vector V1) and Line(Start P2, Vector V2)
Same $V1 \times V2 = 0, (P1 - P2) \times V1 = 0$
Parallel $V1 \times V2 = 0$, Vertical $V1 \bullet V2 = 0$
On same plane $((P1 - P2) \times V1) \bullet V2 = 0$

Figure 20.4 Example of characteristic between elements

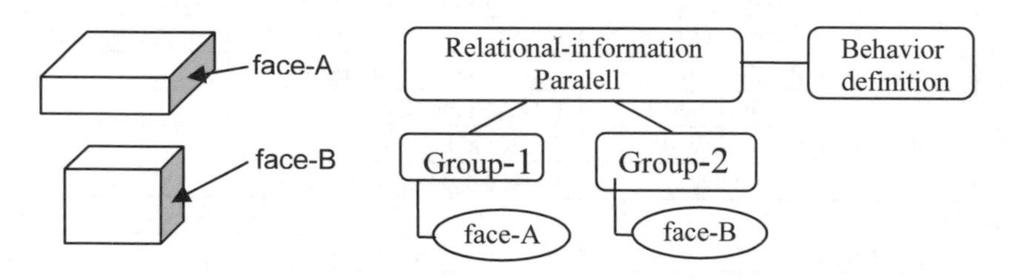

Figure 20.5 Relational design information image

```
<Relational Design Inf. update Relational Key="contact area">
  <characteristic value between elements control
                                   editing method="message output">
    <characteristic value between elements >
         <characteristic value>contact area</characteristic value>
         <Comparison ope > <!CDATA[<]]></Comparison ope >
              <Comparison val> 1.0 </Comparison val >
    </characteristic value between elements>
       <message> The contact area is less than 1.0. </message>
  </characteristic value between elements control>
</Relational Design Inf. update>
```

Figure 20.6 Example of relational behavior definition

20.4 Consideration for Break-down Design Process

In Section 20.2, we assume during the initial design stage, that there are several lines, work planes, sheets and simple solids. However, each geometric element has important meaning. To treat the Break-Down design process, there are several important items.

- The sub-composition should be able to be made from the whole composition; therefore, it is possible to make a sub-assembly from the whole assembly.
- In the upstream design stage, the geometric elements are very simple. However, reflect the design intention, such as restrictions or requirements for the sub-assemblies and the parts.
- It is possible that a single sheet denotes a sub-assembly or a single line denotes a part.

Based on the above-mentioned analysis, we describe the important points to support the Break-Down design process.

Create Sub-assembly

If there are several elements, the element group should be able to be defined as one sub-assembly. Figure 20.7 is an example. By using this function several times, it is possible to make the assembly structure of the arbitrary hierarchy. Creating the assembly structure means that the whole assembly is automatically generated, if an individual sub-assembly is completed.

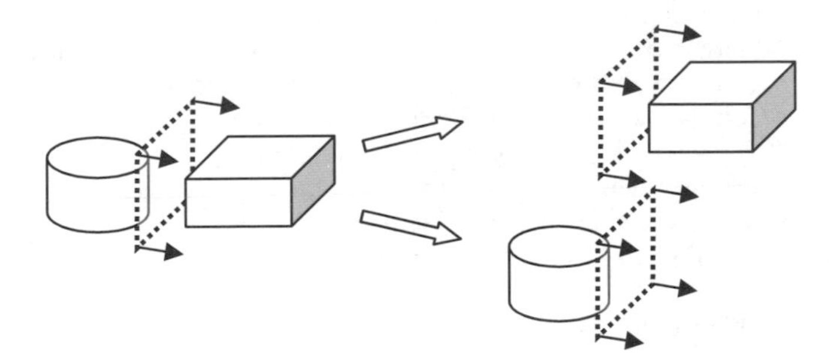

Figure 20.7 Example of the decomposition to sub-assemblies

Region Boundary

One of the typical design information in the upstream design stage is a layout for sub-assemblies. This is to decide the existence region of each sub-assembly. It is very important for designers to transmit this region information. To achieve this requirement, we introduce the region boundary for sub-assemblies and for specific elements as shown in Figure 20.8.

Through region information it is possible to express this by using the relational design information, as shown as Section 20.4. For example, this is the relational design information between a sub-assembly and the elements that show boundaries. Figure 20.9 shows an example of the behavior definition.

It is possible to use the distance in the behavior definition in Figure 20.9 to treat the contact and minimum distance constraint. In the developed system, we are able to use solids, sheets and work plane as region boundaries.

It is possible to use the distance in the behavior definition in Figure 20.9. Thus we can treat the contact and minimum distance limitation. In the developed system, we are able to use solids, sheets, and work plane as the region boundary.

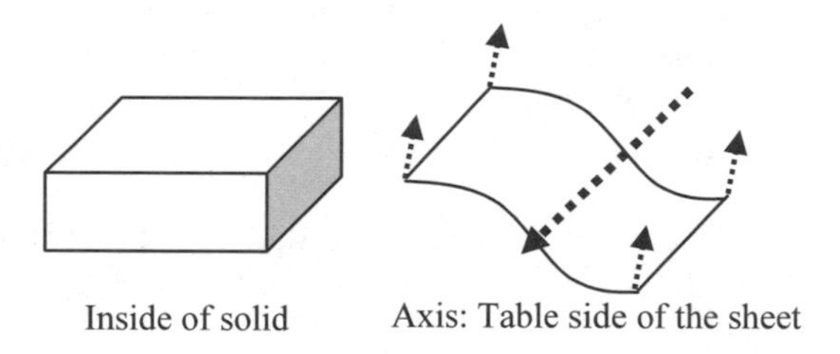

Figure 20.8 Example of the region boundary

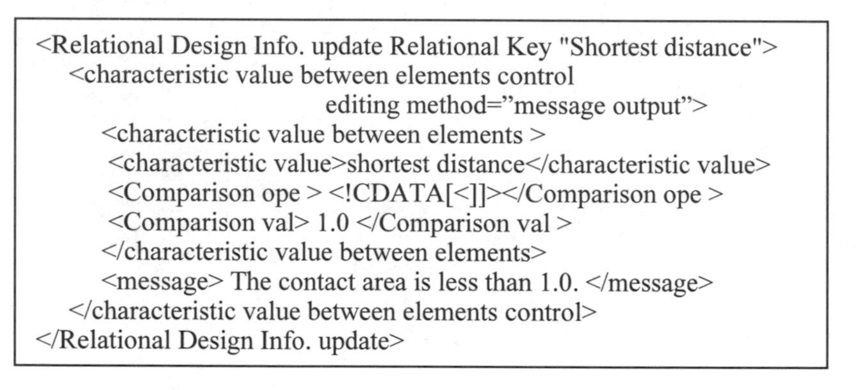
<Relational Design Info. update Relational Key "Shortest distance">
<characteristic value between elements control
editing method=”message output”>
<characteristic value between elements >
<characteristic value>shortest distance</characteristic value>
<Comparison ope > <!CDATA[<]]></Comparison ope >
<Comparison val> 1.0 </Comparison val >
</characteristic value between elements>
<message> The contact area is less than 1.0. </message>
</characteristic value between elements control>
</Relational Design Info. update>

Figure 20.9 Example of behavior definition for the region

Spread Information

In the initial design stage, we can define the single solid model as a sub-assembly and set the weight limitation. In addition, we can set accuracy information to each line, set the parallel to two lines, and define each line as the sub-assembly. There are three consideration points:

- The weight limitation:
  The weight limitation can be defined as the single design information for the sub-assembly.
- The parallelism:
  The parallelism information is the relational information between two lines.

When the geometric model is materialized by using these lines, the axes of this model should be added as a group member of the relational design information.

- The accuracy information:
  The accuracy information should spread to some faces when the geometric model is materialized by using this line as shown in Figure 20.10.

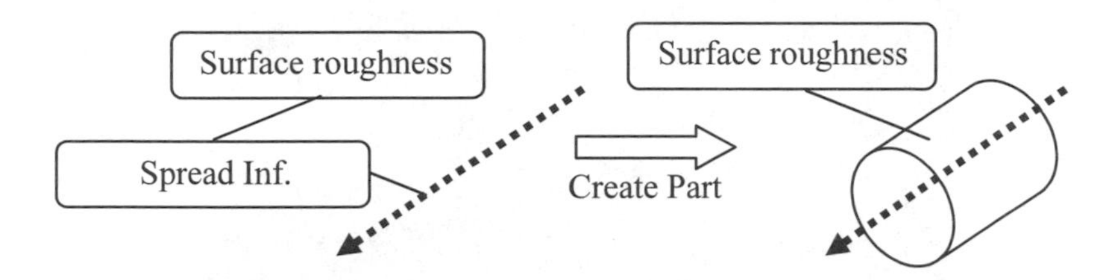

Figure 20.10 Image of spreading design information

20.5 Application

In this section, we apply the integrated model to the Break-Down design process by using the developed system. The following examples show the decomposition, the region boundary and the spreading information.

- Step 1:We consider the design process of the printer. First, the chief designer decides the paper flow and arranges the main components. At this stage, chief designer can define the region boundary or the spread of design information for the elements using the developed system.
- Step 2:Specify the elements where the elements include the region boundary, axes and sheets, the developed system makes each component. Thus, the chief designer makes the structure of an assembly.
- Step 3:The component designer advances the design. By changing the drum in the component, the drum extends beyond the region boundary. The system sends the alarm messages.(Figure 20.11)
- Step 4:To avoid this state, designer changes the dimension value. The system does not issue alarm messages.
- Step 5:The component designer models the roller using the axis. This axis has the spread design information. Therefore, the roller also has the design information such as surface roughness and material (Figure 20.12)
- Step 6:By repeating the operations like step 3, step 4 and step 5, the designer advances the design process. If each individual component is completed, the product is automatically completed because the arrangement of each component has already been decided in step 2.

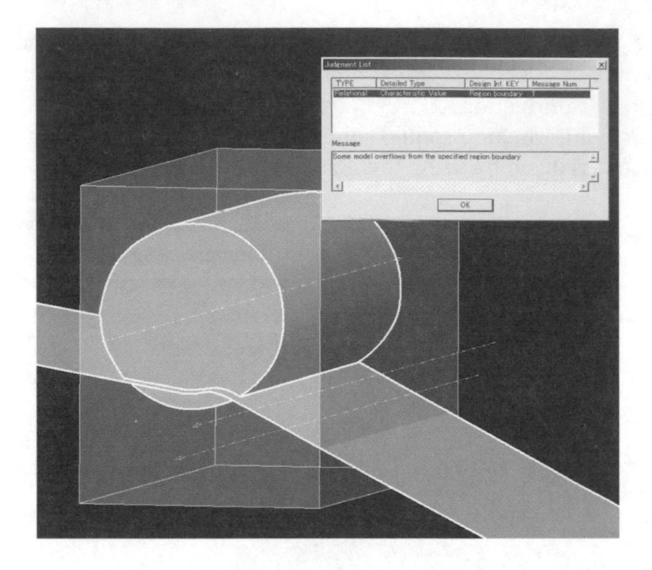

Figure 20.11 Example of system alarm

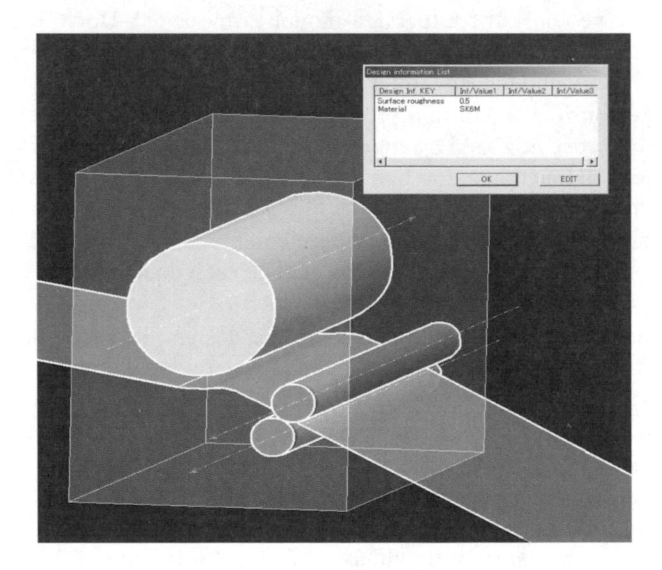

Figure 20.12 Confirm the spread design information of the roller

20.6 Summary and Conclusion

The integrated model for transmission of the design information and intention we discussed. We proposed an architecture consisting of the design information, the behavior definition, and the target. We analyzed the requirements of Break-Down design process and proposed the important design information. Finally, we applied this integrated model to Break-Down design process by using the developed system and showing its effectiveness. In the actual design process, it is very important to transmit design information and intention from the upstream design stage to the detailed design stage. The presented integrated model is one of the effective approaches to support the design process.

20.7 References

[1] Yoshikawa, H., and Tomiyama, T., (ed.), 1889 1990, *Intelligent CAD*, Asakura-syoten, Tokyo, Japan.

[2] Pahl, G., and Beitz, W., 1988, *Engineering Design Systematic Approach*, Springer-Verlag, Berlin, Germany.

[3] Arai, E., Okada, K., and Iwata, K., 1991, "Intention Modeling System of Product Designers in Conceptual Design Phase," *Manufacturing Systems*, Vol. 20, No. 4, pp. 325-333.

[4] Umeda, Y., Ishii, M., Yoshioka, M., Shimomura, Y., and Tomiyama, T., 1996, "Supporting Conceptual Design Based on the Function-Behavior-State Modeler," *Artificial Intelligence for Engineering Design, Analysis, and Manufacturing*, Vol. 10, No. 4, pp.275-288.

[5] Stone, R.B., Wood, K.L., 2000, "Development of a Functional Basis for Design," *Journal of Mechanical Design, and*, Vol. 122, pp. 359-370.

[6] Arai, E., Akasaka, H., Wakamatsu, H., and Shirase, K., 2000, "Description Model of Designers' Intention in CAD System and Application for Redesign Process," *JSME Int. J. Series C*, Vol. 43, No. 1, pp. 177-182.

[7] Chakrabarti, A., (ed.): *Engineering Design Synthesis - Understanding, Approaches, and Tools*, Springer-Verlag, London, UK.

[8] Takeuchi, K., Tsumaya, A., Wakamatsu, H., Shirase, K., Arai, E., 2003, "Expression and integrated model for transmission of design information and intention," *Proc. 6th Japan-France Cong. on Mechatronics*, pp. 83-88.

Part V

Design Management

21

Management of Engineering Design Process in Collaborative Situation

Vincent Robin, Bertrand Rose, Philippe Girard, and Muriel Lombard

Abstract: Product development cycles are greatly shortened and subjected to a growing competitive pressure. In parallel, product and process complexities are increasing. This situation requires new organizational concepts in order to satisfy evolutionary market demand. The various design actors, provided with diverse expertise and culture, are therefore invited to collaborate more closely, in order to perform an effective product design. It is then, that the collaborative design process re–groups actors which have to achieve a common objective: develop a product via interactions, information and knowledge sharing, along with a certain level of co-ordination of the various activities. This paper will show how organization and co-ordination of projects are possible, thanks to the use of design environments, which are adapted to each design context. We will focus particularly on the study of various collaborative forms and collaborative knowledge to manage design environments.

Keywords: Collaborative design, collaborative knowledge, conflict management

21.1 Introduction

The product design phase has been a main research field for many years, due to its influence on enterprise performance. The design process is considered as a set of activities, to satisfy the design objectives and product definition. However, it is not sufficient to focus on product definition only, because the design objectives are constrained by the enterprise organization [1] and by the design steps. Furthermore, they are influenced by technologies or human and physical resources [2]. Design is mainly a human activity and is very complex to understand the activities carried out by designers [3]. Many design models have been proposed [4]. The study of these design models points out that, according to the design type, the design objects are different. When the resolution steps are known (routine design process) the project is structured according to different activities, which transform the product knowledge. In the other cases, design could be considered as a creative or innovative process, and activities do not structure the project. Design must be identified as a process that supports the emergence of solutions [5]. In this case, the design project is organized to favour the collaboration between the process actors and the project manager strives to create design situations that facilitating the emergence of solutions. He/she decides on the adapted organization, favouring collaborative work and supporting the sharing of information and knowledge. This paper focuses on the study of collaborative knowledge, required by the design actors of the design environment, and is implemented to respond to a need for collaboration. We will show how exchanged knowledge during the design process allows a project manager to control the evolution of this design environment. Lastly, we will analyse the influence of this capitalized knowledge, to increase the performance of the resources allocation process during design projects.

The first part of the paper analyses collaboration in design and underlines the importance of the exchanged knowledge during the design process, to increase efficiency of collaborative work. The second part defines the different knowledge exchanged during the co-design process. The third part presents the control of the design environment. This control is based on collaborative knowledge analysis, according to the design environment evolution during the design project progress. The last part of this paper, presents an example illustrating the proposed concepts.

All concepts proposed in this paper are developed through the IPPOP project - "Integration of Product Process Organization for Performance improvement in Design"- http://www.opencascade.org/IPPOP. This project is supported by the French Government, as part of the RNTL program ("Réseau National des Technologies Logicielles").

21.2 Collective Work Analysis

Collaborative design process is not prescriptive, even though a nominal process could be defined. During collaborative work, the designers' tasks are performed in parallel and their results should be convergent to satisfy design objectives. These objectives could be refined as the design project progresses. Therefore, it is important to understand the collaborative design process to be controlled. Rose *et al.*, [6] propose to study the various works performed by co-designers and their occurrence during collaboration. Three main collaborative works are identified: decision-making in collaboration, information in collaboration and management of conflicts during collaboration (Figure 21.1).

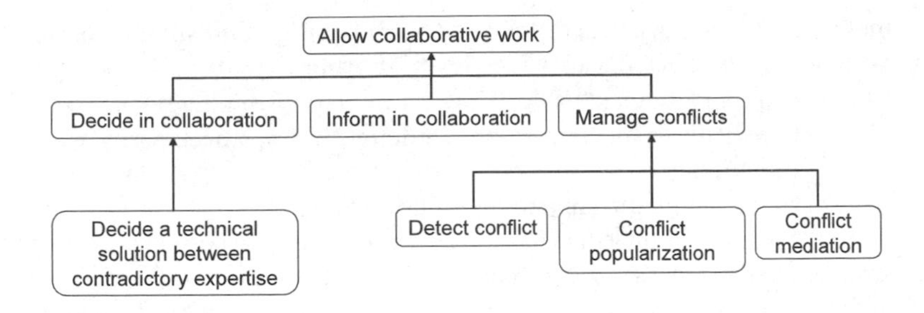

Figure 21.1 Typology of collaborative work

Efficiency of each collaborative work depends on the actor's capabilities to collaborate. Therefore, it is necessary to be able to analyse progress of these situations, according to collaborative context and design objectives evolution. Girard *et al.* [7] have shown how it is possible to encourage collaboration, thanks to an adapted collaboration form (Figure 21.2).

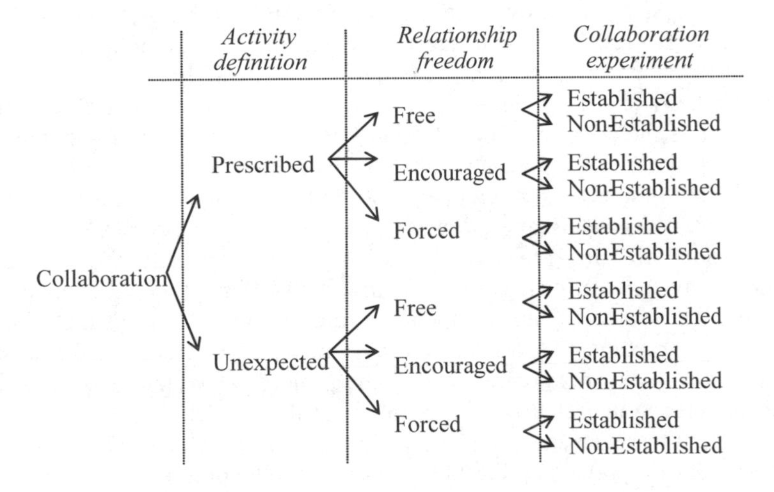

Figure 21.2 Taxonomy of the collaboration

This taxonomy permits the evaluation of the collaborative work according to the activity definition, the relationship freedom and the collaboration experiment of the actors. Consequently the project manager identifies which characteristics they could act on to increase collaboration in order to satisfy design objectives. For each type defined in this taxonomy, some action levels are identified to inform the project manager on their capabilities to change the design context.

Nevertheless, Girard's taxonomy is generic and does not clearly take into account the shared knowledge during collaborative design processes, whereas it is very important to succeed in collaborative design activities. Indeed, Traum and Dillenbourg [8] emphasised that collaborative situations depend on whether participants are at similar levels of knowledge and ability, and whether or not they share common goals. They discussed the level of symmetry of:

- action, participants able to and allow to perform the same tasks,
- knowledge may be at a similar level, but not necessarily within the same field,
- status within the community of collaboration.

Therefore, shared knowledge has to be considered to correctly manage the collaborative design process. We propose to extend the notion of the collaboration experiment, proposed by Girard. First of all, it is necessary to study what "knowledge" is, in a collaboration situation, in order to define precisely the collaborative knowledge. The following section presents a description of the exchanged knowledge during a collaborative design process.

21.3 Exchanged Knowledge During Collaborative Design Process

We can define collaborative knowledge as being the support of a partial and superficial exchange of knowledge among various actors and software tools involved in a project. This exchange authorizes the collaboration among these various participants, coming from different professional horizons, each with a different past, by sharing models or common references in order to perceive a global vision of the problem. Collaborative knowledge has to be considered with the product, process and organization visions. This knowledge is distributed in the context of which the actors are evolving and could appear under a heterogeneous, imprecise and incomplete shape. All actors are supposed to store this knowledge of "popularization" in the application field of the project, allowing a common coherence between various expertises involved during the design process. Therefore, some prerequisite components are requested to characterise this knowledge. Characterization occurs by a common project culture and a common language for each expertise represented in the project [9], which is based on basic knowledge inherent in each discipline involved in the project.

To create this language, it is necessary to have capitalised the pertinent information during previous projects to prescribe collaboration, as well as during the current project, in order to be reactive when a need for collaboration appears. If this has been efficient, the design activities have to be traced to capitalize the design process. Knowledge and collaborative experiment of each actor also have to be taken into account. Consequently, the actor is a predominant factor for the performance of the design process. Knowledge of each actor could be defined as being the meeting of both in-depth knowledge and collaborative knowledge. They re-group all their expertise into one or several given domains. In each situation, collaborative knowledge could be structured in:

- Popularization knowledge acquired by the actor, coming from the other members of the group.
- Popularization knowledge distributed to other actors of the design project. It supports problem resolution.

- Knowledge-being, used by each actor when they have to initiate communication with the other actors. It can be seen as interface ports to reach the other actors of the surrounding context.
- Synergy knowledge, implemented to carry out and maintain the intra-group knowledge exchanges. It's a support of communication.

Depending on the situation, it is necessary to use this kind of knowledge to perform efficient design collaboration. For instance, during a design activity, one actor uses their in-depth knowledge to solve a given problem. They also use some pre-requisite information to accomplish the task. The actor dealing with the design task could solve the problem alone or with someone. In this case, the initial actor must communicate the problem data to the second actor by using synergy knowledge that enables the communication and popularization knowledge to explain it. The resolution of the problem goes through a succession of popularization and mediation actions. Each actor respectively uses their popularization knowledge to communicate with the other and their in-depth knowledge to find a solution to the problem [6]. Moreover, each actor uses different synergy knowledge and pre-requisite information according to the situation. At the end of the collaborative resolution process, a knowledge set is generated, related to the retained solution and the historic resolution. This generated knowledge not only contains product information (structural definition, calculation results, machining process, *etc.*.) but also information about the process and the organization adopted to solve the problem. Figure 21.3 shows different knowledge types involved during collaborative design. Among the various kinds of knowledge identified above, it is essential to capitalise the production of the generated knowledge regarding a given design project, by structuring the exchanges of popularization knowledge.

The following section proposes integrating the collaborative knowledge in the management of the collaborative design process.

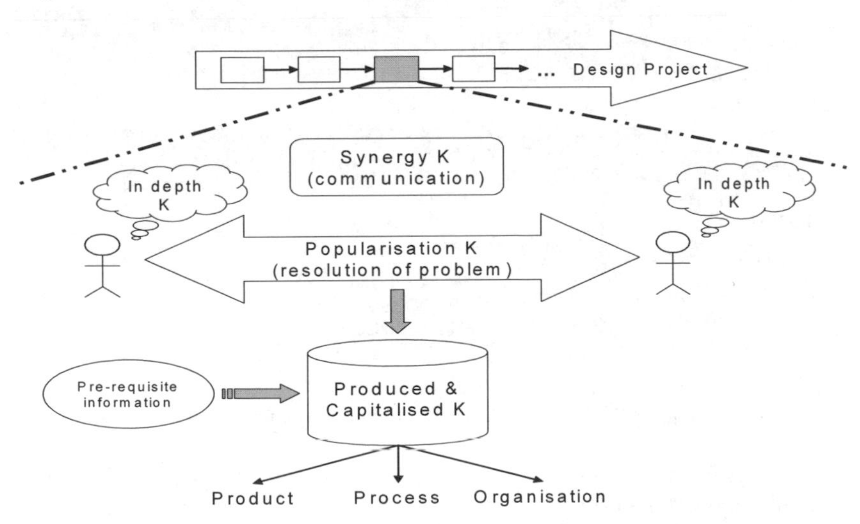

Figure 21.3 Exchanged knowledge during collaborative design process

21.4 Control of Collaborative Design Process

The design process has to respond more and more to restrictive cost, delay and quality objectives; designers are increasingly dislocated through the extended enterprise; technologies are more and more integrated. In this context, design control should be more reactive and take into account external constraints. Therefore, the collaborative design processes control requires an understanding of the context, in which those processes take place [10] in order to modify them to facilitate the actors' work. The GRAI model [11] offers a framework to control the creation, the deployment, the follow-up and the evolution of the adapted design context to improve collaboration. The GRAI reference model [12] describes the engineering design system as composed of 3 subsystems: the decision system, the technological system and the information system. The project manager's decisions to organize the technological system are structured according to time criteria (Horizon-Period), defining the strategic, tactical and operational levels (lines), and to functional criteria defining products or project-oriented decisions (columns) (Figure 21.4). In this structuring, intersections between lines and columns represent a decision centre and the biggest arrows (vertical or horizontal) represent decision frames. A decision centre describes the way a project manager will take to make decisions.

At a specific decision-making level, decision centres control the technological system broken down into design centres. Each design centre receives a design frame from the decision centres to specify its design context. A design centre is a local organization and is responsible for a set of design objectives. The structuring in a design centre and the definition of the design frame are decided by the project manager (or a group of people) who is responsible for the decision-making level. A design centre is the place for collaborative work.

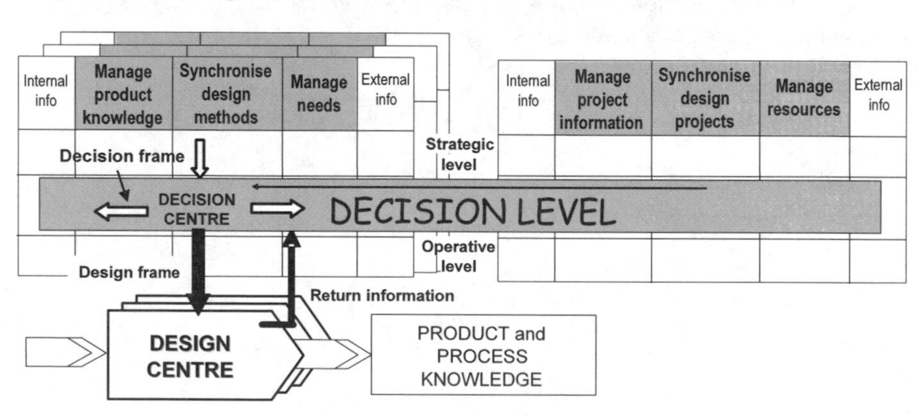

Figure 21.4 The GRAI model to control design centres

The design context may change as the design project progresses. Therefore, a design environment is defined as the context in which the project manager decides to place design actors in order to achieve the assigned objectives.

Design Environment

Robin *et al.* [13] have defined a design environment as the actors' context of work, developed in order to optimise performances relative to customers' or enterprises' expectations. This permits the project manager to promote collaboration between actors, during the progress of the project. Creation and deployment of a design environment obliges the analysis of the design situation it has to optimise. A design situation is defined as: the state of the technological system at a specific point of time [14]. A design environment will be defined as a combination of many parameters, which will evolve during the project's progress, according to the design system situation. The management of design environments consists of a continuous phase of adjustment and evolution of the environment according to the design situation.

The dynamic of design environment's management is based on four main phases [15]:

1. Identification of the need for collaboration,
2. Description of the as-is design situation,
3. Analysis and the comparison of this as-is situation with the objectives of the design system in order to make decision using action levers,
4. Implementation of the new adapted design environment to change efficiently the design context.

Nevertheless, it is the quality of the knowledge exchanged between the design actors that will influence the evolution of the design environment. Indeed, results of the collaborative design activity directly depend on the relationships between the design actors.

Analysis of Exchanged Knowledge to Control Design Environment

Integration of exchanged knowledge in the design environment model could be made into the description of the design situation and during the control of the evolution of the design environment.

The description of the as-is design situation is used to develop a design environment. This description takes into account the actors' experience concerning similar projects, their knowledge and their socialization. We suggest that popularization knowledge, coming from the other members of the group and is distributed to the other actors of the project, has also to be integrated into a general human resources description. It permits defining collaborative knowledge, which is described in Section 21.3. It will subsequently be possible, after many projects' capitalization, to suggest pertinent information in the design environment to favour creation of a common culture and of a common language for each expertise represented in the project. This information will enable a creation of collaboration between actors. We therefore propose to complete the description of the as-is design situation with characteristics concerning knowledge:

- Actors, particularly their roles and their uses in the design process. Their experience(s) concerning similar projccts, their knowledge and their socialization have to also be taken into account. The objective is to properly adapt their work environment according to their needs.

- Product, according to its nature, its complexity, its status in the process and its interfaces.
- Process and in particular the design approach, the design type (routine, innovative or creative) and the collaboration type.
- Material and financial resources (business premises, computers, budget,…).
- Constraints of the enterprise and particularly, constraints of its environment.
- Popularization knowledge, knowledge-being and finally, synergy knowledge, which permit to help the decision-maker choose an adapted collaboration form and help actors communicate and collaborate.

Previous work on the conflict resolution domain enables us to define a dynamic protocol for conflict management in product design [6]. This dynamic protocol was improved by proposing a data model to depict the various states of the product [16]. Nevertheless, this protocol and data model could be generalized to every type of situations in collaborative design. The dynamic protocol presented in Figure 21.5, which takes place in the design environment, is divided into three (3) sequences:

- First, an initialization sequence, corresponding to the implementation of the design environment.
- Second, the main phase, based on a stage of popularization/mediation activities. It corresponds to the decision-making phase of problem solving. It consists of explaining the current problem by using elements and arguments from popularization knowledge for the popularization stage; proposing and arguing about alternative solutions to fit the problem at hand for the mediation stage.
- Third, the closing phase consists of informing the various interested actors involved in the solving process, if a solution is commonly accepted after *n* iterations. Otherwise, in case of failure, this means that the design environment is not well adapted. Consequently, the problem is brought to the design centre in order to inform the decision maker of the necessity of a new design environment.

In order to run efficiently the popularization/mediation protocol, a subscription list of potential interested actors must be set up during the definition phase of the design environment. This list is obtained by matching the availability matrix of each actor with a competence matrix and a responsibility matrix. The competence matrix selects the actors of the projects by their effective skills and competencies in a specific domain. The responsibility matrix selects the responsible person for a part of the product to design and specify the users. Those users can access this entity and the authorizations that have been granted to them (create, modify, delete or just read). The matrices are based on the information collected during the capitalization phase of previous projects. In this case, the follow-up and the capitalization of the process described in Figure 21.5 are very valuable. They allow the project manager to progressively complete each matrix by applying pertinent performance indicators on actors' work, in order to precisely define the information contained in the description of the collaboration experiment of each actor (Figure 21.2).

The following of the dynamic protocol of popularization and mediation process, subsequently permits:

- to follow the evolution of the design environment,
- to complete the different matrices established, to define actors' availability, competencies and responsibilities,
- to refine information about the collaboration experiment of each actor and about the collaboration in the teams,
- to build-up and empower more efficient work teams in an adapted design environment together with an adapted collaboration form.

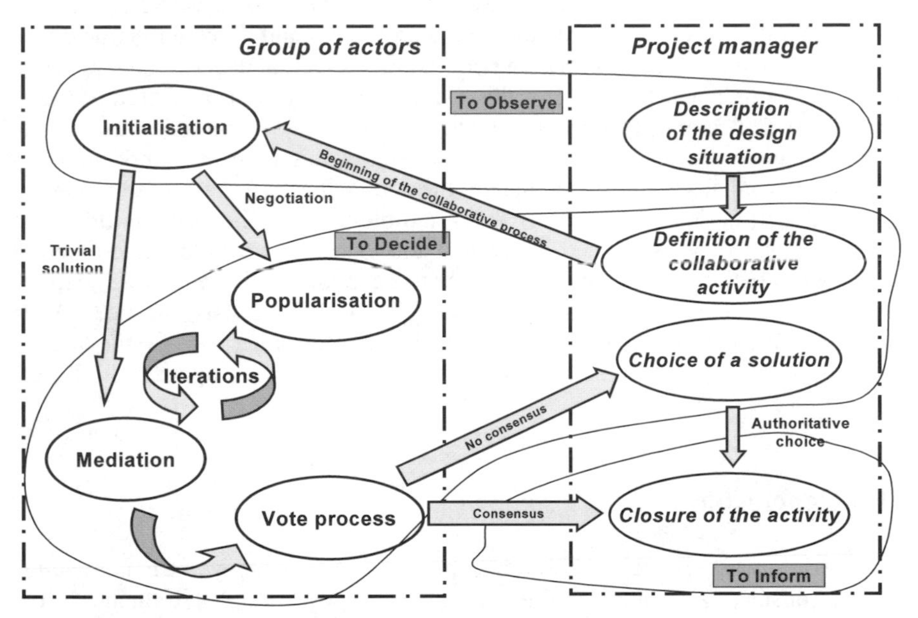

Figure 21.5 Dynamic protocol of popularization/mediation process

21.5 Example

The example that illustrates the proposed concepts is developed in the IPPOP project. It concerns the design of the fixing between the rack-and-pinion and the support of a translation system of a bowl on a mixer. This design phase is representative of the needs for collaboration that could appear during a design project. Three design actors are involved to co-design the fixation: a design expert, a manufacturing expert and a material expert. In our example, IPPOP identifies that the third expert modifies data, which concerns the material of the fixing system. As this data is critical to the manufacturing methods engineer, a conflict could appear between two experts and IPPOP notifies the project manager. He/she creates and deploys a design environment that groups the three experts together, in order to develop a dynamic of collaboration between them to solve the conflict. Before the creation of the new design environment, the design process was sequential and the design activities' sequence was not precisely defined. Now, thanks to IPPOP, a project manager can follow the evolution of the design process, in the design environment frame, and they are able to capitalize this process. IPPOP collects and distributes information concerning product, process and organization. Consequently, the project manager could complete a different matrix concerning actors and refine their knowledge of the actors. Thanks to the dynamic protocol of popularization and mediation process, they can complete knowledge on the collaborative experiment of each one and satisfy needs for information. The last iteration summarises the various stages of the solving process, while embedding the collaborative knowledge used to solve the problem and agree on the proposed alternative solution. This information permits the project manager to be reactive to the group's needs and to capitalise this experiment to reuse it in a similar future project. At this moment, since the manager will have to optimise the design process, the contents of the design framework and the team composition by considering capitalised information about the actors, they will be able to propose a more efficient design environment. This example suggests interests of IPPOP to put into evidence the need for collaboration and to supply a detailed description of the design situation. IPPOP permits the user to be more reactive and more adaptive in the creation phase of the design process, in order to satisfy the need(s) for collaboration between actors. This also increases the reactivity of the project manager towards a potential conflict, and/or when a conflict appears during the design process.

21.6 Conclusion

Engineering design processes are very complex, and now-a-days, it is not enough to only consider the design activities' results to improve their performances. It is necessary to manage and to capitalise on relationships as well as to exchange knowledge between actors and more generally, the design process as a whole, with a particular attention to the organization that was set up to satisfy the objectives. Therefore, the organization has to integrate aspects centred on the actors, in order to be reactive and efficient, considering the design process evolution.

21.7 References

[1] Mintzberg., H., 1989, "Le management : voyage au centre des organizations," *Les Editions d'Organisation*, Paris, France.

[2] Wang, F., Mills, J.J., Devarajan, V., 2002, "A conceptual approach managing design resource," *Computers in Industry*, Vol. 47, pp. 169-183.

[3] Gero, J.S., 1998, "An approach to the analysis of design protocols," *Design Studies,* Vol. 19, No. 1, pp. 21-61.

[4] Love, T., 2000, "Philosophy of design: a meta-theoretical structure for design theory," *Design Studies*, Vol. 21, No. 3, pp. 293-313.

[5] Tichkiewitch, S., 1994, "De la CFAO à la conception intégrée," *Revue internationale de CFAO et d'infographie*, Vol. 9, No. 5.

[6] Rose, B., Gzara, L., Lombard, M., 2003, "Towards a formalization of collaboration entities to manage conflicts appearing in cooperative product design," *Int. CIRP Design Seminar*, May 12-14, 2003, Grenoble, France.

[7] Girard, Ph., Robin, V., Barandiaran, D., 2003, "Analysis of collaboration for design coordination," *10th ISPE Int. Conference on Concurrent Engineering: Research and Applications (CE'03)*, July 26-30, 2003, Madeira, Portugal.

[8] Traum, D.R., Dillenbourg, P., 1998, "Towards a Normative Model of Grounding in Collaboration," In working notes, *Workshop on Mutual Knowledge, Common Ground and Public Information (ESSLLI-98)*.

[9] Midler, C., 1998, "*The Automobile that Never Was: Project Management and Corporate Transformation,*" Inter Editions, Paris, France.

[10] Chiu, M.L., 2003, "Design moves in situated design with case-based reasoning," *Design Studies*, Vol. 24, pp. 1-25.

[11] Girard, Ph., Eynard, B., 1999, "Proposal to control the systems design process: application to manufactured products," In: *Integrated Design and Manufacturing in Mechanical Engineering*, J.L. Batoz *et al.* (Ed.), Kluwer Academic Publishers, pp.537-544.

[12] Girard, Ph., Doumeingts, G., 2004, "Modelling of the engineering design system to improve performance," *Int. J. of Comp. & Ind. Eng.*, Vol. 46, No. 1, pp 43-67.

[13] Robin, V., Girard, Ph., Barandiaran, D., 2004, "A model of design environments to support collaborative design management," *5th Int. Conf. on Integrated Design and Manufacturing in Mechanical Engineering (IDMME 2004)*, April 5-7, 2004, Bath, UK.

[14] Eder, W.E., 1990, "A typology of designs and designing," *International Conf. on Engineering Design (ICED'03)*, August 19-21, 2003, Stockholm, Sweden.

[15] Robin, V., Girard, Ph., Barandiaran, D., 2004, "Performance evaluation of the collaborative design process," *11th IFAC Symp. on Information Control Problems in Manufacturing (INCOM2004)*, April 5-7, 2004, Salvador Bahia, Brazil.

[16] Gzara, L, Lombard, M., 2003, "Cooperative design of mechanical product: specification for a conflict management support system," *Computational Eng. in Systems Applications (CESA'2003)*, July 9-11, 2003, Lille, France.

22

Requirements Management for the Extended Automotive Enterprise

Rajkumar Roy, Clive I.V. Kerr, and Peter J. Sackett

Abstract: The evolution of product requirements in an automotive extended enterprise often involves numerous time consuming interactions between the vehicle manufacturer and their suppliers. It is necessary to manage these interactions and the associated design information in order to ensure transparency such that engineering designers are informed about any changed requirements. One of the avenues currently taken by the automotive industry is the implementation and integration of requirements management into the product development process. This paper will present an electronic requirements management framework that will represent the next step in this digital environment.

Keywords: Product development, Web-based collaboration, Requirements management

22.1 Introduction

The automotive industry relies heavily upon the application of digital product development technologies. The use of such technologies has greatly contributed to the reduction of lead-time. For example Audi has reduced their lead-time, from styling freeze to start-of-production, down from five years to less than two years [1]. However, market forces are driving the necessity to reduce this lead-time further. In order to compress the product development time it is therefore necessary that the entire process be managed better and integrated fully with suppliers since upwards of 70% of the development is performed by the supply chain.

When working collaboratively in an extended enterprise the key enablers are a clear and shared understanding of the requirements at the start of a project and an in-built flexibility for handling changes in these requirements. However, the area of requirements management has been overlooked with the advances made in the digital product development environments utilising CAD/CAM/CAE and PDM.

To address this issue Nissan Technical Centre Europe, Johnson Controls, EDS and the Society of Motor Manufacturers and Traders (SMMT) are sponsoring the electronic Requirements Management (e-RM) project at Cranfield University on 'Integrating requirements in digital product development for the automotive industry.' The aim of the e-RM project is to improve the business capability to develop, capture and manage requirements for engineering designs in the automotive extended enterprise using a digital process. The benefit to industry will be the visibility of requirements during competitive tendering, design, development and validation testing. This will provide both the vehicle manufacturer and their suppliers with an ability to react more effectively as problems arise and to eliminate many changes during the trial build phase. This paper will present the e-RM framework for the integration of requirements generation, dissemination and amendments within an extended enterprise through a web-enabled architecture. Such a tool will allow the vehicle manufacturer to control the evolution of a product requirement.

22.2 Product Development in the Automotive Extended Enterprise

In today's global economy, vehicle manufacturers are facing increased pressure to satisfy market demands. One avenue for gaining significant competitive advantage, in such a competitive arena, is through extensive collaboration with suppliers. Automotive companies are thus adopting the extended enterprise model as depicted in Figure 22.1. Toyota and DaimlerChrysler are two such organisations that have successfully implemented and realised the benefits of an extended enterprise [2]. This offers the vehicle manufacturers great possibilities for gaining access to specialist knowledge and capabilities, to spread and share costs and risks, and to better exploit the expertise of their suppliers [3-5]. For the total costs involved in automotive manufacture, it is estimated that 60% are those of suppliers and service industries [6]. Vehicle manufacturers not only outsource the production of parts to their suppliers, but also the associated development of those parts and increasingly the development of complete subsystems or modules.

The automotive industry has come a long way in implementing digital product development across the supply chain. There have been great efforts made in the harmonisation of CAD systems, the utilisation of simulation tools and the realisation of the digital mock-up (DMU) for a complete vehicle. PSA Peugeot-Citroen, for example, use their "Co-Conception" application for digital modelling [7]. This application allows PSA's suppliers to discuss the impact of design modifications graphically; thus, reducing the time taken for checking designs, which then results in improved productivity in their design conception process. Additionally, Renault uses their intranet portal with their suppliers for the real-time sharing and consultation of the DMU [8]. Designers can conduct space analysis and fitting simulations to determine component distances and potential clashes. For example, Chrysler's Crossfire project used no physical prototypes for development support [9]. Instead, there was the intensified use of simulation tools such as Ramsis (for ergonomics), the DMU for collision examinations and verification of assembly and disassembly, and FEM simulations for functional configuration. Initial tests were only made with vehicles from the pre-series' production [9].

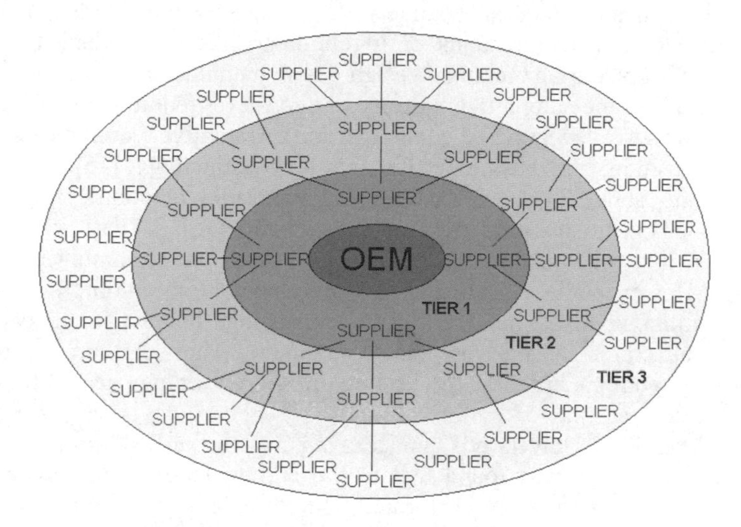

Figure 22.1 An automotive extended enterprise

Design engineers have numerous computer-aided tools for modelling, simulation and analysis that support their efforts at the detailed design stage. Design information management requires similar development and automation. Requirements management is still very manual and time consuming. It is necessary to formalise and automate the process to reduce the vehicle development time and reduce cost. This will provide much needed support to the engineering designers during conceptual design. Verma & Wood [10] state "computer support for future product development must focus on the conceptual design stage where informed decisions make the most impact on design." It is of course the product requirements that drive the development and product design [11]. Neelamkavil & Kernahan [12] have identified that one of the limitations inherent in almost all the current CAD/CAM/CAE systems is the difficulty in mapping and managing of requirements data resulting in only restricted support in the early phases of product development. Additionally, although PDM systems have the advantage of facilitating data access across the supply chain such systems have virtually no requirements management support features embedded into them [12].

The Management of Product Requirements

One of the most critical aspects of the design and development process of any product, not just a motor vehicle, is the effective management of the requirements and the associated communication of the requirements between the stakeholders, whether it be just internal departments or an entire extended enterprise. Requirements are "an elaboration, expansion and translation of the problem definition into engineering terms" [13]. This engineering definition must then be realised into a delivered product. However, requirements evolve throughout the lifecycle of the product to reflect the changing needs of the operational environment and associated stakeholders. For example, the product may have adapted due to new requirements or to eliminate mistakes, whether these be omissions or errors, during the initial design before manufacturing [14].

With regard to the automotive industry, the top-level requirements for a motor vehicle amounts to a few hundred qualitative and quantitative requirements relating to market position, the business case and new technology [15]. These are translated into about 2,000 measurable and verifiable functional requirements relating to characteristic properties of the vehicle [15]. These can then be decomposed into requirements for individual systems and communicated to the suppliers. However, the evolution of a requirement in an automotive extended enterprise often involves numerous time consuming interactions between the vehicle manufacturer and their associated suppliers. Thus, the current processes for managing requirements are experiencing difficulty keeping pace with the drive towards shorter lead-times.

From Cranfield University's undertaking of the e-RM project for the automotive industry it was found that in the current requirements management activities there is a lack of a formal and structured representation of the requirements. In the preparation of the requirements, the vehicle manufacturer is reliant on their employees' knowledge on where the requirements data is and how it should be compiled. There lacks a single source acting as a central information warehouse for the preparation and processing of the product requirements. Additionally, the majority of the requirements documentation is paper-based and in some instances the changes made to a requirement are not easily identified let alone tracked. For example, the supplier will have to go through a complete specification to identify the changes by making comparisons to the original paperwork. Also due to the nature of the paperwork, changes are usually sent out individually and serially. This results in suppliers receiving changes in rapid succession since the paperwork associated with individual requirement changes are not combined and issued as a package. Further, the paperwork tends to lag and when it 'catches up' it is usually academic since the work is already completed or, in the worst case, been superseded by a change. There is a lack of a visible audit trail through the supply chain since there is no real mechanism for readily identifying where the requirements documentation is and its current status. Thus, the primary problem for requirements management in the automotive industry is the administration and dissemination of product requirements through the extended enterprise.

Engineering design research for the aerospace industry as report by Eckert *et al.* [16] has identified complementary findings. There was a reported lack of status information and a lack of awareness of information history. For example, designers "often didn't know where items of information such as specifications and parameter values" came from. Eckert *et al.* [16] also acknowledged that the tracking of information was especially difficult across organisations. A requirements management case study at an aerospace OEM reported that personnel stored copies of the requirements and specifications on their personal computers resulting in engineers working with different versions [17].

22.3 Electronic Requirements Management

"Engineering design is an information intensive activity" [18]. Directing the right information to the right person at the right time is a crucial issue and this is especially critical when dealing with the requirements of a product. According to Thomson [19], a key benefit of requirements management is the establishment of traceability from the original need and product specification through the lifecycle to the completed deliverables and acceptance criteria. According to Zhang *et. al.* [20], web-based product information sharing is a foundation for collaborative product development. Thus, one avenue to achieve improvements in the automotive product development process is the e-enabling of requirements management using a collaborative web-based architecture.

An e-RM enabled platform is the next evolutionary step in digital product development for the automotive extended enterprise. The aim of such a platform is to provide a fluid and seamless tool for the handling and communication of product requirements thus allowing their greater availability, distribution and sharing in the extended enterprise. The perceived benefits of an e-RM platform are:

- Greater consistency and integrity of requirements since the associated data is no longer stored in separate locations, across various platforms and in different formats.
- Greater visibility of product requirements, since all stakeholders in the extended enterprise have access to, and can retrieve, information in a timely and accurate manner.
- Greater ownership of requirements data since the origin of the requirements is recorded and easily identified.

With regard to the state of the e-RM project, the AS-IS and TO-BE models have been completed. The project work is currently focused on developing a prototype of the system for testing.

The e-RM Framework

To achieve the benefits of electronic requirements management, a framework is being developed in order realise the e-RM platform for utilisation in the automotive industry. Figure 22.2 provides an overview of the web-based integration framework for e-enabling requirements management in the extended enterprise. At the centre of the framework is the vehicle manufacturer who houses the 'global' requirements management system. This system is the one single source of product requirements information for a vehicle programme or project and is based on an ontological structured 'global' requirements repository. Thus, the structuring and documenting of the requirements is in a common standardised electronic format. The vehicle manufacturer then allows their suppliers authorised access limited to the modules or subsystems to which they are responsible for producing. Thus, the suppliers have a portion of these requirements data housed in their own 'local' clients for internal dissemination in their respective organisations. A secure web-based front-end is used by both the vehicle manufacturer and suppliers for the uploading, browsing and downloading of the product requirements for specific parts. Using the requirements repository as the single source of information, requirements changes can be automatically updated through the extended enterprise to the affected suppliers. This automatic provisioning of only the actual updated change, as opposed to the whole specification, will reduce the manual collection effort and workload through the use of a front-end reporting tool. An automated set of business process procedures will control the aspects of creating and maintaining agreement of the requirements together with the associated decision-making and communication in the extended enterprise. Instead of individual changes being issued serially, the requirements processing capabilities of the e-RM system will highlight how several changes can be concurrently compiled and released as single combined issue. Additionally, there will be links in the electronic requirements documentation for the CAD and styling data. Interfaces to CAD/CAM/CAE and PDM systems will be used in order to facilitate a complete encapsulated requirement information package containing 2D drawings, 3D models, FEA data, bill of materials and even test data. The requirements repository is central to the e-RM framework (Figure 22.2) and the next section will discuss how this can be realised such that a common understanding of the requirements is achievable between the vehicle manufacturer and their suppliers.

Ontology-based Requirements Repository

According to Toye *et al.* [21] design occurs as a result of reaching a "shared understanding" of the design problem, the requirements and the process. However engineering designers, from the vehicle manufacturer and the supplier organisations, bring their own language and perspective to a project. This is termed "design identities" by Kilker [22] and may result in ineffective collaboration. In the domain of requirements management this is apparent in the lack of a formal and structured representation of the product requirements. Agouridas *et al.* [23] also remark the lack of defined structures in requirements and a lack of a rigorous means for their classification. Additionally, there are a number of commercial requirements management tools available such as Telelogic's DOORS and EDS's SLATE however they do not have a mechanism for controlling the language that is used to represent the requirements. In the automotive extended enterprise, this results in a lot of manual effort being spent searching, interpreting and transforming product requirements data. To address these issues in requirements management, the proposed e-RM framework as presented in Figure 22.2 is based on an ontological structured 'global' requirements repository. This 'global' repository is housed and maintained by the vehicle manufacturer.

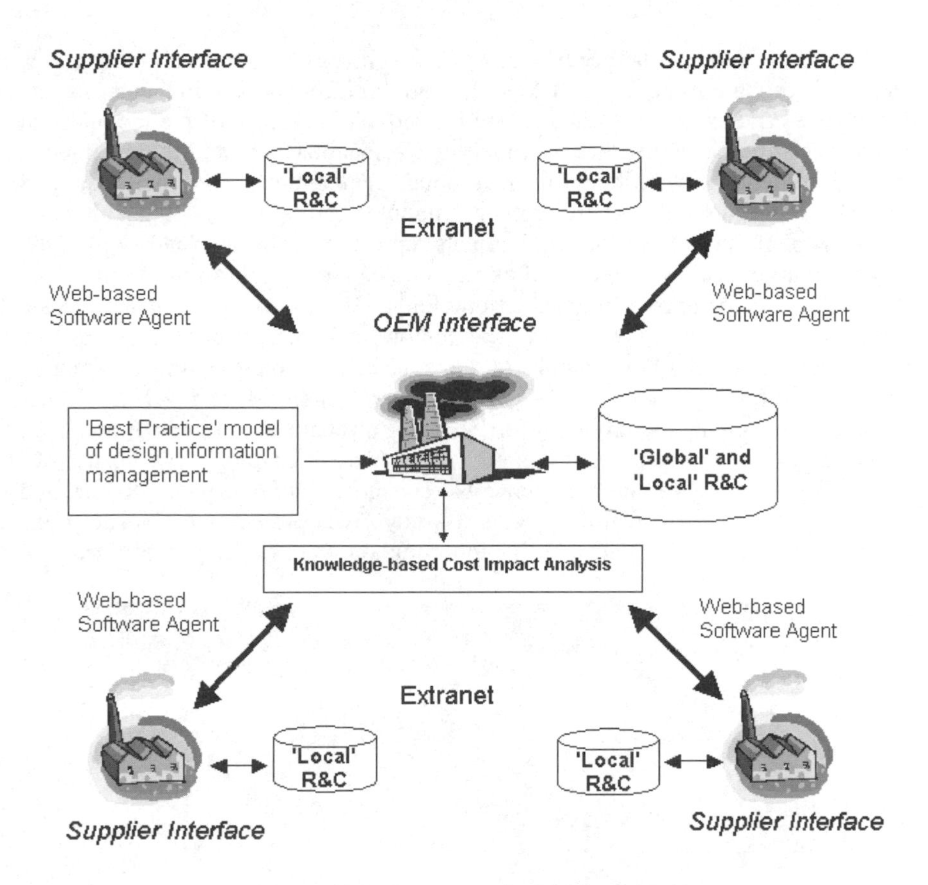

Figure 22.2 The proposed e-RM framework

With regard to utilising ontology in the e-RM framework the most applicable definition in the literature, which is also the most referenced, is Gruber's [24]: "An ontology is an explicit specification of a conceptualisation", where a conceptualisation is a set of definitions that allows one to construct expressions about some physical domain [25]. An ontology is therefore a content theory about the sorts of objects, properties of objects, and relations between objects that are possible in a specified domain of knowledge [26]. Details for the application of ontology to the industrial domain can be found in [27] and the work of Bailey [28] provides an example of how the ontological approach can be applied for the formalised structuring of information at the preliminary design stage. In the e-RM framework, the automotive product requirements will be based on a set of definitions of formal vocabulary that provides potential terms for describing the knowledge about each vehicle module or sub-system. The aim of the ontological structured requirements repository is to make the requirements knowledge sharable by encoding domain knowledge using standard vocabulary based on 'pluggable' ontologies.

A motor vehicle can be decomposed into a number of modules or sub-systems (*e.g.* body-in-white, engine, cockpit [9]), and the product requirements can be decomposed and assigned to each respective module. For each of the modules, an ontology can be developed between the vehicle manufacturer and the supplier in order to provide a definition of the product requirements for elicitation and documentation purposes. These ontology modules can be made 'pluggable' into the e-RM platform to form the requirements repository. This approach will allow easy integration and development of extra modules into the platform. Figure 22.3 illustrates an example of a 'pluggable' ontology. This ontology is for the vehicle seating assembly and it provides a common and shared definition for the requirements of a seat. For example, in terms of the functionality of a seat from an occupant positioning perspective, a seat can recline, slide, lift and swivel. Under the slide function, the requirements for the slider mechanism are inclination angle, travel length and travel pitch (Figure 22.3). Thus, an ontology can be used to design the structure of the requirements repository for a vehicle module and 'plugged' into the e-RM platform. Then for any given project, the repository can be populated with domain data and shared throughout the extended enterprise.

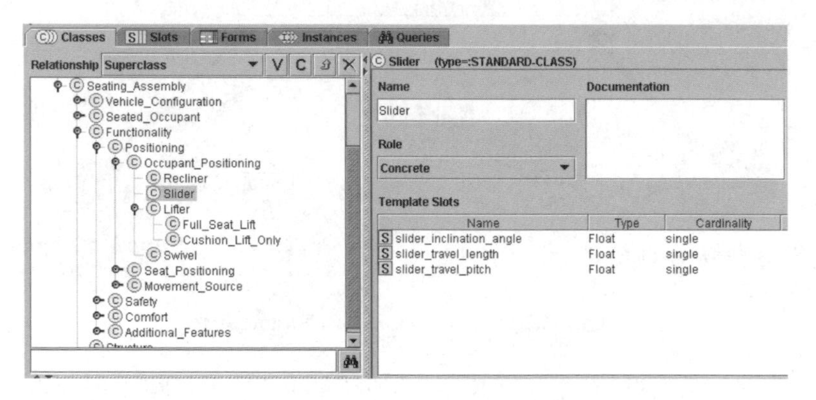

Figure 22.3 Seating assembly ontology

22.4 Conclusions

In the digital product development environment of the automotive industry, requirements management activities are still predominately performed manually through paper-based quotations, proposals, specifications and change requests. These current processes are experiencing difficulty in keeping pace with the drive towards shorter lead-times. The primary problem for requirements management is the administration and dissemination of product requirements through the extended enterprise. One avenue to achieve significant improvements in this area is the e-enabling of requirements management using a collaborative web-based architecture. The aim of such a platform is to provide a fluid and seamless tool for the handling and communication of product requirements thus allowing their greater availability, distribution and sharing in the extended enterprise; In essence, the systematic management of product requirements. The electronic requirements management (e-RM) framework is based on an ontological structured 'global' requirements repository. The aim of this approach is to permit the structuring and documenting of the product requirements in a common standardised electronic format. The motor vehicle is decomposed into a number of modules or sub-systems and 'pluggable' ontologies are developed for each respective module to form the requirements repository. Then for any given project, the associated product requirements are assigned to each module and the repository can be populated with actual specification data and shared throughout the extended enterprise.

22.5 Acknowledgements

The authors would like to acknowledge the UK Engineering and Physical Sciences Research Council (EPSRC); Nissan Technical Centre Europe, Johnson Controls Automotive, EDS and the Society of Motor Manufacturers and Traders (SMMT) as the e-RM project sponsors; Kamal Sehdev, Chrysanthi Makri and Patrick Oduguwa; and the Decision Engineering team at the Enterprise Integration department.

22.6 References

[1] Hermann-Krog, E., 2003, "Logistic Challenges," *Odette International Conference*, September 16-17, 2003, Paris, France.

[2] Dyer, J.H., 2000, *Collaborative Advantage: Winning through extended enterprise supplier networks,* Oxford University Press, Oxford, ISBN: 0-19-513068-5.

[3] Douma, M.U., 1997, "Strategic Alliances: Fit or failure?" *PhD Thesis, University of Twente*, The Netherlands.

[4] Quinn, J.B., Hilmer, F.G., 1994, "Strategic Outsourcing," *Sloan Management Review*, Vol. 35(4), pp. 43-56.

[5] Littler, D., Leverick, F., & Bruce, M., 1995, "Factors affecting the process of collaborative product development: A study of UK manufacturers of information and communications technology products," *Journal of Product Innovation Management,* Vol. 12(1), pp. 16-32.

[6] Millyard, T., 1996, *Communication at the UK TEAM-IT Seminar: Knowledge Engineering*. DTI, London, UK.

[7] Arozamena, C.M., 2003, "A strategic gamble with ENX," *Odette International Conference*, September 16-17, 2003, Paris, France,.

[8] Jordan, A., 2003, "Consultation and sharing of digital mock-ups," *Odette International Conference*, September 16-17, 2003, Paris, France.

[9] Marotz, D., 2003, "Digital Mock-up," *Odette International Conference*, September 16-17, 2003, Paris, France.

[10] Verma, M., & Wood, W.H., 2003, "Functional Modeling: Toward a common language for design and reverse engineering," *ASME 2003 Design Engineering Technical Conferences and Computer and Information in Engineering Conference*, September 2-6, 2003, Chicago, Illinois, DETC2003/DTM-48660.

[11] Hooks, I.F., & Farry, K.A., 2001, *Customer-centered Products: Creating successful products through smart requirements management,* AMACOM, New York, ISBN: 0-8144-0568-1.

[12] Neelamkavil, J., & Kernahan, M., 2003, "A framework for design knowledge reuse," *ASME 2003 Design Engineering Technical Conferences and Computer and Information in Engineering Conference,* September 2-6, 2003, Chicago, Illinois, DETC2003/CIE-48215.

[13] Shefelbine, S.J., 1998, "Requirements capture for medical device design," *MPhil. Thesis, Engineering Department, Cambridge University*, UK.

[14] Lindemann, U., & Reichwald, R., 1998, *Integriertes Änderungsmanagement,* Springer, Berlin.

[15] Sunnersjo, S., Rask, I., Amen, R., 2003, "Requirement-driven design processes with integrated knowledge structures," *ASME 2003 Design Engineering Technical Conferences and Computer and Information in Engineering Conference*, September 2-6, 2003, Chicago, Illinois, DETC2003/CIE-48218.

[16] Eckert, C.M., Clarkson, P.J., Stacey, M.K., 2001, "Information flow in engineering companies: Problems and their causes," *13th International Conference on Engineering Design*, August 21-23, 2001, Glasgow, UK.

[17] Kritsilis, D., 2003, "Requirements management within the aerospace industry: A case study," *MSc. Thesis, Cranfield University*, UK.

[18] Dong, A., Song, S., Wu, J.L., & Agogino, A.M., 2001, "Automatic composition of XML documents to express design information needs," *13th International Conference on Engineering Design*, August 21-23, 2001, Glasgow, UK.

[19] Thomson, G.A., 2001, "Requirements engineering - Laying the foundations for successful design," *13th International Conference on Engineering Design*, August 21-23, 2001, Glasgow, UK.

[20] Zhang, S., Shen, W., & Ghenniwa, H.H., 2003, "A framework for internet based product information sharing and visualization," *ASME 2003 Design Engineering Technical Conferences and Computer and Information in Engineering Conference*, September 2-6, 2003, Chicago, Illinois, DETC2003/CIE-48270.

[21] Toye, G., Cutkosky, M.R., Leifer, L.J., & Glicksman, J., 1993, "SHARE: A Methodology and Environment for Collaborative Product Development," *Proceedings of IEEE Infrastructure for Collaborative Enterprises*.

[22] Kilker, J., 1999, "Conflict on Collaborative Design Teams," *IEEE Technology and Society Magazine*, Fall, pp. 12-21.

[23] Agouridas, V., Baxter, J., McKay, A., & de Pennington, A., 2001, "On defining product requirements: A case study in the UK health care sector," *ASME 2001 Design Engineering Technical Conferences and Computer and*

Information in Engineering Conference, September 9-12, 2001, Pittsburgh, Pennsylvania, DETC2001/DTM-21692.

[24] Gruber, T., 1993, "A translation approach to portable ontology specifications," *Knowledge Acquisition*, Vol. 5(2), pp. 199-220.

[25] Schreiber, G., Wielinga, B., Jansweijer, W., 1995, "The KACTUS view on the 'O' word," *JCAI Workshop on Basic Ontological Issues in Knowledge Sharing, International Joint Conference on Artificial Intelligence*, August 19-20, 1995, Montreal, Canada.

[26] Chandrasekaran, B., Josephson, J.R., & Benjamins, V.R., 1999, "What are ontologies, and why do we need them?" *IEEE Intelligent Systems*, Jan./Feb., pp. 20-26.

[27] Roy, R., 2001, *Industrial knowledge management: A micro-level approach,* Springer Verlag, London, ISBN: 1852333391.

[28] Bailey, J.I., 2003, "Cutting tool design knowledge capture and reuse," *EngD Thesis, Cranfield University*, UK.

23

Federated Product Data Management in Multi-company Projects

Henk Jan Pels

Abstract: An approach for enabling concurrent engineering between companies by providing a proper collaboration platform, as developed in the VIDOP project is proposed. Apart from security, information status is an important element of trust. However, status-coding schemes differ much between companies and are deeply anchored in local engineering culture. An abstract life cycle model that enables comparing and relating different life cycle conventions is presented. A federated PDM architecture is proposed that enables interface with the local PDM systems in a loosely coupled, but yet effective way.

Keywords: Product Life Cycle Management (PLM), Product Data Management (PDM), document life cycle, collaborative engineering

23.1 Introduction

The Need for Collaborative Engineering

Where in the past decennium the strategy of companies to outsource the production of their components has lead to the supply chain, at present companies tend to outsource the design of their non-critical components [10]. This tendency leads to new processes like collaborative engineering [7].

Product Data Management (PDM) is an important technology to enable collaborative engineering and has been introduced in the past 15 years in most of the larger companies as "... *the discipline of making the right product and process related data available and accessible to the right parties at the right time in the product lifecycle in order to support all business processes that create and/or use this data."*[4]. Since 2000 the term PDM is gradually being replaced by the term

Product Life Cycle Management (PLM) that broadens the scope from product development to the whole product life cycle [2]. Although PLM is a more 'modern' term than PDM, we will use to the more traditional, but better-defined term PDM.

This paper analyses the problems of optimising inter-company engineering processes. Most of the research behind this paper has been funded by the VIDOP project.

The VIDOP Project

Manufacturers or turnkey suppliers of production facilities plan, design and build production facilities (*e.g.* spot-welding lines, assembly lines) from sub-systems (*e.g.* robots, machines, cells), which are created and supplied by sub-suppliers located throughout Europe. The sub-systems themselves are built from components and are partly planned, designed and optimised in a *virtual world*. Building and maintaining a complete model of the whole production facility is difficult because of the evolving character of the sub-models. Objective of the VIDOP (Vendor Integrated Decentralized Optimisation of Production facilities) (EC GRD1-2000-25705) project is to define *an Infrastructure for Vendor Integrated Decentralized Modelling* (IVM) for all phases in the life cycle of a production facility. Technische Universiteit Eindhoven is the responsible partner for the model management system in the IVM [1]. This model management system is in essence a prototype for an inter-company PDM system. This paper focuses on the design of this system.

23.2 How PDM Supports Engineering Processes

PDM systems build upon relational database systems and thus inherit features like optimised query languages, secure concurrent access, distributed storage and reliable back up and recovery. To this they add typical functions like data vault and document management, work flow and process management, product structure management, project management, product classification and a set of utilities for data translation, image services and data administration. These functions can be presented in a more structured way in the PDM function matrix [5] (see Figure 23.1).

Figure 23.1 shows how PDM functions can be ordered in a 3X3 matrix. Note that below this structure a relational database system is assumed for efficient and secure storage. The rows correspond to three levels of data management:

1. **Object repository management**: every entity in the PDM world is represented by a versioned object, such that data on these objects can be easily retrieved or manipulated,
2. **Object structure management**: relationships between objects can be created, browsed, manipulated and maintained,
3. **Object life-cycle management**: different lifecycles for different types of objects can be defined with the associated life cycle processes (work flows). Objects are guided through their life cycles.

The columns make the distinction between functions for document, product and process/project management. In practice engineers easily mix up these concepts, for example, by using the item number (product-id) as drawing number and intermixing project phase and product status. This is dangerous because documents, products and processes have different life cycles, use different status codes and make life cycle steps at different points in time.

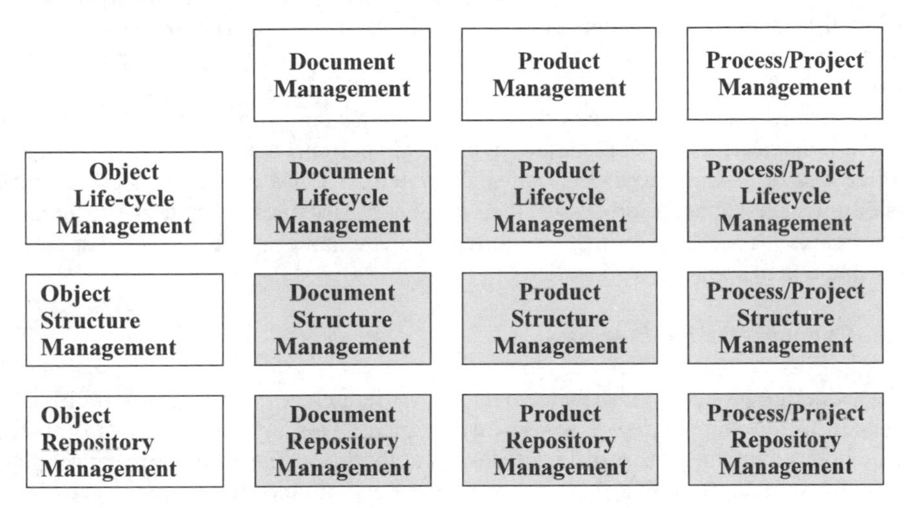

Figure 23.1 PDM function model

The elements of the PDM function model enable improvement of the product development process in several ways. The object repository management provides shared secure storage (data vault) with sound mechanisms for identification, check-in/check-out and versioning that ensure that users will always get the right version. The object structure management enables building and maintaining relationships between documents and product items. By enabling the user to browse these structures in a simple visual way, it becomes much easier to find the right documents in relation to products and activities.

The object life cycle management is crucial for not only monitoring the progress of the project, but also to enabling real concurrent engineering by supporting the exchange of preliminary design information [5]. The problem with preliminary information is that it is possibly incomplete, immature and likely to change in the future. A real danger is that very heavy commitments are made based on this information, which incurs high cost for future changes. Therefore, a well understood status scheme is required that indicates precisely for what purposes unreleased information may be used in what stage of the process. Consequences of exchange of preliminary information are that more versions will emerge and that the number of information exchanges increases drastically. If no PDM system would be used to automate these exchanges, then concurrent engineering would cause a dramatic increase in document handling cost.

23.3 Inter Company Collaboration

Intra Company PDM Systems

In order to be effective and acceptable by the users, a PDM system must support the local versioning and status coding conventions. These conventions often exist for several generations of engineers, are part of the engineering culture and, therefore, difficult to change. The adaptation of the PDM system to the local culture is responsible for an important part of the customisation cost. Also a PDM system is not a personal workstation, like a CAD system, that can be tuned to the preferences of a single user. Like an ERP system, a PDM system is a corporate system, which supports processes that involve many users, often in different departments. It is known that corporate systems take much more effort to implement than personal workstations.

Inter Company Design Processes

When a design process is distributed over two different companies, a new problem is added to the complexity of design management. This problem is that in most cases both companies have different design cultures, so that the engineers have difficulty in understanding how to react on version and status codes on the drawings of the other party. This causes misunderstandings and errors. The project managers who are responsible for the coordination of the processes try to minimise these problems by delaying the exchange of documents until the content of these documents has reached sufficient stability. This means that communication is minimised, which is the opposite of collaboration. The design process takes much longer than necessary and many opportunities for design optimisation are missed. Introducing collaboration, and thus concurrency between design processes in both companies would require a well defined and well understood status scheme the implementation of which might take a lot of time because it would require integration of different design cultures.

Inter Company PDM Systems

When improvement of intra company design processes requires a well implemented intra company PDM system, then it is very likely that for optimising inter company design processes a shared PDM system will be beneficial [6]. However implementing such a system will be problematic for at least two reasons.

1. A PDM system must be configured to the engineering culture of the users. The complexity of configuring one PDM system to the different cultures of collaborating partner companies is an order of magnitude higher than that of implementing an intra company PDM system. Implementation would therefore require unacceptable long preparation,

2. A shared PDM system must have one company as owner/administrator. The other companies will have to leave administration of their intellectual property to an external party, which is not an acceptable option for many companies.

In current practice we see that the biggest partner in the consortium (in most cases the OEM) builds a web based PDM installation outside his firewall and fills it with documents that must be available for suppliers. These so called project portals contain only released information and are owned by one party. Therefore, they cannot solve the problems mentioned above. In order to overcome these problems the VIDOP project has developed a concept of Federated Model Management (FMM). The purpose of FMM is to provide shared document and workflow management, while enabling partners to continue working in their own engineering culture with their own PDM system. Since real process improvement requires a shared convention for versioning and status coding, a well-founded theory on document version and status would be very helpful to analyse the real differences between local conventions. The next section presents such a theory. The VIDOP Federated Model Management approach, which using this theory, will be presented in Section 23.5.

23.4 A Document and Product Life Cycle Theory

The idea of product and document life cycle is that the form and function of the product as well as the contents of the documents evolve from incomplete and uncertain to complete and definitive. During this process different versions will emerge and be used by different parties in the development process. Status is used to indicate how complete and certain the specifications or contents are and what may or must be done with it and by whom. Version and status are key concepts in the management of design processes [3]. For a proper understanding the concepts of product, document, version and status must be clearly distinguished. In this paper object oriented conceptual modelling (UML) is used as a definition language and in this language generic properties of the concepts are defined.

PDM Objects

Figure 23.2 shows a UML static structure with the main PDM concepts. PDM objects (PDMObject) represent the entities that are managed by PDM and can be either a document or a Manufacturing Object (MfgObject). Manufacturing objects are the products, processes and equipment that are specified in the documents. Each individual object is identified by a PDM object identifier (PDMObjId) and has a meaningful name (PDMObjName). To enable discussion of PDM in multi-company situations, each company is represented by a Company object. Companies are identified with organisation identifier (OrgID). Each PDM object is linked to the creator company. In order to make companies independent in assigning PDM object identifiers, the OrgId is part of the key of PDMObject. The meaning of the other attributes and procedures is explained later.

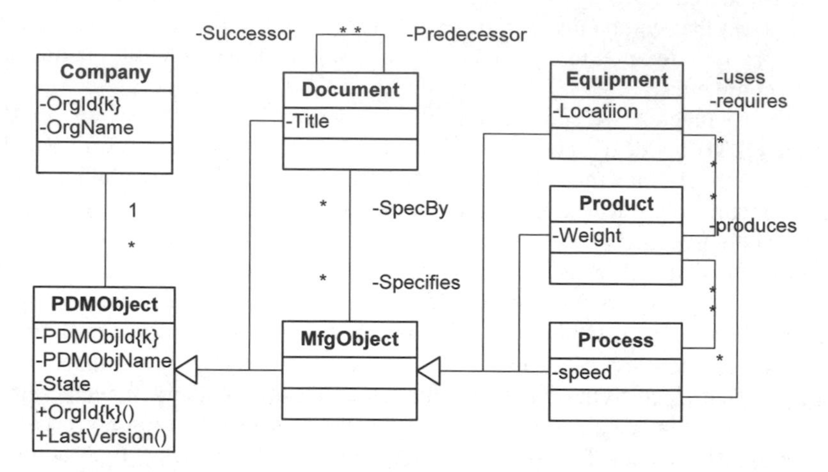

Figure 23.2 PDM Objects

PDM Object Life Cycles

Natural language often does not make a clear distinction between document and document version. In order to be able to compare and relate document life cycles in different companies, a simple but precise model of life cycles is needed. A document is a kind of abstraction of its different versions. Data modelling uses three forms of abstraction: classification, generalization and aggregation. Classification is the proper form: a document can be conceived as a class of versions.

In order to be able to model this, objects must be allowed to be classes: it must be possible to represent a specific PDM entity (*e.g.* a document or product) with an object and allow representing each version of this entity with a different object that is an instance of the PDM object [8, 9]. This construct is modelled in Figure 23.3 with the dotted arrow from Version to PDMObject. It means that every Version object must be an instance of a PDM object. It also means that each version object inherits all attribute values from its PDM object. In other words: a document version inherits the ID and name of its document. Versions are identified with a version number (VersionNr) that is unique per PDM object. Every version object is linked to a data object (DObject) holding the content of the version. Data objects are uniquely identified per Company. Different companies may have their own replicate of a data object. For documents the content is a readable file, while for manufacturing objects the content will be a record with values for attributes like price, weight, supplier *etc.*.

A PDM object can be created as soon as the need for it in a design project is identified. The State attribute of a PDM object can have one of the values {Checked-in, Checked-out}. Checked-out means that a new version is in preparation. The main function of the Check-in/out mechanism is to prevent that two new versions being prepared in parallel.

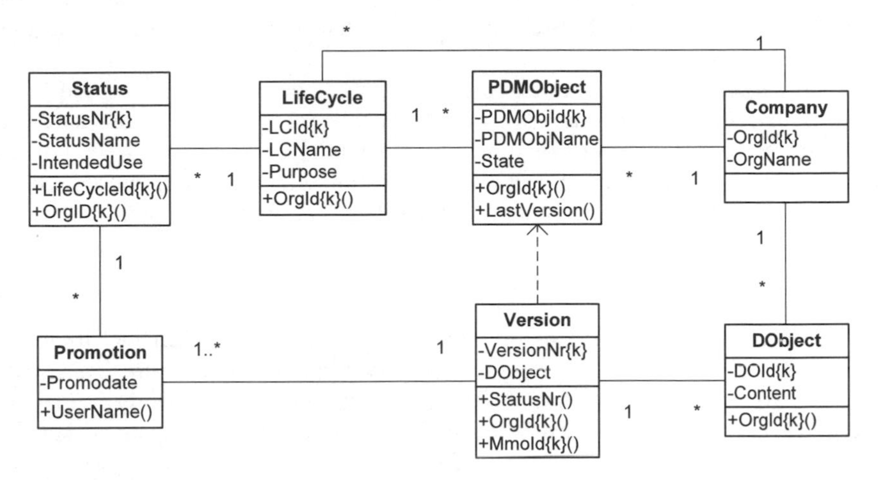

Figure 23.3 Object life cycle

The function of the version number is twofold: (1) it identifies the versions of one object and (2) it defines the sequence of versions. This means that in principle any ordered set of symbols can be used to identify the versions of an object.

Each PDM object is assigned to a Life Cycle. A life cycle consists of a sequence of statuses. The essence of a release procedure is that the object goes through a sequence of release steps [11]. The simplest life cycle has two statuses: {Concept, Final}. Engineering methods often require a formal review and approval steps and therefore use a four-step life cycle like {In-Work, For-Review, For-Approval, Released}. During the life cycle of a version its contents do not change. What changes is the probability of future changes. This means that more commitments can be made when the status increases. This characteristic of status makes it a very important control for concurrent engineering. The status For-Review could allow the Purchase department to use the document for supplier selection. Status For-Approval may be used to allow documents that used this one as input, to be submitted for Review. If status is used for controlling concurrent actions, then it is important that status is never decreased because this would take away the justification for commitments that may already have been made.

On basis of the considerations above, we may conclude that:

1. Status codes must be an ordered set,
2. A higher status allows implicitly all commitments of lower statuses,
3. There must be a clear description of what commitments can be made on basis of each status,
4. The status of a version may only be increased.

A generic way of coding status is to use consecutive numbers starting from number 1. Important for concurrent engineering is that work requiring a higher status can continue on older versions as long as newer versions still have a lower status. If for instance a new version is created because of a change proposed by the reviewer, then the old version will keep status For-Review and remain in use by the purchaser to continue the selection process. Only when the new version receives status For-Review, one may check whether the change could affect the selection criteria.

Since making commitments on basis of PDM object version is allowed only from the date that the required status has been assigned, the promotion date for every status must be known. Therefore in the model of Figure 23.3 Version is associated with Status via the object class Promotion with attribute promotion date (Promodate). The creation of a version is modelled as its first promotion.

Conclusion on PDM Object Life Cycles

The described model of PDM object life cycles is simple but sufficient to cover most of the version and status coding schemes used in practice. Its value lies in that it enables comparing and relating different coding schemes. Especially it enables describing how different schemes of different cooperation companies can be linked. This will be described in Section 23.5 below.

23.5 A Distributed, Federated Solution

In the VIDOP project a collaboration solution with a single shared PDM system was totally unacceptable for a number of partners, for the reason that they could not accept that another company would manage their data. In answer to this requirement, two solution principles were adopted:

1. A fully distributed peer-to-peer software architecture,
2. A federated PDM architecture.

The peer tot peer software solution called *Infrastructure for Vendor Integrated Decentralised Modelling* (IVM) is based on the principle that each partner has his own independent installation of a set of web services for collaboration and data management. In this environment projects can be defined and other partners, who also have an IVM installation, can be invited to participate in a project. Within this project users can be assigned roles that give access to specific PDM objects. Access is controlled on basis of user identification and authentication and data transfer between IVM sites is protected by encryption. Thus a safe environment for sharing data is created where each partner keeps full control over the protection of his data, until it has been delivered to the partner.

The federated PDM architecture enables each partner to perform his contribution to the project in his own local engineering environment. A federation of PDM systems supports the project work, where the shared system is a kind of federal PDM system that manages PDM objects and their life cycles only as far as they are shared between two or more partners. A simplified example is presented below to illustrate the concept.

Example:
Company OEM defines a project for a new production line. Company TKS, a turnkey supplier of production equipment, is invited as participant in the project.

The deliverables in the project are a requirements specification RS1 by OEM and a functional design FD1 to be delivered by TKS. These documents have as a life cycle: (1) For-Information, (2) For Acceptance, (3) Accepted. When RS1 version 3 receives internal status For-Review in the local PDM system, it is checked-in in the Federal system as version 1 with federal status For Information. TKS is notified, checks-in RS1 in its own local PDM system. TKS engineers inspect the document and propose a few improvements. These changes are communicated to the OEM so that they can be implemented together with the changes proposed by the internal reviewers. Finally the internal version 6 is released locally in OEM and checked-in in the federal system as version 2 with status For-Acceptance. The TKS project manager downloads this version in his local PDM system, has it accepted by his engineers and promotes the federal status to Accepted. Then he releases the design order for FD1 and a similar communication follows around this document until OEM accepts it.

The above example has been restricted to a two-company case. This is sufficient to illustrate the principle. However the federated approach serves also multi-company collaboration situations. For instance, a three-company extension of the example, where TKS out sources part of its work to a sub-supplier SP, can be considered. Then there are two options to configure the project: either SP can be invited in the existing project, so that OEM can have access to documents shared between TKS and SP and monitor the progress, or TKS can create a new project in which it invites SP. In the last case OEM cannot see that work has been sub-contracted.

The federated PDM system can be implemented in short time for a single project. Essential is that the parties agree on the federal life cycle. This life cycle enables OEM to present a preliminary version of the requirements specification to TKS, so that review processes in both companies can be executed in parallel. In the current practice the OEM project manager would not be allowed to send the requirements to the supplier before full release, because of the risk of confusion and liability problems. More detailed analysis shows that this simple parallelism can reduce the elapsed time of the project by 40%. Refining the federal life cycle can increase parallelism further.

23.6 Conclusion

Lack of trust, or the non-readiness to make information available via a public medium such as the Internet, is an important reason to stick to paper or CD-ROM as communication medium. The VIDOP project revealed that trust has at least two aspects: (1) security and (2) status of information. Security must ensure that the wrong persons do not use information; status must ensure that information is not used for the wrong purpose. This paper addresses the status issue. Status is defined in information (document) life cycles. Companies have quite different internal conventions for coding these life cycles, which makes it very difficult to integrate them between companies. PDM systems have an important role in supporting and enhancing document life cycle processes.

This paper proposes an abstract life cycle model that enables comparing and relating different life cycle conventions. A federated PDM architecture is proposed that enables interfacing the local PDM systems in a loosely coupled, but yet effective way and thus makes it easy to implement collaboration environments. The integration is not just technical, but also organisational: the federalized approach allows defining version numbers and status codes between the partners, without having to adapt the local systems. The federal system acts as glue between the local processes.

Future research will be directed to the question how to establish the best federal life cycles in order to find optimised matches between existing local engineering cultures.

23.7 Acknowledgements

The authors thank the European Commission for supporting this work in the context of the project contract n°: g1rd-ct2000-00301 "Vendor Integrated Decentralised Optimisation of Production Facilities" (VIDOP).

23.8 References

[1] Caskey, K.R., Rouibah, K., Pels, H.J., 2002, "Managing Vendor Supplied Models of Production Facilities," *Proc. CARS-FOF conference on CAD/CAM, Robotics and the Factory of the Future*, July 3-5, 2002, Porto.

[2] 'Product Lifecycle Management – Empowering the Future of Business'; *A CIMdata Report*; http://www.cimdata.com/publications/PLM_Definition_0210.pdf; 2002.

[3] Hamer, P. v.d., Lepoeter K., 1996, 'Managing Design Data: The five Dimensions of CAD Frameworks, Configuration Management and Product Data Management,' *Proceedings of IEEE*, Vol. 84, No. 1.

[4] Harris, G., Cantrell, S., 2002, 'Enterprise Solutions Integration Technology,' *Next Generation Enterprise Solutions, Accenture - Institute for Strategic Change*; Issue Three, March 16, 2002, http://www.line56.com/research/download/accenture_isc_ent_int.pdf.

[5] W.Helms, R., 2002, "Product Data Management as enabler for Concurrent Engineering; Controlling the flow of preliminary information in product development," *PhD Thesis, Technische Universiteit Eindhoven.*

[6] Huhtinen, H., Lehtonen J.M., 2003, "Implementing a PDM Application in the Supply Chain: Key Requirements from the Communication Perspective," *Proc. ICE 2003 Conference of Concurrent Enterprising*, 16-18 June 2003, Dipoli Congress Center, Espoo, Finland.

[7] Kumar, R., Midha, P.S., 2001, 'A QFD based methodology for evaluating a company's PDM requirements for collaborative product development,' *Industrial Management & Data Systems*; MCB University Press; Volume 101, Issue 3, pp. 126-131; ISSN 0263-5577.

[8] Pels, H.J., 2002, 'Een Voorbeeldige Classificatie', *Data Base Magazine*, Issue 2002, nr 8, Array Publications, Alphen a/d Rijn, Netherlands, December 13, 2002 (in Dutch)
[9] Pels, H.J., 2004, "Classification hierarchies for Product Data Modelling," Internal discussion paper, http://is.tm.tue.nl/staff/hpels/classification4.doc.
[10] Rezayat, M; 2000, 'The Enterprise-Web portal for life-cycle support,' *Computer-Aided Design*; Elsevier Science; Vol. 32, pp. 85-96.
[11] Rouibah, K., Caskey, K.R., 2003, "Change management in concurrent engineering from a parameter perspective," *Computers in Industry*, Vol. 50, pp. 15-34.

24

STEP PLCS for Design and In-service Product Data Management

Rohit Sharma, and James Gao

Abstract: The Product Life Cycle Support (PLCS) [1] initiative is beginning to move from the conceptual phase to implementation. PLCS is part of the ISO 10303, which is the Standard for Exchange of Product Data. It has recently become an International Standard. The Standard covers a wide range of product design through to in-support activities. An effort to drive the standard forward through a practical implementation is reported here.

Keywords: ISO 10303, PDES STEP, PLCS, PDM, XML

24.1 Introduction

As products become increasingly complex, the tools required to design and support them become more numerous and specialised. This problem can be tackled by reliance on an overarching suite of products from a single vendor or by providing an assured mechanism of exposing information held in disparate native systems to those who require it. The former option exposes through life information users to risk as it can result in proprietary lock-in. When considering the life cycle from concept to disposal, products from a single vendor will inevitably have gaps in functionality or weak areas. The cost of licensing an overarching solution from a single vendor, for all information users throughout the life cycle of a complex product is often prohibitive. The result is that many users are unable to quickly and easily access information that would make their day-to-day tasks easier and more productive. The latter option is difficult to implement due to the increasing complexity of the disparate native systems. Keeping pace with these developing systems can be an extremely specialised task. Therefore, the cost of developing a point-to-point interface between each system becomes prohibitively time consuming and expensive.

The objective of the PLCS Product Information Explorer project was to develop a flexible and adaptable approach to exposing information from disparate sources, federated through a single, web-based interface utilising the PLCS information model [1] as a common context. The PLCS Product Information Explorer application then acts as a portal for users to access the information they require without the need to invest in run-time licenses for native systems or develop numerous point-to-point interfaces. The ability to gain access to information and the transformation of native descriptions to a common vocabulary allows the application to support the evaluation of information quality and consistency.

The PLCS Product Information Explorer supports information users needs by providing a method of accessing information from any native database or file export source, whether it be an as-required, as-designed, as-built or as-maintained configuration. Previously, this has only been possible through the use of an overarching solution or through funding the creation of many point-to-point interfaces.

24.2 Background

The problem of supporting the product over its entire life cycle, ironically, gets more complex with technology. The main problem is the various types of proprietary design data. For example a typical large product has:

a) Product structures and breakdowns;
b) Attribute data applied to elements of product structure;
c) Computer-Aided Design data;
d) Other sources of design information, such as cabling, piping or common equipment data;
e) Catalogue parts and manufacturers library parts; and
f) Many sources of in-service configuration and logistics data.

These disparate sources of data require many applications to enable effective use of the information. Many users simply do no have access to the required software. This means that the value added by the use of digital design, management and support tools cannot be effectively leveraged and ends up locking the information rather than releasing it to the users.

The problem can hence be summarised as:

a) The source applications effectively generate vast amounts of useful yet inaccessible information.
b) With no common context to provide a backbone to integrate this information to a coherent whole;
c) Therefore, it is impossible to provide a view on the quality and consistency of this information, let alone give any sort of guarantee;
d) An average product life is increasing. Although this is emerging as a problem in simple consumer products and cars, it is a chronic problem in case of defence products like ships, submarines and aircraft where the life of

a product can be anywhere between 25 to 50 years and possibly more. How is this data going to be managed cost effectively for this period of time?

e) The problem takes on a different level if the product is manufactured and delivered by a sub-contractor. Build-and-handover contracts require the technical datum pack to be defined. This datum pack (or information about the product as manufactured) is delivered to the end customer in the electronic format.

f) The configuration of the product changes over time due to maintenance and modification/upgrade. Any product life cycle support system needs to cater to this.

Literature Review

Few publications have reported on practical implementation of PLCS [2]. Although STEP [3] has been under development for the past two decades, PLCS is one of the youngest Application Protocols. The Product Life Cycle Support (PLCS) initiative is supported by both industry and national governments, with the aim of accelerating the development of a new international standard for the exchange of product support information. The initiative is being undertaken within the framework of the International Organization for Standardization (ISO) and aims to produce a full international standard. The standard will be the mechanism to ensure product and support information is aligned with the evolving product definition over the entire lifecycle, from design to disposal. The standard will be an extension to the existing proven exchange capability of the Standard for Exchange of Product Data, also known as ISO 10303, STEP. The PLCS Standard will be published as an Application Protocol to the STEP standard and will be known as ISO 10303, AP 239.

The PLCS Standard (AP 239) will enable the exchange, sharing and archiving of support data. The compatibility with STEP will enhance the utility of PLCS in enterprises where STEP already supports design, analysis and manufacturing, *e.g.* automotive and aerospace industries.

The ultimate aim of the PLCS initiative is to service three significant business requirements for owners of complex engineering assets such as aircraft, ships, and power plants:

a) Reduction in total cost of ownership of such assets
b) Protection of investment in product data through life
c) Increased use of the asset to deliver enhanced business performance

The PLCS Process Model defines the product support activities and associated information flows throughout the product lifecycle from concept design to disposal. The process model is captured in the IDEF 0 format and was developed by professional engineers drawn from a wide cross-section of design, manufacturing and logistic support backgrounds. It aims to be generic process model applicable to any complex, high value product where: Typical products would include aircraft, ships and power generation equipment.

In the past, implementation of STEP/ EXPRESS based data-models has been difficult and limited to using a few advanced CASE tools. This has changed in the recent past with most data persistence and modeling moving to XML/ XSD based standards. STEP models are traditionally represented in EXPRESS. A new recent initiative of the STEP community has been in the development of STEP Part 28. This part of ISO 10303 specifies means by which schemas specified using the EXPRESS language (ISO 10303-11) and data governed by EXPRESS schemas can be represented as an XML document (Extensible Markup Language W3C Recommendation).

The advantage of using XML for representation of STEP schemas and data is primarily portability. Manipulating EXPRESS schemas and instances required expensive specialist software. XML allows the user to select from a vast choice of software tools including a lot of freeware and shareware tools. Another advantage of using XML is the ease of data transformation using XSLT/ XSL.

24.3 Methodology

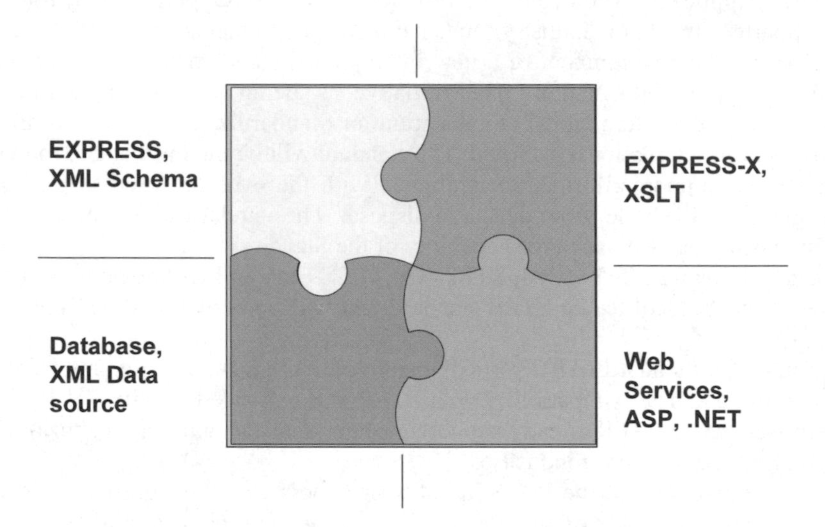

Figure 24.1 4M Architecture

The PLCS Product Information Explorer has been built on the 4M architecture shown in Figure 24.1. The architecture has four distinct parts which fit together:

a) Models – This includes the PLCS product model and the proprietary model;
b) Metadata – This includes the standard reference data and the product breakdown;
c) Mappings – This is the mapping from one model to another. Typically this is a mapping from the proprietary model to PLCS model implemented using XSLT;
d) Middleware – This is the software element of the architecture and encompasses such technologies like DOM (Document Object Model) and SAX (Simple API for XML), which are used for implementation.

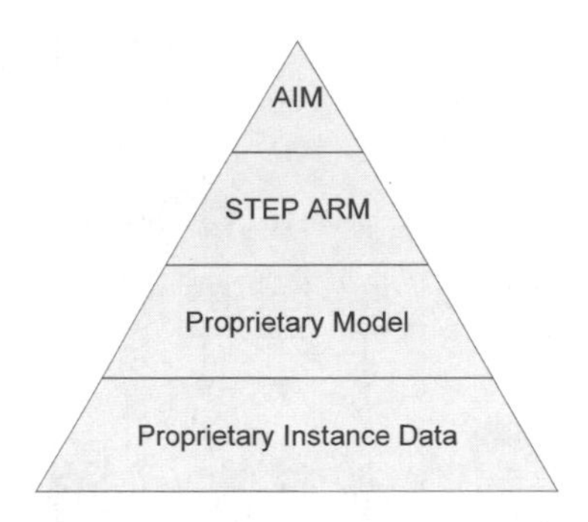

Figure 24.2 Data Model Pyramid

The mappings are used to transform the models. The model transformation pyramid is shown in Figure 24.2. The lowest form of the exchangeable model is the proprietary model of the application. The instance of the proprietary model is mapped on to the STEP PLCS ARM (Application Reference Model). The ARM model is then mapped on to the AIM (Application Interpreted Model) which is the implementation model according to the PLCS standard.

24.4 Implementation

A schematic diagram of the architecture is shown in the Figure 24.3. The first web service allows the user to browse through and select any of the registered data sources. The second web service transforms the selected data source into a PLCS-XML compatible file. This web service maps the proprietary data model of the source application onto the PLCS standard data model using XSLT. The knowledge of the data mapping is encapsulated in the XSLT files and therefore easy to change and adapt for any future needs. The output XML is validated against the PLCS schema before being used as the data source for the Product Information Explorer application.

The exact sequence of events are enumerated below:

a) First, the database source is queried using a SOAP (Simple Object Access Protocol) service, according to a set of SQL statements held in a text file. This allows easy modification of the SQL queries if required;
b) This query process returns a bunch of XML record sets.
c) The record sets are then processed using XSLT. This processing integrates the multiple record sets to a single file and transforms the source data into a PLCS representation
d) The end of the chain is the Product Information Explorer application which queries the XML database using XPath. The query process is done whilst the user is browsing the data to improve system performance.

The multi-tier architecture works in real-time and the PLCS-XML data source presented to the user is derived in real-time when selected by the user. The information browser is an ASP.NET application running on a web server. The front-end presents a graphical view of the information and allows the user search and navigates the information using any web browser.

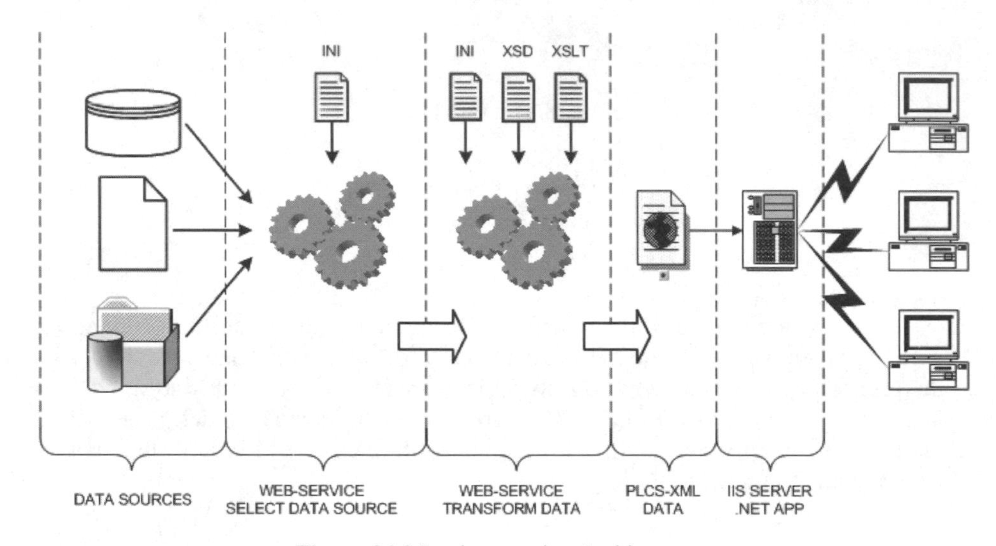

Figure 24.3 Implementation Architecture

24.5 Case Study

The main requirement was to develop an architecture that did not require access to expensive runtime licences of proprietary systems and at the same time allowed exchange of information using the PLCS standard.

A screenshot of the current system is shown in Figure 24.4. The current system is capable of combining information from a number of separate sources into a single related view. The integrated systems include:

a) Product structures defined in CADDS5 CAMU (Concurrent Assembly Mock Up);
b) Attributes data in both CAMU and Optegra CM (Configuration Master);
c) CAD data defined in CADDS5;
d) Other systems such as Manufacturers Parts Items Database to complete the design;
e) And configuration and logistics information supplied by in-service configuration master of the product.

Additionally, we can input breakdown definitions allowing us to attach elements of product structure to an element of a breakdown structure. For example we can define a hierarchical functional breakdown for a ship, for example;

a) Electrical system, under which we have general lighting, emergency lighting, auxiliary power, broadcast and communication *etc.*.

b) Fresh water system, under which we have chilled water, hot water, grey water *etc.*.

The product breakdown is also defined as metadata using XML. The system can attach elements of product structure, from multiple sources, to these breakdown elements based on some metadata attribute belonging to the product structure element. It is important to note that these breakdown definitions are defined in the PLCS data model and implemented in XML, thus maintaining the benefits offered by the architecture as a whole.

The "Plug and Play Data Access Architecture" allows literally any PDM system to be plugged in as long as some information is available about its source data and appropriate mappings can be developed in XSLT to map this data to PLCS. The same can be said of other database type applications, such as the in-service systems. The system can also read in simple files such as spreadsheets, text files or CSV files and map them to the PLCS model. All these sources can be accessed remotely by the client system, using a web browser.

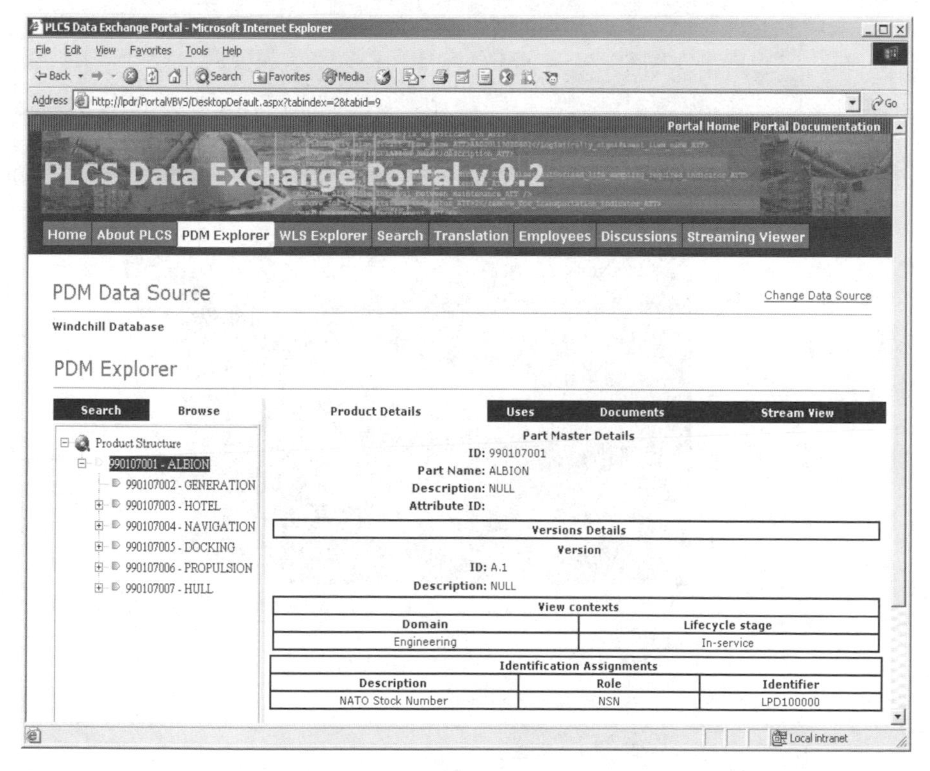

Figure 24.4 Application interface

The PLCS Product Information Explorer application has proven that the PLCS model is entirely suitable for the federation and integration of information from the many sources through a single, web-based interface. Using the PLCS information model as the common context to describe the contents of a native database or export file makes comprehension of the bigger picture far easier for information users. The technology used in implementing the PLCS Product Information Explorer is well understood by IT professionals facilitating improved maintenance opportunities.

24.6 Step and CAD Models

Not only does the system use PLCS, but it also uses STEP for representation and archival of CAD geometry. The current work is focused on CADDS5 and ACIS translation to STEP AP203 and AP214. The viewing capability has been implemented using a CAD streaming technology which allows the streaming of STEP CAD data to internet desktop clients. Figure 24.5 shows a screen shot of the application with a 3D CAD model opened in the viewer.

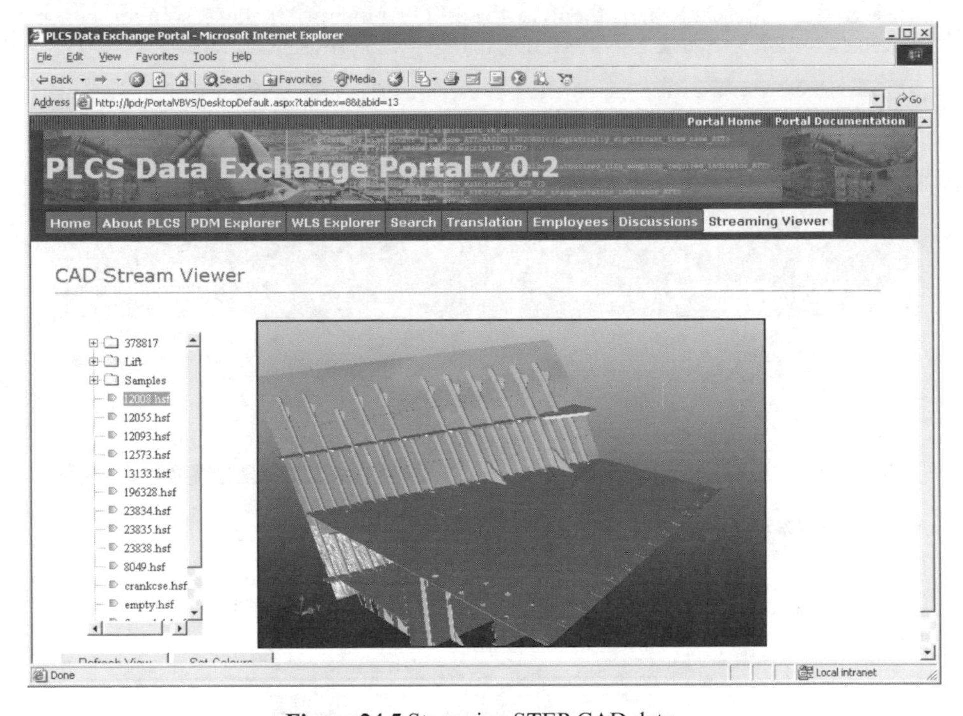

Figure 24.5 Streaming STEP CAD data

24.7 Summary

This approach is based on the federation of information. The integration data is integrated from multiple disparate sources through a common interface. The approach is flexible, allowing new forms of information to be incorporated at a later date without the need to modify huge chunks of complied code. The approach also uses a common context for representation of product information. That allows information from disparate sources that may have very different native descriptions, to use a common vocabulary. For this purpose, we use PLCS to supply the backbone, to integrate the information effectively as a cohesive whole.

The implementation architecture is also based on Standards; in this case W3C Standards such as XML, XML Schema, XSLT (the XML transformation language), XPath (a query language for XML documents) and Web Services (that is technologies such as SOAP and WSDL). Using these Standards has advantages such as improved maintenance opportunities and interoperability with other systems based on these Standards due to sensitive nature of the implementation, real data has not been used for illustration the discussion and the case study

Whilst this is still a proof of concept at this stage, we hope to effectively demonstrate the power of the PLCS standards-based approach based on its obvious merits.

24.8 Acknowledgements

The authors acknowledge the support of LSC Group, UK.

24.9 References

[1] PLCS Inc. Website : http://www.plcs.org/.

[2] King, T.M., 2002, "Early practical realisation of Product Life Cycle Support," In: *11th Symposium on Product Data Technology Europe 2002 – PDT Europe*, May 2002, Turin, Italy.

[3] Bloom, H.M., 1992, "STEP - Standard for the Exchange of Product Model Data," *National Institute of Standards and Technology*, USA.

25

Value Chain Structure and Correlation Between Design Structure Matrices

Marco Cantamessa, Maurizio Milanesio, and Elisa Operti

Abstract: An empirical study of the relationship between product architecture and industry structure is discussed. Product architecture is modeled by using Design Structure Matrices (DSMs) representing three different types of inter-component relationships: technological homogeneity, functional interaction and assembly process contiguity. The DSM models may be used to explain firms' specialization choices within an industry. Moreover, the same models can provide a rough-cut forecast of the impact that modular and architectural innovation may have on industry structure. The method is then applied to the automotive industry, using empirical data on automotive suppliers located in the province of Turin, in Northwestern Italy.

Keywords: design, manufacturing, industrial, Design Structure Matrix

25.1 Introduction

Literature generally defines product architecture as the set of components that make up a product, together with the way with which they interface with each other. It is generally taken for granted that architecture has a significant impact on the organization of the manufacturing firm and, more generally, of industry structure. For instance, Ulrich [14] observed "*there is some evidence that the organization of the firm and the architecture of the product are inter-related*", and suggested conducting "*an empirical study of the elements of difference in product architecture.*"

It is generally acknowledged that a modular architecture enables a more flexible approach to organizing and managing product development, by making it easier to distribute design tasks across the value chain. For instance, the producer of the final good may assign design tasks to component manufacturers and coordinate their work. By contrast, the design process associated to an integral architecture is more complex, and this may hinder the outsourcing of component design. The tight relationship that exists between product architecture on one side and organizational structure and routines on the other has also been observed by Henderson and Clark [5], who have shown that architectural innovation may find a significant barrier in the inertia with which organizations adapt to new relationships between components.

A number of motivations have been proposed to explain alternative architectural choices and to understand the factors that determine how manufacturing and design activities are distributed in the value chain. Some authors focus on technology and knowledge [4, 11], since the benefits of specialization have traditionally led firms to organizing themselves around technical domains. Others [1, 10] suggest that the web of functional interdependence between components defines a pattern of communication flows within the design process. In turn, this determines the way design and manufacturing activities are organized. Finally, the assembly process may have a significant impact on defining product architecture as well [2, 3]. For instance, this latter perspective is the one currently used by the automotive industry in its use of the term "modules", which essentially are large component subassemblies that incorporate heterogeneous technology and technical functions. The influence of these three perspectives, namely technological homogeneity, functional relationships and assembly processes, on the organization of individual firms and on industry structure is usually taken for granted although, to the authors' knowledge, empirical evidence is quite scarce.

Section 25.2 of this paper proposes a method that uses Design Structure Matrices (DSMs), [10, 13] to test the degree to which product architecture and industry structure are correlated, and which of the three above-mentioned perspectives has the most significant impact. If demonstrated, such correlation between product architecture and the organization of activities in the value chain may have a direct application in forecasting the impact of architectural innovation on the firms operating within in a value chain. Given the previously mentioned problems that organizations encounter when tackling architectural innovation, a forecast of this kind can be quite relevant to industry. This is particularly true in the current production environment, in which product development is seldom performed within a single vertically integrated company, but is ever more frequently carried out within a network of distinct and often geographically dispersed firms. Though it hasn't yet been subject to empirical verification it may be assumed that, due to lack of centralized control, firms operating in value chains of this kind may find architectural innovation even more confusing and difficult to tackle. In greater detail, managers may use a forecast of the impact of architectural innovation in order to:

- understand the main challenges lying ahead for their company (for instance, whether there is a greater need to become confident on a new emerging technology, or whether their product development process ought to be redesigned),
- compare such challenges to the innovative capabilities owned by their firm,
- Benchmark their firm's position in the "challenges ahead" versus "readiness" space against other firms belonging to the same value chain.

Similarly, policy-makers operating at industry-wide level may use the same results to design actions aimed to support the industry regarding the architectural innovation process.

In Section 25.3, the proposed method is applied to a sample of 300+ automotive suppliers located in North-Western Italy. Statistical tests show a strong correlation between the DSMs that respectively describe the product architecture and the industry structure. Companies are then mapped in order to forecast the impact of architectural innovation (in this case, the introduction of hybrid powertrains and of Drive by Wire technology) on the supply chain. Finally, the results are used to evaluate industry-level readiness with respect to architectural innovation and to suggest policies for design management at a strategic level.

25.2 Method

This section describes the method for comparing and analyzing product architecture with respect to the organization of design and manufacturing activities in the value chain. The approach involves six main steps (Figure 25.1).

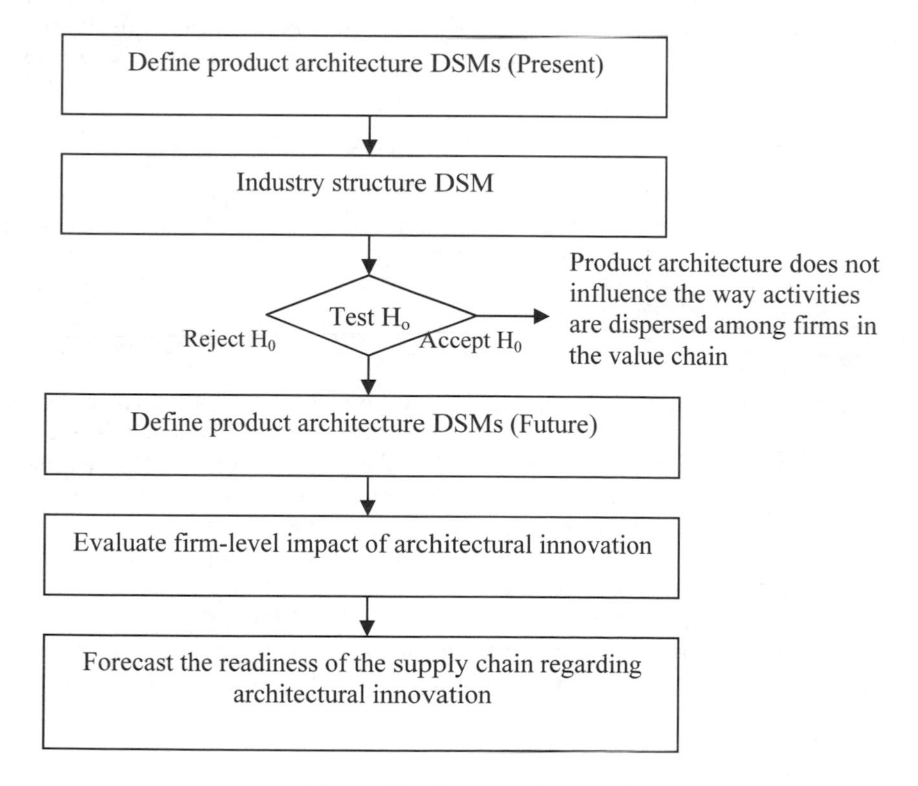

Figure 25.1 Conceptual approach

First, the Design Structure Matrix is used in order to study the current product architecture. The DSM method is an analytic method [13, 10] used to document product decomposition and component inter-dependence. In order to build the DSMs that describe product architecture it is necessary to decompose the product at a sufficiently high level of detail and assess, through expert evaluation, the intensity of pairwise interactions between components from the three perspectives of technological homogeneity (DSM^t), functional relationships (DSM^f) and assembly requirements (DSM^a). A five-point Likert scale can be used to capture the level of criticality of each type of dependency (Table 25.1).

Since it may be useful to have a single evaluation of the relationship existing between each component pair, it is necessary to define a distance measure aggregating the three perspectives. It has been decided to use a simple element-by-element summation of the matrices (equivalent to a city-block distance measure on each component pair), though a number of alternatives are of course available (such as Euclidean, maximum vector component, *etc.*.). This summation, therefore, yields a fourth matrix that will be used as a first-cut evaluation of the overall relationship between components (DSM^o).
Industry structure is described by a second matrix, DSM^i that measures, for each component pair, the number of suppliers whose design and manufacturing activities are associated with both components. This DSM measures the firms' production choices and the way with which design and manufacturing activities are dispersed in the value chain.

Table 25.1 Level of criticality of dependencies between component pairs

Level	Description
0	Components not tied up
0.25	Weak relationship between components
0.50	Significant relationship between components
0.75	Strong relationship between components
1	Components are extremely dependent between one another

Given the architectural DSMs and DSM^i, it is possible to evaluate whether there exists a statistically significant correlation between them. Correlation between distance matrices can be measured through either, the Mantel test [8] and the PROtest [7, 6]. Firms are reasonably expected to produce a set of components that reflect strong architectural links related to some combination of technology, function or assembly between them. Hence, we expect to find empirical support for *rejecting* the following set of hypothesis:

- H_{0a}: *There is no correlation between DSM^t and DSM^i* (by rejecting, one states that firms are expected to specialize in one technology and exploit economies of scope across multiple components that use the same technology),
- H_{0b}: *There is no correlation between DSM^f and DSM^i* (by rejecting, one states that firms are expected to choose to design and manufacture sets of components that enable to minimize coordination costs in design),

- H_{0c}: *There is no correlation between DSM^a and DSM^i* (by rejecting, one states that firms are expected to choose to manufacture sets of components that may facilitate assembly operations),
- H_0: *There is no correlation between DSM^o and DSM^i* (by rejecting, one states that firms are expected to choose to design and manufacture components according to the joint influence of the three reasons outlined above).

The fourth step of the method consists of estimating what might happen in a new technological scenario. Expert knowledge may be elicited in order to define a new set of architectural DSMs, which we label *newDSM^t, newDSM^f, newDSM^a* and *newDSM^o*, and to evaluate modular (component-level) change, m_j. A five-point Likert scale can be used for this latter purpose (Table 25.2).

Table 25.2 Innovation within sub-systems

Value	Description
0	Component essentially unchanged
0.25	Component needs minor adjustments
0.50	Component needs to be modified
0.75	Component is significantly transformed
1	Component is subject to radical change

The fifth step now allows evaluating the firm-level impact of innovation, thanks to the knowledge of which components are designed and manufactured by each firm. The measure of modular innovation for firm i is given by:

$$MI_i = \sum_j p_{ij} m_j \tag{25.1}$$

where m_j measures component-level innovation within component j and $p_{ij} = 1$ if firm i produces component j, and = 0 otherwise.

The measure of architectural innovation for firm i under the technological perspective is given by:

$$AI_i^t = \sum_{j,j'>j} p_{ij} \left| newdsm_{j,j'}^t - dsm_{j,j'}^t \right| \tag{25.2}$$

where $dsm_{j,j'}^t$ is the element in column j and row j' of DSM^t.

A similar measure can be used for functional, assembly and overall perspectives, leading to impact indicators AI_i^f, AI_i^a and AI_i^0. These indicators can be used to assess the relationship between modular and architectural innovation.

The final step allows comparing these firm-level forecasts of the impact of innovation against indicators of innovative competencies owned by each firm (IC_i). This amounts to defining a mapping tool that allows identifying four types of firms, as shown in Figure 25.2:

- Competent firms: impact of innovation will be high, and the firm's innovative capability is adequate,
- Inadequate firms: innovation can be disruptive, because it will lead to a significant impact, while the firm's innovative capability are inadequate,
- Unchanged firms: have low innovative capability, but will be subject to little impact,
- Overqualified firms: whose innovative capability exceeds the requirements associated to the foreseen innovative impact.

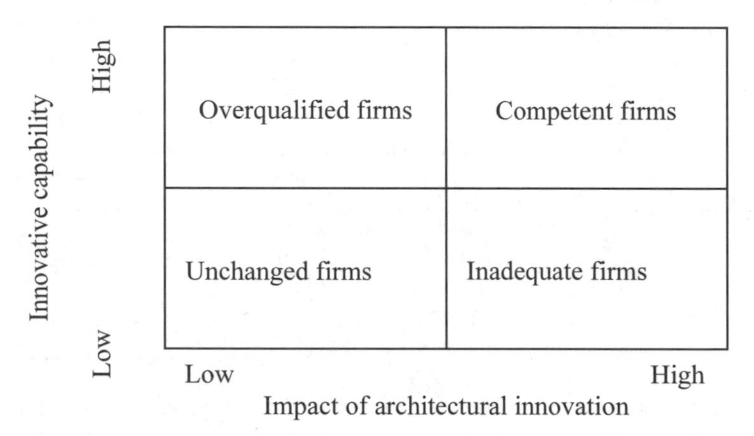

Figure 25.2 Innovation Map

25.3 Case Study

Empirical data from the automotive industry is used to provide an application of the described method. The study is particularly interesting because of the economic and engineering significance of the automotive industry, and because passenger cars nowadays are on the verge of facing significant innovations (*e.g.*, weight reduction, "by-wire" technology, alternative propulsion systems, modular production, *etc.*.). The study is based on a sample of 300+ automotive suppliers located in the province around Turin, in Northwestern Italy. Data has been collected by a survey aiming to promote the local automotive cluster [12]. Although these records have not been specifically collected for applying the method discussed in this paper, they can be considered the most relevant and updated census of the local automotive supply chain.

The product architecture DSMs were based on 28 sub-systems or major components that, due to the complexity of a passenger car, include a number of lower-level components. The strength of links between sub-systems was measured from the three perspectives (technological, functional and assembly) by using expert evaluation. The industry DSM was built by counting, for each pair of sub-systems or major components, the number of suppliers whose design and manufacturing activities were associated with both sub-systems. In this way it was possible to perform the study on both tier-one (module and sub-system level) and tier-two (component level) suppliers.

A statistically significant correlation between DSM matrices has been found, as Table 25.3 reports. This finding is confirmed by the PROtest result, that rejects H_0 with a statistically significant level of confidence ($p < 0.001$), and shows a goodness of fit $R^2 = 0,4347$. Moreover, the results in Table 25.3 show that the strongest correlation is found when comparing the industry structure DSM to the DSMs that represents technological homogeneity and functional interdependence (DSM^t and DSM^f). Therefore, while firms' production choices are mainly associated with functional interdependence (as suggested by Ulrich, 1995), technological homogeneity, and to a lesser degree ease of assembly, also are important determinants. The finding that, within a value chain, all of the three perspectives may be significant is hardly surprising. In fact, functionally coupled components are often based on the same technology and it is quite wise to locate them nearby in the product layout, but it is important to stress that functional interdependency does appear to be the main determinant of firms' production choices. It may also be interesting, though it is beyond the scope of this paper, to investigate whether firms being similarly influenced by the three perspectives (individually or in combination) also exhibit similarity in other features as well (*e.g.*, company size).

Table 25.3 Result of statistical test: Mantel Test

Hypothesis	Conclusion according to Mantel's test	r
H_{0a} (no correlation between DSM^t and DSM^i)	Reject with $p < 0,001$	0,465
H_{0b} (no correlation between DSM^f and DSM^i)	Reject with $p < 0,001$	0,433
H_{0c} (no correlation between DSM^a and DSM^i)	Reject with $p < 0,002$	0,158
H_0 (no correlation between DSM^0 and DSM^i)	Reject with $p < 0,001$	0,509

We can therefore conclude that product architecture, and the way with which activities are organized among firms, are closely related. Moreover, the higher correlation associated with technological and functional interdependencies, suggests that these are fundamental in defining the structure of the value chain.

Based on this result, the method has been used for evaluating the impact of a future emerging technological scenario. The main innovations that have been considered are the introduction of alternative propulsion systems (in our hypothesis, hybrid powertrains) and the implementation of Drive-by-Wire systems. Impact within sub-systems (modular innovation) and between sub-systems have been evaluated by experts and firm level impact has been measured by computing MI_i, AI_i^t, AI_i^f, AI_i^a and AI_i^0.

It is interesting to notice a linear relationship between indicators associated with innovation within individual sub-systems and indicators measuring architectural innovation (Figure 25.3).

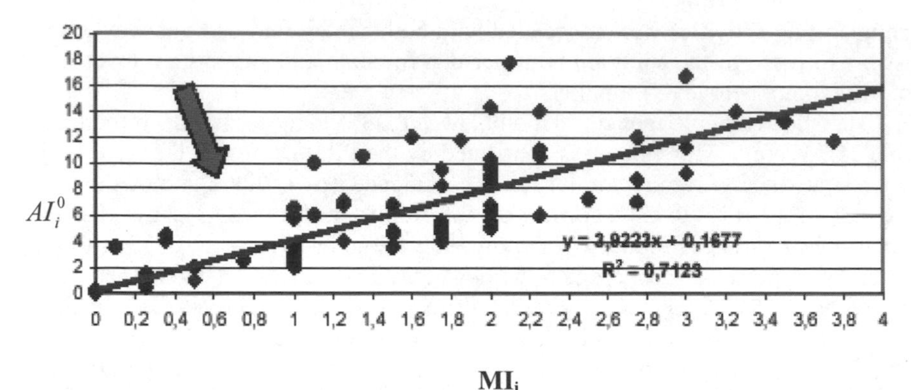

Figure 25.3 Modular innovation vs. architectural changes

If the architectural changes are decomposed according to the three perspectives of technological homogeneity, functional interdependence and assembly requirements, we realize that the major impact of modular innovation is on functional interdependences. Accordingly, firms which are subject to significant change in the components being designed and manufactured will also be involved in product architecture innovation, affecting especially the design process and its relationship with other firms.

Firms have also been classified according to their innovative capability, which has been measured by an indicator (IC_i) developed considering variables such as:

- existence of an R&D unit or, at least, of a design office,
- involvement in international R&D projects
- patents registered within the last five years
- technological tools (CAD, CAM, CAE)
- size and employment policy of R&D unit.

The result is shown in Figure 25.4; the horizontal axis captures the impact of architectural innovation in firm *i*, while the vertical axis represents the firms' innovative capability. The bubble size represents company size. According to the classification shown in Figure 25.2, some of the companies within the sample can be considered as "unchanged" or "overqualified". Although the situation appears encouraging at a first glance, the graph brings to light a small group of "inadequate" firms, whose innovative capability appears to be substantially lower than the required capability. Further analysis on this group of firms has revealed that they are characterized by common features:

- average size is lower than sample average (about 63,7% of "inadequate" companies have less than 50 employees)
- companies are either independent or belonging to small groups. Conversely, none of the firms in the sample that are subsidiaries to large multinational groups fall in this category (Figure 25.5).

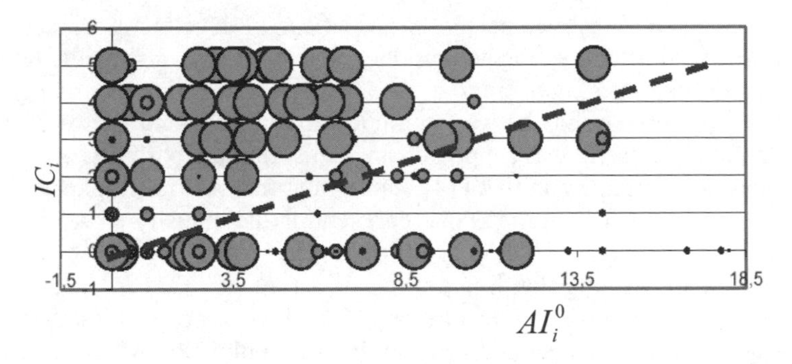

Figure 25.4 Architectural innovation vs. innovative capability (and firm size)

Finally, if the architectural changes are decomposed according to the three perspectives of technological homogeneity, functional interdependence and assembly requirements, the impact of changes in functional interdependencies appears to be stronger and, therefore, harder to tackle. Hence, it is possible to foresee that a major problem for firms facing architectural innovation will consist of creating and managing new communication channels, filters, and product development strategies with which they may interface with other players. In contrast, it may be expected that changes in specific technical fields will not be as disruptive (since technology can be outsourced), nor will be changes associated to assembly processes (as production can more easily be outsourced).

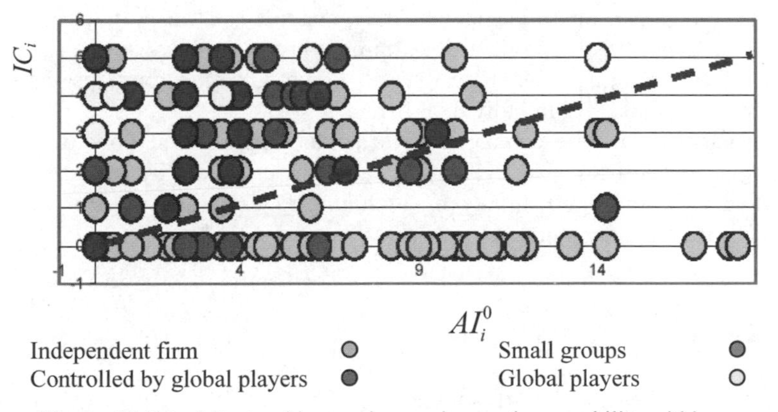

Figure 25.5 Architectural innovation vs. innovative capability within groups

25.4 Conclusions

The influence of product architecture on the organization of individual firms and of value chains is usually accepted and taken for granted. This relationship has been explored, from an empirical perspective, with the aim of developing a practical tool for the management of innovation and product development at a strategic level.

The overarching message, from a theoretical perspective, is the empirical validation of the relationship between product architecture and the allocation of activities among firms in the supply chain. This finding partially answers the research questions raised by Ulrich [14] concerning this relationship. Specifically, the correlation found between the DSM representing the industry structure and the functional interdependence DSM confirms the major role of functional interdependence in defining the organization of activities. This is true, at least in the case of a product whose architecture exhibits a low degree of modularity, such as a passenger car. It is reasonable to expect that the application of this method to a different kind of product, whose architecture exhibits a higher degree of modularity, such as personal computers, might lead to highlight the importance of other perspectives, such as technological homogeneity. Hence, the application of the method to another industrial sector seems worthy of further study.

The method can also be used by individual firms in order to evaluate the impact of innovation and to estimate the need to renovate production choices and technical competencies. The tool can be applied by companies in order to structure and organize architectural knowledge, bringing to light and quantifying the effects of architectural innovation.

From a policy maker's point of view, the method can be applied to the entire value chain, in order to forecast the industry level effects of products architectural innovation and to identify supporting policies for those firms whose capabilities are inadequate to meet the requirements set by technological innovation. For instance, the role of functional inter-dependence may suggest the need to pay greater attention and support to the product development process. This can be done, for example, by encouraging the redesign of communication patterns between suppliers and OEM and within individual firms.

Further studies could be conducted in order to support these conclusions; by studying other industries, by evaluating innovative capability through more sophisticated indicators, and, finally by including factors such as firm size and geographic location in the study.

25.5 Acknowledgements

The authors thank the Chamber of Commerce of Turin for having provided the data used in Section 25.3 of the paper.

25.6 References

[1] Andersson, S., Nilsson, P., Malmqvist, J., 2003, "Exploring Requirements Management in the Automotive Industry," *Proceedings of ICED'03*, Stockholm, Sweden.

[2] Camuffo, A., 2001, "Rolling Out a 'World Car': Globalization, Outsourcing and Modularity in the Auto Industry," *Working Paper, IMVP (International Motor Vehicle Program), available at:* http://imvp.mit.edu/papers/0001/camuffo1.pdf

[3] Cantamessa, M., Rafele, C., 2002, "Modular products and product modularità – implications for new product development," *Proc. DESIGN2002, 7th Int. Design Conference*, May 14-17, 2002, Dubrovnik.

[4] Cebon P., Hauptman O., Shenkar C., 2001, *Industries in the Making: Product Modularity, Technological Innovation and the Product Lifecycle*, Melbourne Business School.

[5] Henderson R., Clark K.B., 1990, "Architectural innovation: the reconfiguration of existing product technologies and the failure of established firms," In: *Administrative Science Quarterly*, No. 35, pp. 9-30.

[6] Jackson, D.A., 1995, "PROTEST: a PROcrustean randomization TEST of community environment concordance," In: "*Ecoscience*", Vol. 2, pp. 297-303.

[7] Kruskal J.B., Wish M., 1978, *Multidimensional Scaling*, Sage Publications, Beverly Hills, CA.

[8] Mantel, N., 1967, "The detection of disease clustering and a generalized regression approach," In: "*Cancer Research*", No. 27, pp. 209-220.

[9] Novak S., and Eppinger, S.D., 2001, "Sourcing by design: product complexity and the supply chain," *Management Science*, Vol. 47, No.1, pp. 189-204.

[10] Pimmler T.U., Eppinger, S.D., 1994, "Integration Analysis of Product Decompositions," *Proc. ASME Design Theory and Methodology Conference*, September 1994, Minneapolis, MN.

[11] Sanchez R., Mahoney J.T., 1996, "Modularity, Flexibility, and Knowledge Management in Product and Organization Design," *Strategic Management Journal*, Vol. 17, Winter Special Issue, pp. 63-76.

[12] *STEP Economics*, 2003, "La mappatura della filiera autoveicolare in Piemonte", final report of the "Dall'Idea all'Auto" project (in Italian).

[13] Steward D., 1981, "The Design Structure Matrix: a Method for Managing the Design of Complex Systems," In: *"IEEE Transactions on Engineering Management,"* Vol. 78, pp. 71-74.

[14] Ulrich, K., 1995, "The Role of Product Architecture in the Manufacturing Firm," In: "*Research Policy*", Vol. 24, pp. 419-440.

26

Integration of Cost Models in Design and Manufacturing

Nicolas Perry, Magali Mauchand, and Alain Bernard

Abstract: Cost control in the early phase of the product life cycle became a major competitiveness asset for the companies, due to the world competition. After defining the problems related to these control difficulties, an approach using a concept of cost entity related to the activities of the product to be designed and realized is presented. This approach is applied to the fields of the sand casting foundry. The enterprise modelling difficulties, limits of a global cost modelling and some specifics limitations of the tool used for this development, as well as the limits of a generic approach will be highlighted.

Keywords: cost management, enterprise - product - cost modelling, cost entity

26.1 Objective and Brief Overview

In the early nineteen-seventies, studies in the United Kingdom and in the United States highlighted the strategic role of the design activities. The conclusions lead both companies and authorities towards new approaches in order to improve the economic performances of the companies. At the end of the Eighties, the paramount role of quality in design was reinforced in the United States by the Made in America report from the MIT Commission on the Productivity. These conclusions were confirmed in 1991, by the Improved Engineering Design: Designing for Competitive Advantage report, from the United States National Research Council "Engineering Design Theory and Methodology". According to Perrin [1], the design phase is the key factor of the product development process. The ability to produce new products with high quality, low cost and which fit the customer requests is fundamental to improve the nation competitiveness [2-3]. Consequently, the costs, and cost management from the early design to the end delivery, become as important as the other technical requests.

Due to the global market and the worldwide competition, reactivity and agility are the only way to maintain the enterprise competitiveness. This can be characterized by the ability to change its products and/or processes in a very short time and at minimal cost. The cost control, at the early stage of design, becomes a key factor of success, since it is at this phase that an average up to 70 to 80% of the end product costs are fixed depending to the kind of production.

Moreover, the costs distribution (direct and non-direct) is changing: more time and services dedicated to the studies for smaller product batches and shorter product life. The former fees sharing out methods, the analytic cost accounting or by analogical method, no longer give efficient results. Then, thanks to studies from CAM-I (Computer Aided Manufacturing-International) and authors like Johnson and Kaplan, the increasing gap between "traditional methods" of cost estimation and the new management requirements were highlighted.

All these works lead to new approaches integrating the complete cost and spread accounting methods based on the enterprise activities (ABC for instance). French economist, since the sixties, also developed a method based on a single cost indicator identification through all the steps of the product development process (Added Value Unit method). We implemented such a costing management in a French sand casting foundry in order to allow a several level management, based on indicators linked with the exact costs of the product to be delivered [4]. In this Ph.D. thesis we validated not only the concepts but also the methodology needed through a complete numerical traceability.

The work presented in this paper, is linked with this former study and uses a concept called cost entity [5]. It includes several concepts, the cost indicators from the activity based accounting methods, the features from the CAD and the homogeneity from the analytical cost accounting. Consequently, in order to define a cost entity, it is necessary to fill in several attributes linking technical and economical variables. The product model uses the concept of manufacturing feature. The cost is evaluated on the basis of knowledge and reasoning models with the tool "Cost Advantage" (from Cognition Europe), giving costs information to the CAD model until any semantics related to the cost are empty. This model (called costgramme) makes the expertise of the manufacturing cost available to the designer.

Some models, dedicated to the sand casting production of primary parts, were created with the wish to evaluate the limit to be reached in order to make a meta-model that could be deployed in all the sand casting industries. Thus, the goals of this study are to create the model related to the sand casting application to be as generic as possible, and to determine up to what point they are transposable from a company to another (or from a production line to another). Therefore, we will have to define and discuss the limits of the concepts from the triptych product/process/cost, and what level of dctail is necessary to implement in the most industrial environment.

26.2 The Cost Entity Concept and the Modelling Logic

The aim of our study is to carefully manage the costs (direct and indirect) during the production of sand casting parts. As illustrated previously, it is imperative to give a tool to the engineers of the engineering and design departments with an aim of controlling the costs of the parts design. In collaboration with the company Cognition Europe, and on the basis of the tool Cost Advantage, we work on the costs models to apply in the case of the foundry sands steel parts. We, based on a preceding work, propose an approach integrated for the sand foundry, realized within the framework of a thesis in partnership with company SMC Colombier Fountain (France) of group AFE Métal. This work, formalized the base of trade knowledge necessary to the control of the product life cycle in a foundry company. In addition, we validated an approach, a methodology and a deployment leading to ensure an exact knowledge of the parts costs and their impact on the output of the company [6].

Cost Entity Concept

A Cost Entity is a grouping of costs associated with the resources consumed by an activity (Figure 26.1) [7, 8]. The general condition is due to the homogeneity of the resources, which makes it possible to associate a single indicator the entity cost [9]. The model allows the expertise formalization, knowledge capitalization and to have, at the early design phase, information about the production step. Moreover, it helps the communication between several collaborators during the product life cycle.

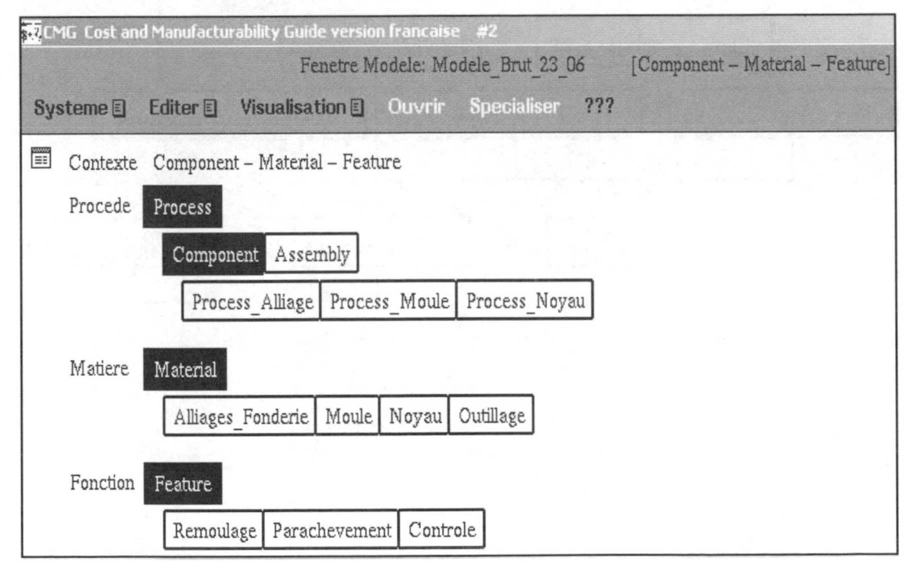

Figure 26.1 Sand casting base components (Process, Material, Feature)

Context / Instantiation

The contexts specify the definite entities in three levels in our model. The first is defined in a process level, the second on a material level and the last is directly related to the feature. This context is a cross between a process, a material and a feature, connected to an environment (Figure 26.2). Complete realizations are specified, depending to the exact process chosen our forecast for the part.

Based on this analysis we have created a generic model using Cost Advantage Software. The first step is to closely define the production process dedicated to this industry. The master parameters acting upon the product cost must be identified and used to enrich the cost semantic of the model.

Calculations are simple, taking into account volumes of material, rates of production, losses and the machine and labor costs. Put aside the difficulty in knowing the exact parameters, the rules of calculations are simple; there is not the problem of modeling which positions more on hierarchy problems and of model organization.

The rules of calculation then implemented will make it possible for the future user to provide only the relevant data about its study. Indeed, only the operational process, rates, dimensions, numbers of cores (*etc..*) will be required (or deduced directly in a CAD software) to allow an automatic calculation of the cost of the part according to its particular characteristics.

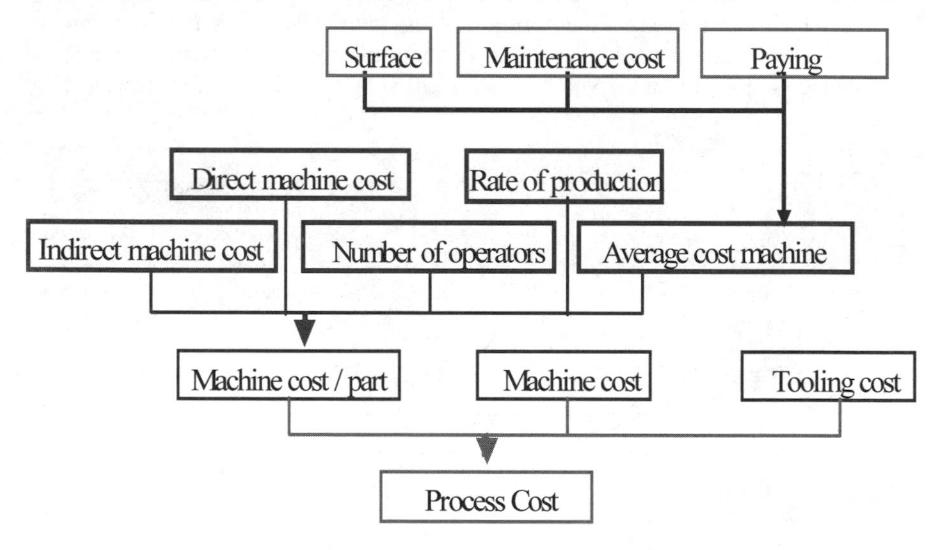

Figure 26.2 Process cost structure

26.3 Sand Casting Modeling

We created a generic model using Cost Advantage Software after a sand casting foundry process analysis. First, we started with the production process definition dedicated to this industry. We also identified a master parameter acting upon the product cost in order to master the cost of the product. This approach highlights the problems of the contexts characteristics. How to define a significant cost indicator for one part (a batch, the global production…) and with which level of detail (in order to be generic)? We will not answer this question, but we will present the paths or solutions we used.

This model is based on the SMC Colombier Fontaine Foundry (France), from AFE Metal group [10]. We will focus on the production phase, from the sand elaboration, the tooling machining and the parts perfecting and limit in this area. We took into account the several physical compounds (raw material, tooling…) and the elements needed to manufacture a part linked with the major indicators dealing with the final cost (loss, scrap ratio, production rate…).

Figure 26.3 represents a transposition under the concepts of Cost Advantage of this model gathering the three levels of entities defined in the software. For example, with the mould, the tooling and the cores, which are components needed to carry out the assembly (the moulding) by the operation (feature) of weating (positioning and assembly). In this approach, it is necessary to define the final part, to carry out the two assemblies, first the moulding (realization of the mould), then the casting.

In terms of model design, the functional view identifies the assemblies needed, it is then necessary to define the components and choose and define the related operations. An ascending step must be practiced, starting with the components up to the definition of the assemblies. The costs are calculated according to Figure 26.2. The implemented data structure is shown prior in Figure 26.4. Calculations are simply taking into account volumes of material, rates of production, losses and the machine and labour costs.

The implemented rules of calculation will make it possible to the user to only fulfil the relevant data about its study. Indeed, only the operational process, rates, dimensions and numbers of cores will be taken into account (or extracted from a CAD software) to allow an automatic calculation of the cost according to its characteristics.

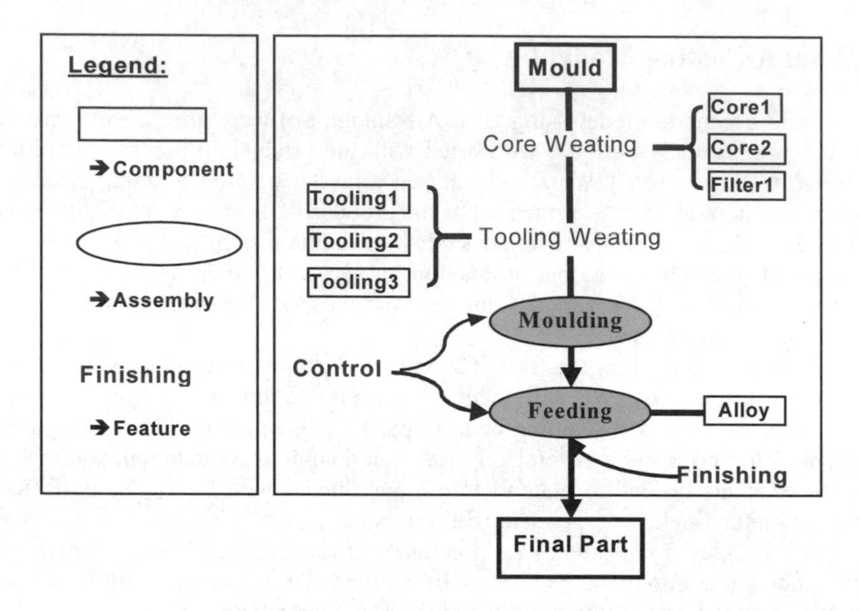

Figure 26.3 Cost advantage modeling example at the assembly level

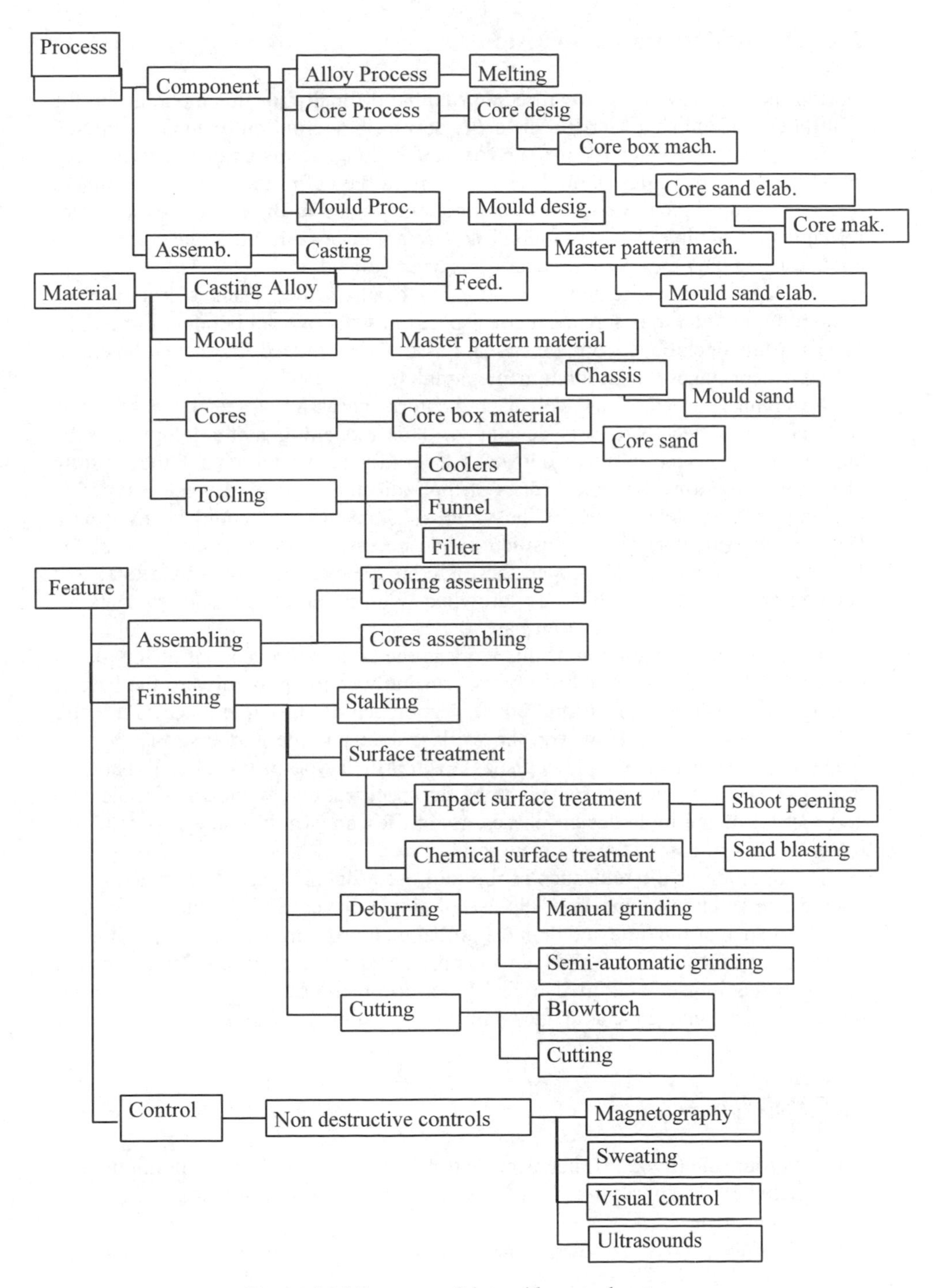

Figure 26.4 Structure of data with cost advantage

26.4 Discussion and Conclusion

During this work, we have identified a principal difficulty for this modeling in the multiplicity of the characteristic elements and in the definition of their hierarchy. Even if a manufacturing process, such as sand casting, seems simple, it uses many components (alloy, cores, mould…). We limited the definition in terms of model refinement since each one of these components could be the subject of a further refinement modeling. A basic minimal skeleton, transposable from one company to another using the sand casting process, has been defined.

The logic of cost-oriented modelling has been clarified using the concept of cost entity. In order to ensure a generic aspect our work, we deliberately limited the details of the operations, components and assemblies. Indeed, the development of these elements takes into account many parameters.

A minimal structure, as well as indicators, necessary to evaluate all costs without the indirect part, was developed. This modelling methodology will be applied to other sand casting companies to configure the model for the existing processes and define the exact values of the indicators. This will make it possible to compare the effectiveness of the various companies and could be used as a Benchmark evaluator. The acquisition of the necessary information is one of the foreseeable difficulties. Moreover, these factors are often managed by a total cost accounting system and thus are drowned within indicators and systems of management that are not very transparent.

A significant continuation of this work is the taking into account of the global costs related mainly to the indirect parts. Our introduction puts forward the lack of management of these aspects and our first approach did not give place to a better control of these factors. However, the work is done and the workers must be paid (designer, maintenance, buyers, logistics) even if their work is not as well managed through a cost management system. A better specification (by means of indicators and metric) of the tools design phases, tools lifespan, would make it possible to integrate the real cost of the complete series.

The question of the relevance of the tool used for this type of approach arises then. Some solutions come from the use of single or very limited number of cost indicators such as the time and define a global enterprise minimum cost per hour to balance its financial objectives. Such an approach allows a multi level management of the parts, impact and give a real-time information to asses the enterprise objectives and manage the strategic tactic and operational decisions.

26.5 References

[1] Perrin, J., 1996, "Cohérence, pertinence et évaluation économique des activités de conception," *Cohérence, Pertinence et Evaluation, ECOSIP, Economica*.

[2] Ostwald, P.F., 2000, *Construction Cost Analysis & Estimating*, ISBN 0-13-083207-3.

[3] Giannopoulos, N., Roy, R., Taratoukhine, V., Sarasua-Echeverria, A., 2003, "Embedded systems software cost estimating within the concurrent engineering environment," *CE03: the Vision for the Future Generation in Research and Applications*, ISBN 90 5809 623 8, pp. 352-357.

[4] Bernard, A., Perry, N., Delplace, J.C., Gabriel, S., 2002, "Optimization of complete design process for sand casting foundry," *Proceedings of IDMME'2002 Conference*, Clermont-Ferrand, France.

[5] Perry, N., Bernard, A., 2003, "Cost objective PLM and CE," *CE: The Vision for the Future Generation in Research and Applications*, ISBN 90 5809 622 X, pp. 817-822.

[6] Delplace, J.C., 2004, "L'ingénierie numérique pour l'amélioration des processus décisionnels et opérationnels en fonderie," *PH.D. Thesis, IRCCyN – Ecole Centrale Nantes & University of Nantes*, Nantes.

[7] Liu, Y., Basson, A.H., 2004, "Case Study Using COM (OLE) to Link CAD and Manufacturing Cost Estimation Software," *COMA'04, International Conference on Competitive Manufacturing*, Stellenbosch, South Africa, ISBN 0-7972-1018-0, pp. 168-174.

[8] H'Mida, F., 2002, "L'approche entité coût pour l'estimation des coûts en production mécanique," *Ph.D. Thesis, LGIPM-ENSAM Metz-Univ.*, Metz.

[9] Bernard, A., Perry, N., Delplace, J.C., Gabriel, S., 2003, "Quotation for the value added assessment during product development and production processes," *2003 CIRP Design Seminar Proc.*, May 2003, Grenoble, France.

[10] Perry, N., Mauchand, M., Bernard, A., 2003, "Modèles de coûts en fonderie sable: les limites d'une approche générique," *CPI 2003, 3ème Colloque Int. en Conception et Production Intégrées*, Octobre, Ref 044, Meknes, Maroc.

Part VI

Product Life Cycle

27

Life Cycle Product Support in the Digital Age

Jörg Niemann, and Engelbert Westkämper

Abstract: In order to master the constraints of an effective life cycle management, all kinds of data and information concerning the actual machine behavior need to be available. Latest developments in information and communication technologies allow a "look inside" the machine to visualize its actual status. This paper will establish a performance controlling system, including a fleet management system, to gain competitive advantages in manufacturing by focusing on the entire product life cycle.

Keywords: Life cycle management, life cycle controlling, production management, PDM

27.1 Introduction

The rising globalization of companies combined with the potentials of modern information and communication technologies have combined to make the world a global village with global competitors. Global competition also means benchmarking one's performance to the worldwide best of class. Therefore, competitive manufacturing strategies have to focus on efficient machine utilization, reliable processes, and an effective overall performance controlling even the fringe ranges of technological potentials. Immediate reactions to performance deviations are crucial for keeping the budget lines and for securing calculated profit ratios. Therefore, we need transparent manufacturing systems, which provide all kinds of data necessary to monitor online key indicators of machine performance. Industrial corporations generally direct their strategies at economic targets manufacturing technical products. Their main business is developing, producing and operating products either for individual customers or for complete sectors of the market. Service and maintenance are necessary for many companies in order to attain lasting business relationships with customers and to generate added value [1, 2, 3].

27.2 Reliable Data for Transparent Life Cycles

The critical factor for success in these developments is the management of data and information. The volume of information is exploding and industry needs actual and reliable information on the state–of–the–art [4].

Data Management

Life cycle management offers the opportunity to maximize the benefits of each product in all its phases. To do this, lifelong information, up–to–date in each situation, is required. Most of these data are generated during the production phases and up to the end of the ramp-up phase. In the usage phases, in particular in the maintenance phases, the basic data about the actual condition of the product change permanently. As with the assembly process, service, diagnosis and disassembly need actual data and a background of operation and programs coming out of planning processes [5].

Today there are no system solutions, which have a potential for the future. It seems to be advantageous to open the PDM systems for data management. PDM systems are centralized systems and their application for life cycle management is reduced by their characteristic functionality, which is oriented to support engineering. PDM systems may be applied for managing the digital data of shops and assembly or disassembly systems. They support distributed engineering with common standards and communication functions.

Life Cycle Platform

A future development has to take into account the possibilities for implementing all basic products data into their internal information system. This would help to support all operations done with the product, and surrounding activities, with documentation [2, 6, 7, 8, 9, 10, 11].

Basic standards for management and exchange of data are available. A key problem is the validity of data and the protection of know-how. In the automobile industry's world standards for the exchange of data for life cycle and managing services, including logistics, are developed. The machine tool industry should start with basic product and process models to find solutions for standardization and the application of new services in the life of technical products.

In the future, there will, of course, be a high potential for product–oriented knowledge management. Knowledge is needed to optimize the production and products finished by assembly. But facing the flow of costs and the efficiency of products, it seems necessary to support the usage phases especially in critical situations, like breakdown, change of usage, or change of configuration by disassembly and assembly [12, 15, 16, 17].

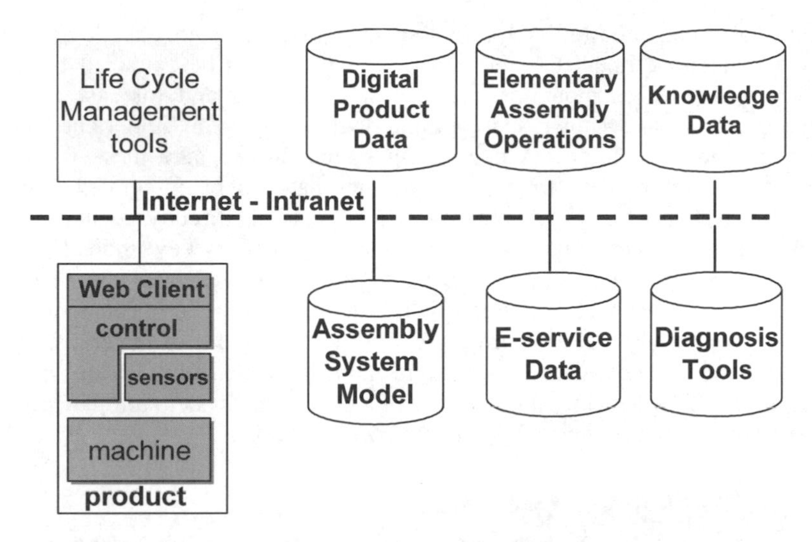

Figure 27.1 Platform for the integrated management of products life cycle

27.3 Monitoring and Evaluation of System's Performance

Structural Framework for System Monitoring

Different data from various sources are needed in order to control the cost of a manufacturing system. Obviously the master control of system behaviour requires machine and machining data. Some of these data can be easily acquired from the machine control system [18, 19].

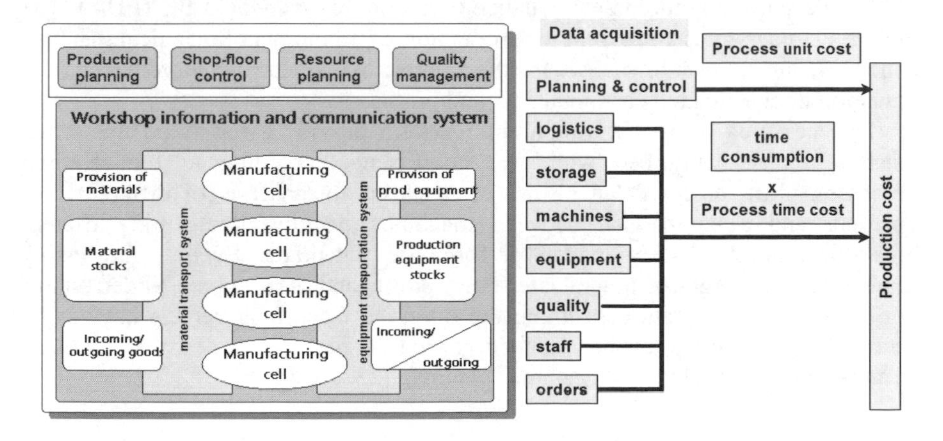

Figure 27.2 Data acquisition in manufacturing segments

This offers the opportunity for remote machine access to data logging via internet or telephone lines. The relevant machine data can be extracted from the data flow and serve as input for in-situ cost monitoring and forecast. Various research projects have shown that optimal logistics play an important role in avoiding performance losses. A controlling system has to take these facts into account and therefore data from parts logistics have to be integrated into the supervision system [4, 14]. Another group of data is directly related to the machine's environment. Figure 27.2 shows the possible key indicators and improved planning data for production management, derived from the described model [20, 21]. Order, size, required quality, number of workers, calculated lead times, *etc.*. can be taken directly from the work schedule, bill of materials, or the order management. These data are static and can be extracted from various internal sources. Figure 27.3 describes the structure of a controlling system implemented on a precision machining center at IFF, University of Stuttgart.

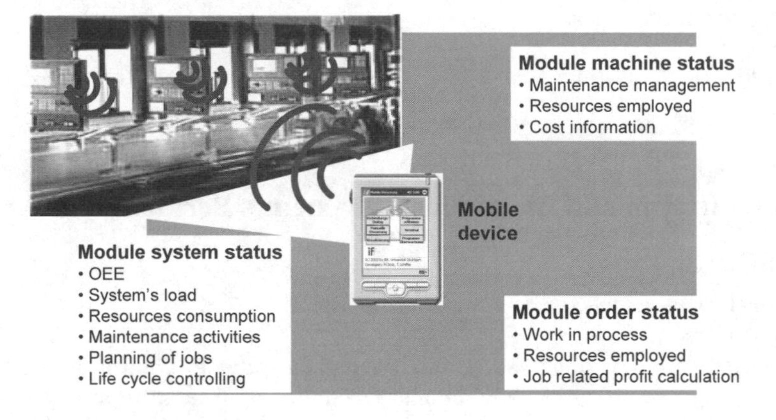

Figure 27.3 Online manufacturing data for a controlling system of manufacturing segments

The data are monitored and visualized via a mobile handheld PC (PDA). The mobile PDA serves as a platform for production staff in terms of technical machine control (failures, breakdowns *etc.*.) and economic manufacturing surveillance (*e.g.* deviation from estimated cost, total cost and profit…).

The measured data of the monitored system provide a report on the actual machine status. Multiplied with cost coefficients according to the required processes, profit analysis can be made. A sensitivity analysis of different cost positions and a comparison between machine operation times and different breakdown times identify hidden performance potentials. Even a forecasting module can be integrated to simulate future profits and performance under "status quo" conditions. All data concerning the observed machining centers have to be accumulated on the top level of production program planning to derive key actions in mid-term performance and resources planning [22, 23, 24].

Permanent System Performance Evaluation

Process quality can be evaluated by monitoring the actual process at different times (process monitoring) and compressing and comparing real data with planned data (process controlling). Process information systems provide information support for this task. At the moment, the main challenges are increasing process orientation and implementing the necessary organizational and information tasks.

The relevant processes that mainly contribute to the result need to be identified from the wide range of involved company processes. In order to obtain sound process controlling information, the real data needs to be compared with the planned data (Figures 27.4, 27.5). Process information systems are much more than process cost calculation systems assisted by data-processing, because data mainly concerned with value are determined rather than quantifiable data, such as cycle times or the time factor of a process organizational unit [21].

The permanent machine data acquisition also allows the evaluation of overall equipment effectiveness (OEE). This measurement has its origin in the philosophy of total production management. The OEE measures all losses occurring during machine operation. All sources of losses are combined into one % factor– the OEE factor. This number ranges from 0 to 1 meaning that a factor of 1 (or 100%) describes the optimal machine performance.

The OEE factor consists of six major loss categories. The data that are necessary to analyze and aggregate these categories can mainly be acquired from machine and production operation. Therefore, it is possible to analyze and optimize the machine cost by these controlling systems. The losses identified by the OEE analysis can be interpreted as lost profits (or added cost, "opportunity cost") because for inappropriate machine operation, parts cannot be sold and additional labor and material costs are increased. The measured equipment effectiveness losses can be expressed as a coefficient and transformed into economical values by linking them to resource process cost rates. The expression of this "performance loss" usually represents an enormous and often underestimated value adding potential in production, according to research findings by the Institute for Industrial Manufacturing and Management (IFF) and the Fraunhofer IPA, Stuttgart. The retrograde analysis and cumulated analysis of production data also show the main cost drivers and "expensive" work steps. This knowledge is useful for re–organization planning, re–engineering, and long–term technology planning purposes [24].

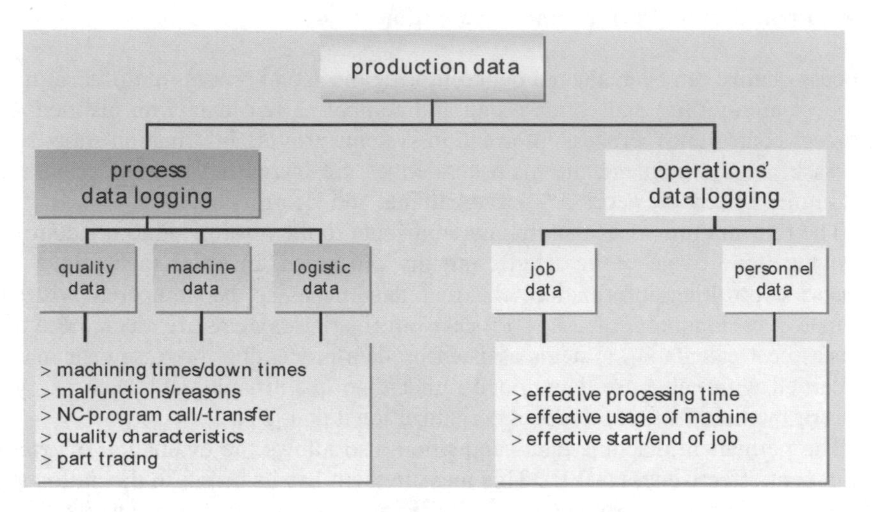

Figure 27.4 Framework for manufacturing data acquisition

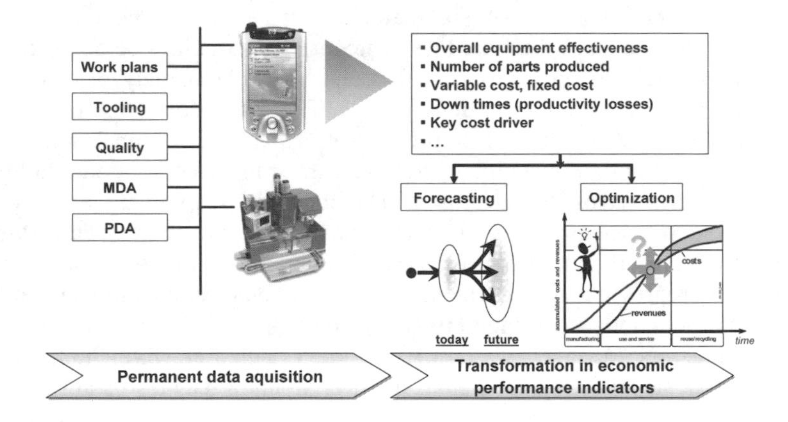

Figure 27.5 Economic machine performance controlling

27.4 Root Cause Analysis and Risk Management

The method of Life Cycle Controlling (LCC) comprises a comprehensive life cycle calculation, which takes technical and organizational parameters into account. The method focuses on maximizing the overall usage of a product throughout its life cycle. All expenses and returns in all phases of a product's life cycle are compared with one another and balanced. A structured classification assists in compiling a catalogue of positions, which is as complete as possible. It is clear that a catalogue intended to encompass the entire life cycle of a product is neither absolutely complete nor monetarily accurate. For controlling to be consistent, the calculation must be continuously updated in order to recognize deviations as quickly as possible and to take counter–measures.

Cost and profit positions arise over a long period of time. The total balance is therefore based on the method of dynamic investment calculation. Here the capital method has asserted itself, with the investment discounted from the net present value. Thus, profitability can be assessed against alternative investment decisions (*e.g.* other equipment). Once the cost and profit catalogues have been completed, the positions can then be depicted as a graph in an aggregated form. This representation can be seen as a "blueprint" or benchmark and is filed for analyzing later deviations. The graphical depiction also enables a strategic analysis to be performed with regard to the main costs and profit drivers occurring during the product life cycle.

The "system life cycle" of complex investment goods and products with system characteristics serves to create and analyze models as part of company-specific and problem-specific analyses. Thus, concrete life-cycle related decisions can be prepared. Specific life cycle calculations serve as operands into which life cycle costs are incorporated and then recorded as returns, profits or deposits. Models are formed for the model-assisted planning, recording and analysis of life cycle costs or other monetary factors. When making purchasing decisions, all procurement/acquisition costs, subsequent costs, and any existing differences in quality or performance, can also be adequately included. Quantitative and qualitative outputs such as profits are also taken into consideration in a model-related way. The system unit of life cycle control thus makes a model analysis which is adapted to the special features of the life cycle object possible. The aim is to derive information of strategic importance to corporate decisions. Therefore, each object has a major influence on costs - especially in the early stages - and interactions exist between costs associated with various phases of the life cycle [20, 21, 24, 25, 26, 27].

The concept of a continuous profitability calculation throughout the life cycle of an object should enable statements to be made concerning the economical effects of technical and organizational measures within the scope of successful technology management. Due to such a long-term viewpoint, LCC analyses reveal both hidden cost drivers and profit potentials during a life cycle. So the analysis also supplies coefficients for outsourcing strategies up to and including modern full-service concept calculations and contracts. The identification of cost drivers permits these processes to be analyzed more accurately and, where necessary, to achieve a more efficient provision of services by redistributing performance bundles between value-adding partners. An LCC analysis of the manufacturing portfolio (resources used/performance output) may thus point out unused potentials and business risks very early on. To summarize, LCC analyses can be used for the following purposes:

- Comprehensive investment calculation of assets (both investment costs and running costs)
- Budgeting
- Trade-offs analysis between investment and running cost
- Identification of cost and profit drivers

- Impact of outsourcing decisions
- Cash-flow analyses, ROI
- Analysis of "what if" scenarios (prognosis)
- Optimum point in time for implementing machines
- Pro-active budget planning
- Designing and shaping life cycle cost contracts
- Optimum point for machine replacement
- Machine's upgrade decision

However, as critical profitability values are only revealed by the analysis, it is not possible to analyze equipment in this way as far as its overall profitability is concerned. The impact of these values can be analyzed in more depth by carrying out sensitivity analyses. As a result, by accepting alternative monetary values for sensitive positions, forecasting uncertainties can be reduced. Alternative scenarios (*e.g.* worst-case; best-case) provide an indication of the degree of risk involved in an investment. It may be that risky positions are introduced from external service providers. Life cycle controlling thus supports financial planning (cash flow, capital demands) and enables far-sighted budget planning.

Customers today increasingly require security with regard to future operating costs. They are also increasingly demanding equipment manufacturers to give them contractually fixed guarantees related to future maximum operating costs. Such life cycle cost contracts limit cost risks and involve equipment manufacturers in product responsibility, as illustrated in Figure 27.6.

Corresponding analyses serve to increase the transparency, identification and examination of interactions between costs in various phases and also serve to evaluate alternative actions, thus paving the way for life cycle-related decisions. Life cycle controlling is comprised of cost management, cost controlling, and additional management activities associated with life cycles. By analyzing product life cycles, numerous spatial disparities between the fixing of costs and profits and how they occur during a life cycle (*i.e.* trade-offs) become clearer. The traditional method of calculation is often inadequate for evaluating equipment in the long-run with regard to their overall profitability. With life cycle controlling, however, such interactions can be revealed and utilized to enable the future-oriented, long-term management of costs and profits.

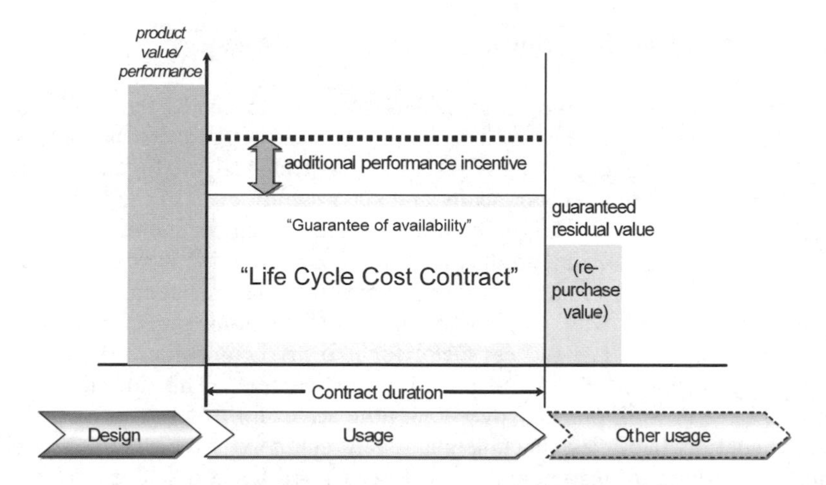

Figure 27.6 Life cycle cost contract

27.5 Learning from Fleet Management

Modern information and communication technologies not only enable equipment status to be read out in-situ, but also to make the acquired data available over long distances. Thus, a system operator is able to network with customers worldwide. Should the networking of system management not prove possible, he/she is at least able to analyze centrally and individually the production data obtained from all the systems in use, which have been distributed to customers [5]. Subsequently, fleet-assisted learning curves can be generated with an incline much steeper than those of competitors as a result of the scaling effect. The broadness of the database also allows benchmarking activities and the identification of best-practice solutions.

27.6 State of Implementation

Many industrial companies think about linking all their different digital data sources to a life cycle–oriented information system. Up to now the different standards and data formats do not support this. It is generally accepted that the overall analysis and data provision provides a huge potential for performance improvements.

Manufacturing units are designed for one specific task and will be dissolved after the product is taken out of the market. Simulation is used in the initial phase to design and plan such systems. Only rarely is it used for performance evaluation. Most of the single elements of thc proposed model are already available as "single solution". The boosting impact can be reached by combining them into a holistic management information system.

27.7 Summary and Outlook

The development of modern products is decisively influenced by the application of technologies that contribute to increasing efficiency. Products nowadays are complex highly integrated systems with internal technical intelligence, enabling them to be used reliably, economically and successfully even in the fringe ranges of technology. As a result, business strategies are aiming for increasingly perfect technical systems, optimizing product utilization and maximizing added value over the entire lifetime of a product. In this context, the total management of product life cycles associated with the integration of information and communications systems is becoming a key success factor for industrial companies. Manufacturers today have to guarantee process reliability by contract. This implies that the machines have to work properly over long time scales for different work tasks. To meet the contract liabilities, machine manufacturers monitor their facilities, collect all manufacturing information and try to forecast and boost machine performance by intelligent process optimization. Modern IT also allows adding more data sources and experiences from other machines all over the world.

27.8 References

[1] Seliger, G., 1997, "More use with fewer resources – a contribution towards sustainable development," In: Krause, F.-L., Seliger, G., *Life Cycle Networks: Proceedings of the 4th CIRP International Seminar on Life Cycle Engineering*, 26-27 June 1997, Berlin, Germany, Chapman and Hall, London.

[2] Zülch, G., Schiller, E.F., Müller, R., 1997, "A disassembly information system," In: Krause, F.-L., Seliger, G., *Life Cycle Networks: Proceedings of the 4th CIRP International Seminar on Life Cycle Engineering*, 26-27 June 1997, Berlin, Germany, Chapman and Hall, London.

[3] Brissaud, D., Tichkiewitch, S., 2001, "Product models for life-cycle," In: *CIRP Annals Manufacturing Technology*, Vol. 50(1), pp. 105-108.

[4] Seliger, G., Grudzien, W., Zaidi, H., 1999, "New Methods of Product Data Provision for a Simplified Disassembly," In: *Proceedings of the Life Cycle Design '99*, June 1999, Kingston, Canada.

[5] Feldmann, K., 2002, "Integrated Product Policy - Chance and Challenge," *9th CIRP International Seminar on Life Cycle Engineering*, April 9-10, 2002, Erlangen, Germany, Bamberg: Meisenbach.

[6] Tichkiewitch, S., Brissaud, D., 2001, "Product models for life-cycle," *CIRP Annals Manufacturing Technology*, Vol. 50/1, pp. 105-108.

[7] Feldmann, K., Trautner, S., Meedt, O., 1999, "Computer Based Design for Recycling and Disassembly Planning," *Proc. of 6th International Seminar on Life Cycle Eng.* June 21-23 1999, Queen's Univ., Kingston, Canada.

[8] Jackson, P., Wallace, D., Kegg, R., 1997, "An analytical method for integrating environmental and traditional design considerations," *Annals of the CIRP*, Vol. 46/1, pp. 355-360.

[9] Gu, P., Hashemian, M., Sosale, S., 1997, "An integrated modular design methodology for life cycle engineering," *Annals of the CIRP*, Vol. 46/1, pp. 71-74.
[10] Birkhofer, H., Schott H., 1995, "Development of Environmentally Friendly Products – Methods, Material and Instruments," In: Jansen, H., Krause, F.-L., *Life cycle Modelling for innovative Products and Processes, Proceedings on life cycle modelling for innovative products and processes*, November / December 1995, Berlin, Germany, Chapman & Hall, pp. 432 – 443.
[11] Anderl, R., Daum, B., John, H., Pütter, C., 1997, "Cooperative Product Data Modelling," In: Krause, F.-L., Seliger, G., *Life Cycle Networks: Proc. of the 4th CIRP Int. Seminar on Life Cycle Eng.*, 26-27 June 1997, Berlin, Germany, Chapman and Hall, London, pp. 435 – 446.
[12] Krause, F.-L., Kind, C., 1995, "Potentials of information technology for life-cycle-oriented product and process development," In: Jansen, H., Krause, F.-L., *Life Cycle Modelling for Innovative Products and Processes, Proc. on life-cycle modelling for innovative products and processes*, Berlin, Germany, November/December 1995, Chapman & Hall, pp. 14 – 27
[13] Kimura, F, 2000, "A Methodology for Design and Management of Product Life Cycle Adapted to Product Usage Modes," *The 33rd CIRP Int. Seminar on Manufacturing Systems*, 5-7 June 2000, Stockholm, Sweden.
[14] Anderl, R., Daum, B., John, H., 2000, "Produktdatenmanagements zum Management des Produktlebenszyklus," In: *ProduktDatenManagement 1*, pp. 10-15.
[15] Arai, T., Aiyama, Y., Maeda, Y., Ota, J., 2000, "Agile Assembly System by "Plug & Produce"," *Annals of the CIRP*, Vol. 49/1, pp. 1-4.
[16] Berger, R., Krüger, J, Neubert, A., 1998, "Internet-basierter Teleservice," In: *Industry Management,* Vol. 6/98, GITO-Verlag, Berlin, Germany.
[17] Niemann, J., 2003, "Life cycle management- das Paradigma der ganzheitlichen Produktlebenslaufbetrachtung," In: *Bullinger*, Hans-Jörg (Hrsg.) u.a.: *Neue Organisationsformen im Unternehmen: Ein Handbuch für das moderne Management*. Berlin, Springer, pp. 813-826.
[18] Brussel, H. van, Valckenaers, P. (Hrsg.), 1999, "Intelligent Manufacturing Systems," *Proc. of the 2nd Int. Workshop on Intelligent Manuf. Systems*, September 22-24, 1999, Katholieke Universiteit Leuven, Leuven, Belgium.
[19] Westkämper, E., 2000, "Technical Intelligence for Manufacturing," In: Cochran, David S. (Hrsg.) u.a., *CIRP u.a.: The Third World Congress on Intelligent Manufacturing Processes & Systems: Proc.*; June 28-30, 2000, Cambridge, Massachusetts, USA, pp. 411-420.
[20] Kemminer J., 1999, *Lebenszyklusorientiertes Kosten- und Elösmanagement,* Gabler Verlag Wiesbaden.
[21] Westkämper, E., Niemann, J., 2002, "Life Cycle Controlling for Manufacturing Systems in web-based environments," In: *CIRP u.a., CIRP Design Seminar Proceedings*, 16-18 May, 2002, Hong Kong, China.
[22] Dowie, T., Simon, M., Fogg, B., 1995, "Product disassembly costing in a life cycle context," *Proc. of the Int. Conf. on Clean Electronics Products & Tech. (CONCEPT), IEE Conf. Publication.*

[23] Trender, L., 2000, “Entwicklungsintegrierte Kalkulation von Produktlebenszykluskosten auf Basis der ressourcenorientierten Prozesskostenrechnung,” In: *wbk Forschungsberichte aus dem Institut für Werkzeugmaschinen und Betriebstechnik der Universität Karlsruhe*. Bd. 98, Karlsruhe.

[24] Westkämper, E., Osten-Sacken, D.v.d., 1998, “Product Life Cycle Costing Applied to Manufacturing Systems,” *Annals of the CIRP*, Athens, Greece, Vol. 47.

[25] Kirk, S.J., Dell’Isola, A.J, 1995, “Life Cycle Costing for design professionals,” In: Siegwart, H., Senti, R., *Prod. Life Cycle Management*, Die Gestaltung eines integrierten Produktlebenszyklus, Schäffer Poeschel Verlag, McGraw-Hill, Inc.

[26] Schimmelpfeng, K., 2002, *Lebenszyklusorientiertes Produktionssystemcontrolling*, Deutscher Universitäts-Verlag GmbH, Wiesbaden, zugl. Habilitationsschrift Universität Hannover.

[27] Niemann, J., 2003, “Ökonomische Bewertung von Produktlebensläufen – Life Cycle Controlling,” In: Bullinger, Hans-Jörg (Hrsg.) u.a.: *Neue Organisationsformen im Unternehmen: Ein Handbuch für das moderne Management*. Berlin u.a., Springer, pp. 904-916.

28

Total Quality Management and Process Modeling for PLM in SME

Umberto Cugini, Andrea Ramelli, Caterina Rizzi and Marco Ugolotti

Abstract: Product Lifecycle Management (PLM) as a business strategy is becoming a must not only for big companies but also for small and medium enterprises SME's that consider product development a core competency. However, a PLM solution deeply impacts the business process and requires the analysis and, if necessary, the re-engineering of the process itself. This paper presents the application of process modeling and simulation techniques for the implementation of a PDM/PLM system within a SME using Total Quality Management procedures to integrate process analysis. Two As-Is models have been realized: the first, extracting the process knowledge from Total Quality Management procedures, and the second, interviewing a company's staff. A gap analysis has been carried out to identify first, which aspects of the process could be modified and improved introducing a PDM system, and then to forecast a complete extension to the PLM paradigm. A re-engineered process, described by the To-Be model, has been designed and compared with the As-Is models using a discrete events simulator. On the basis of simulation results, considerations have been drawn related both to the new process asset and its future evolution to full implementation of the PLM paradigm.

Keywords: Process Modeling, Process Simulation, PLM-Product Lifecycle Management, Total Quality Management.

28.1 Introduction

Product Lifecycle Management - PLM is increasingly becoming a must for those enterprises that consider product development a core competency [1].

The various definitions have been proposed to explain and synthesize the PLM concept [1-4]. Herein, we consider a PLM system as a set of tools and methodologies to manage the evolution of a product during its life cycle from conception to disposal. It is mainly a business strategy and one of its aims is to enhance integration and collaboration during definition, sharing and using of engineering data, *i.e.*, all the information needed throughout the product life cycle. It implies the coordination and integration of processes and applications used to define and manage the so-called virtual product with those used to manufacture and maintain the physical product. Therefore, it can be considered the *glue* to connect such environments (Figure 28.1).

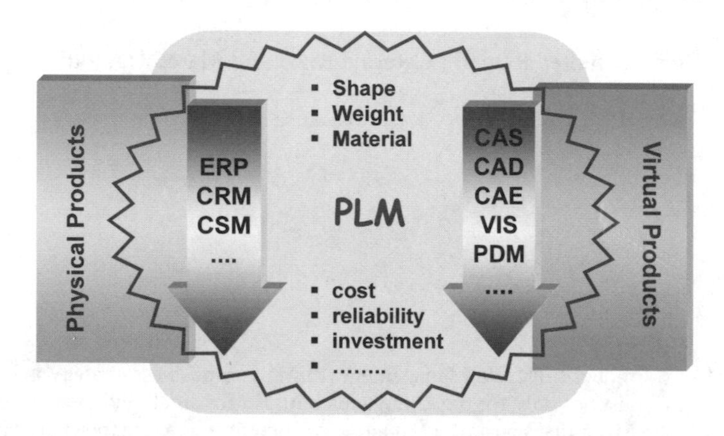

Figure 28.1 PLM as an integrator between the physical and virtual products

Generally speaking, a PLM implementation process can be considered from two different points of view:

- Consider PLM as an extension of the engineering data and process management environment crossing the boundary with manufacturing and providing integration with this environment (namely ERP system);
- Consider PLM as an extension of the ERP environment involving engineering documents and process management.

In both cases, PLM requires an understanding of the whole product development process, and of the interface (in terms of information exchange and flow) between the two main processes involved with the PLM: engineering and manufacturing.

A PLM solution consists not only of a mere introduction of new technologies, but it is also a new organizational paradigm that requires a deep analysis of the company business processes. In this paper, we describe the application of process modeling and simulation techniques for the implementation of a PDM/PLM system within a SME company using Total Quality Management (TQM) procedures to integrate process analysis.

28.2 Product Lifecycle Management and Business Process Re-engineering

The design and implementation of a PLM solution requires adequate analysis of the business processes to be supported; therefore, often such implementation is coupled with Business Process Re-engineering (BPR) to better deploy technologies and/or methodologies [5]. In BPR activities, process modelling plays an important role and is used to understand:

- How the current process is working (As-Is model);
- How to exploit innovation and how new tools and/or methodologies can improve the process (To-Be model).

A Methodology for Process Analysis and Modeling

In the case of a complex integration effort, such as a PLM implementation requires, it is important that the employed process modeling techniques represent all business aspects both to provide a complete picture of the main process and to go into detail when necessary [6]. A methodology is required, which provides guidelines and tools to correctly represent the complete business process [7, 8]. We have adopted a methodology developed by the KAEMaRT Group (www.kaemart.it) [9] that integrates modeling and simulation techniques. It gives the technicians a structured framework that provides a step-by-step roadmap, techniques and tools for technological innovation and BPR as defined in [10]. In particular, it is a methodology to analyze and represent the so-called *As-Is* and *To-Be* processes. First, the As-Is process (*i.e.*, the process currently carried out at the company) is analyzed, collecting knowledge through interviews with the experts of the process. During this activity, the knowledge regarding both the product and the process, usually used among the technicians and company departments, is acquired and formalized. It permits process problems and possible improvements to be highlighted, and constitutes a term of comparison to evaluate quantitatively the effectiveness of the new organizational paradigm. The second phase consists of modeling the new process that implements the new technological solutions. The main objective of this phase is to permit the process experts to highlight advances and changes with respect to the As-Is process and to quantitatively evaluate the new scenarios, such as a PLM solution.

In literature, we can find different techniques to represent process knowledge: ARIS [11-12], IDEF [13-14], UML [15], *etc.*. Studies on business modeling came to the conclusion that there is not a universal tool; the challenge is to find the right tool for the considered problem [16]. Thanks to our experience carried out in several industrial contexts on product-development processes reengineering, we adopted IDEF techniques. For both models, we considered IDEF0 for activities and data flow representation, and IDEF3 for execution flow and states assumed by product data. They represent process knowledge with graphical languages, simple and easy to be used and understood by people without a technical background. This facilitates the communication between work teams with different competencies and from different company departments. It is particularly important to validate the process model with process owners and to ensure that the collected information (process and/or product knowledge) has been correctly formalized.

The following steps are performed for each process model (As-Is and To-Be):

- *Capture process/product knowledge* through interviews with company experts, and technical documentation;
- *Represent and formalize the process knowledge* using IDEF0 techniques. It leads to the definition of the process of the *functional model* and identification of what the current system does right and what it does wrong.
- *Review* the IDEF0 model with the process experts and consolidate the model that correctly captures the current status of the process in order to define a common view of the process agreed upon by all technical persons involved in the process.
- *Define* the IDEF3 models that show the process execution flow. Precisely, IDEF3 Process Flow (IDEF3 PF) model permits identifying and formalizing *decisional processes* embedded within the business process itself, while the IDEF3 Object State Transition Network (IDEF3 OSTN) model allows the representation of object states along their lifecycle.
- *Collect data* about *execution times* and *probability* that *mistakes* can occur, leading to a partial or total re-execution of activities/sub-processes.
- *Define and execute* the simulation model to *gather quantified sample data* that will be used to evaluate and compare different process assets (As-Is vs. To-Be).

Extracting Knowledge for Process and Data Modelling from Total Quality Management Procedures

Process knowledge capture and related sources are critical issues among the mentioned steps. Companies using a Total Quality Management system keep a repository of procedures, which describe their process and product data. They contain a complete description of the processes, but are typically written in plain language using a business administration terminology [11], which, because of its nature, is not suitable to specify information systems. However, their examination could be a good starting point for process analysis. In the presented work, we considered the possibility of using Total Quality Management procedures for the definition of the As-Is model. In particular, the main goals have been:

- To evaluate the possibility of extracting all necessary information directly from Total Quality Management procedures, such as activities, resources, execution times, and data flow necessary for process modeling;
- To study the possibility of using a semantic processor to extract automatically the modeling concepts from the procedures text.

28.3 Study of a PDM/PLM Solution for a SME

The work has been carried out in collaboration with a SME producing hydraulic systems. The company holds an ISO9001 certificate Total Quality Management system that guides the design phase of product development. As mentioned previously, one of the main goals has been to extract the As-Is model of the design process from Total Quality Management procedures and to use it as a reference for the implementation of a PDM/PLM system. The main company's need was a better coordination and data exchange among engineering, manufacturing and test labs in order to facilitate testing and prototype activities and to improve the product development lead-time. In fact, documents and data flow were evolving from a 2D drawing-based system to a 3D model-based system. Basically, this means that instead of circulating only electronic files representing 2D technical drawing, 3D CAD models were exchanged and used, where possible, directly for all the activities downstream from the engineering department. In such a context, the introduction of a PDM/PLM system was envisaged to leverage best practice based on collaboration.

Therefore, the main objectives of the process modelling have been:

- Build a new model for the product development process to support and monitor PDM/PLM implementation;
- Evaluate the possibility of using Total Quality Management procedures to enhance modelling activity;
- Evaluate the suitability and effectiveness of IDEF tools to be used as aids in Total Quality Management environments to improve dissemination and understanding of procedures.

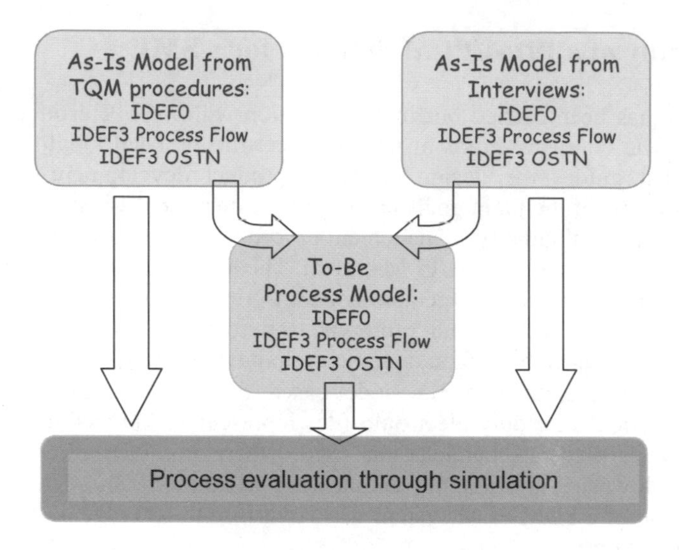

Figure 28.2 Process modelling methodology

Figure 28.2 shows steps carried out according to the adopted methodology and above-mentioned objectives.

As-Is Process Modelling

Two different As-Is models (in the following named *ISO9001 As-Is* and *As-Is*) have been realized using two different knowledge sources: the procedure ISO9001 "development and check of the planning", and interviews with the company's staff. This permitted evaluating and comparing which method is more suitable to capture each kind of knowledge necessary for process modeling.

In order to have a complete overview of the design process, we considered all design process typologies described in the Total Quality Management procedures. Each type of design process was subdivided into four main phases: *Concept generation*, *Preliminary design*, *Detailed design*, *Product engineering*. The *As-Is* model has been built by gathering information only from interviews with design process owners. Figure 28.3 and 28.4 show two IDEF0 diagrams of the As-Is model; precisely, the first (A1 diagram) describes the four main phases and the latter (A13) a shot of the detailed design phase.

As-Is model comparison and knowledge capture methods

The two As-Is models have been compared and differences, particularly for activity flow and synchronization aspects, have been highlighted. There are essentially two main reasons.

First, people aseptic from the process have accomplished the procedures review. They interpreted the procedures literally, and so introduced in the resulting process model many checkpoints and gates that are not present in the real life processes. This also depends on the language used to write the procedures-typically a business language suitable for good interpretation by people with experience and insight in the process, but which can give ambiguous responses to software experts reading it with the purpose of extracting a workflow suitable to automation [11].

Secondly, differences between the process actually carried out and Total Quality Management procedures are caused by the dynamics of innovation inside the process; namely, the transition from 2D-drawing-based information to 3D Man-based information exchange was already started in the company. Therefore, in many cases, the process is no more the drawing based flow described in procedures, but a hybrid process using partially the 2D drawing based flow, and starting to deploy the advantages of the 3D CAD model exchange. This is not due to a scarce maintenance or respect of Total Quality Management procedures, but instead it relies on different dynamics of the process and procedures evolution. This is particularly true for small and medium-sized companies where the steps for changing the process are not so well planned to allow their correct inclusion in the procedure from the first stage. They are often introducing new solutions, testing and fitting them in the current process and making on-line runtime the adjustments to make things work correctly, and then documenting the procedures.

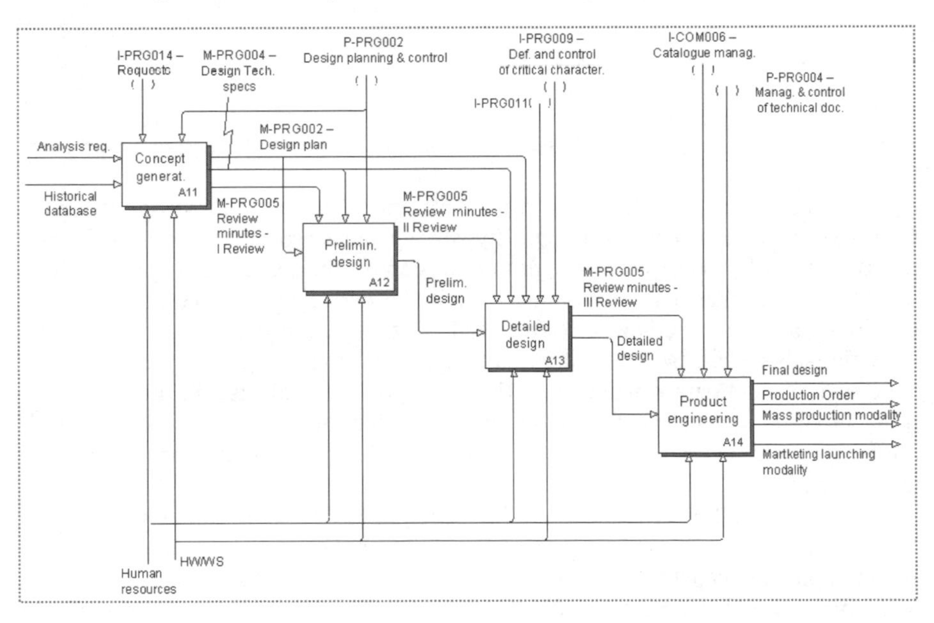

Figure 28.3 As-Is model: the IDEF0 Diagram-A1

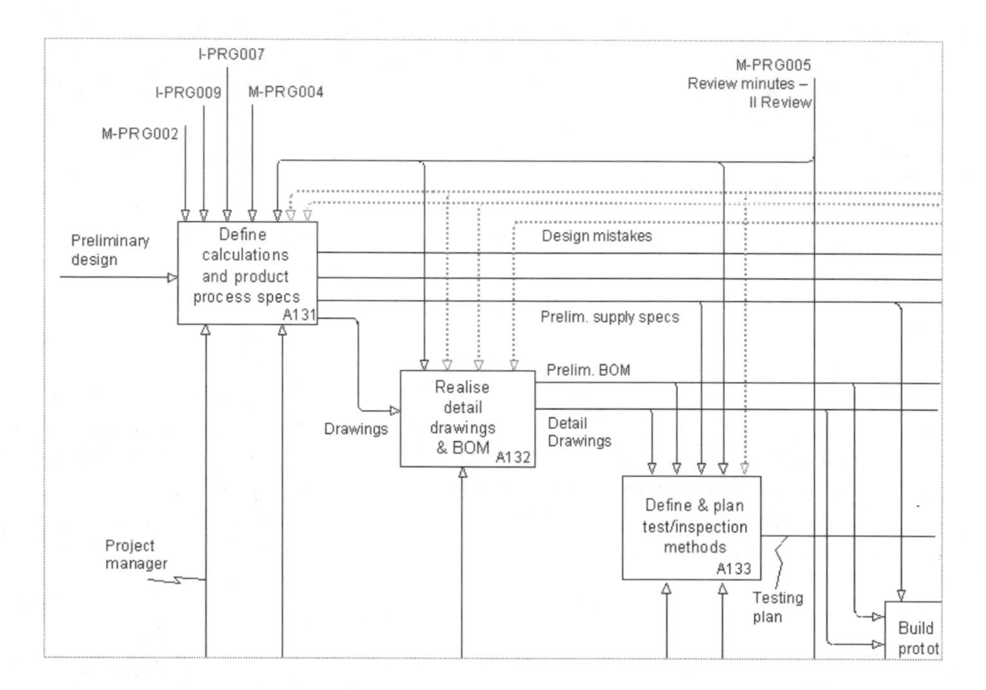

Figure 28.4 As-Is model: A shot of the IDEF0 Diagram-A13: Detailed design sub-process

The following considerations are important regarding knowledge sources. Compared with interviews, Total Quality Management procedures permitted rapid identification of process activities, criteria to control their execution and produced output. On the other hand, it is more difficult to derive precedence and causality relations between activities (sequential activities/sub-processes, concurrent activities/sub-processes, alternative activities/sub-processes), and resources to execute tasks and data management. However process analysts can profitably use these products to gather information on the product development process and, if correctly combined with interviews, they permit the speeding up of process analysis activities.

Concerning IDEF techniques, and in particular IDEF0, they can be useful tools to support procedures comprehension, especially thanks to their graphical nature. In addition, the capability of commercial process modeling software to make available information in a graphical format and in a web-based form allows navigation within the entire process and identification of specific information associated with a single activity.

To-Be Process Modeling

A gap analysis using the two As-Is models has been done to harmonize what was perceived by the analysts and to study which aspects of the process could be improved introducing a PLM system. According to company requirements, the re-engineering activity has been concentrated on two phases of the design process:

Preliminary and Detailed design. This means that changes and improvements have mainly affected these two sub-processes. The re-engineered process, described by the To-Be model, has been designed calling for:

- The introduction of CAE tools, already in use in the R&D area, to assist the designer and involve technological experts during identification of possible problems for the product;
- The introduction of rapid prototyping systems;
- The implementation of a PDM system to improve information flow, document production, and recovery, making them more secure and rapid.

The re-engineering process leads to the definition/representation of the engineering data flow, object transition states and processes/activities, which use and manage the engineering documents. Figure 28.5 portrays a shot of the IDEF0 Diagram-A13 of the To-Be model related to the *Detailed design* phase where the sub-process has been re-designed (see Figure 28.4) and a PDM system has been considered. Figure 28.6 shows a shot of an IDEF3 OSTN diagram describing the data flow and evolution within the entire design process.

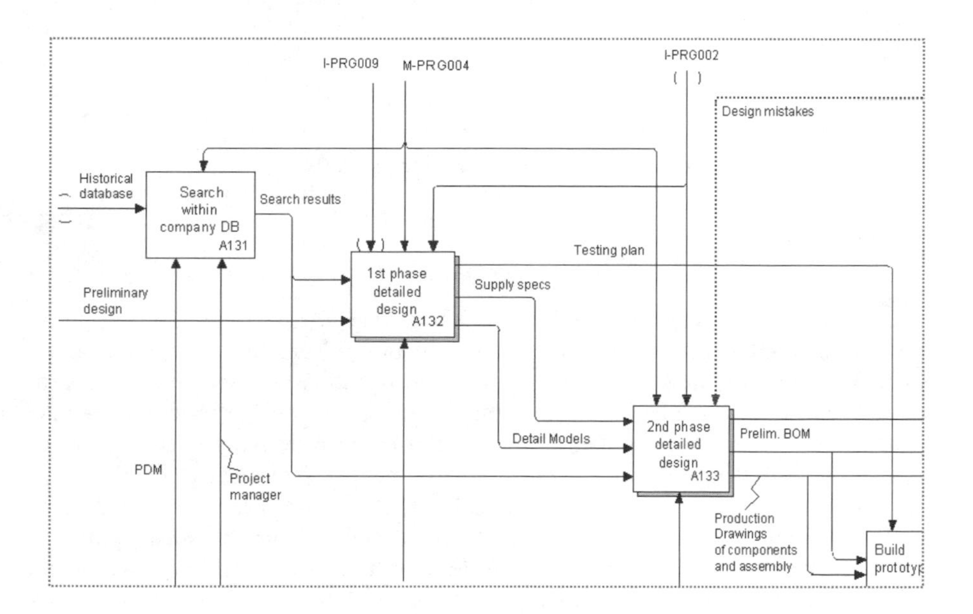

Figure 28.5 To-Be model: A shot of the IDEF0 Diagram-A13 - Detailed design sub-process

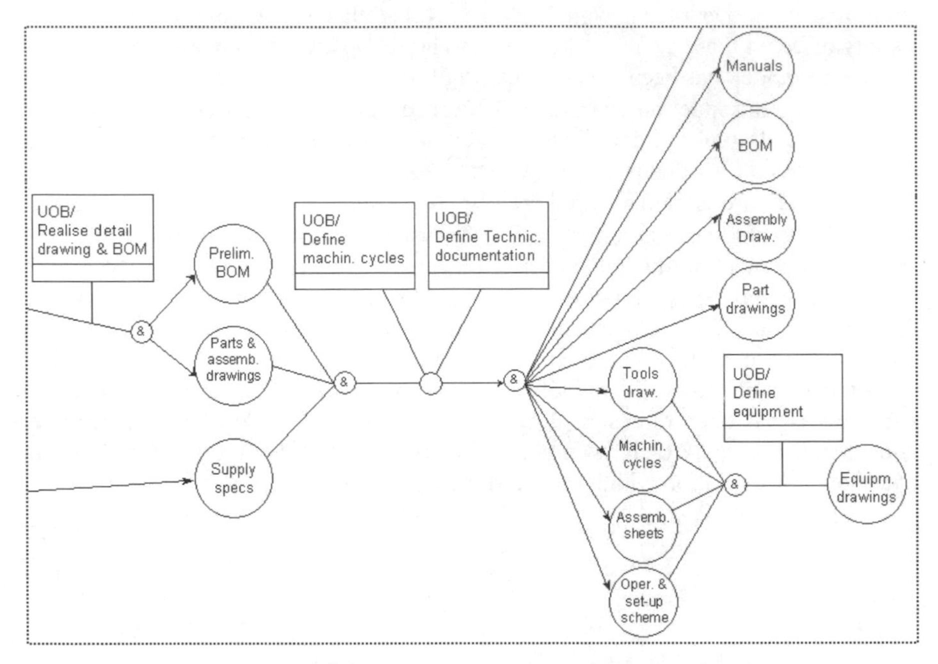

Figure 28.6 To-Be model: An IDEF3 OSTN diagram

Simulation: As-Is vs. To-Be

To compare the As-Is and To-Be Data for As-Is process simulation (*e.g.*, execution times of activities, required resources and availability) have been acquired through interviews with the process owners; simulation results have been compared with those recovered from company documents related to the design process of a new product. To-Be process simulation has been performed deriving execution times and *probability* that *mistakes* can occur for the new activities (*i.e.*, not present in the As-Is process) from information provided by the process experts. A reduction of the execution times have been taken into account either for those activities affected by the introduction of a PDM system, or those related to the development of physical prototypes using rapid prototyping. Comparing results obtained by simulating the As-Is and To-Be processes the following conclusions can be drawn. A reduction of about 35% the development time has been estimated as shown in Figure 28.7. Note that the reduction mainly concerns the *Preliminary and Detailed design*; this is because changes (*e.g.*, PDM system) have been mainly introduced in these two phases as mentioned in paragraph *To-Be Process Modeling.* As far as the project manager utilization is concerned a reduction of about 32% has been estimated (Figure 28.8).

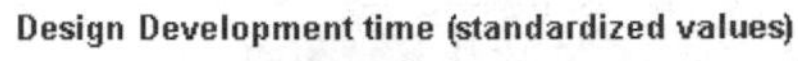

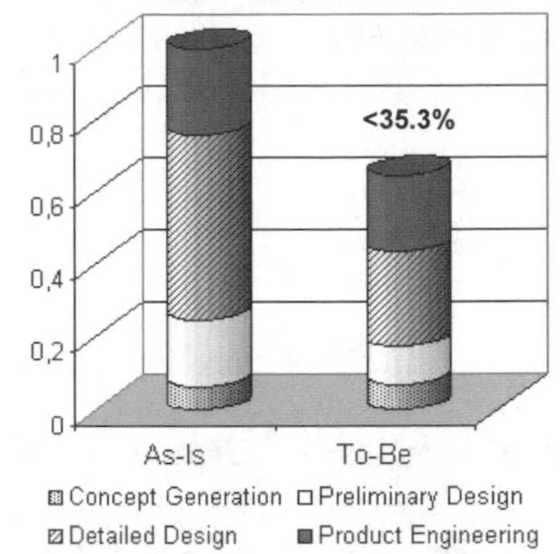

Figure 28.7 As-Is vs. To-Be: Design Development time

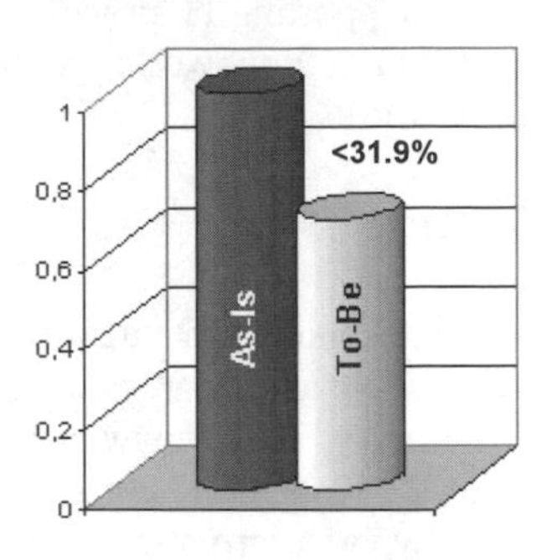

Figure 28.8 As-Is vs. To-Be: Project manager utilization

It is important to note that at this stage the influence of a PDM system has been evaluated only for two phases of the design process. Its implementation will surely influence positively all the company business processes, in particular those phases involving a frequent data exchange within and outside the company.

28.4 Conclusions

Modelling and simulation techniques permitted the capturing and the formalizing of the process aspects necessary to study and guide the implementation of a PDM/PLM solution.

The As-Is model provides a view of the current process - *where we are now* - and an understanding of changes required. The As-Is process model cannot be a pure translation of Total Quality Management procedures; nevertheless, these procedures can be a useful tool to capture process knowledge if properly integrated with interviews of process owners or other information sources, such as project management systems to collect historical data on activity duration. An envisaged future research direction is the possibility of using automated text analysis software based on semantic processor technology to extract information from text-based business procedures. The To-Be model was used as a decision making tool and has permitted the evaluation of how a new business paradigm, such as PLM, can improve the process, the investments needed, and provided information about how new tools can leverage industry best practices.

Finally, the production of structured process documentation formalizing As-Is and To-Be process knowledge permits the spreading and the sharing of knowledge among people with different roles, competencies and working in different departments, that are in some cases, geographically distanced.

Future work will be based on the evaluation of suitability of certain business modelling frameworks to represent all aspects needed to keep up-to-date procedures and supporting IT processes oriented to PLM implementation systems. The first candidate frameworks to be evaluated are the ARIS [11], and the Zachman frameworks [17].

28.5 References

[1] http://www.technologyevaluation.com.

[2] "Product Life-cycle Management "Empowering the Future of Business", *CIMdata Report*, 2002 (www.cimdata.com).

[3] Bacheldor B., Kontzer T. 2003, "What PLM is and is not", *Information Week*, July 2003 (www.informationweek.com).

[4] "Understanding Product Lifecycle Management", *Datamation Limited Report,* n. PLM-11, rev. 1.0, September 2002 (www.datamation.co.uk).

[5] Miller E., 1997, "The synergy between PDM and Business Process Re-engineering", *Computer Aided Magazine*, March 1997.

[6] Vernadat, F.B., 1996, "*Enterprise modeling and integration: principles and applications",* Chapman & Hall, London, UK.

[7] "First Method to Implement", 1998, *Report BRITE DMU BP D2.0*, September.

[8] "A Refined Method To Implement DMU-BP, Part 1 – Introduction To The Methodology And The Road-Map", *Report BRITE DMU BP D3.0,* 20 September 1999.

[9] Cugini, U., Ramelli, A., Ruozi, D., 2001, "A product development process re-engineering oriented to the introduction of new DMU technologies", *5th IEEE Int. Conf. on Intel. Eng. Sys. (INES 2001)*, Sept. 16-18, Stockholm, Sweden.

[10] Mayer, R.J., deWitte, P.S., "Delivering Results: Evolving BPR from Art to Engineering", http://www.idef.com/Downloads/free_downloads.html (to be published in book: *Business Process Reengineering*, Ed. Kluwer).

[11] Sheer, A.W., 1998, "*Business Process Frameworks",* Springer-Verlag, Berlin.

[12] *"ARIS Methods Version 4.1"*, July 1999, Ed. IDS Sheer AG.

[13] "Integration definition for function modelling (IDEF0)", 1993, *Draft Federal Info. Processing Standard Publication 183 (FIPAPUB183),* FIPA, USA.

[14] Mayer, P., Menzel, C., De Witte, P., "IDEF3 Process Description Capture Method Report", *Information Integration For Concurrent Engineering (IICE),* www.idef.com.

[15] http://www.omg.org/uml.

[16] http://www.cit.gu.edu-au/~noran/cit_6114.

[17] Zachman, J.A., 1987, "A Framework for Information Systems Architecture," *IBM Systems Journal*, Vol. 26, No. 3, pp. 276-292.

29

Selection and Evaluation of PLM Tools for Competitive Product Development

Matteo Benassi, Monica Bordegoni, Umberto Cugini, and Gaetano Cascini

Abstract: PLM solutions propose methodologies and tools aiming at improving the product development process and competitive engineering. In particular Engineering Knowledge Management (EKM) has proved to be a key enabler for reducing lifecycle cost and time, improving quality and helping to ensure safe products. Nevertheless the selection of the most proper tools for a given product development process is not a trivial task. The application of a methodology for the selection and evaluation of new PLM technology to be adopted for improving product development processes is presented. The focus of this work is illustrated in two different study cases that apply and test the developed methodology: one belonging to the consumer products sector, the other belonging to the machine tool sector.

Keywords: Product development process, PLM, engineering knowledge management, integrated knowledge-based environment, product development modeling and simulation

29.1 Introduction

Product development processes include several critical aspects. Among those, today costs and time are increasingly constraining for companies' competitiveness. The introduction of CAD/CAM systems has contributed to the reduction of time spent for design and manufacturing. By means of Finite Elements and Multibody analyses it is possible to perform virtual tests of the proposed solutions with minimal expenses. PDM solutions allow a better management of data that are stored and retrieved in a more rational and reliable manner. Recent PLM solutions propose methodologies and tools aiming at improving the product development process and competitive engineering. They support a more closely integrated management of engineering activities of product lifecycle with process planning and manufacturing aspects. Within this broader view of product development and integration of its various aspects, Engineering Knowledge Management (EKM) has proved to be a key enabler for reducing lifecycle costs and time, improving quality and helping to ensure safe products [4].

In fact, the analysis performed by Moenaert, *et al.* [3] on several European multinational corporations aimed at identifying the requirements that determine the effectiveness and efficiency of communication in international product development teams, revealed that the most relevant factors are network transparency, knowledge codification, knowledge credibility, communication cost and secrecy.

Knowledge-related issues are considered as critical aspects for companies, since Knowledge is recognized to be one of the major assets, but it is often not explicitly expressed and often exists as personal and tacit know-how of people. In current collaborative environments, Knowledge is managed not only within individual companies, but also across supply-chain relationships. Therefore, it is even more strategic to adopt methods and tools that support Knowledge Management, where Knowledge is also distributed in collaborative environments.

By investigating the practices of 281 new product teams from around the world, Lynn *et al.* [2] elicited the most significant factors impacting a team's ability to acquire and use knowledge to reduce cycle time and improve their probability of success: 1) documentation of project information, 2) storage and retrieval systems for project information, 3) information reviewing practices, 4) vision clarity, 5) vision stability, and 6) management support of the project.

Boston *et al.* [1] demonstrated that standard supplier literature is as well a key source of design information within the early phases of new product development where the cost and quality are largely defined. Besides, an extensive investigation into the way this information source was organized and handled within a typical engineering organization revealed that an array of deficient "systems" were used for classifying its content, and there were no formal procedures in place for its life-cycle management, with corresponding consequences for the effectiveness of the design operation.

Several other works confirmed these results, but none of them aims at supplying engineering teams with methods to identify knowledge gaps in the product development cycle and most of all criteria for selecting the most proper tools for overcoming such knowledge lacks.

Two issues have to be considered for the adoption of new EKM methods and tools within the product lifecycle: 1) how to select appropriate and effective methods and tools; and 2) how to estimate benefits and impacts before adopting and/or integrating those methods and tools. The first issue is related to the identification of the best technology that meets the requirements for improving of a specific product development process. The second issue deals with the critical aspect related to the estimation of the benefits deriving from the eventual adoption of the identified new technological solutions, before actually making investments and changes within the company organization.

The authors are running a research project that proposes a solution to these two issues. The aim of the project is defining and validating an integrated environment for studying and evaluating the adoption of knowledge and innovation management tools within product lifecycle (www.kaemart.it/ike). The outcome of the project is a roadmap that consists of guidelines for the adoption and integration of new technologies within the product development process [5].

29.2 The Methodology

The methodology developed in the research project proposes a sequence of activities aiming at supporting companies in providing answers to the issues described in the previous section.

In order to *select the most appropriate methods and tools* (issue 1) the following activities are planned (Figure 29.1):

1. modeling the product development process, selected as candidate process, as it is currently implemented in a company (As-Is process modeling);
2. identification of critical issues related to EKM that underlines the necessity for improvements;
3. analysis of current and emerging technologies for knowledge and innovation management.

Major details about the above-cited activities can be found in Bordegoni M., et al., 2003 [5].

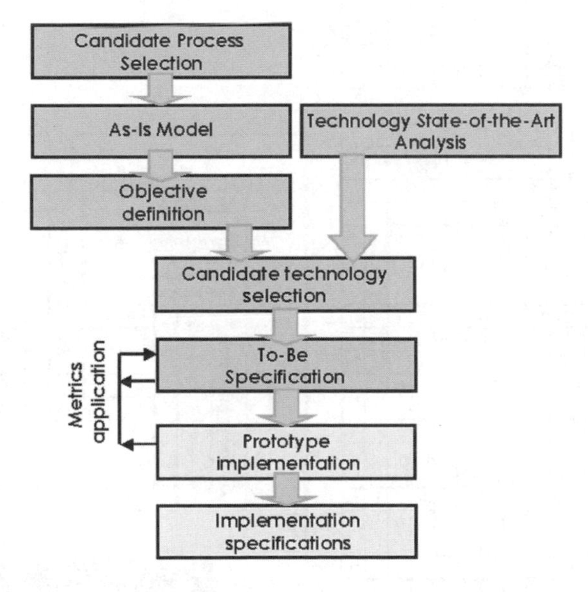

Figure 29.1 Roadmap for the selection and evaluation of new EKM solutions to be adopted in product development processes

The last step is the application of a method - based on the well-known Quality Function Deployment method (QFD) - to select the most appropriate technologies to be adopted, and/or to be integrated into the actual product development process.

After first activities, two matrices are obtained (Figure 29.2):

– *Matrix 1* that reports the importance of generic knowledge management activities with respect to the process task of the selected product development cycle;
– *Matrix 3* that links PLM and EKM tools available on the market with a comprehensive set of functionalities related to knowledge management.

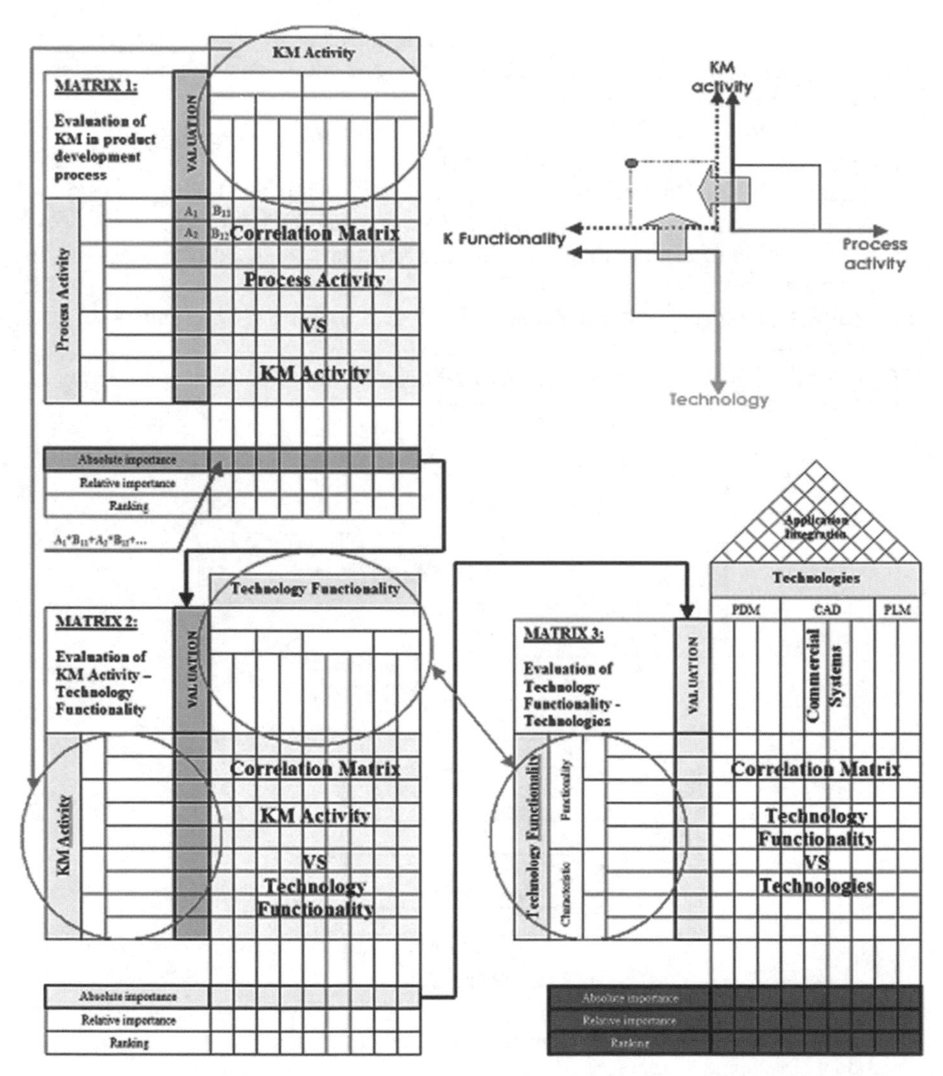

Figure 29.2 Method for candidate technology selection, based on the use of the three correlated matrices

An intermediate matrix has been defined, between *Matrix 1* and *Matrix 3* for correlating EKM activities and K functionalities (Figure 29.2). This matrix, named *Matrix 2,* is filled in by process experts together with technology experts (the former are involved in defining the importance of EKM activities for their specific process).

Moreover, the roof of the third matrix allows taking into account the level of integration of the examined technologies. This is a crucial aspect when several technologies require to be integrated in order to provide a comprehensive solution and/or when the selected new technology has to be integrated with ones already in use.

Furthermore, according to the proposed roadmap, in order to evaluate benefits, impacts and cost related to the adoption of new and innovative solutions (issue 2) the following activities should be performed:

1. definition of an evaluation metrics;
2. definition of a candidate new model of the product development process integrating the selected technologies (To-Be model);
3. application of the metrics to the As-Is process model, to the To-Be process model, and their comparison.

29.3 Application of the Methodology

The research project (funded by MIUR - Italian Ministry for University and Research) is carried out by four academic partners (Politecnico di Milano, Università Politecnica delle Marche, Università di Firenze, Università di Udine) and involves four companies: two of them belonging to the consumer products sector, and two of them belonging to the machine tool sector, assumed as representative cases for the definition and the validation of the proposed methodology.

In order to evaluate the effectiveness and benefits of the newly defined product development process, including new methodological and technological solutions, it is necessary to define metrics capable of measuring if there is and which is the degree of improvement. The metrics, applied to both the As-Is and To-Be process models, takes into account different aspects at two levels:

- An *operational level*, where relevant aspects to be considered and measured, refers to design activities. Indicators (like execution time, capability of proposing innovative solutions, *etc.*.) are defined according to the analyzed test case and process.
- A *strategic level*, where judgments given by several characters within the company organization (company's managers, design managers, senior designers, junior designers, *etc.*.) are analyzed. Those judgments consider the effectiveness of both new tools adopted for the specific problem solution, and also of the deriving advantages in terms of increased know-how within the company, that can be fruitfully exploited in subsequent activities.

Only one test case for every industrial sector is introduced to analyze the methodology flexibility in different application fields.

First Case Study: JOBS (Italy)

The first study case is the one of JOBS in Piacenza (Italy). JOBS designs and produces 3/5-axes operating machines and automated milling systems for the aerospace, car industry, general mechanical and energy sectors.

According to the steps of the proposed methodology, first the candidate process has been selected: the design process of the tool-changer of a machine tool.

Interviewing the designers involved in the machine development has preceded the As-Is analysis. The collected data have been organized and modeled using IDEF0 diagrams for the identification of critical aspects and requirements within the current design processes, related to issues concerning Knowledge Management:

- simplify the reuse of existing projects;
- transfer singular experiences to the whole work-group;
- better manage design knowledge (better availability and quality of information);
- limit errors and loops, especially during preliminary phases where they can play a more critical role;
- introduce better support for junior designers.

In order to identify the technological solutions that satisfy the requirements (objectives), the method for identifying the candidate technological solutions has been applied. The application of the method has suggested that the system RuleStream (a KBE application capable of capturing and sharing design practices in order to avoid mistakes and reduce time consumption) would be the most appropriate solution to satisfy the process requirements. The decision about which technology to adopt has been taken by the company designers and managers, also taking other relevant aspects into account such as the applications already in use in the company, the usability and user-friendliness of the new technology.

The To-Be IDEF0 models, considering the identified technology, have been subsequently defined. RuleStream allows simplifying product configurations and introduces new product development processes in which senior and junior users have to be identified. These users play different roles in KM activities: senior users are delegated to introduce properties, rules and procedures that will drive juniors during the final configuration process.

Once the To-Be process has been defined, the implementation of new technology has been analyzed using the evaluation metrics. Particularly, at operational level, times and errors are studied during the processes: these two indicators allow a more accurate analysis (particular in cases where decreasing time is less significant than decreasing errors). During the first implementation period the time of KM activities, in As-Is and To-Be processes, are comparable, due to learning new technology and introducing properties, rules and procedures. Nevertheless, after the full-implementation, process activities duration decreased (Figure 29.3). As a result of the knowledge formalization, RuleStream allows faster product configuration, faster viewing, use and capture of knowledge and significant reduction of typical errors (like missing rules and unknown procedures). In Figure 29.3 the errors are represented with points in 2D space, where frequency

(F) and gravity (G) lie on two axes, then As-Is and To-Be best fitting curves are created for the comparison between the two processes.

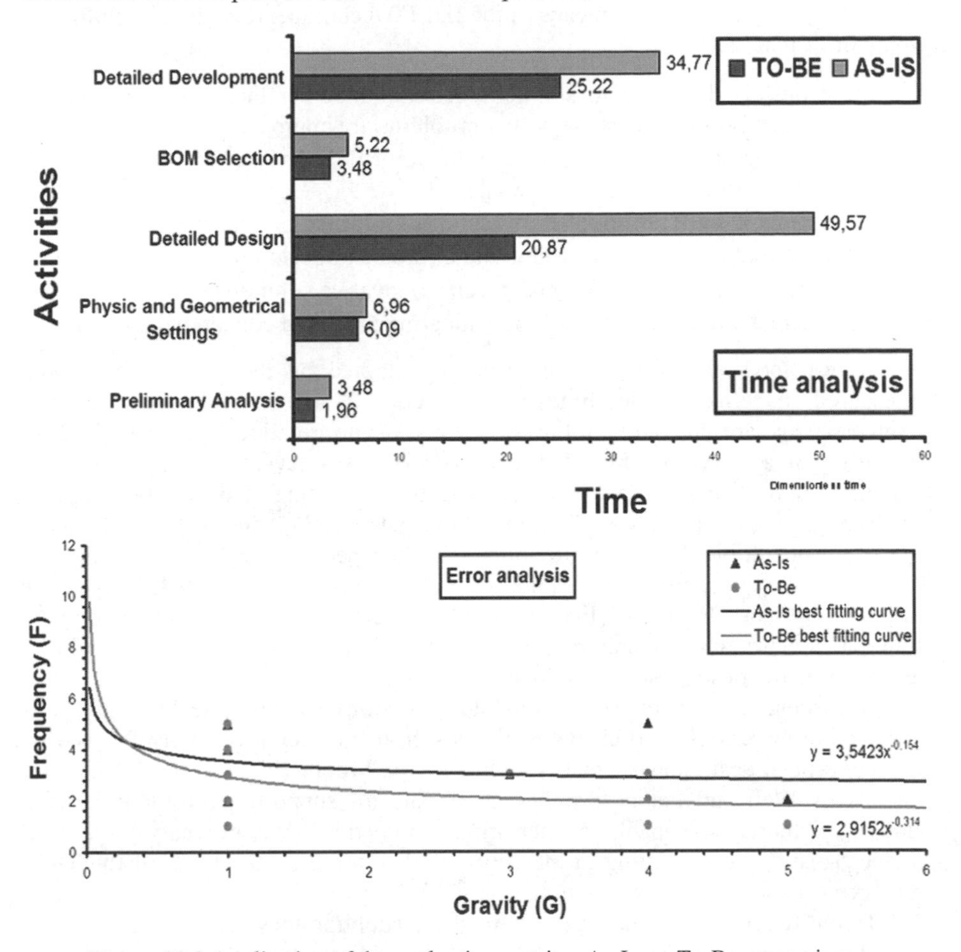

Figure 29.3 Application of the evaluation metrics: As-Is vs. To-Be comparison.

Second Case Study: Whirlpool (Siena Plant)

One of the case studies dedicated to the consumer product sector has been developed in the Siena plant of Whirlpool Europe s.r.l. Such a plant is mainly dedicated to the production of chest freezers (more than 700 thousands per year) engaging about 700 employees. Compared with other plants of the same corporation, a relevant peculiarity is that the whole product cycle is managed by Siena employees autonomously and only few secondary services are provided by the main R&D center in Cassinetta (VA, Italy).

Several product innovation tasks have been approached aimed at solving functionality problems of current products and/or reducing manufacturing costs.

Moreover a major product revision project is still in progress with the goal of increasing productivity and quality of icemakers. The analysis of the current product cycle, performed by means of the IDEF0 technique, revealed the following main critical aspects:

- more time is dedicated to solutions generation rather than problem analysis (consequently, sometimes the wrong problems are approached);
- trial and error solutions without a theoretical basis are often tested with consequent time and investment wastes;
- no systematic tools are adopted for technical information retrieval;
- informal conversations have a key role in product development;
- designers sometimes fear proposing very innovative solutions (!);
- no systematic tools are adopted for evaluating proposed conceptual solutions.

A test about the source of information adopted by the product designers, measured in terms of the time spent (Figure 29.4), revealed that informal conversations are the most followed means of information retrieval (31.3%), resulting in a very low knowledge reusability. A relevant role is played by suppliers (8.8%) and competitors' products benchmarking (5.0%). The usage of Internet is negligible (2.8%), Technical Journals (2.0%), and Patents Database (0.3%) is negligible. It is worth to notice that their perceived efficiency (evaluated by means of a questionnaire) does not follow such a rank (Figure 29.4, below): the normalized score highlights that informal conversations are less effective than Internal Reports and Laboratory Test Results, while greater relevance should be given to suppliers suggestions and benchmarking activities.

According to the proposed methodology a set of candidate technologies was selected to increase the efficiency of the design team: they are mainly focused on the conceptual design phase, that is tools for Text Mining, in order to obtain better results by Web and Patent searches, and tools for supporting Problem Solving activities, therefore capable of performing Functional/Value Analysis, Cause-Effect Analysis and pointing to the conceptual solution on the basis of the TRIZ theory.

Two different approaches were defined for adopting these tools in the product development process (TO-BE model): since major efforts in terms of employees education are necessary in order to achieve the expected results by means of the selected tools and methodologies, a basic training was given to the whole design team for every-day activities so that the authors contribution, in these cases, has been limited to the role of facilitator in "strategic" technical meetings. On the other hand, a major product revision has been approached with the authors operating as skilled consultants and a direct cooperation has been established with the designers. Both the activities provided successful results with minor limitations as described below; the major revision has driven a very cheap and effective solution for frost reduction actually patent pending.

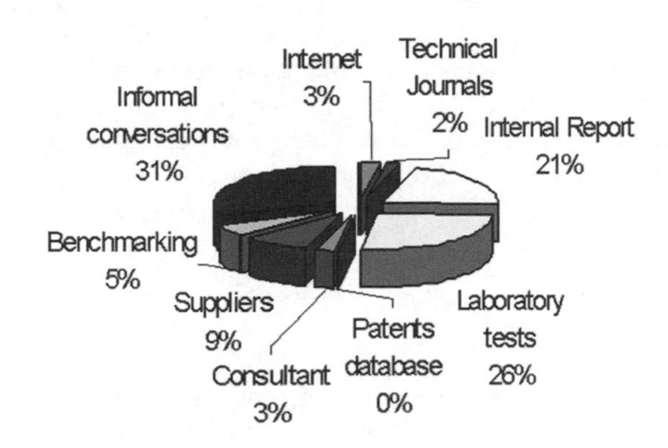

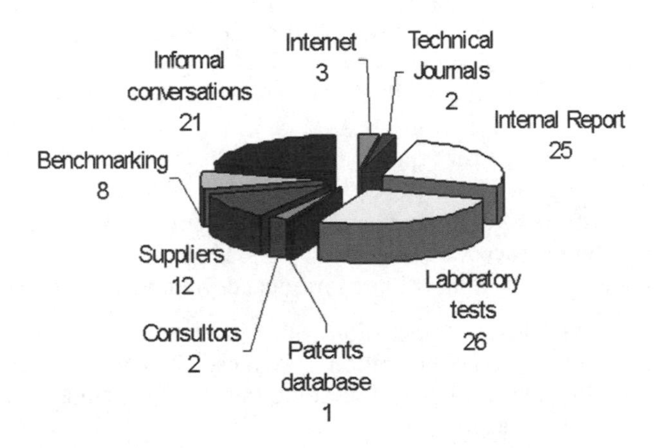

Figure 29.4 As-Is product development cycle in Whirlpool Siena Plant: sources of design information (above) and their perceived efficiency (below)

Table 29.1 Overview of the comparison between AS-IS and TO-Be models in Whirlpool tests

Evaluation Parameters	TO-BE vs. AS-IS
Number of Proposed Solutions	>>
Time effort in individual activities	> =
Time effort in team activities	=
Early identification of possible failures	>
Perceived Efficiency of Proposed Solutions	>

The metrics application (Table 29.1) revealed a greater number of generated solutions. It is usually effective, even if no time reduction of the design activities elongated times have been observed in some cases. Since the attention has been focused on the conceptual design phase, it is hard to identify "errors", while it is worth highlighting that satisfactory solutions have been developed, and that even discarded ideas constitute coded knowledge enrichment for future activities. Major advantages have been obtained in dealing with complex problems and in collaborative design tasks since the proposed methodologies and tools allow the accomplishment of comprehensive and detailed analyses systematically.

29.4 Discussion

Evaluating the efficiency of the proposed roadmap with quantitative means is not a trivial task since it is not possible to make a direct comparison with analogous methodologies. Moreover, it is intended as a general set of guidelines for any product development process. Nevertheless, the presented study cases have provided positive results that validate the methodology application. Of course a comprehensive evaluation of the proposed TO-BE models must also take into account medium-long term effects, since all the adopted technologies give better results with increasing skill of the users.

The implementation of the methodology has allowed the achievement of important goals:

- identification of KM critical aspects in different product development process using a methodological approach;
- identification of best technologies (selecting from those available in the market) to overcome such limits;
- evaluation of a set of PLM tools in real product development processes.

Moreover, in methodology application, other aspects about PLM tools have been analyzed, such as systems integration with those already in use in examined study cases; usability of such tools and learning time; strengthening of innovation capability; and economic benefits.

With these results, the methodology demonstrates flexibility not only in different application fields, but also in different processes:

- in the first case the methodology is applied in a design process, where the management of project rules and information are preponderant;
- in the second case it is applied in a product innovation process, where KM activities are concerning problem analysis and concepts generation.

Further works are in progress for applying the methodology to new product development processes with remarkable technology integration requirements. In this case the role of top-roof of matrix 3 that takes into account integrability of systems belonging to different technological classes, must be enhanced.

29.5 Conclusions

The paper presents the application of a methodology for evaluating the adoption of PLM tools in product development processes. The research work has developed a roadmap consisting of a sequence of activities for the evaluation of the impacts and costs related to the adoption of new and innovative technologies for knowledge and innovation management, within currently implemented companies' product development processes.

The roadmap plans a sequence of activities where initially a candidate product development process is selected and analyzed. The analysis points out issues related to Knowledge Management that might require improvements. A method, based on the Quality Function Deployment method, has been set up for the identification of technological solutions that may improve the performances of the process. The effectiveness of the identified solution is proved through a simulation of the To-Be process, where product development processes are re-defined introducing the use of new EKM technologies. Metrics are used for measuring the performance of the To-Be processes, compared to As-Is ones. The roadmap has been evaluated through its application to some selected case studies.

The application of the metrics allows the evaluation of the effectiveness of the proposed methodology. The effectiveness and the benefits of the To-Be process can be further evaluated by implementing some prototypes of the To-Be process. This type of evaluation is currently under development.

From a general perspective it is quite evident that EKM technologies can increase companies' efficiency and competitiveness, but very often they require start-up costs both in terms of software acquisition and employee training, that are not negligible. The availability of a methodology for performing the best process analysis and technology selection, with benefits evaluation means, is the greatest added value of the presented work.

29.6 Acknowledgements

The authors would like to thank D. Pugliese, M. Pulli and M. Ugolotti from Politecnico di Milano, M. Cambi and D. Russo from Università di Firenze, S. Filippi from Università di Udine and F. Mandorli from Università Politecnica delle Marche for their contributions to the research. Special thanks are dedicated also to Dr. Schiavi and Mr. Caminati from JOBS and Dr. G. Rosi and Mr. A. Braccagni from Whirlpool for their support in test cases implementation.

29.7 References

[1] Boston, O.P., Culley, S.J., McMahon, C.A., 1999, "Life-cycle management of supplier literature: the pertinent issues," *Journal of Product Innovation Management*, Vol. 16, No. 3, pp. 268-281.

[2] Lynn, G.S., Reilly, R.R., Akgun, A.E., "Knowledge management in new product teams: practices and outcomes," *IEEE Transactions on Engineering Management*, Vol. 47, May 2000, pp. 221-231.
[3] Moenaert, R.K., Caeldries, F., Lievens, A., Wauters, E., 2000, "Communication flows in international product innovation teams," *Journal of Product Innovation Management*, Vol. 17, No. 5, 2000, pp. 360-377.
[4] Cugini U. and Wozny N., (Eds.), 2002, *From knowledge intensive CAD to knowledge intensive engineering*, Kluwer Academic Publishers.
[5] Bordegoni M., Cascini G., Filippi S. Mandorli F., 2003, "A methodology for evaluating the adoption of Knowledge and Innovation Management tools in a product development process", *ASME Int. Design Eng. Technical Confs. & Computers and Info. in Engineering Conf.*, 2003, Chicago, IL, USA.
[6] QFD, URL: www.qualisoft.com
[7] IDEF0, 1993, "Integration definition of functional modeling," *Deaft Federal Info. Processing Standard Publication 183 (FIPAPUB183)*, FIPA, USA.

Part VII

Collaborative Engineering Design

30

Efficient Product Data Sharing in Collaboration Life Cycles

Frank-Lothar Krause, Haygazun Hayka, and Bernhard Pasewaldt

Abstract: The efficiency of the collaboration is a decisive factor for successful product development and production. The processes of the Collaboration Life Cycle should be supported with powerful IT tools to increase efficiency. This paper introduces some concepts of the Collaboration Life Cycle and discusses aspects of collaboration. To overcome the integration problem in product development collaborations standard-based solutions including methodology and tools, which have been achieved in the joint project "PDM Collaborator" are presented. The focus is on the components Collaboration Services and the Federation Services, which create one virtual PDM system that contains all the product data, needed for the collaboration process.

Keywords: Collaborative Design, Collaboration Life Cycle, Product Data Management, Federation, Integration, Data Mapping, PDM Collaborator

30.1 Introduction

The creation of a product requires nowadays the collaboration of OEM (Original Equipment Manufacturer), numerous system and component suppliers and sub-suppliers. This is indispensable for creating attractive as well as high-quality products and for reaching time to market objectives. In the past, the entire know-how for the development and production of a product laid in the hands of the OEM. In the meantime this knowledge is distributed over multiple partners along the entire value creation chain and very intensive dependencies arise between collaborating partners. Therefore, the efficiency of the collaboration determines the success of a product. New methods and tools are needed to increase this efficiency. Different aspects of collaboration processes are the subject of various research and development projects worldwide. At the same time, global players of production industries, such as the automotive and aviation industry are developing proprietary solutions to make the collaboration with their partners more effective [1]. Different aspects of collaboration should be regarded to increase its effectiveness. Important aspects of collaboration are *e.g.* processes, communication, project management and product data sharing. Additionally, the intensity of the cross company integration of processes, projects and product data should be enhanced to improve the success of the collaboration.

30.2 Collaboration Life Cycle

Stages of the Collaboration Life Cycle

Collaboration processes usually follow a unique scheme starting with the decision to collaborate. The next stages of collaboration are the selection of partners, set-up of the collaboration and the actual collaborated product development or production activities. Collaboration ends with disconnection of the infrastructure as well as the evaluation of the success of the collaboration in order to use the experiences made for further collaborations. Regular monitoring or controlling activities accompanies all these stages. The chain of these processes can be referred to as "Collaboration Life Cycle" (CLC) (Figure 30.1).

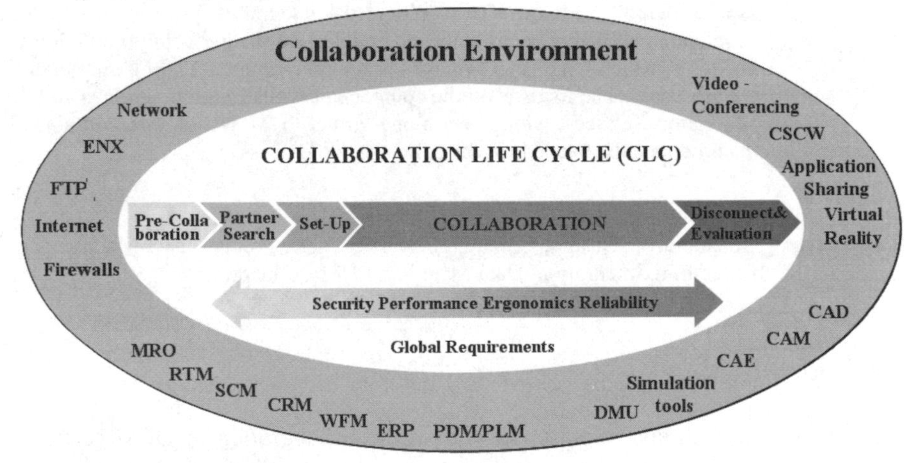

Figure 30.1 Stages of the Collaboration Life Cycle

Within the "pre-collaboration" activities the product is specified and make or buy decisions are made for systems or components of the product. For the parts of the product to be outsourced the selection of appropriate partners is performed next. To support the process of partner selection, the companies have to create their collaboration profile, which describes all the characteristics of the company regarding such things as manufacturable products, IT infrastructure, machines and competences. It also contains information about the company's development systems, data formats and schemata, communication protocols or encryption methods. When the appropriate partners are found, all participants in the collaboration have to agree on technical and organizational matters. Usually these matters, such as responsibilities, deliverables, milestones and IT infrastructure to be used, are specified in the collaboration contract.

Before starting the collaboration the necessary infrastructure has to be built up, especially IT-systems, hardware and software. It is also necessary to structure the collaboration processes. Therefore, a "collaboration model" should be defined which contains generic roles, deliverables, tools or milestones. This definition can be done based on "collaboration templates" that can be easily adapted to a specific collaboration situation. A certain methodology is needed, specified in the form of checklists or guidelines, and on how to instantiate a collaboration using these models and templates. In addition, the collaboration profiles and templates can be used to support the actual set-up of the technical infrastructure. This includes configuring cross company access control in accordance with the defined roles and responsibilities, firewalls, collaborative PDM (Product Data Management) and connecting workflows - in particular for engineering change management. Only a highly effective set-up process enables flexible, efficient and secure collaboration.

Once the collaboration is set-up, the actual collaborative product development or production processes can start. An integrated collaboration environment gives the user access to all engineering tools needed to fulfil the tasks. In this environment product data management systems are connected over the "virtual enterprise" forming one virtual product data management system, lasting during the collaboration as realized in the PDM Collaborator project [2, 3]. In case of changes all affected partners have to be determined. The affected parts and the cost caused by change in the actual stage of product creation should be identified. Data consistency has to be assured in all stages of the change process. Workflow coupling, decision support systems and the virtual PDM system support collaborative change processes.

When the goal of the collaboration is reached, all resources that were shared during the collaboration have to be disconnected. Even more important, the experiences of the actual lifecycle have to be evaluated and incorporated into the models, profiles and templates to make the next collaboration lifecycle more efficient. During the whole lifecycle, controlling tools should evaluate the status and development of the process and support systems should help the users avoid mistakes and improve their collaboration skills.

Aspects of Collaboration Life Cycle

The activities within the Collaboration Life Cycle can be regarded from different points of views. For an efficient collaboration all aspects should be jointly considered and supported with powerful IT tools. The most important aspects for collaboration are:

- Process-related aspects,
- Communication-related aspects,
- Project- related aspects and
- Product data-related aspects.

Process-Related Aspects

Most of the companies spend great effort to specify and control their internal product creation processes. The harmonisation of cross company processes is still a difficult task that requires new methods and tools.

Many processes within the product creation phase are highly dynamic and undetermined and therefore cannot be formalized sufficiently. To handle these processes some methods and tools have already been developed [4]. But processes highly relevant to product creation such as release or change processes can be modelled and ensured to run with a defined result. The models of such processes can be built in workflow management components, which should have capabilities to support cross company needs.

Communication-related Aspects

Communication-related aspects focus on how participants in collaborations communicate with each other. They include the involved users and their IT capabilities, communication channels, the applied communication tools, exchanged information or the reasons why communication has taken place. These aspects were originally covered by research areas such as CSCW (Computer Supported Cooperative Work) or tele-cooperation. Typical tools used are e-mail, video-conferencing and application sharing. Actual technical progress has enabled first approaches for advanced communication capabilities within collaborations. Technologies such as VR (Virtual Reality) and 3D computer visualisation could make many collaborative tasks easier or even possible. A vision for the future could be a conference in a collaborative design space in VR including real size images of the conference participants. The question of mobile access and participation is another important point.

Lack of security for cross-enterprise communication is one of the most important barriers against the extensive use of advanced communication techniques or tools in cross-enterprise collaboration.

Project-related Aspects

In collaborative engineering projects organizational problems have to be mastered beside system-oriented hurdles. The organization of distributed projects causes great challenges to the enterprises in the form of multi-project management, cross enterprise workflows, distributed security management and product creation reviews.

To get multilateral collaborations under control the management of cross company projects should be improved. For this purpose collaboration models should be specified, which take roles and responsibilities of the companies into consideration. The existing project management systems are mostly sufficient for company needs. They must be extended with powerful collaboration functionalities to handle the requirements of multiple cross company project management.

Product Data-related Aspects

A substantial factor for an efficient collaboration is the availability of information-that is the supply of the correct data at the right time at the right place. To manage the product data PDM systems are utilized. In medium and large-sized companies,

PDM systems have already become the central source for all product-related data. Yet, the integration of product data is still not realized sufficiently between companies or sometimes even between departments. Besides technical problems, each company has its own guidelines, workflows and semantics regarding the product creation process. Integration and collaborative use of product data halts at the borders of the companies, often even at the borders of the departments of the same enterprise. Therefore, product data is spread over various PDM systems during the collaboration. To provide information about the latest status of development to all concerned project partners, the actual data is replicated over the collaborating companies. Because of the large amounts of data, this replication takes place off-line. Each replication of data means a big effort: data have to be extracted and imported, checked and probably converted. This uncontrollable replication leads to inconsistencies and waste of time.

The product developer should be able to access heterogeneous data sources as one logical system, which delivers all needed data consistently independent of the specific system and its location. In this way a minimization of the collaboration expenses and information losses can be achieved. However, intelligent methods and systems to realize these demands have been missing to-date. The integration problem does not only exist between different companies, but also between different departments of large enterprises, since single departments often use their own process structures and PDM systems. Even small suppliers without any central data management system have to share data and information with their OEM or system supplier. Otherwise they break the integration chain.

The PDM Collaborator project has been launched to create solutions for the outlined requirements, especially to support project and product data-related aspects of collaboration. The PDM Collaborator system supports tools for the management of collaboration projects and acts due to its federation component as the "glue" between the heterogeneous PDM systems of the single collaborating companies. It creates one virtual PDM system containing all product data of the collaboration for the duration of a product development project. Through this, each partner has access to the data he needs and is authorized to access, without worrying about system details.

30.3 The PDM Collaborator System

Overview

The PDM Collaborator system can be seen as the business tier of a three-tier architecture. The PDM Collaborator system serves clients as one virtual data management system. It accesses the data-tier made up of the data management systems that hold information about the product creation.

The functionality of the PDM Collaborator is realized by four main components [2]:

- Basic Services,
- Application-specific Services,
- Collaboration Services and
- Federation Services.

The Basic Services of the PDM Collaborator system fulfil tasks such as user or session management and user mapping. Application-specific services contain CAD-format conversion or off-line data exchange. But the very heart of the PDM Collaborator system is the collaboration and federation services. While the collaboration services support a cross-company project and workflow management, the federation services are responsible both for data model mapping and the linking as well as distribution of data in the connected heterogeneous data management systems.

The system architecture is based on a neutral product data model and a neutral, generic API for accessing data management systems [5]. The neutral product data model is the so-called PDTnet schema [6], which is a subset of STEP AP 214. It covers the PDM-relevant conformance classes CC6-CC8 and is represented in form of a XML schema. The neutral API was defined during the project and is going to become an OMG (Object Management Group) standard as part of the PLM Services V1.0 specification of the OMG's MANTIS (Manufacturing Technology & Industrial Systems) task force [7]. Adaptor and Connector components adapt API and data model of the PDM Collaborator system to those of the data management servers and clients.

Project – and Process-related Aspects of Collaboration – The Collaboration Services

Collaboration services of PDM Collaborator deliver tools and functionalities covering project- and process-related aspects of collaborations and are by that the basis for product-related collaboration aspects. A project management module supports the cross- enterprise project management. For actions that are relevant to other users a notification is sent automatically. The information needed by the project management module is stored in a so-called cooperation model which also includes the project structure [2].

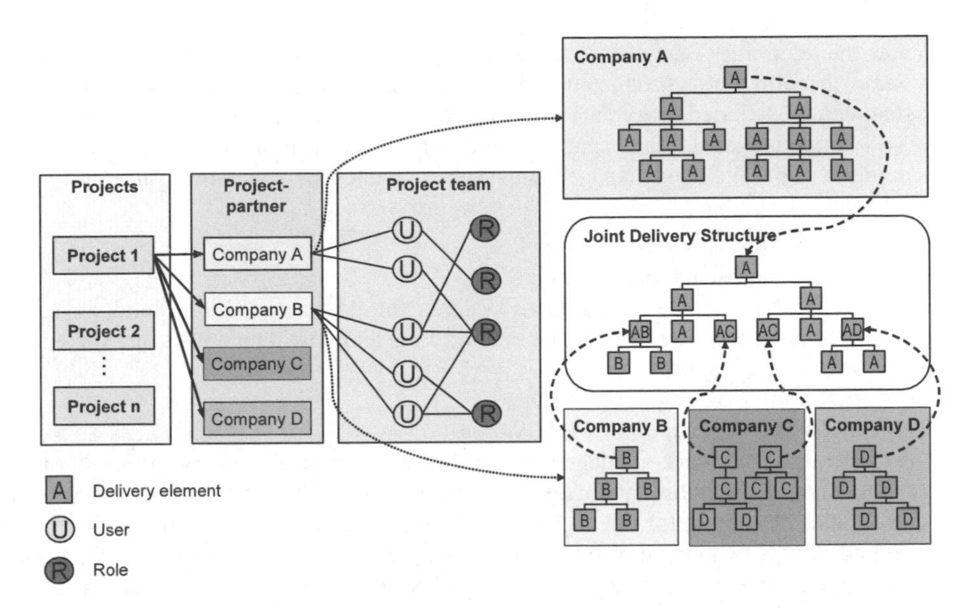

Figure 30.2 Structure of the Cooperation Model

The cooperation model can be specified interactively at the beginning of a project. This model forms the framework for the collaboration between different partners and roles as well as specifies the project structure (Figure 30.2).

The first activity for the definition of the cooperation model is the specification of a project. The companies taking part in the collaboration are added to the project from an existing pool. Persons working on the project in each company and their roles as well as access rights are further information to be attached. Additionally, what components have to be supplied by which company in the context of collaboration is defined? This results in a delivery structure and is modelled using delivery units and delivery elements, to which the project partners can assign their development documents or even links to complete product structures through the other components of the PDM Collaborator. Workflow management has been developed for the integration along the supply chain a cross enterprise. For fast and simple configuration workflow templates are supported. The redefinition and customization of these templates can be performed graphically.

Product Data–related Aspects of Collaboration – The Federation Services

Federation of Data Management Systems

In the case of a collaboration of several organizational units during product creation, the term "federation" describes their temporary cooperation fulfilling the purpose to develop and manufacture a product. During product development and manufacturing processes, larger companies nowadays use product data management systems to manage the product data created, processed or interchanged during product creation. The federation of data management systems means the temporary integration of several organizational units' data management systems to form one virtual, temporary data management system. This virtual system contains all necessary data for product creation that is provided and/or used by all organizational units participating in the development collaboration. Federation of product data management systems enables the following concepts:

- data object linking and
- data object distribution as well as
- data model and semantics mapping.

The federation of data management systems can be classified in several ways, such as how long the federation or the method lasts and how data objects are linked [2].

Concepts for the Federation of PDM Systems

The federation of data management systems is realised by two types of components: the federation and mapping services. The federation service acts as a central point of contact for clients and implements the central management of the federated data management systems, the linking and distribution of data objects. Furthermore, it manages the relationships of data management systems to participating companies, users and development projects of the collaboration services. All information is stored in the so-called federation repository. It is important to mention that besides the configuration information, no product data is held in the central federation repository. Thus, the federation approach goes beyond existing approaches for collaborative data management, that use central data repositories to which all participants have access, as for example the "Share-A-Space" approach [8].

The federation of data management systems requires that all product data is described using one neutral data model. This is assured by the application of mapping rules through mapping services. The working principle of the federation service is the following: any request of a user to the federation engine has to be split onto all affected data management systems (Figure 30.3). This may include splitting up data objects. The splitting process is controlled by federation rules describing the distribution of product data over the systems, are stored in the federation repository.

When all results of the request have been sent back to the federation engine, they are joined or merged to one consistent product structure, again guided by the federation rules stored in the federation repository. It is important to mention that fine-grained access rights to product data are not managed by the federation engine, but by the data management systems themselves. The user initiating a request must be allowed to do so in the connected data management systems, otherwise the request is not processed. Nevertheless, the federation engine ensures that change or creation of product data is usually only allowed in the data management system(s) of the organisational unit to which the actual user belongs. This special data management system(s) is (are) called the "HomeDM" of the user. Creation requests are normally forwarded to the HomeDM for the user, if he has not explicitly chosen another data management system.

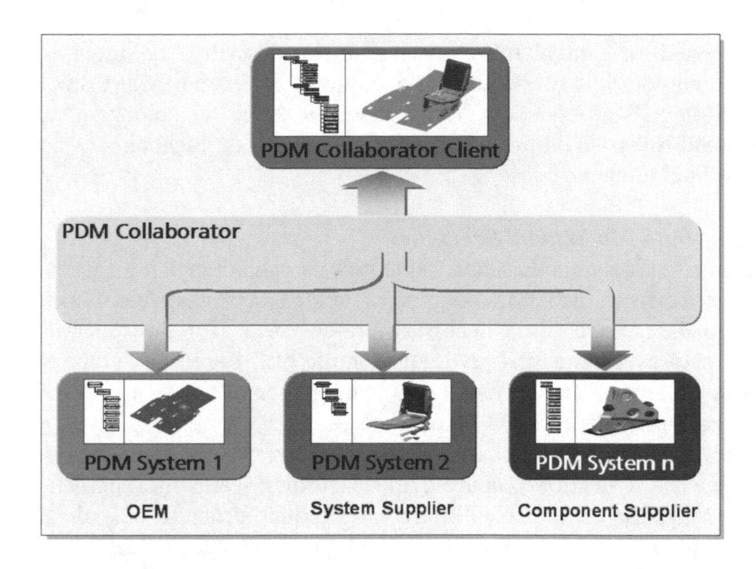

Figure 30.3 Federation of product data management systems

Configuration of the Federation Services

The configuration of federation requires the following information:

- federation rules about the data object relationships and distribution of data objects over the different systems,
- mapping rules between the data models of the different PDM systems and semantics of different companies.

Only the configuration of data object relationships and the distribution of product data over the data management systems participating in the collaboration are considered. The configuration of the federation is based on the described cooperation model.

There are two main approaches for describing data object relationships and distributions:

- Federation based on linking rules between data objects and
- Federation based on general rules.

We now focus on the first approach. Links between data objects residing in data management systems at different organizational units are formed by interface elements of the product structure, which are contained in more than one data management system and by that redundant from the cross-company viewpoint. The actual linking information between those redundant data objects can be managed centrally by the federation services or its managements can be decentralized by the connected data management systems. The linking information may easily be changed while collaboration is going on, to incorporate new suppliers for example. The idea of linking different aspects of a product model has been presented previously, for example in the ULEO (Universal Linking of Engineering Objects) approach [9]. There, the focus is on linking data objects of different disciplines of product views rather than on different distributed data management systems. An approach based on general rules rather than the described linking rules is more complex. It could allow rules such as "documents of System A at Company X and items of System B at Company Y that have the same ID belong to each other". These general rules would provide powerful distribution mechanisms but are very difficult to implement.

Mapping of Data Models and Semantics

The mapping of data models and semantics is mandatory for the exchange of information between different companies and data management systems. The mapping of data models is a necessary pre-condition for the federation of part product structures. Figure 30.4 gives an example of differences in data models and data model contents of two different data management systems implemented at two different companies [5].

Another important problem to be solved by the mapping services is the different levels of detail, comparing the customer's and the supplier's product structure. While the customer will need the product structure only detailed to the level of spare parts, the supplier will detail his product structure to the level of spare parts of his supplier. The mapping rules responsible for the reduction of granularity may for example be a simple filter removing all data objects not labelled as spare parts [5].

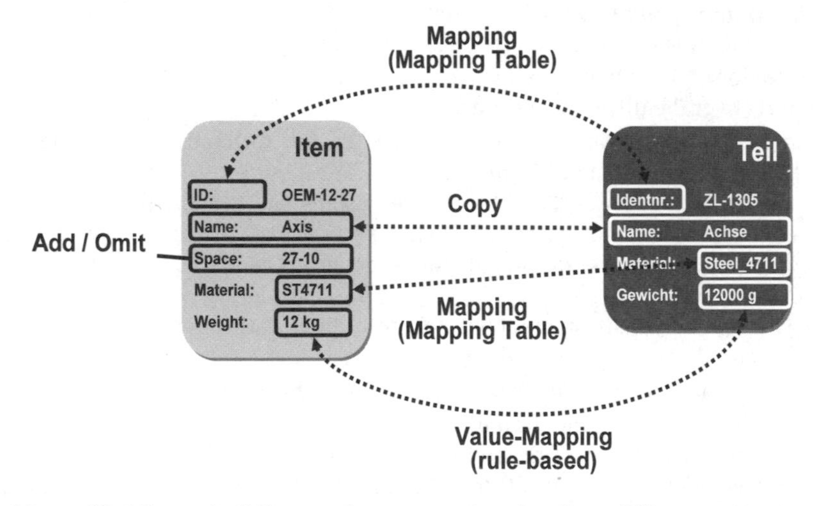

Figure 30.4 Semantic differences between product data from different companies

30.4 Conclusion

Concepts of the Collaboration Life-Cycle (CLC) and its single stages were presented. Important aspects of the CLC have been described including process-, communication-, project- and product-data-related ones. Following, the PDM Collaborator system has been presented that supports project- as well as product data-related aspects of the CLC. The corresponding services of the PDM Collaborator have been presented in more detail: the Collaboration and Federation Services. Beside these aspects, which are efficiently supported by the PDM Collaborator system, there are still a lot of issues that have been solved only partly or isolated from related ones. Security, holistic integration of all collaboration aspects or the usages of future technologies as VR conferencing, to name a few, are still major issues.

30.5 Acknowledgements

The presented results have been partly developed during the joint project PDM Collaborator. The project was funded by the German Federal Ministry of Education and Research (bmb+f) under the label "2PL1004/2". Nine partners were participating in this project. The Fraunhofer IPK Berlin has coordinated the project.

30.6 References

[1] Tseng, M.M., Kjellberg, T., Lu, S.C-Y., 2003, "Design in the New e-Commerce Era," In: *CIRP Annals 2003 – Manufacturing Technology,* Vol. 52/2/2003, 'Technische Rundschau', Edition Colibri Publishers, Wabern, Switzerland.

[2] Krause, F.-L., Hayka, H., Pasewaldt, B., 2003, "Supporting Collaborative Processes by the Federation of Product Data Management Systems," In: *Proc. of the ProSTEP iViP Science Days 2003*, Vol. 8/9, Germany.

[3] Hayka, H; *et al.*, 2003, "Final Report of the Joint Project PDM-Collaborator" (www.pdm-collaborator.de).

[4] V. Ende, A., Helmke, M., u.a., 2002, "Adaptives Prozessmanagement für verteilte Produktentstehungsprozesse," In: Krause, F.-L.; Tang, T.; Ahle, U. (Eds.): *Leitprojekt integrierte Virtuelle Produktentstehung – Final Report*, pg. 33.

[5] Krause, F.-L., Hayka, H., Pasewaldt, B., 2003, "Produktdatenbasierte Kooperation in der Produktentstehung," In: Adam, W.; Pritschow, G; Uhlmann, E.; Weck, M. (Eds.): *Datenmodelle in der Produktion. Fortschritt-Berichte VDI Reihe 2*, Nr. 633, Düsseldorf: VDI Verlag.

[6] N.N.: "PDTnet Implementation Guide," Vol. 1.5, November 2002 (http://www.pdtnet.de/de/results/wg2/guide/).

[7] N.N.: "Homepage of the OMG's MANTIS DTF," 2003.

[8] Rosén, J., 2001, "Collaborative engineering, a tool for the extended enterprise based on an open information model," In: *Proc. of the Int. CIRP Design Seminar – Design in the New Economy*, 6-8 June 2001, KTH, Stockholm, Sweden.

[9] Zimmermann, J.U., Haasis, S., van Houten, F.J.A.M., 2002, "ULEO – Universal Linking of Engineering Objects," *CIRP Annals 2002 – Manufacturing Technology,* Vol. 51/1/2002, 'Technische Rundschau', Colibri Ltd., Uetendorf, Switzerland.

31

Design Iterations in a Geographically Distributed Design Process

Toufik Boudouh, Daniel-Constantin Anghel, and Olivier Garro

Abstract: Iteration is an inherent component of any design process. It is a very important characteristic since it influences product development cost and time. In this paper, an experiment is used to make observations about iterations in a geographically distributed design process. Our objective is to understand how and why iterations occur in the design process. This investigation will help us in the classification of iterations in order to distinguish useful iterations from negative ones. The results of such work could be used to improve assumptions adopted to develop engineering design models, which is very helpful in design planning. We provide a brief review of some design models integrating the iterative aspects of engineering design. After describing the experiment environment, the research method is presented, and observations are then analyzed.

Keywords: design experiment, iteration, geographically distributed design

31.1 Introduction

In this paper, a laboratory experiment is used to make observations about iterations in a geographically distributed design process. Iteration is the process by which a design solution is approximated step by step [1]. It is also defined as the repetition of design tasks due to the arrival or discovery of new information [2].

Iteration has been used as an important issue in several research works for developing models of the design process. These models are used to make important managerial decisions in planning design processes. However, the models do not consider direct observations in order to understand the iteration process, but are based on intuition about their occurrence.

Engineering design is a source of competitive advantage for manufacturing companies. Reducing the development time is a key factor in the successful completion of the product development process. Understanding and mastering the iteration process is a way to achieve this goal. The design process is complex and dynamic. It requires a large variety of approaches to be understood. Therefore, the emergence of a great number of design process models is well justified, and we have to use them in a complementary way rather than considering them as competing. As mentioned above, the models are generally based on intuition. The aim of this work is to provide experimental data based on direct observation to develop models integrating iterative design behaviour in detail.

The remainder of the paper is organized as follows. Section 31.2 provides a literature review of design process models integrating the iterative aspect of design processes. Section 31.3 presents the research method used in this study, with a description of the experimental environment. In Section 31.4, the experimental data and observations, and their analysis, are presented. We conclude in Section 31.5 with a summary of our work and a perspective of future developments.

31.2 Iteration in the Design Process: A Literature Review

Design is an information intensive activity. Because of the complex information dependencies that exist between design tasks, we cannot perform the design process as a once-through procedure. Therefore, iteration is fundamental to resolving design problems.

Understanding and controlling iterations can improve the design process, and reducing them would have positive effects for the product development cycle time. In a study involving semiconductor design, Osborne [3] reported that iterations accounted for 13 to 70% of total development effort for nine projects. It is therefore important to consider the iteration aspects of design tasks when developing models of the design process for managerial issues.

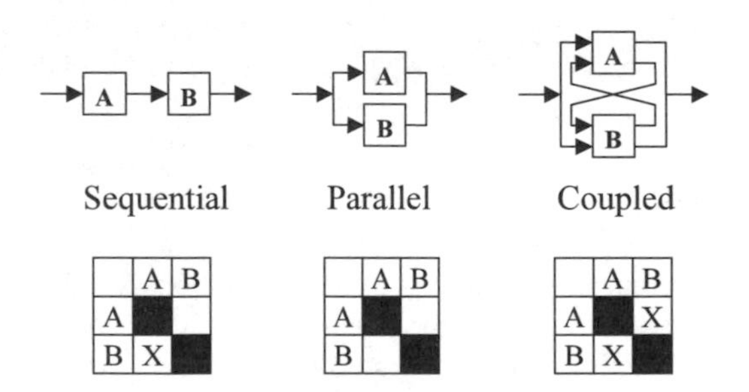

Figure 31.1 Task relationships in a DSM

Several models of the design process that account for iterations are derived from the design structure matrix (DSM) method. In DSM, design tasks and their information dependencies are encapsulated within a compact matrix representation. The DSM matrix is square with one row and one column per task. The tasks are listed in a roughly chronological sequence of execution. In a simple DSM, diagonal elements of the matrix are not used. Off-diagonal elements indicate information flows between design tasks. Reading across a row indicates all of the tasks whose output is required to process the task corresponding to the row, whilst reading down a column indicates all of the tasks, which receive output from the task corresponding to the column. There are three types of sequence relationship that link design tasks: serial (or independent), parallel (or dependent), and coupled (or interdependent). Figure 31.1 shows the three relationship types and their DSM representations.

Upper diagonal elements of a DSM depict the existence of cyclic information flows. Matrix elements are manipulated in an attempt to eliminate or minimize the number of upper diagonal elements, a process known as partitioning. The remaining information cycles are resolved through iteration within the design process.

The original DSM method [4] does not contain quantitative information about the strength of interaction between design tasks. Several extensions have been developed to allow quantitative analysis. One of these models is the sequential iteration model developed by Smith and Eppinger [2], in which they use the linear systems theory to identify controlling features of iteration in a coupled development process. The authors also developed the work transformation matrix model [5], which has a fully parallel structure. In this model, the tasks are re-sequenced and an expected duration for each sequence is determined. While both of these models are useful in characterizing the two extreme cases of product development (parallel and sequential iteration) for any number of tasks, they do not model intermediate scenarios where overlapping might be more appropriate.

Carrascosa [6] developed a mathematical model based on characterizing the information exchanged between tasks in terms of probability of change and impact. DSM has been used in this model as an interface to graphically represent the information flow. The model provides a basis for determining the appropriate task sequence and degree of concurrency to minimize development time and cost.

Belhe and Kusiak [7] proposed a different matrix based methodology to define and sequence tasks by mapping their technical relationships. The developed model includes probabilistic OR and XOR (exclusive OR) relationships between tasks. In this model, the probabilities of executing one or more of the OR/XOR paths are dependent on the iteration number, and these probabilities are fixed in advance. In this model, the length of each task is fixed and deterministic and does not change as the iteration process progresses.

In addition to these models, other models of the iteration process in design do not use the DSM approach. The model developed by Aitsahlia, Johnson and Will [8] presents the effects of parallel scheduling. Doing tasks in parallel allows the completion of the design process more quickly. On the other hand, doing tasks in series leads to fewer tasks requiring repetition (design iterations), and therefore the development cost is lessened. The objective of the model is to determine a trade-off between much parallelism and greater development cost.

Krishnan, Eppinger and Whitney [9] developed a model for overlapping nominally sequential tasks in order to reduce development lead-time. In their model, the downstream activity begins with preliminary upstream information and incorporates subsequent upstream design changes in future iterations. They present a framework to determine how to disaggregate design information and overlap consecutive stages based on the evolution and sensitivity properties of the information exchanged.

All of the models presented above require significant restrictive assumptions. Because of the complexity of engineering design and because of the high level of human intervention in design, it is difficult to develop models with the level of sophistication and application reached in production modelling, for example. The basic assumptions of the presented models are based on general beliefs about how design processes occur, rather than any specific observations. However, we believe that design experiments are a useful way to improve modelling assumptions and therefore to improve engineering design models.

31.3 Research Method: The Design Experiment

To understand the design process, it is useful to use experimentations in order to observe how the designer's design progresses. Design is a social activity [10], and then observations are necessary to depict the complex human behaviour in engineering design.

Design could be accomplished in different situations: by an individual designer, or by a design team or several teams. The members of a design team could work synchronously or asynchronously, in the same space, or in a geographically distributed environment.

Those direct experimental observations that examine the work of individual designers are appropriate for understanding cognition, creativity, and innovation in the design process. Examining the work of design teams provides understanding of the design process in an organizational setting.

Our intention in this study is to observe iterations in a design activity performed by a geographically distributed team of designers. Video-based observational techniques were used in this experiment to provide a useful record of the design process that is then used by different researchers in different ways to study different issues.

The experiment described in this paper was undertaken in the context of the GRACC projects (Cooperative Design Activity Research Group). Four French research laboratories from the Universities of Belfort, Grenoble, Nancy, and Nantes formed the GRACC GROUP.

The experimental subjects were four undergraduate mechanical engineering students, one from each university. Each participant was given a distinct role. There were roles for a project manager, a frame designer, a link designer, and an ergonomics specialist.

The experiment lasted for one month. There were four weekly design sessions conducted synchronously by the four designers. These sessions were recorded in order to fully capture the sights and sounds of the activity. Each session required about two hours to complete. In addition to these sessions, the designers performed their own tasks individually and asynchronously, but could send messages to each other via the project manager. In this case, the designers were asked to document their activities to get the most complete information about the design process.

During the synchronous sessions, the experimental subjects used Netmeeting and speakerphones for personal communication, and FTP (file transfer protocol) for computer file exchanges. Solidworks was chosen as the application to create part and assembly drawings. Moreover, a shared whiteboard was used to allow real time sharing of notes, drawings, and sketches.

The goal of the design experiment was to redesign a trailer pulled by a mountain bike. There were three main parts in this design problem: (1) the design of the chassis or the trailer frame performed by the frame designer; (2) the design of the tow bar, used to link the chassis to the bike and the wheels of the trailer, performed by the link designer; (3) and the design of the hood performed by the ergonomics designer. During the four synchronous sessions, the designers cooperated in order to perform together the design of the different parts of the trailer.

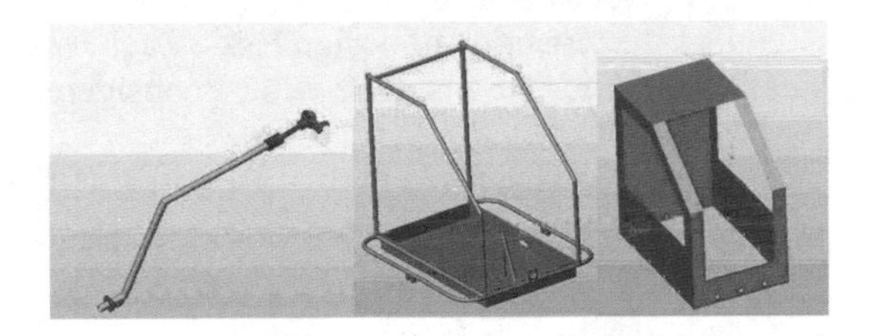

Figure 31.2 The three parts of the trailer

31.4 Observations, Analysis and Results

The four synchronous design sessions were videotaped for later analysis and coding. No design process was externally imposed on the experimental designers; rather they were free to choose the process necessary to develop an appropriate design result.

The first design session was the first meeting of the design team, so it was dedicated to the presentation of team members and the design problem to be resolved. The roles of each designer were also fixed in this session. A part of the last session was intended for the project termination. In these phases, no iterations were observed since the designers were just sharing and exchanging information, and performing the first solution developments.

The results and observations presented in this section concern the third session, which was the most information intensive phase and, therefore, the phase where most of the iterations occurred.

Task and Iteration Identification

In order to identify iterations in the design process, we have first identified the design tasks performed by the experimental designers. Task identification was performed using three means.

1. *Product decomposition.* Decomposing the product from its different parts allows us to identify some of the tasks performed by the designers.
2. *Intermediate objects.* These are the results of the different designers' actions. They could be drawings or calculus notes.
3. *Videotapes.* Using the first two means and the videotapes of the recorded design sessions, we were able to identify both of the design tasks and the information flow between these tasks.

The tasks identified in the third design session are represented in the DSM shown in Figure 31.3. The DSM indicates also the task relationships. In our analysis, iterations are considered as the repetition of design tasks. Each new switch to a task after designers have executed it for the first time will be considered a new iteration.

		1	2	3	4	5	6	7	8	9	10	11	12	13	14	15	16	17	18	19	20	21	22	23	24	25
Problem presentation	1									1																
New chassis shape	2	1																								
Chassis-bike link analysis	3		1												1											
Tow bar solution analysis	4			1			1	1	1																	
Chassis link analysis	5				1																					
Tow bar tightening solution	6				1	1				1																
Detailed tightening solutions	7				1		1																			
New link solution	8				1																					
Tightening solution validation	9				1			1																		
Chassis design analysis	10	1										1	1													
Chassis assembly solution	11										1		1													
Seat fixation solution	12										1	1					1	1	1							
Chassis material selection	13										1															
Chassis detailed design	14													1												
Cost evaluation	15			1													1			1					1	
Hood shape proposition	16															1		1			1	1		1	1	
Comfort evaluation	17										1		1			1	1				1	1				
Seat selection	18										1															
Hood material selection	19												1			1	1									
Water tightness evaluation	20																1			1						
Opening ease evaluation	21																1	1		1			1			
Hood shape optimisation	22																					1				
Safety evaluation	23																1									
Hood detailed design	24																1			1						
Hood design validation	25																								1	

Figure 31.3 The DSM of the observed design process

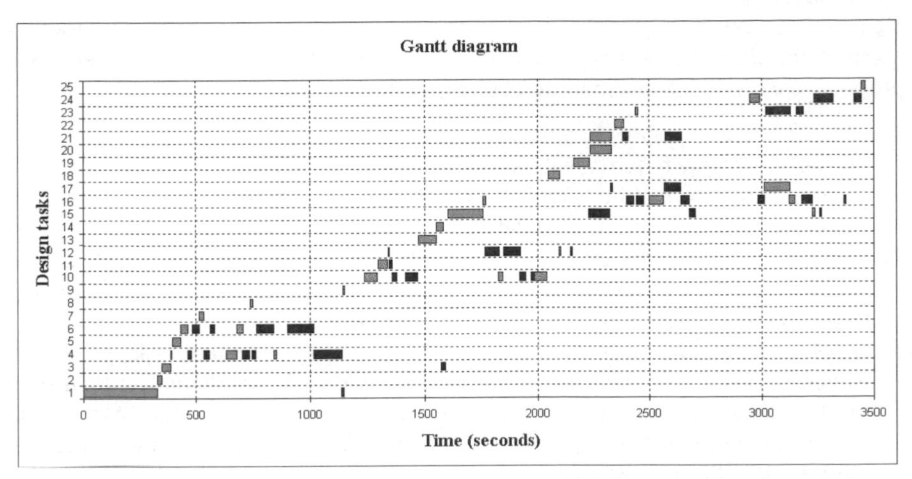

Figure 31.4 The Gantt diagram of the observed design process

In the iteration process, the rework may concern the entire task or just a part of it. Figure 31.4 shows the Gantt diagram of the design process as it was observed. The execution of each task is indicated using different shades for the different iterations.

Iteration Types

Iterations were classified into three typologies: expected and unexpected, short and long, and positive and negative.

1. *Expected and Unexpected iterations.* The first type describes the planned iterations. They are typically performed in the case of coupled tasks to converge to a desirable solution. The second type of iteration results from unexpected failure of the design task to meet the specifications. Upon failure, some or all of the design tasks need to be repeated. Expected iterations accounted for 67% (13.7 minutes) and the unexpected iterations for 33% (6.8 minutes).
2. *Short and Long iterations.* This classification allows us to determine if the iteration process is resolved rapidly or not. It concerns also the number of tasks involved in the iteration process. In the DSM method, to minimize the number of iterations we should minimize the number of non-zero elements of the upper diagonal matrix. The remaining elements should be closer to the diagonal so the number of iterated tasks is minimized. In our case, short iterations represent 71% (13 minutes) of the total iteration time. Long iterations account for the remaining 29% (7.5 minutes).
3. *Positive and Negative iterations.* In this category, any iteration that can be eliminated or avoided without loss in value is considered negative. This is the case of iterations caused by design errors. In the experiment, 23% (4.5 minutes) of the iteration time is lost in negative iterations. Positive iterations accounted for the remaining 77% (16 minutes).

Iteration Origins

Design iteration implies rework by returning to previously achieved tasks to account for changes. This rework could have different causes. In this study, we propose to classify the origin of iterations into three types.

1. *Specification changes.* In this case, design objectives and requirements are unstable, or new requirements are added. It is particularly so when tasks use preliminary information supplied by un-finished upstream tasks. Changing specifications was the cause of 24% (5 minutes) of the iteration time.
2. *Design errors.* Design errors and mistakes could have negative impact on the design process. Generally, failures in achieving design tasks are due to human errors. Design errors represent 25% (5 minutes) of the total iteration time.
3. *Task interdependence.* This is the case of mutually dependent tasks for which several iterations are necessary to reach an acceptable solution. The total time of the design process depends on the initial scheduling of the tasks. Task interdependence is the cause of the most of time spent in the iterative process with 51% of the total iteration time.

31.5 Conclusion

Engineering design is a complex activity and experiments are a useful means for obtaining relevant data to help understand it better. In this paper, an experiment involving a geographically distributed engineering design team was used to study the iteration process in engineering design. Iteration is an important characteristic of the design process and its analysis is a key factor in reducing product development cost and time.

We do not claim that the observations provided by this experiment hold for design iterations in industrial projects, but this experiment is an appropriate first step toward understanding iteration mechanisms in the design process.

The experiment presented in this paper was not specifically intended to study the iteration issue in the design process; rather, it was used by several researchers for different research purposes.

It is possible that the geographically distributed aspect of the design team may have influenced the behaviour of the design team members, but we believe that this influence, if it exists, is minimal. Moreover, the analysis performed for iterations concerns the third session, so the designers were familiar with the geographically distributed environment. In further research, we will attempt to compare the assumptions of existing models of the iteration process with the observed mechanisms in order to improve them. The perspective of this work is to perform a quantitative evaluation of the impact of iteration in the design process in terms of cost, time, and quality.

31.6 References

[1] Pahl, G., Beitz, W., 1996, *Engineering Design: A systematic approach*, Springer-Verlag.

[2] Smith, R.P., Eppinger, S.D., 1997, "A predictive model of sequential iteration in engineering design," *Management Science*, Vol. 43, pp. 1104-1120.

[3] Osbrne, S.M., 1993, "Product development cycle time characterization through modelling of process iteration," *Master Thesis, MIT*.

[4] Steward, D.V., 1981, "The design structure system: A method for managing the design of complex systems," *IEEE Transactions on Engineering Management*, Vol. EM-28, No. 3, pp. 71-74.

[5] Smith, R.P., Eppinger, S.D., 1998, "Deciding between sequential and concurrent tasks in engineering design," *Concurrent Engineering: Research and Applications*, Vol. 6, No. 1, pp. 15-25.

[6] Carrascosa, M., Eppinger, S.D., Whitney, D.E., 1998, "Using the design structure matrix to estimate time to market in a product development process," *ASME Design Automation Conference*, Atlanta 98-6013.

[7] Belhe, U., Kusiak, A., 1996, "Modelling relationships among design activities," *Journal of Mechanical Design*, Vol. 118, No. 4, pp. 454-460.

[8] Aitsahlia, F., Johnson, E., Will, P., 1995, "Is concurrent engineering always a sensible proposition?" *IEEE Transactions on Engineering Management*, Vol. 42, No. 2, pp. 166-170.

[9] Krishnan, V., Eppinger, S.D., Whitney, D.E., 1997, "Model-based framework to overlap product development activities," *Management Science*, Vol. 43, No. 4, pp. 437-451.

[10] Bucciarelli, L.L., 1988, "An ethnographic perspective on engineering design," *Design Studies*, Vol. 9, No. 3, pp. 159-168.

32

A Cluster-based Approach for Collaborative Design Process Analysis

Reza Movahed Khah, Egon Ostrosi, and Olivier Garro

Abstract: In this paper, we propose an analysis approach for the collaborative design process. That is centered on the messages generated by the actors. It is structured with two levels: the first level concerns the analysis of interactions between actors inside each discussion, whereas the second level concerns the analysis of relationships between different discussions. The analysis distinguishes between three concepts: the formation of micro-groups; the articulation of design process around one or several key actors; and the types of interactions. The application of the proposed approach in collaborative design experience allowed the identification of three properties of this process: the auto-organization, the dynamics and the auto-similarity.

Keywords: collaborative design process, teamwork, experimental analysis, knowledge sharing.

32.1 Introduction

Organization of work in companies is increasingly based on computerized environments, such as Groupware, *CSCW (Computer Supported Collaborative Work)* and Workflow. The concept of collaborative design appears both as an effect of globalization and as a prospective tool for enabling this new business approach [1]. Collaborative design is an activity that requires the participation of individuals for sharing information and organizing design tasks and resources. The purpose of design collaboration is to share expertise, ideas, resources, or responsibilities [2]. For a clear understanding of collaborative design, many researchers have built up different experiences. Associated with modelling tools and with theoretical development, which came both from DAI (*Distributed Artificial Intelligence*) and from *CSCW,* these experiences show that designers use a major part of their time to create, manipulate, discuss, interpret and transform the texts, graphs, calculations, digital models, diagrams, physical model, *etc.*. During the design process, the traces and supports of design actions are produced or used. They relate to tools, procedures, and actors [3]. These objects are called *Design Intermediate Objects* (*DIO*) [4, 5, 6].

One of the most important Design Intermediate Objects emerging from a *collaborative design process* is the meeting known as *corpus* [7]. *Corpus* represents the complete traceability of verbal message exchanges during the collaborative design process. Furthermore, during the collaborative design process, the actors communicate when they want to cooperate, to coordinate their actions, and to realize tasks in common. Communication established in this way between the actors is expressed in the form of interactions in which the dynamic relations between the actors can be expressed via indicators. Once interpreted, these indicators can produce effects on the *cooperation between the actors*.

The problem of cooperation can help determine "who makes what, when, where, by what means, in what way and with whom?" Within the framework of the collaborative design process, *cooperation* expresses the resolution of various sub-problems: *collaboration* through the tasks distribution, action *coordination* and *conflict resolution* [8]. Therefore, *cooperation* can be seen as an effective and concrete articulation of the actors around a common action. The action coordination during the collaborative design process represents a set of functioning rules established by one or several actors in order to accomplish a task in a group [9].

Analysis of the collaborative design process during communication between actors is an important task that can allow the development of pertinent tools to assist the design process. For design process analysis, we propose an approach based on the interaction between actors. Our approach is centred on the communicative traces, or messages, generated by the actors. Therefore, we use the notion of corpus that represents communication traceability.

In this paper, our objective is to analyze the *collaborative design process* during meetings between the actors. Through the corpus, we study various forms of interaction between the actors and we try to express the dynamic relations during these interactions by co-operation indicators. These indicators, interpreted by the actors, can produce changes for the various elements of the cooperation, such as: *collaboration*, *action coordination* and *conflict resolution*. In the second section, we develop two levels of our analysis approach for the collaborative design process. Finally, we present conclusions and perspectives.

32.2 Development of the Approach

In the research, many works based on systemic, axiomatic, psycho-cognitive analysis, socio-technique, and administrative analysis approaches contributed to defining the design process of a product. Aiming to study collaboration (or cooperation if it takes place) through communication, our goal is to develop an approach for design process analysis centred on the communicative traces or messages generated by the actors. The message represents the *communication unit* between the actors. Consequently, our study is based on the analysis of messages inside a corpus, at the time of communication between actors during the collaborative process of design. The formal representation of the analysis of interactions between actors is achieved by means of the corpus decomposition into a set of sub-corpus, called discussions. Then, the proposed approach is divided into two levels:

Level 1: Analysis of interactions between actors inside each discussion;

Level 2: Analysis of relationships between different discussions.

Analysis of Interactions between Actors inside Each Discussion

The analysis is done in the following phases:
1. Decomposition of the corpus into discussions;
2. Representation of interactions for every discussion;
3. Interactions Analysis.

Decomposition of the Corpus in Discussion

During the observation of the design experience, and after re–reading the corpus, we found that the design experience could be represented as a discussion set. In fact, actors exchange messages that represent blocks of knowledge linked to their own register of knowledge. In every message, we can distinguish different concepts defined as a set of key words. From these identified concepts, we can represent the corpus as a chain of discussions $C= (D_1, D_2,...,D_p)$. So, the analysis of the interactions between actors consists of studying, representing, and interpreting the interactions in every discussion and between discussions. For example, the corpus C_3 of the third meeting of a design experience is decomposed into eight discussions (Figure 32.1).

N°	Subject	Interventions Interval
D_1	Meeting Organization and tools setup	46 – 114
D_2	Informative discussion on the relationship between link and trailer	115 – 184
D_3	Fast tightening, choosing of the pitch and the screw	202 - 280
D_4	Make a study of FEM results to select materials	281 - 377
D_5	Discussion on the specifications and prices of baby seat on bicycles	378 - 653
D_6	Hood – design	654 - 835
D_7	Hood – choosing of materials	835 - 880
D_8	Hood- welding the bar interior of chassis	889-950

Figure 32.1 Decomposition of a design experience

Representation of Interaction for Every Discussion

The representation of interaction between actors depends on the definition of *communication unit* between these actors. Interactions are realized by the interventions of actors. During the design process, an intervention contains one or several messages. Each message, considering it's semantic, represents the communication unit that is defined as an exchange between two actors. Therefore, in that analysis, we define a message as an exchange between actors if, and only if, this message concerns the registers of knowledge of these actors. For example, the following extract (Figure 32.2) shows the exchanges during the considered collaborative design meeting:

Intervention N°	Actor	Emission message	Response message	Conversation
N	A_3	A_2		Can you draw with Paintbrush?
N+1	A_2	A_1 A_4	A_3	Yes, I will try. I will redraw the tow bar. Do you see it?
N+2	A_1		A_2	Yes, I see it.

Figure 32.2 Extract of a conversation between several actors

If an exchanged message between actors can contain information and questions, *etc,* relative to a subject, then, we call this type of message an *emission message.* In the same way, the information, the responses, *etc.,* related to an *emission message* is called a *response message.* In addition, an intervention could contain two types of messages: *response* and *emission.*

For example, the intervention n+1 of the actor A_2 contains the *response message* to the actor A_3 and the *emission message* to actors A_1 and A_4 (Figure 32.2).

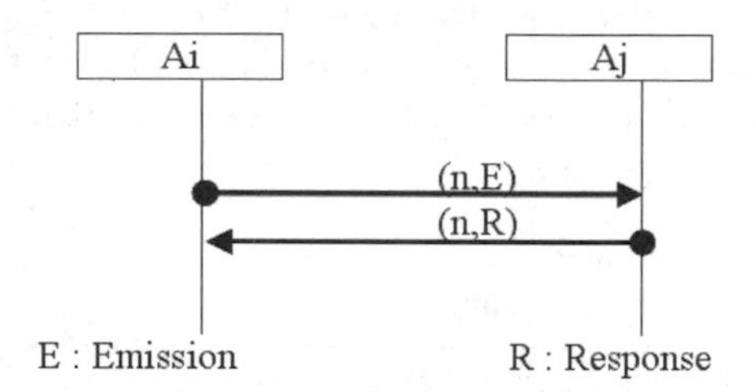

Figure 32.3 Diagram of interactions

Given discussion D in corpus C, then the interactions between two actors could be represented by the diagram of interactions, where:

- A_i represents the set of nodes, where each node represents an actor;
- L is the set of edges, where each edge represents an interaction between the actor A_i and A_j. An interaction between two actors exists if, and only if, a message concerns these actors;
- F is an attribute $F=(n, X)$ where *"n"* represents the intervention number and *"X={R, E}"* represents the message. If the message is a response, then $X=R$, otherwise $X=E$ (Figure 32.3).

For example, in Figure 32.4, the diagram represents the interactions between three actors during a design experience. It shows that actor A_2 emits the message during the intervention *130* in favour of actor A_3. Then, actor A_3 interferes with intervention *131* and emits the message in response to A_2. In this case, we can find that the interventions *130* and *131* contain repetitively the emission and the response messages. In another case, actor A_2, in intervention *160*, sends the message for actor A_1 and in the same time through this message, he replies to the actor A_3. Therefore, intervention *160* is composed of two different types of messages: *emission* and *response*. Thus, through this example, we show the concept of the message used in this *analysis*.

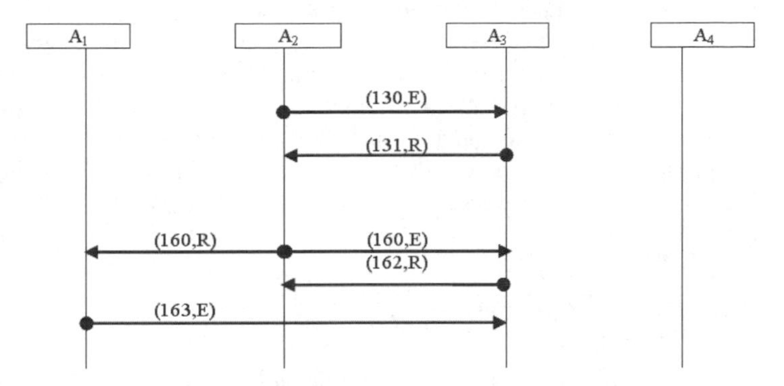

Figure 32.4 Interactions between three actors

Interaction Analysis

The interaction analysis follows these steps:

1. Matrix representation of interactions for every discussion;
2. Matrix representation of interactions for the set of discussions (corpus);
3. Interaction analysis for every discussion.

E	A1	A2	A3	A4
A1	-	{e12}	{e13}	{e14}
A2	{e21}	-	{e23}	{e24}
A3	{e31}	{e32}	-	{e34}
A4	{e41}	{e42}	{e43}	-

a) Emission Matrix

R	A1	A2	A3	A4
A1	-	{r12}	{r13}	{r14}
A2	{r21}	-	{r23}	{r24}
A3	{r31}	{r32}	-	{r34}
A4	{r41}	{r42}	{r43}	-

b) Response Matrix

ER	A1	A2	A3	A4
A1	--	{er12}	{er13}	{er14}
A2	{er21}	--	{er23}	{er24}
A3	{er31}	{er32}	--	{er34}
A4	{er41}	{er42}	{er43}	--

c) Emission-Response Matrix

Figure 32.5 Matrix representation of interactions

Matrix Representation of Interactions for Every Discussion

For analyzing the interactions between actors, we propose the construction of three matrixes called the *emission matrix E* (e_{ij}), the *response matrix R* (r_{ij}), and the *emission-response matrix ER* (er_{ij}). These matrices are constructed from the diagram of interactions. Thus, each row i, $i=1, n$ and corresponding column j, $j=1, n$ of each matrix represent actor A_i, and A_j respectively. An element e_{ij} of the matrix E represents the actor's emission messages A_i toward the actor A_j; an element r_{ij} of the matrix R represents the actor's response messages A_i to the actor A_j; an element er_{ij} is defined as the union of emission messages e_{ij} and response messages r_{ij}.

For example, let us consider the 6th discussion. This discussion, which we call *hood-design,* occurs between the interventions *653* and *834*. It is represented by the three matrixes, respectively, the *emission matrix E* (e_{ij}), the *response matrix R* (r_{ij}), and the *emission-response matrix ER* (er_{ij}) (Figure 32.5). The following table represents two elements of matrixes e_{ij}, r_{ij}, respectively, E (e_{ij}) and R (r_{ij}).

e_{12}= {685,675,799,786,716,704}
r_{42}= {687,709,771,773,796,811,813}

(32.1)

Here, the elements er_{ij} could be defined as the sum of the respective elements of matrix E (e_{ij}) and the respective elements of matrix R (r_{ij}). To measure interactions between actors through emission messages and response messages, we transform each matrix into quantitative matrixes. So, an element e_{ij}*, respectively r_{ij}* of the quantitative matrix $E^*(e_{ij}^*)$, respectively $R^*(r_{ij}^*)$ is defined as $e_{ij}^* = card\{e_{ij}\}$, respectively $r_{ij}^* = card\{r_{ij}\}$.

The transformation of matrixes from Figure 32.5 is represented in Figure 32.6. We can find, for example, that element
$e_{12}= \{685,675,799,786,716,704\}$ is represented by $e_{12}^*=6$.

E	A1	A2	A3	A4
A1	-	6	19	11
A2	3	-	21	14
A3	2	17	-	16
A4	2	20	30	-

R	A1	A2	A3	A4
A1	-	0	1	2
A2	0	-	5	11
A3	2	7	-	17
A4	6	7	4	-

ER	A1	A2	A3	A4
A1	-	6	20	13
A2	3	-	26	25
A3	4	24	-	33
A4	8	27	34	-

Figure 32.6 Quantitative matrix $E^*(e_{ij}^*)$, $R^*(r_{ij}^*)$ and $ER^*(er_{ij}^*)$

Matrix Representation of Interactions for the Discussion Set

Therefore, the diagonal *discussion* (D_i)-*discussion* (D_j) matrix can represent corpus C. For example, corpus $C=(D_1,..., D_8)$ is represented by the discussion matrixes. A zoom on this matrix shows in detail the interactions in each discussion (Figure 32.7).

Interaction Analysis for Every Discussion

Two types of analyses have been performed on the considered discussion:
1. Searching of design "*micro-groups*";
2. Searching of "*key actors*";

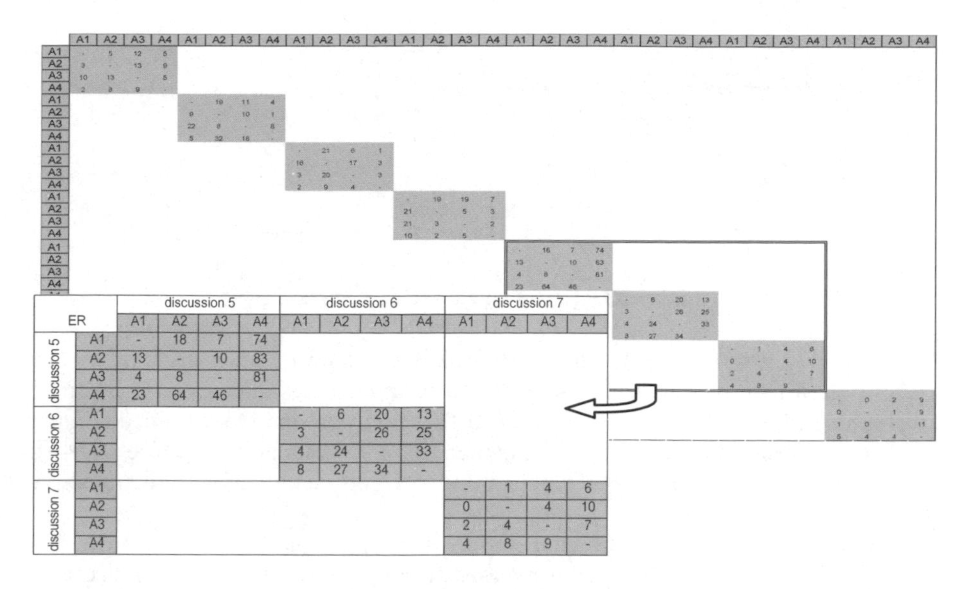

ER		discussion 5 A1	A2	A3	A4	discussion 6 A1	A2	A3	A4	discussion 7 A1	A2	A3	A4
discussion 5	A1	-	18	7	74								
	A2	13	-	10	83								
	A3	4	8	-	81								
	A4	23	64	46	-								
discussion 6	A1					-	6	20	13				
	A2					3	-	26	25				
	A3					4	24	-	33				
	A4					8	27	34	-				
discussion 7	A1									-	1	4	6
	A2									0	-	4	10
	A3									2	4	-	7
	A4									4	8	9	-

Figure 32.7 Matrix Z (z_{ij}) of the discussions

Searching for Design "Micro-groups"

Matrix Z (z_{ij}) where $z_{ij}=er_{ij}^{*}+er_{ji}^{*}$ was used for searching for micro-groups. This matrix is transformed into a fuzzy matrix where each element is defined according to relation $z_{ij}=z_{ij}/max\ (z_{ij})$. For example, decomposition of the fuzzy matrix $Z(z_{ij})$ (Figure 32.8) shows the formation of two micro-groups in the 6th discussion of the collaborative design process. The micro-groups are the following: $\{A_2,A_3,A_4\}$ and $\{A_1\}$.

zij	A1	A2	A3	A4
A1	-	0.13	0.36	0.31
A2		-	0.75	0.78
A3			-	1.00
A4				-

Figure 32.8 Fuzzy Matrix Z(z_{ij}) of D_6

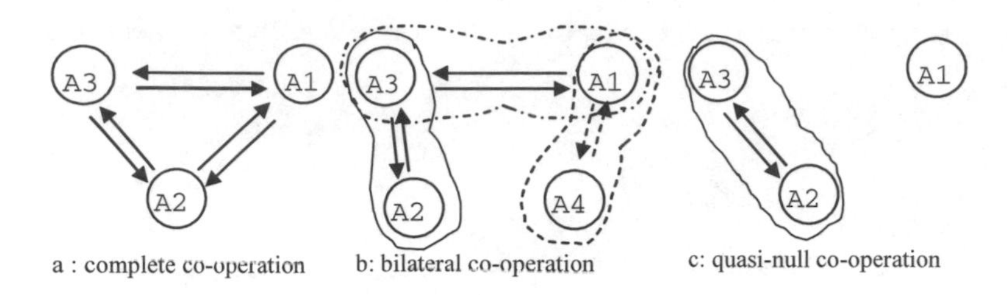

Figure 32.9 Representation of co-operation graph

Figure 32.9 shows the design micro-groups obtained for each discussion. The results show that, depending on the interactions, three types of co-operation can be distinguished: cooperation is *quasi-null*, if the micro-groups are disjoined (Figure 32.9-c); cooperation is *bilateral*, if interactions are symmetrical (Figure 32.9-b); and cooperation is *complete*, if interactions are both symmetrical and transitive (Figure 32.9-a).

We have observed that the cooperation is efficient in each discussion if this cooperation is complete. The opposite cases emerge when the cooperation is quasi-null. For example, in the 6th discussion, the cooperation is complete inside the micro-group (A_2,A_3,A_4) and is quasi-null between the micro-group and the actor A_1. Figure 32.10 shows types of cooperation for each discussion.

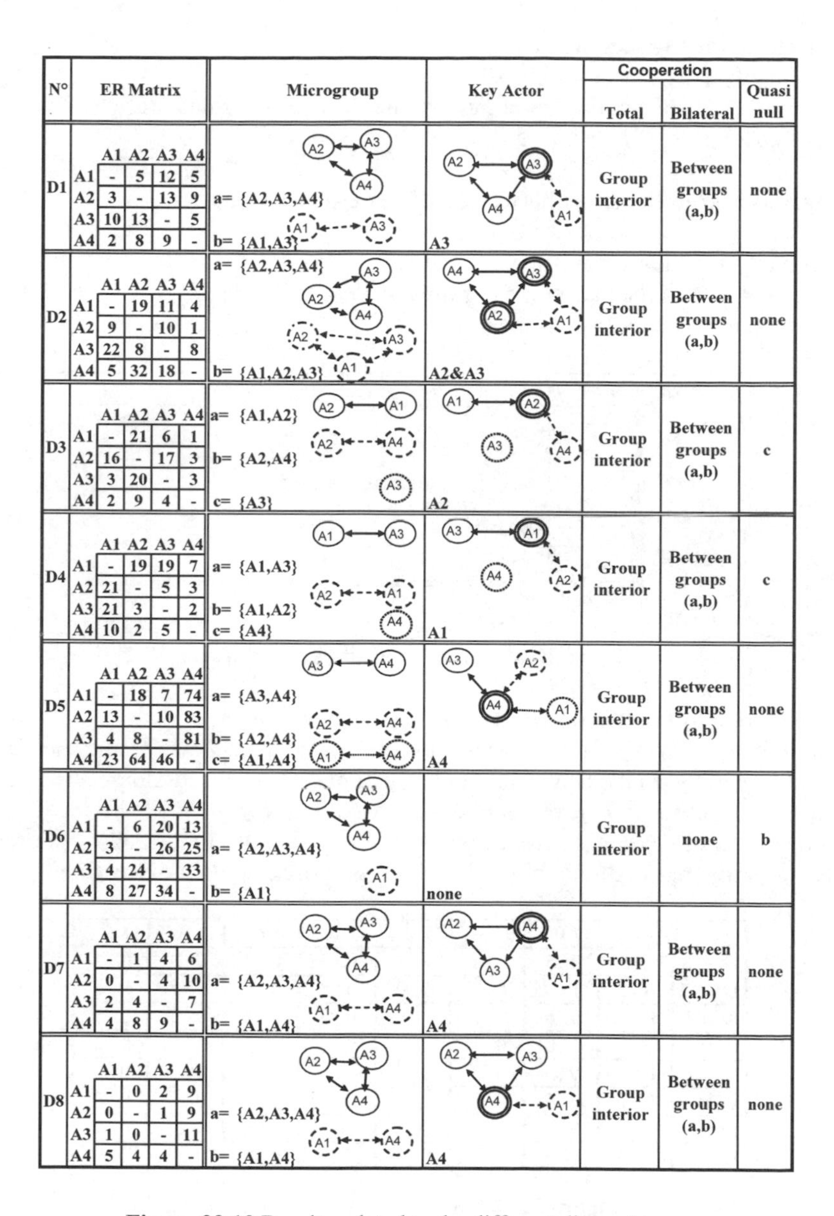

N°	ER Matrix	Microgroup	Key Actor	Cooperation: Total	Cooperation: Bilateral	Cooperation: Quasi null
D1	A1 A2 A3 A4 A1: - 5 12 5 A2: 3 - 13 9 A3: 10 13 - 5 A4: 2 8 9 -	a= {A2,A3,A4} b= {A1,A3}	A3	Group interior	Between groups (a,b)	none
D2	A1 A2 A3 A4 A1: - 19 11 4 A2: 9 - 10 1 A3: 22 8 - 8 A4: 5 32 18 -	a= {A2,A3,A4} b= {A1,A2,A3}	A2&A3	Group interior	Between groups (a,b)	none
D3	A1 A2 A3 A4 A1: - 21 6 1 A2: 16 - 17 3 A3: 3 20 - 3 A4: 2 9 4 -	a= {A1,A2} b= {A2,A4} c= {A3}	A2	Group interior	Between groups (a,b)	c
D4	A1 A2 A3 A4 A1: - 19 19 7 A2: 21 - 5 3 A3: 21 3 - 2 A4: 10 2 5 -	a= {A1,A3} b= {A1,A2} c= {A4}	A1	Group interior	Between groups (a,b)	c
D5	A1 A2 A3 A4 A1: - 18 7 74 A2: 13 - 10 83 A3: 4 8 - 81 A4: 23 64 46 -	a= {A3,A4} b= {A2,A4} c= {A1,A4}	A4	Group interior	Between groups (a,b)	none
D6	A1 A2 A3 A4 A1: - 6 20 13 A2: 3 - 26 25 A3: 4 24 - 33 A4: 8 27 34 -	a= {A2,A3,A4} b= {A1}	none	Group interior	none	b
D7	A1 A2 A3 A4 A1: - 1 4 6 A2: 0 - 4 10 A3: 2 4 - 7 A4: 4 8 9 -	a= {A2,A3,A4} b= {A1,A4}	A4	Group interior	Between groups (a,b)	none
D8	A1 A2 A3 A4 A1: - 0 2 9 A2: 0 - 1 9 A3: 1 0 - 11 A4: 5 4 4 -	a= {A2,A3,A4} b= {A1,A4}	A4	Group interior	Between groups (a,b)	none

Figure 32.10 Results related to the different discussions

Searching for the Key Actors

Figure 32.11 shows the variation of the key actor roles during this design experience.

Analysis of Relationships between Different Discussions

This analysis follows these steps:
1. Constitution of the discussions *"similarity matrix"*;
2. Searching for "*similar sub-groups*".

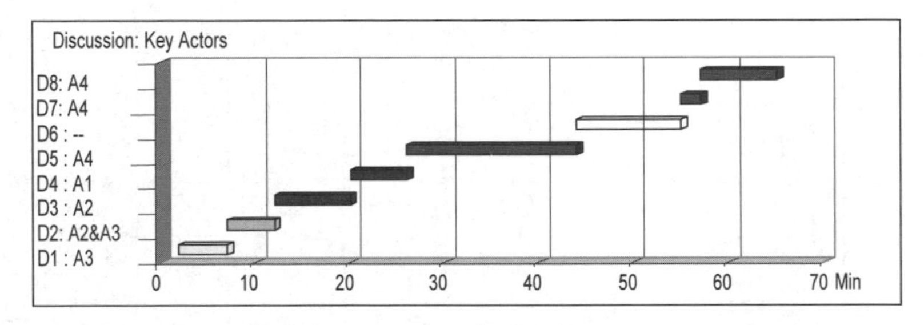

Figure 32.11 Variation of the key actor roles

Constitution of the "Similarity Matrix" of Discussions
The similarity matrix is represented by the matrix $S\ (s_{ij})$ where the elements s_{ij} could be defined as the degree of similarity between two discussion matrices $Z\ (z_{ij})$. Thus, each row i, $i=1, n$ respectively each column j, $j=1, n$ of the matrix $S\ (s_{ij})$ represents a discussion D_i respectively D_j. For example, Figure 32.12 shows that the similarity matrix of the third meeting is composed of eight discussions:

	D 0 1	D 0 2	D 0 3	D 0 4	D 0 5	D 0 6	D 0 7	D 0 8
D 0 1	1 , 0	0 , 5	0 , 7	0 , 5	0 , 3	0 , 7	0 , 8	0 , 5
D 0 2	0 , 5	1 , 0	0 , 5	1 , 0	0 , 5	0 , 5	0 , 7	0 , 3
D 0 3	0 , 7	0 , 5	1 , 0	0 , 5	0 , 3	0 , 7	0 , 5	0 , 5
D 0 4	0 , 5	1 , 0	0 , 5	1 , 0	0 , 5	0 , 5	0 , 7	0 , 3
D 0 5	0 , 3	0 , 5	0 , 3	0 , 5	1 , 0	0 , 0	0 , 2	0 , 2
D 0 6	0 , 7	0 , 5	0 , 7	0 , 5	0 , 0	1 , 0	0 , 8	0 , 8
D 0 7	0 , 8	0 , 7	0 , 5	0 , 7	0 , 2	0 , 8	1 , 0	0 , 7
D 0 8	0 , 5	0 , 3	0 , 5	0 , 3	0 , 2	0 , 8	0 , 7	1 , 0

Figure 32.12 Matrix S (s_{ij}) of extract meetings

Searching for "Similar Sub-Groups"
For the searching for similar sub-groups, we used the matrix $S\ (s_{ij})$ that resulted from the previous step. The hierarchical clustering algorithm (complete linkage) is applied on the similarity matrix $S\ (s_{ij})$ [10, 11].

Final results of hierarchical clustering are presented in the form of a dendrogram. On the x-axis of the dendrogram, the indices of clustered objects (or variables) are displayed, whereas y-axis represents the corresponding linkage distances (or an adequate measure of similarity) between the two discussions, which are merged. We found six sub-groups (Figure 32.13).

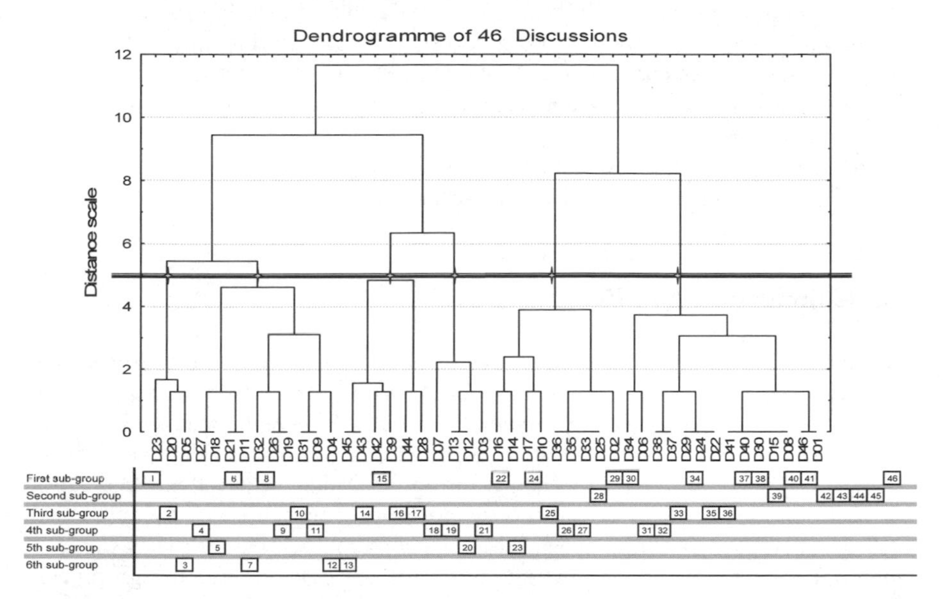

Figure 32.13 Hierarchical Classification Result Shows the Creation of Several Similar Sub-Groups (Similarity Degree of Approximately 60%)

Discussion: From the interaction analysis between actors, we can define some interesting notions related to a design experience:

First, we notice that in a collaborative design process, the actors organize themselves automatically to solve a particular problem. This auto-organization is obviously a consequence of the emergence of the design micro-groups.

Second, we also notice that this auto-organization, during the collaborative design process, is dynamic as a result of variation of the micro-group formation. Therefore, for each new discussion, new micro-groups emerge, and the design process will articulate around one or several new key actors.

Third, the relations within each micro-group and between micro-groups permitted distinction between the different types of co-operation. It is interesting to emphasize that we have a complete cooperation inside a micro-group. This permits qualitative classification of a design process. The associated indicators and these qualitative measures remain to be defined.

Fourth, we notice that the design discussions are qualitatively similar. Therefore inside each discussion, we can find: (a) *the formation of the micro-group;* (b) *the articulation around the key actors;* and (c) *the types of co-operation.*

On the other hand, from the interaction analysis between discussions, the following concepts are inferred:

First, the similarity between discussions permitted classification of the similar sub-groups. Then, it can be seen that several convergences and divergences occurred by moving the discussions to similar sub-groups. In this experiment, we noted that actors diverge first and converge at the end of the design process.

Second, we notice also that the auto-organization, during the collaborative design process, is dynamic. The dynamism is the result of the variation of the sub-group formation in each discussion. Therefore, each new discussion can change its place in the sub-groups.

Third, we notice that in a collaborative design process, in the majority, each discussion is a component of the similar sub-group of discussion.

32.3 Conclusions and Perspectives

We have analyzed a collaborative design experience (*GRACC*: *Groupe de Recherche sur l'Activité de Conception Coopérative.*). The design experience that we have carried out is mainly *remote design experience* which implements commercial computer software (such as CAD, MS-Office, video conference, sharing of applications...) for mechanical product design, where the specification list of the product and the documents of pre-prototyping are already defined. A work group with four actors consisted of a *project manager*, a *form designer*, a *frame designer* and a *link designer*. Analysis of the collaborative design process during communication between actors is an important task that allows us to develop pertinent tools to assist the design process. In order to analyze this design process, we propose an approach developed at two levels: the first level concerns the analysis of interactions between actors inside each discussion whereas the second level concerns the analysis of relationships between different discussions that are performed in the corpus. Our approach is centred on the communicative traces or messages generated by the actors. Therefore, we use the notion of corpus, which represents the communication traceability.

The formal representation of the interactions between actors in this analysis is achieved by decomposing the corpus into a chain of discussion sets. The diagram of interactions represents the interactions between actors for every discussion. We transform this diagram into different types of matrices. Their analysis permitted us to set up micro-groups, key actors and similar sub-groups of discussions. Through the different types of interactions inside each micro-group, and between the micro-groups, we distinguished three types of co-operation: complete cooperation, bilateral cooperation, and cooperation quasi-null.

A data analysis approach is developed to analyse relationships between different discussions. The concept of the similarity between two discussions is used to understand the dynamics of organization and the knowledge of sharing and developing.

The interpretations of the results of a real collaborative design experience enabled us to observe some interesting properties of the design process such as: the auto-organization, the dynamism, the auto-similarity and the convergence/ divergence of discussions.

32.4 References

[1] Ostergaard, D., Summers, D., 2003, “A taxonomic classification of collaborative design”, *ICED'03*, August 19-21, 2003, Stockholm, Sweden.

[2] Chiu, M., 2002, “An organizational view of design communication in design collaboration,” *Design Studies*, Vol. 23, Issue 2, pp. 187-210.

[3] Boujut J-F., Laureillard P., 2001, “A co-operation framework for product–process integration in engineering design,” *Design Studies*, Vol. 23, Issue 6, pp. 497-513.

[4] Jeantet, A., 1998, “Les objets intermédiaires dans la conception éléments pour une sociologie des processus de conception,” *Sociologie du travail*, Vol. 3, pp. 291–316.

[5] Laureillard, P., Boujut, J.F., Jeantet, A., 1997, “Conception intégrée et entités de cooperation”, *dans les actes de Design, “Les objets en conception”*, coordinateurs B. Trousse et K. Zreik, pp. 119-134, Edts Europia.

[6] Vinck, D., and Jeantet, A., 1995, “Mediating and commissioning objects in the socio-technical process of product design: a conceptual approach,” Dans Vinck, Maclean and Saviotti (eds), *Design Network, Strategies,* COST Social Sciences Series, CCE, pp. 111–129.

[7] Perry, M., Sanderson, D., 1998, “Coordinating joint design work: the role of communication and artefacts,” *Design Studies,* Vol. 19, Issue 4, pp. 273-288.

[8] Ferber, J., 1995, « les systèmes multi-agents : vers une intelligence collective, » *Interéditions*, Paris, France.

[9] Legardeur, J., Merlo, C., Franchistéguy, I., 2003, “Coodinattion et Coopération dans les processus de conception,” *AIP-PRIMCA*, Cd-Rom, La Plagne.

[10] Benzécri, J.P, 1973, *L'Analyse des Donnèes*, Paris: Dunod.

[11] Hartigan, J.A, 1975, *Clustering Algorithms*, New York, Wiley.

33

Workspaces and Cooperation Notions in the Design Process

Ezio Pena, Denis Choulier, and Olivier Garro

Abstract: Understand the way the actors organize their activities is needed for better managing the resources used in distance collaborative design. We introduce here the notion of design workspaces and distinguish them according to the number of actors (individual versus collective) and the type of activity (communication versus cooperation, related to the presence or absence of Intermediary Objects - IO - on these spaces). A method for analysing a design meeting in order to delimit and then qualify design phases is presented. A specific device was used to delimit the different workspaces. Audio and video records allow for a precise observation of glances, gestures, and exchanged words. From these observations, we coded: actions on the IO (we distinguish draw, point, annotate, and handle), attentions from the glances (to another actor, or to an IO, noting its workspace), and design acts which are requests and propositions of information, solution, criteria definition and evaluation. Several graphs representing the different data versus time are proposed and used for identifying and qualifying phases. Analyzing a short (1 hour) experiment with 4 designers gave two main results. First, a significant modification of the designers' attention is revealed to be a good indicator for phase shift detection, especially when an IO appears in the collective space thus six phases were identified. Second, the type of actions and design acts used for each phase show important differences between the three main phases that were analyzed. These results are promising and show relevant indicators for segmenting, and qualifying design phases. Repeating such analyses should lead to activity models onto which new design tools could be proposed.

Keywords: empirical study, user observation, design phases, workspaces

33.1 Introduction

The development of tools for collaborative design is nowadays a main challenge. Design activities involve immense communication and interaction between individuals and groups in more or less complex social settings [1]. In order to improve the performance of these tools we must consider many features or issues. Moreover, the issues that we could considerer trivial in a face-to-face situation could turn out to be a major challenge in a distributed situation [2]. Therefore, for better defining these resources we must try to understand the way in which actors organize their activities. Thus, computer tools must be constructed on activity models. We propose to identify these models from the analysis of face-to-face design experiments, observed on real situations or in laboratory [3]. Design activities are led largely by the designers thanks to the utilization of both verbal exchanges and actions made onto the different manipulated objects. These objects are called Intermediary Objects (IO) [4]; for instance: graphs, diagrams, bills of materials, calculation spreadsheets, geometric and numeric models, and even a physical object, providing it plays the role of mediator during the actions. The design process is complex, and it is composed of a succession of different phases, often delicate to identification. We have a particular interest in defining these phases, in order to consider tools adapted to each phase. The objective of this article is to present tools of analysis for face-to-face design meeting. They must allow for the identification of phases, considering the actions made on the intermediary objects, their positions in the workspace, and the verbal exchanges during communication or cooperation instances. The method of analysis we are working on includes a first stage of identification of phases, for which we use the notion of workspaces detailed in the next section. The second stage depends (for *each* phase) on the analysis of actions done on the IO, and on the verbal exchanges (acts of design). Its objective is to qualify the different phases, and the "tools" used to support every phase.

33.2 Workspaces in the Design Process

Typology Definition

The notion of workspace in design process seems to be little approached in the literature [5], except for computer interfaces that consider the best placement of windows into a screen [6, 7]. In a previous work, we established for the collective work a distinction between communication space, and cooperation space(s) [8]. The definition of spaces must at least consider the intermediary objects and their position, as well as actions done onto and around these objects. Features of a space for each instant are therefore its spatial features - a focal point, and a zone into a space, a plan or within a screen, the presence of an IO (for cooperation), and the attention (or the action) reached by at least one actor. We distinguish here, apart from a communication space without IOs, the individual spaces (mobilizes the attention of only one actor) from collective space of cooperation in which several actors - or even all actors - interact. Of course, an IO can forward from an individual space towards a collective space, and vice versa. Making this distinction among spaces permits evoking several aspects of work, in particular, the imposed constraints (rightly or wrongly) by the computer tools, confidentiality, the division and sharing of activities (parallelism of activities) individual consultations.

Face-to-face and Remote Workspaces

During remote work, the distinction between individual and collective spaces can be applied to different zones of a screen. Indeed, it is possible to share a screen as well as applications, or windows - as many initial choices which can force the use of the tools thereafter. Moreover, each actor frequently uses printed documents or blank sheets for consulting or capturing notes in a space which is in fact individual, and tools of communication. In face-to-face situation, when the actors are "simply" invited to work around a table, a whiteboard or a screen, the workspaces are observable thanks to the gestures, the conversation (words), and from handling or making the actions on an IO. It is, however, delicate for cooperation spaces, in particular it is difficult to dissociate – individual or collective - spaces whose statuses are not specified and which, moreover, permanently evolve and overlap. We thus designed a device that allows more rigorous observations, while permitting the actors the possibility to work without many additional constraints.

Figure 33.1 View on one of two cameras showing the partitioned table

A Device for the Observation of Workspaces

The device consists of separate spaces by partitioning a table as shown in Figure 33.1. These partitions are open in the lower portion, thus, permitting the actors to have access to the central collective space and even to look at each other's individual spaces. In addition, we have placed two camcorders, for recording and subsequent analysis of the glances and the gestures, and a microphone above the middle of the table for the verbal exchanges. The cameras were placed on two opposite sides of the table. Figure 33.1 corresponds to the sight of one of these cameras. The use of a space is observed by the direction of the glances or the attention given to an object or an actor. Spaces can be collective (central) or individual (the sides of the table). The communication space is defined on top of this device (we did not analyze these spaces of communication). Here for instance, actors A and B use the communication space. For actor D, the position of his eyes enables him to observe his individual space, such as actor C observes the collective space.

Type of Data Collected

The glances, gestures and exchanges are raw data which should be interpreted. As the final objective is to specify then develop tools, we do not take into account the semantic content of the interactions. For the verbal exchanges, we use the acts of design. They make it possible to describe the linguistic interactions from a pragmatic and intentional point of view (action). Design acts show the participation of a designer in the group during a design meeting [13, 14] by reference to language acts [9].

Table 33.1 Example of data collected

Time	Actor	Glance	Action	IO	space
00:28:37	A1	A2			
	A2	A1	speak		
	A3	OI	drawing	7	Collective
	A4	OI	annotate	8	Individual

We translate gestures into actions. We have defined the following typology. To point: when an actor points a part of an IO (finger or pen). To draw: when an actor creates or modifies an IO. To handle: when an actor changes the position of an IO in a space. To annotate: when an actor adds explanatory or descriptive text on an IO. The granularity used also corresponds to an analogy with the actions allowed by computer tools. Approximately we can draw, point, manipulate (move, rotate, zoom, hide…), and make annotations. It is possible to make a finer analysis of the actions of the type draw, for example to create a new typology. However, that leads to finely analyze the drafts and drawings, which in this work not considered.

Lastly, a glance indicates the attention. We noted in an Excel file (extracted Table 33.1) the moment of the observation, and for each actor: its attention, either on an object, or on another actor, the action carried out, the number of the IO, and the space where this object is.

Each time that an attention changes or that an action is carried out, we define a new moment. This table is used as the basis for the quantitative analysis.

Tools for Data Analysis

Partition of the Meeting

First, we must find a manner of partitioning this experiment which has two objectives: first, to find homogeneous subsets making it possible to identify phases, then, this partitioning will enable us to compare these phases. The identification of the phase shifts must be attached onto indicators, which can be of various natures. However, the quantity of data collected is important as illustrated in the example. They are moreover interacting with each other, and it is difficult to distinguish an order at first glance. Initially, we considered several types of data individually: for instance, a type of action (to draw, to annotate, to handle, to point), or the appearance of a type of act of design. This approach appeared unfruitful. For example, acts P2 (proposal for a solution) appear very often at the beginning of an experiment, but this does not necessarily mean that we are facing a phase of solution searching. The failure of these attempts is certainly in relation with the distinction that we must make between individual acts and collective tasks. We have also tried to plot this information in the form of graphs according to time (IO/spaces, attention/actions, actions/spaces, acts/actions). At last, a graph representing the attentions related to the IO by the actors and the position of these IO in spaces proves to be relevant, provided that one distinguishes the type of space in which this IO that is individual, or collective. It highlights in particular the simultaneous use of several intermediate objects on the same space, and the simultaneous use of several spaces, reveals tasks occurring in parallel. Thus, a modification of the attention of the actors can reveal a phase shift. In particular, the moment at which an object appears in the collective space seems an obvious indicator. It can be either a new IO, or an IO which comes from an individual space. We thus consider that the appearance of an object in the collective space is used to start new actions and exchanges between the actors. When these moments are defined, a second graph representing for each phase the attention related to the IO makes it possible to check the importance of the role of mediator played by the considered IO (Figure 33.2).

Qualification of Task of Design

First, the analysis of identified phases includes the type of actions carried out on the IO. In this analysis we have gathered all of the actions done - individually - by the actors: to draw, to annotate, to point and to handle. A graph representing for each phase the relative importance of the various actions is used as illustrated by the example in Section 33.3. The presence and the importance of the actions "to draw" and "to annotate" seem to be a relevant data. For instance, a phase of search for solutions must have a strong presence of actions of the type "to draw". On the other hand, a phase of evaluation must comprise mainly actions of type "to point". The second element that we have considered for qualifying these phases is an analysis of the acts of design. For this analysis we transcribed then coded the entirety of the speeches and exchanges carried out by the actors. Two types of information are extracted. Initially, we determine the type of acts and their distribution in a graph representing, for each phase, the relative importance of each type of act, graph similar to the previous one (Figure 33.2). The interpretation of these graphs is mainly related to the relative importance of the P2 acts (proposal of solution) and P3 acts (proposal for an evaluation). In general, P2 is more present in phases of research for solutions; on the other hand a phase of evaluation will have to comprise mainly acts of the evaluation type P3. This analysis will then consider the bonds between the exchanges, or sequences of acts. An act is often a response to act preliminary emitted. For example, it is the case of the acts D2 (request of solution) and P2 (proposal) or of the acts P2 (proposal of solution) and P3 (proposal for an evaluation). We use here the concept of structure of acts as a way for visualizing these interactions in the form of graphs, described in [10].

33.3 Illustrative Example

Description of the Experiment

The subject for the experiment is relatively known: to design a device allowing an egg to touch the ground without damage after a fall of 10 meters. It was rewritten in order to include a significant number of constraining rules, which represented the many criteria the device must satisfy. Therefore, the subject is already much documented and it prohibits the actors to questions about the limits of the design. The instructions given to the four actors are, starting from the specifications, to imagine principle of solutions for the device, and then to select one candidate for manufacturing a prototype, justifying the choice. We did not propose a step by step methodology to be followed. Their work thus comprises the entirety of the "conceptual design" stage of the Pahl & Beitz model [11], with a preliminary clarification of the problem (reading of the subject). The duration is one hour, with possibility of exceeding this time if necessary. Concerning the partitioned table, we asked the actors to considerer that the partitions are not obstacles. They can move and look at all spaces (no confidentiality). The only strict order is to use the central space when in working common.

Analysis of the Actor Attention on the Intermediary Objects (IO)

The observation allowed us to retain 980 moments, that is to say 3920 data (4 actors). In these 3920 pieces of data, 609 corresponding to attentions on the OI were noted in individual spaces and 1706 in the collective space. The difference corresponds to the conversation. Figure 33.2 represents the whole of the attentions on each intermediary object, on collective or individual space. The meeting partition revealed 6 phases. Phase #1 contains 6 IO, of which objects 1 to 4 are the subject. These objects stay most of the time in individual spaces. Object #8 is an object created by an actor, but this object remains throughout the experiment in its individual space. Phase #1 contains only one object in the collective space, a physical object (IO # 15), which forms part of the system to be designed (an egg). This object remained in the collective space, throughout the experiment. Phase #2 began with the appearance of object # 6 in the collective space. It is an object created by actor #2, and it represents a first solution. We also note the appearance of object # 7 in an individual space. Phase #3 is the most important one, in term of time, number of actions and diversity of acts. This phase starts when object # 7 becomes collective. This object represents new solutions. Then, the attention returning on the object "subjects" (1 to 4), we decided to define phase #4.

The main characteristic of phase #5 is an important presence of three objects in the collective space, the objects # 7, 11, 16 (plus IO #15), and object #10 in an individual space; The latter triggers phase #6 when it passes in the collective space. The most important phases (expressed as a percentage compared to the total) are the phases #1 (22 %), #3 (37%) and #5 (17%).

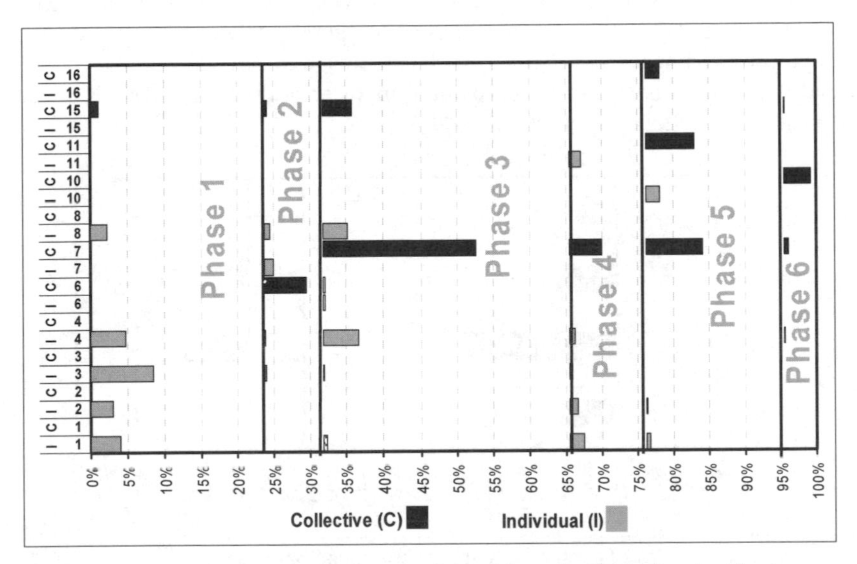

Figure 33.2 Phases and attentions onto various objects (individual and collective spaces)

Analysis of the Types of Actions Done on the IO$_s$

Table 33.2 shows a summary of the actions onto IOs in individual and collective spaces. We can see that actions onto IOs are most important in the collective space. Moreover, the principal actions are to point and to annotate.

Table 33.2 Summary of actions onto IO

	draw	point	annotate	handle
individual	16	11	95	0
collective	74	303	132	38

Figure 33.3(a) gives the relative proportions of actions (to point, to draw, to annotate, and to manipulate) for each phase. First, we note the absence of actions during phase #1 (individual reading of the topic, however there are attentions on

objects 1 to 4). The most important phases are phase #2 (16%), phase #3 (48%) and phase #5 (18%).

Phase #2 mainly includes actions of type "to point", and a very few actions of type "to draw".

Phase # 3 is composed by almost the totality of actions. Actions of type "to draw" and "to annotate" have a very similar occurrence. Actions of type "to handle" are here more numerous than for the other phases. However, acts of type "to point" are even important.

During phase 4, actions "to annotate" and "to point" are present in equal proportions. On the other hand, actions of type "to draw" are rare. These impressions are nearly the same for phase #5, therefore, we could think about merging these two phases, but Figure 33.4 has shown a difference.

Finally, phase #6 is characterized by actions of annotation, and drawing, and a weak presence of actions "to point".

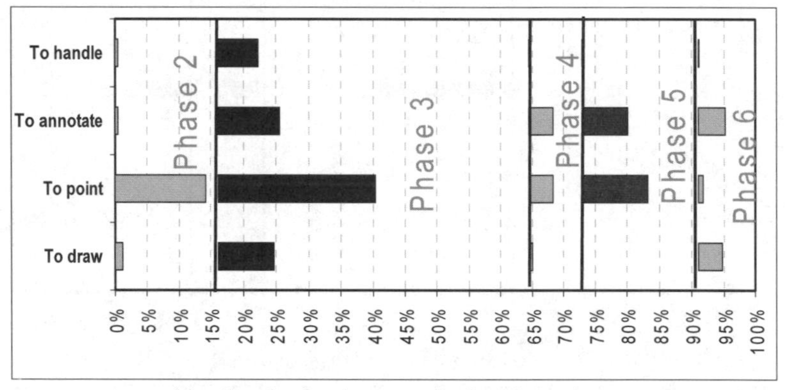

(a) Graph of action onto the found out phases.

	To draw	To point	To annotate	To handle
phase 2	6	74	2	2
phase 3	44	126	49	32
phase 4	3	20	20	1
phase 5	1	54	38	1
phase 6	20	4	22	1

(b) Total number of actions onto the found out phases

Figure 33.3 Type of actions made on the whole of the OI, collective space

Analysis of Acts of Design

The third type of analysis corresponds to the acts of design (Figure 33.4). Differences between phases are considerable. For instance, Phase #1 includes an important number of types of act: nearly the totality of acts is presented there, except for D2 acts (requests for solutions). The phase #3 includes a large number of P2 type acts. Here the difference with phase #1 is a lower quantity of P3 acts, a more important number of evaluations (P3+ and P3 -), as well as a more important number of D2 acts. Otherwise, phase # 4 presents a reduction of P2 acts (proposition of solution), in relation to the whole of evaluation acts (P3, P3 -, P3+). This difference is again more important in phase #5, with a disappearance of P2 acts (proposition of solution). P1 and P2 acts concern the design process, not the product.

The second type of analysis concerns sequences of act. Some less frequence sequences are not taken into account. Phases #1 and #3 are presented.

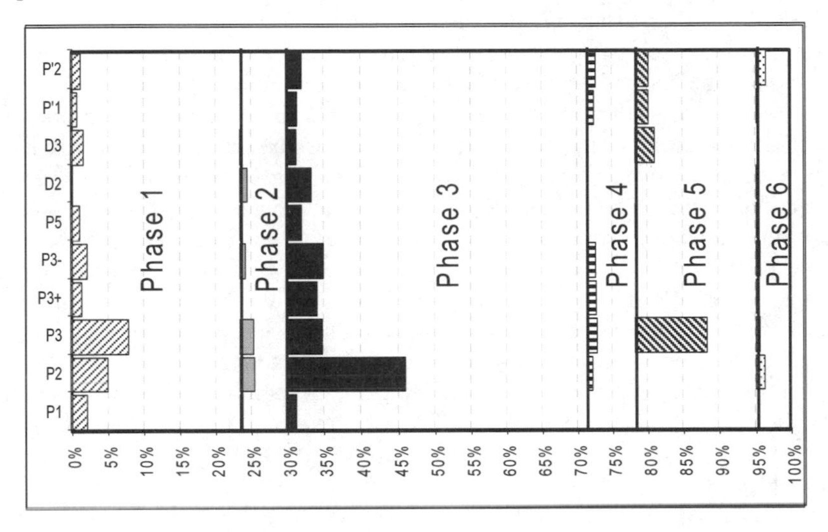

Figure 33.4 Types of acts of design for each phase

Figure 33.5 shows very different sequences of action from one phase to another. Thus, for example during phase #1, P3 (proposal for criterion) is the most important act, in interaction with P2 (Proposal for solution). For phase #3, we note a preponderance of the P2 acts, in interaction with the proposals for criteria, and evaluations (positive, and negative). This characteristic is connected to the emergence of solutions [12].

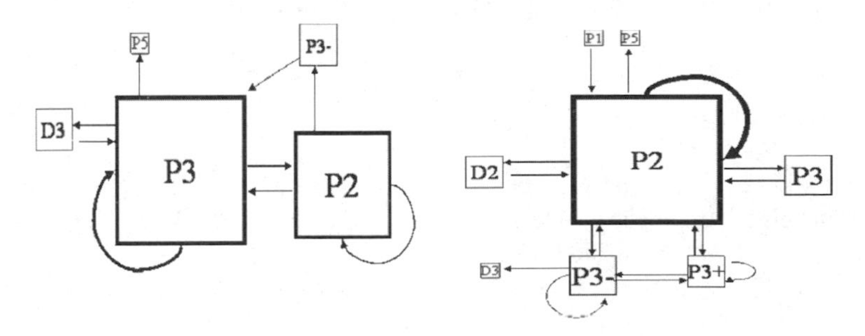

Figure 33.5 Sequence of acts for phases #1, and #3

33.4 Conclusions and Prospects

This article presents tools of analysis tested in a mechanical design meeting. The results presented here follow upon several abandoned attempts made to partition this experiment and previous ones into phases: unfruitful, too delicate or difficult. It was indeed delicate to identify phases starting from only one type of information. We tried the acts of design, the sequences of acts, the actions on the IO, or the spaces alone. To identify phases, we used the notion of spaces where these IO are. This supposes an experimental device for which spaces are relatively easy to observe, making a clear distinction between collective and individual working areas (one for each actor). The use of a partitioned table offers this possibility, but other types of instrumentation are certainly possible, in particular for the observation of remote design situations where spaces are effectively materialized by windows, and screens. A change of attention on spaces seems well adapted to the detection of phase shifts, in particular when it is induced by the appearance of an IO in the collective space. It is advisable, however, to check that the object in question is indeed used as a support for cooperation, by an analysis of the attentions related by all of the designers on this object. In a second time, to characterize these phases, we propose to analyze the gestures of the actors, and the verbal exchanges between them, coded respectively in actions, and acts of design. This analysis can take the form of a counting of the numbers of acts or actions, or/and stick to the sequences of acts. The two analyses (actions and acts) are complementary: sometimes the analysis of one of them is not sufficient to distinguish phases. The results of the first analyzed experiment are encouraging. Here, 6 phases have been found for a one-hour meeting: 3 are determined, and 3 are less easy as they can correspond to transition phases. The three principal phases present in particular great qualitative differences, which will certainly allow us, by repeating this experiment, to carry out a coherent analysis. It is now necessary to analyze the results on other design meetings; this is why we mainly presented here tools for analysis, and not an analysis of the tasks. The latter should lead to models of activity on which specifications of design tools could be built. It must thus be possible to propose better-adapted tools at every situation for the realization of the various actions. However, we think that computer tools must be able to go beyond the simple realization of the actions - for example, starting from the models of the activity, to provide elements of analysis of the process, or to capitalize [13].

33.5 References

[1] Törlind, P., Larsson, A., 2002, "Support for Informal Communication in Distributed Engineering Design Teams," *Annals of 2002 Int. CIRP Design Seminar*, 16-18 May 2002, Hong Kong, China.

[2] Larsson, A., Törlind, P., Mabogunje, A., Milne, A., 2002, "Distributed design teams: embedded one-on-one conversations in one-to-many", Durling D. & Shackleton J. (Eds.) *Common Ground: Design Research Society International Conference 2002*, UK. ISBN 1-904133-11-8.

[3] Gero, J., McNeill, T., 1998, "An approach to the analysis of design Protocols," *Design Studies 19*, 21~61.

[4] Blanco, E., Garro, O., Brissaud, D., Jeantet, A., 1996, "Intermediary object in the context of distributed design," *CESA'96*, Lille, France.

[5] Nam, T.J., 2001, "The development and evaluation of Syco3D: a real-time collaborative 3D CAD system," *Design Studies,* Vol. 22, pp. 557-587.

[6] Kamel, N., 1999, "A unified characterisation for shared multimedia CSCW workspace designs," *Info. and Software Technology*, Vol. 41, pp. 1–14.

[7] Sungho, W., Eunjoo, L., Tsuyoshi, S., 2001, "The multiuser workspace as the medium for communication in collaborative design," *Automation in Construction,* Vol. 10, pp. 303–308.

[8] Peña, E., Choulier, D., Garro, O., 2002, "Spécifications d'un outil Coopératif de définition de Problème en Conception à partir de l'analyse d'expériences," *IDMME,* 14 - 16 May, 2002, Clermont-Ferrand, France.

[9] Searle, J.R, 1972, *Les actes de langage : essai de philosophie du langage*, Hermann, Paris, (Original English edition 1969).

[10] Choulier, D., Garro, O., Pena, E., 2004, "Design acts, the analysis of a cooperative design experiment," *COOP'04 Scenario-Based Design of Cooperative Systems,* 11-14 May, 2004, Nice, France.

[11] Pahl, G., Beitz, W., 1996, *Engineering Design: A systematic approach*, Springer, 2nd edition, London.

[12] Garro, O., Choulier, D., & Micaëlli, J.P., 2001, "L'émergence, processus clé de la conception inventive: Application à la conception d'une partie d'un robot," *7ème Colloque sur la Conception Mécanique Intégrée (AIP-PRIMECA)*, 2-4 avril 2001, La Plagne, France.

[13] Pena, E., Choulier, D., Garro, O., 2004, "A proposition to capitalize and share the logic of design," *CIRP Design Seminar,* 12-14 May 2003, Grenoble, France, *Methods and Tools for cooperative and integrated design*, Kluwer Academic, ISBN 1-4020-1889-4, (January 2004), pp. 241–257.

[14] GRACC: Groupe de Recherche sur l'Activité de Conception Collaborative, 2001, "Une expérience de conception collaborative à distance," *7ème Colloque sur la Conception Mécanique Intégrée (AIP-PRIMECA)*, 2-4 avril 2001, La Plagne, France.

34

Pitfalls of Engineering Change
Change Practice during Complex Product Design

Timothy Jarratt, Claudia Eckert, and P. John Clarkson

Abstract: The majority of design projects involve adapting a known solution to meet new requirements. Therefore, understanding the issue of engineering change is of vital importance if companies are to deliver product development projects on time and to budget. When a change is made to part of a product, the change is likely to propagate to affect other components or systems. This paper examines the engineering change process within a UK engineering firm and focuses on the issue of change propagation. The findings are compared with an earlier study in the aerospace industry. Four reasons why propagation occurs are proposed and discussed.

Keywords: Design, Management, Engineering change

34.1 Introduction

From a business perspective, changes to a design are “a fact of life” in taking a product from concept, through design, manufacture and out into the field [1]; they are the rule and not the exception in product development processes in all companies and in all countries [2]. As an example of the importance of engineering change, a survey of German engineering businesses found that approximately 30% of all work effort was due to engineering changes [3]; this included rework as well as the adding of functionality to a product. It has been reported that engineering changes in the automotive industry consume between a third and a half of the engineering capacity and account for 20-50% of tool costs [4].

Engineering Change Research

Relatively little work has been published on engineering change and engineering change management. Wright [5] conducted a survey of engineering change management literature published between 1980 and 1995 and found only 15 "core" papers. Other researchers have also commented upon this scarcity (*e.g.* [6]).

Historically the design community has seen change as the responsibility of manufacturing research groups [5] and any alterations made to a product during design were just regarded as normal iterations of the design process. However, the rise of concepts such as concurrent engineering and simultaneous design, plus the influence of business disciplines such as configuration management, has seen production and organizational issues becoming an integral part of the design process and associated research. Thus, engineering change is starting to be featured much more prominently in academic work.

Background and Motivation

The work described in this paper is part of an ongoing research project into the field of engineering change and the engineering change process. An earlier, general study into change processes at an aerospace company (reported in [7]) indicated that potential change propagation was a major pitfall of engineering change management. Propagation occurs when an alteration to a component or system spreads to affect other parts of the product. A few authors (*e.g.* [3] and [4]) have mentioned propagation as a possible effect of implementing a change. There have been studies into change practice in UK [6], Hong Kong [8] and Swedish [9] companies. However, there has never been a study that specifically investigates the issue of change propagation within a company.

The study described in this paper was carried out at a large UK engineering company and was complemented by many informal interviews with engineers from a wide range of other companies. Underpinning this research is the hypothesis that changes potentially can propagate between the elements of a product when an alteration is made to a particular component or system. The purpose of this work was to prove this main hypothesis and to discover how engineers assess the possibility of change propagating, which tools or methods (if any) they use to support their evaluation and what sort of support is required.

34.2 The Engineering Change Process

The formal engineering change process is a critical business process that affects all aspects of product design and development. An engineering change process can be triggered at any point in the product life-cycle once the concept has been selected, because, once the concept has been chosen, information about design decisions starts to be formally released to design teams, suppliers, potential customers, *etc.*. Any changes to this information as the product evolves must be regarded as an engineering change process.

Perhaps the clearest description of the engineering change process is provided by Leech and Turner [10], who state that the process is a mini, highly constrained design process or project and "like any project, is only worth undertaking if its value is greater than its cost."

Figure 34.1 shows a generic high-level engineering change process based upon various processes outlined in literature (*e.g.* [11, 12, 13]). The process is initiated by a change trigger. Eckert *et al.* [7] describes changes as *emerging* from the product (*i.e.* errors) or being *initiated* from outside (*i.e.* customer requests, legislation, *etc.*.).

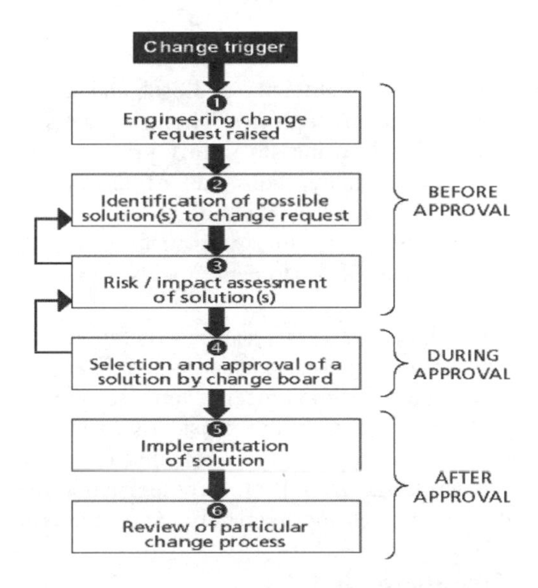

Figure 34.1 A generic engineering change process

The highest risk of the six phases is the third: assessing the impact or risk of implementing each solution. Various factors must be considered: for example the impact upon design and production schedules; how relationships with suppliers will be affected; and will a budget overrun occur. The further through the design process a change is implemented, the more disruption is caused. Several authors refer to a "rule of 10" (*e.g.* [2]), which states that the cost of making an engineering change rises by a factor of 10 between each phase. Thus, a change made during manufacture would be 1000 times more expensive than making the same change during the detail design phase.

There are possible iterations within the process, two of which are marked by arrows in Figure 34.1. For example a particular solution may be too risky for the company to implement and so the process will return to phase 2, so that other possible solutions can be identified. At the approval stage, the Engineering Change Board may feel that more risk analysis is required (maybe in the form of more testing) and so the process will return to phase 3. There are other possible iterative loops, but they are not marked for sake of clarity. The most extreme loop would be when, if during the review phase, it was realized that the implemented solution had been ineffectual or made matters worse. In that instance the process would return to the start with a new change request being raised.

34.3 Details of the Study

This section briefly describes the company involved and outlines the elements of the study.

The Company

The company designs and manufactures diesel and gas engines for two main markets: generator sets and off highway. There is a wide spectrum of customers ranging from famous global businesses to tiny privately owned companies. Large concerns take delivery of many thousands of engines per year, whilst small manufacturers of specialist vehicles may require less than 50. One basic engine can go into a wide variety of applications and be used in a range of environments. Thus, adapting engines to fit customers' needs is a standard activity for the company.

A major issue facing the business is complying with ever tightening environmental legislation, which has caused a significant decrease in product life-cycles and led to huge technological changes. Various aspects of engine performance are being regulated; the most important area is that of exhaust emissions. There is not room for a thorough discussion of this issue but further details are provided by Jarratt *et al.* [14]. New legislation requires the development of a new generation of engines approximately every five years.

Methodology of the Study

The study consisted of four parts: (1) interviews with a range of employees across the company; (2) observations of meetings; (3) shadowing an engineer involved in managing the engineering change process; and (4) structured sessions with engineers filling in a connectivity matrix (see [15] for more details of this last aspect).

Twenty engineers and managers were interviewed for between 45 minutes and two hours in early 2002. The interviewees came from a wide range of roles and functions, and had a range of experience within the company and industry. Just under half of the interviewees had been employed by the company for over 20 years. Each session was recorded on audiotape and was later transcribed. The interviews were semi-structured. After initial questions about the background of each person the interview progressed to focus more on design processes and in particular the engineering change process. During the discussion of engineering change, the interviewees were asked to:

- describe the company's engineering change process and their relationship to it;

- give examples of changes that had been successfully implemented and examples where there had been unexpected difficulties;
- identify those tools and techniques (if any) that were used to support engineering change;
- discuss where the engineering change process could be improved and what support methods / tools they felt were needed to assist them.

Three Design Change Meetings were observed during the interview period to gain an understanding of the issues and trade-offs that are considered when assessing and authorizing engineering changes. Attendance at these meetings also enabled the authors to appreciate the tools and methods used during the risk analysis assessment of each change. The Design Change Meetings are held twice a week and the purpose is to review and authorize engineering changes to the company's product range. A wide variety of people from across the business (manufacturing, after sales support, *etc.*.) attended in conjunction with designers and engineers.

The first author spent a week shadowing the Technical Design Manager. This employee had a number of roles, one of which was being the "owner" of the Design Change Process. Shadowing this person enabled the author to gain a clear understanding of how the engineering change process worked in reality and it helped highlight which areas of the process required support.

34.4 Findings

The company is very successful and on the whole is well organized. A key issue that came out of the study was that all of the company's design and development activities are dominated by engineering change in some form or other. This includes the New Product Introduction (NPI) process, which is used to develop engines to meet new tiers of legislation. When a new engine is required and an NPI process is launched, one of the first acts is to take the current engine and decide which pieces of the architecture and technology can be carried forward into the next generation and which must be altered to meet the new requirements. Diesel engines are evolutionary products with a well-established architecture. However, this is not to say that there is no room for innovation. Over the past decade rapid advances have been made in materials technology and manufacturing processes. Perhaps the biggest area of innovation has been the addition of electronics to engines.

Reasons for Triggering the Change Process

At a high level there are three sources of engineering change at the company: (1) suppliers, (2) customers and (3) internal departments.

Suppliers: either a current supplier is changing a process (*e.g.* manufacturing) or a new supplier is being brought in. These relationships are vitally important as a large amount of the engine is bought in (70-80%).

Customers: the reasons are either to make a change to an existing sales option or to request a new sales option. For example, a customer that designed vineyard equipment requested the development of a new filter head option. Vineyard tractors need to be very narrow and the standard filter head impeded the turning circle of the vehicle.

Internal departments: these changes can be for a wide variety of reasons such as quality and reliability improvements, cost savings, changes to manufacturing processes, improve servicing or a company-wide initiative.

Effects / Impacts of Change

During the interviews it became clear that most of the changes carried out are quite mundane. For example changes are made to paperwork when there is a switch in supplier and the new source's manufacturing process is slightly different to that of the previous supplier's. However, every so often an engineering change can propagate dramatically, as the product engineering manager noted: "On each design project there will be 4-5 things that are major ... things that we did not predict at all". Less dramatic propagation will also occur: "in a lot of cases [propagation] occurs, but only to affect one or two components"

The interviewees gave many examples of past changes to the engines that they had been party to, most of which described situations where the implementation of a change had not been completely smooth. Three of these examples are given below with quotes from the engineers who described them. All show that engineering changes can propagate to affect other parts of the product or other business processes such as after sales support.

Example 1: outlet pipe change – a project was undertaken to replace the metal water outlet pipes in an engine series with plastic ones to reduce overall engine weight and lower manufacturing costs. The water outlet pipes contain a temperature sensor and this was transferred from the old design of pipe to the new one. It was only once the redesigned engine was in production that it was noticed that the sensor no longer functioned: it had been designed to earth through the pipe, which was no longer possible due to the change of material. The solution was to redesign the sensor with a return wire. "Nobody thought about it when they introduced plastic pipes. It was very embarrassing and very expensive."

Example 2: change to bush – a new bush design was suggested for the gears for the power-take-off to reduce cost. As a result, the oil circuit system was altered without any analysis work. The first engine with this change included quickly broke during testing. After much analysis, it was revealed that the change to the oil circuit was the critical factor, which had initially been overlooked. The eventual solution meant changes to four or five other components within the engine. "It was a real systems approach".

Example 3: gear train change – another key requirement (both from customers and legislation) for new engine design is to reduce engine noise. One way to achieve this focused on the backlash between the gears. The gear train was redesigned and some of the gear ratios were altered. Engineers designed a gear train that was perfectly durable, but unfortunately, to get the gears to fit within the same space that had been available before and be located at the same centres, meant that the helix angle had to be altered. These introduced extra loads in the gears and, as the bearing systems were being carried over from common parts; those bearing systems were not adequate. So in order to stop the change from spreading the gears were redesigned again: "otherwise we would have fixed the next problem down the chain and that would have had a knock on effect somewhere else". "You can't just [...] work on your own component – how it interacts with other parts of the engine must be considered".

Tools to Support the Engineering Change Process

During the interviews, when the subject of tools or methods was raised, the interviewees were first asked to describe how they analyzed each engineering change before describing the tools and methods they used.

Analysis of Engineering Changes

When an engineering change is initially evaluated, a group of engineers will meet and use their experience to determine what level, if any, of testing and analysis is required. If the form, fit or function of a component is affected, a Failure Mode and Effects Analysis (FMEA) will be carried out to identify the critical characteristics of the situation: for example whether emissions legislation compliance will be affected. This will be done at three different levels: engine, system and part. People will also consider wider risk issues such as program risk *i.e.* whether the change will affect the whole design program.

Several engineers talked about having a "mental checklist" that they went through with simple components. "For example with flywheels: we have almost standard quotes – the scheme work will be the same, *etc.*. The time to design and detail the new flywheels is quite standard". With customer requests for new options, things are not always so easy: "it comes down to experience. There is no hard and fast way of doing it, there is no checklist as such."

There was universal agreement that a lot of problems with analyzing engineering changes come from oversights and mistakes rather than the unknown: "a lot of the problems come from stupid mistakes – not from horrible ones – the big ones people think about and apply their formidable brains to it, but the little details are overlooked"; "[paradoxically] big changes are less likely to propagate [unexpectedly] than little ones..."

Tools and Methods

In terms of computer tools, the company has an Enterprise Resource Planning (ERP) system, which has a Product Change Control module that supports the workflow aspect of the change process *i.e.* applying a change once it has been authorized. The company also has a Product Data Management (PDM) system, which is integrated with the CAD system and used to manage all the drawings, *etc.*. for each product. This PDM system is also used to populate the company's configurator software, which is used by sales engineers to accurately configure engines for each customer.

When asked to describe which tools and methods are suited to support decision making during the change process, the interviewees mentioned, DFMA, FMEA,

QFD, CAD and solid models. The first three of these can be regarded as "soft" techniques, whilst CAD and solid modelling can be regarded as a hard technology.

Several managers commented that there was a pressing need for support for the risk/ impact analysis phase to avoid simple errors, which have major ramifications as shown with some of the examples (*e.g.* number 3) mentioned above. Others felt that there was a need to capture experience and rationales because so much of understanding possible change propagation "comes down to the experience of the individuals. You can teach a young engineer the basic principles and the application of those principles, but only by experience will the practitioner be able to ask the appropriate questions."

Engineering Change and the Product

Many of the interviewees discussed the complexity of both diesel engines and the company processes that design, manufacture and support them: "one change cascades down into a vast amount of complexity". Important events inside the engine such as combustion occur due to the interaction of many parameters such as injection time, temperature, *etc.*. The interaction with suppliers and customers can be equally involved: "you'd be surprised the number of times a minor change will take place on the engine because maybe a bolt became obsolete and [production] had to put one on that was slightly different and suddenly [a customer] rings up and says "our production line has stopped because this new bolt of yours fouls on so-and-so" so it's horrendously complex in that respect."

In terms of the connections between components that could cause engineering changes to propagate there was almost universal agreement that vibration was the most complicated and the least understood. Even specialist engineers with many years of experience of vibration and dynamic mechanical interactions stated that vibration paths within the engine could still surprise them and "catch them out."

Questions about complexity naturally led on to discussions about how both individual engineers along with design teams gained an overview of the whole product and the complex network of linkages between components. It was agreed that most people lack such an overview and that there were few tools to help people gain such an overview. "Designers have an awareness of the components that they are designing and anything that links in directly into that component. However outside of that area they probably do not have such great understanding – not a good idea of how a change to their component can affect the rest of the engine."

It was also noted that being aware of connectivity did not mean that it was always recognized. "We get caught out – we are aware of [an issue], but we don't always think of it. There are no two ways about it we do miss things." It was felt that most mistakes occurred when the initial assessment of a change was that it should be a relatively simple process. Engineers and designers find it hard to appreciate fully the complexity of linkages between parts that could cause changes to propagate. "We miss the other things that the components are doing because most components are doing several jobs." A manager commented that certain connectivities were only considered now because the firm had "been burnt by them" in the past.

34.5 Discussion

The study showed that although the majority of engineering changes concern amendments to paperwork and instructions, those changes that involve altering the form, fit or function of a component can propagate to impact upon other parts of an engine design.

It was clear from the interviews and observing Design Change Meetings that the risk assessment phase of the engineering change process is strongly dependent upon the experience and knowledge of the engineers involved. This experience is used to examine the change via "hard" technologies like CAD and "soft" techniques such as FMEA. There are no specific tools to assist with the evaluation, especially with the prediction of possible change propagation. When questioned, a large number of engineers and managers felt that there was a need for more support especially with gaining an effective overview of the complexity involved within the product.

The findings from this study were compared with the findings of an earlier study into customization in the aerospace industry [7]. This has lead the authors to propose four reasons why changes may propagate within a product during the engineering change process:

1. propagation due to forgetfulness or oversight because connectivity is known about, but forgotten or discounted due to inexperience (Example 1 given above falls into this category);
2. propagation due to a lack of systems knowledge because the role played by a component in a system is not known or because of a lack of overview of the product (Example 2 falls into this category);
3. propagation due to communication breakdown or failure because different designers are making changes to the same component without informing each other (Example 3 falls into this category); and
4. propagation due to the emergent properties of complex systems (for example issues such as vibration an electromagnetic interference – the "non-linear parts of the problem").

In the company featured in this study very few of the propagation events were due to the emergent properties of the engine, whereas in the aerospace company engineers were frequently surprised this way although all the other reasons occurred too. All the propagation examples provided by the interviewees in this study occurred due to the first three reasons in the above list. The most common reason put forward was a combination of forgetfulness and a lack of product overview.

The findings described above have been discussed with designers and managers in a number of UK and European companies (in the aerospace and automotive industries) and with other academics involved in the field of engineering design. All have provided anecdotal support and encouraged further investigations.

34.6 Conclusions

The study described in this paper has shown that change propagation does occur when alterations are made to a product. It has also shown that current tools and techniques to support the evaluation of engineering changes during the engineering change process are lacking, especially when it comes to the prediction of possible change propagation. There is a need for the development of product models and tools that accurately show engineers the complex network of connectivities that link components together within complex products. Current work by the authors is concentrating on meeting this perceived need.

34.7 Acknowledgements

The authors would like to acknowledge the generous support given by Perkins Engine Company especially the assistance of Richard Weeks, David Andrews and Colin Ingram. This work is funded by a UK Engineering and Physical Sciences Research Council IMRC grant.

34.8 References

[1] Nichols, K., 1990, "Getting Engineering Changes Under Control", *Journal of Engineering Design*, Vol. 1(1), pp. 5-15.

[2] Clark, K.B., and Fujimoto, T., 1991, "Product Development Performance: Strategy, Organization and Management in the World Auto Industry," *Harvard Business School Press*, Boston, MA, USA.

[3] Fricke, E., Gebhard, B., Negele, H., and Igenbergs, E., 2000, "Coping with Changes: Causes, Findings and Strategies," *Systems Engineering*, Vol. 3(4), pp. 169-179.

[4] Terwiesch, C., and Loch, C.H., 1999, "Managing the Process of Engineering Change Orders: The Case of the Climate Control System in Automobile Development," *Journal of Product Innovation Management*, Vol. 16(2), pp. 160-172.

[5] Wright, I.C., 1997, "A Review of Research into Engineering Change Management: Implications for Product Design," *Design Studies,* Vol. 18, pp. 33-42.

[6] Huang, G.Q., and Mak, K.L., 1999, "Current Practices of Engineering Change Management in UK Manufacturing Industries," *International Journal of Operations & Production Management,* Vol. 19(1), pp. 21-37.

[7] Eckert, C.M., Clarkson, P.J., and Zanker, W., 2004, "Change and Customisation in Complex Engineering Domains," *Research in Engineering Design*, Vol. 15(1).

[8] Huang, G.Q., Yee, W.Y., and Mak, K.L., 2003, "Current Practice of Engineering Change Management in Hong Kong Manufacturing Industries," *Journal of Materials Processing Technology*, Vol. 139(1/3), pp. 481-487.

[9] Pikosz, P., and Malmqvist, J., 1998, "A Comparative Study of Engineering Change Management in Three Swedish Engineering Companies," *Proceedings of ASME Design Engineering Technical Conferences*, Atlanta, GA, USA.

[10] Leech, D.J., and Turner, B.T., 1985, *Engineering Design for Profit,* Chichester, Ellis Horwood Ltd.

[11] Dale, B.G., 1982, "The Management of Engineering Change Procedure," *Engineering Management International,* Vol. 1(3), pp. 201-208.

[12] Rivière, A., Dacunha, C., and Tollenaere, M., 2002, "Performances in Engineering Change Management," *Proc. of 4th International Conference on Integrated Design and Manufacturing in Mechanical Engineering*, Clermont-Ferrand, France, CD-ROM.

[13] Maull, R., Hughes, D., and Bennett, J., 1992, "The role of the bill-of-materials as a CAD/CAPM interface and the key importance of engineering change control," *Computing and Control Engineering Journal*, Vol. 3(2), pp. 63-70.

[14] Jarratt, T.A.W., Eckert, C.M., Weeks, R., and Clarkson, P.J., 2003, "Environmental Legislation as a Driver of Design," *Proceedings of 14th International Conference on Engineering Design*, Stockholm, Sweden, CD-ROM.

[15] Jarratt, T.A.W., Eckert, C.M., and Clarkson, P.J., 2004, "Development of a Product model to Support Engineering Change Management," *Proceedings of the TMCE 2004*, Lausanne, Switzerland, CD-ROM.

35

Modeling of Manufacturing Process Complexity

R. Jill Urbanic, and Waguih H. ElMaraghy

Abstract: Gaining momentum in several fields of study is the recognition of the need for a viewpoint that includes the human element as an integral part of the modern production system beyond traditional ergonomics. The "intellectual capital" is as much of a resource as money, materials, software and hardware. A model that considers the human players in tandem with the physical elements is needed to provide insights into the sensitivities of the manufacturing system. Using Systems Analysis and Design methods, a framework has been developed, which is valid for different perspectives and environments, to assess the elements of manufacturing complexity. The manufacturing complexity index allows people with diverse backgrounds to rapidly evaluate alternatives and risks with respect to the product, process or operation tasks. In this paper, the technique for evaluating the process complexity metric is presented. An analysis is performed comparing the relative process complexity for a power steering pump bracket that is manufactured in a CNC machining cell and a dedicated line. The areas of complexity are clearly evident. This provides insight for risk assessment as this systematic approach can be used to "mathematically" show tradeoffs for each important criterion during the design stages.

Keywords: Process Development, Decision-making, Systems Analysis Methodology

35.1 Introduction

In today's increasingly interdependent, volatile, global economy, companies face a combination of unpredictable demand and consumers that expect the introduction of innovative, high quality, low cost new products in a timely fashion [1]. Manufacturing performance must be considered as critical as innovative products [2, 3]. There is a need for both productivity and innovation – and the innovative approach must be expanded to include new processes as well as new products. Focusing piecemeal on each element alone will not suffice. Manufacturing systems need to adapt to new events; hence research into concepts such as transformable factory strategies [4] and reconfigurable manufacturing systems. However to be effective, the system must balance human characteristics, skills and capabilities within the technical and business environment: a multi-faceted human point of view, in conjunction with the development of technical and financial tools, is crucial for long term success.

A framework has been developed that decouples the product, process and operational elements of manufacturing complexity and re-links them in a systematic manner. This manufacturing complexity analysis technique has the scope to effectively define the areas of complexity within the manufacturing enterprise in an objective manner. This allows the players to identify the areas to focus on that could compromise critical performance criteria such as: training and maintenance issues, resources allocation, reduction of process and ramp up lead times, 100% on-time delivery, costs, response time, conformance to specification, product features, *etc.*.

Kjellberg [5] focuses on the overall process efficiency, and narrows in on the fact that the real bottleneck in organizations is the lack of knowledge. Knowledge improvement strategies should be linked to manufacturing strategies to optimize manufacturing performance, or conversely the manufacturing strategies should be linked to the knowledge base where appropriate. The effectiveness of an enterprise depends on the successful exploitation of resources. Various techniques are used to balance product and process design and other manufacturing activities but typically in an unscientific manner. Comprehensive comparisons of different design, configuration and process planning scenarios will help the decision making processes at all organizational levels by highlighting the system sensitivities. Each manufacturing system is unique and to effectively capture this information in a complexity model is a significant engineering challenge.

35.2 Manufacturing Complexity

Introduction

Manufacturing processes consist of highly coupled relationships between the product design, the materials, the production equipment, and support systems. These elements are integrated with activities within all levels of an organization and capturing a relevant perception of complexity can be problematical. There are various approaches to presenting a manufacturing complexity measure. Cooper et al [6] introduce a generic index that does not consider any distinctions between the product or process elements of complexity. Guenov [7] and Kim [8] introduce systems design metrics for a comparison of alternatives. Guenov [7] uses the fundamentals of Architectural Design, Axiomatic Design [9] and Entropy to portray a comparison of alternatives based on cost, value, performance and technical risk or complexity. These methods are applicable to any environment, but are not intuitive to apply. Kim [8] presents several relatively simple metrics that consider the relationship between product variety and system components. The number of flow paths and crossings, the travel distance, the system reliability, setup time, cycle time and so forth are all considered, but the framework does not decouple the elements of product, process and operational complexity, nor does it capture perceptions of complexity, which vary from enterprise to enterprise.

Complexity may be, in part, associated with understanding and managing a large volume or quantity of information, as well as a large variety of information. The general manufacturing complexity model introduced by Urbanic and W. ElMaraghy [10] is a heuristic model that focuses on these elements. The model is composed of three basic components – the absolute quantity of information, the diversity of information and the information content or the "relative" measure of effort to achieve the required results (Figure 35.1).

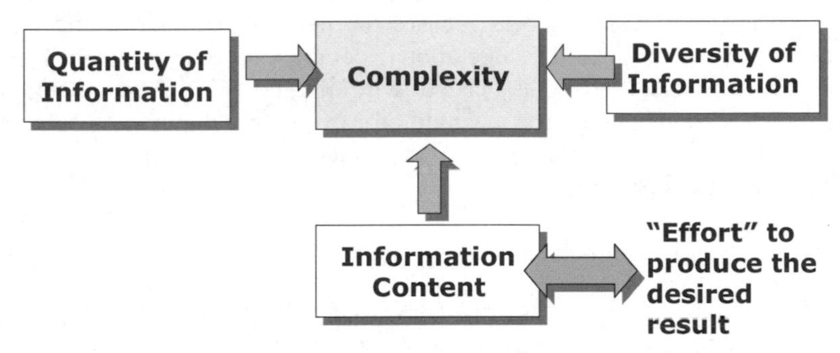

Figure 35.1 Elements of complexity

Although the quantity of information is a factor of complexity, the absolute quantity of information may contain much redundancy. Therefore a compression factor, the information entropy measure H, will be used to represent the quantity of information element:

$$H = \log_2(N+1) \tag{35.1}$$

where N is the total quantity of information.

The measure of uniqueness or the diversity ratio D_R is defined as a ratio of distinct information to total information, as given by:

$$D_R = \frac{n}{N} \tag{35.2}$$

where n is the quantity of unique information and N is the total quantity of information.

Information content is defined here as a "relative" measure of effort to achieve the required result. The higher the effort (*i.e.* the more required stages or tools), the more complex is the feature or task. Each work environment has a different perception of complexity, but it is typically consistent. The complexity index needs to effectively capture this. The relative complexity coefficient, c_j is introduced along with a methodology to determine its value. This coefficient has a value between 0 – 1, complementing the diversity ratio D_R.

The product complexity analysis is performed independently from any process plan, and focuses on the product features and specifications. The product complexity index $CI_{product}$ is a combination of the diversity ratio and the relative complexity, and is scaled by its information entropy. This is expressed as:

$$CI_{product} = (D_R + c_j) * H \tag{35.3}$$

The process complexity analysis focuses on the tools, equipment and systems to manufacture the product. The operational complexity analysis centres on the cognitive and physical effort associated with the tasks that are related to a product/process combination. The technique for evaluating the process complexity metric, which is an extension of the techniques used to provide a product complexity metric, is presented here.

Elements of Process Complexity

The product design, volumes, available capital and the working environment influence the manufacturing process selection (Figure 35.2). In order to balance the process with the human factors, the areas of complexity need to be highlighted in such a manner that people with diverse backgrounds can rapidly assess risks and alternatives. This is required for effective management of the manufacturing system. Complexity affects throughput, reliability and quality, and an area of high complexity has the potential to be a risk unless proactive steps are taken. To generate a value for process complexity, the main constituents of the manufacturing process must be identified and assessed as each factor contributes to the overall process complexity [11]. A sample of the process complexity components in the machining environment are: (i) in-process features; (ii) types of tools, tool holders, and spindles; (iii) fixture setups and product orientations; (iv) machine types and controllers; (v) gauges; (vi) part feeding and material handling devices, and so forth. The example in this paper focuses on machining, but the framework can be extended to any environment.

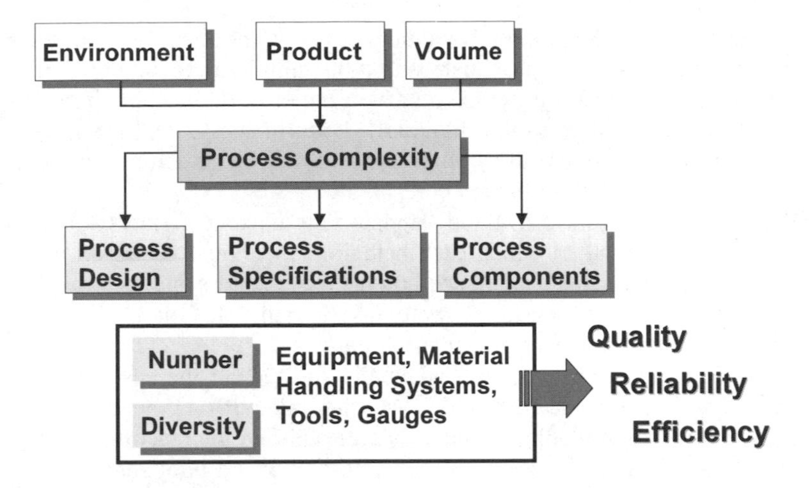

Figure 35.2 Elements of process complexity

Although there are many facets to process complexity, the methodology that was established to generate a product complexity index *CI* (W. ElMaraghy and Urbanic [11]) can be extended to embrace these different process elements. The process complexity index is the sum of the individual constituent complexity values and the product complexity:

$$PI_{process} = \sum pc_x + CI_{product} \tag{35.4}$$

The xth individual process constituent complexity index pc_x is:

$$pc_x = (D_{R\ process,x} + c_{process,x}) * H_{process,x} \tag{35.5}$$

where $D_{R\ process,\ x}$ is the diversity ratio for the xth constituent

$c_{process,x}$ is the relative complexity coefficient for the xth constituent

$H_{process,\ x}$ is the information entropy for the xth constituent.

The methodology to define the product manufacturing complexity coefficient has been extended to evaluate the individual process constituents' complexity; however, for analysis purposes here $c_{process,\ x}$ set to zero.

Example

An example for generating process complexity index is presented for a die-cast aluminum (10% Si. max) power steering pump bracket (Figure 35.3). This product had migrated from a flexible CNC work cell process to a dedicated machine work cell (Figure 35.4). The required volumes of this product had increased by asignificant factor (from approximately 15,000 units/year to 60,000 units/year), and it became cost inefficient to utilize a twin pallet CNC machine based process due to the long cycle time (~ 4.3 min./part). Unfortunately, the "ideal" dedicated line could never be consistently utilized without a CNC machine due to quality issues such as excess flash or broken cores in the casting.

Within the CNC work cell, three standalone tombstone fixtures – two on one machine, one on the other, which each held four locating fixtures or nests, were utilized. The parts were manually loaded on the cast locators and then hydraulically clamped into each nest. Bushings were inserted into the part using a simple pneumatic cylinder press in a separate machine.

For the dedicated line, the part was manually located on the cast locators in an integral fixture in the mill-drill machine. Multi-spindle drilling heads were used to drill all the holes (except hole 2), and a fly cutter milled the machine datum (–A-). The part was manually loaded into the multi-spindle milling machine and was located on the machine datums –A-, -B-, and –C-. Tracer heads were employed to machine the curvilinear surfaces, simple linear stepper motors were used to mill the rest of the surfaces, and hole 2 was drilled. The part was then manually loaded into a tapping machine to tap all the holes. Next the bushings were inserted into the part using the bushing insertion machine. If CNC machining was required, the parts were loaded into a CNC machine (one pallet only) prior to machining on the dedicated line. An analysis of the process complexity is presented for these scenarios.

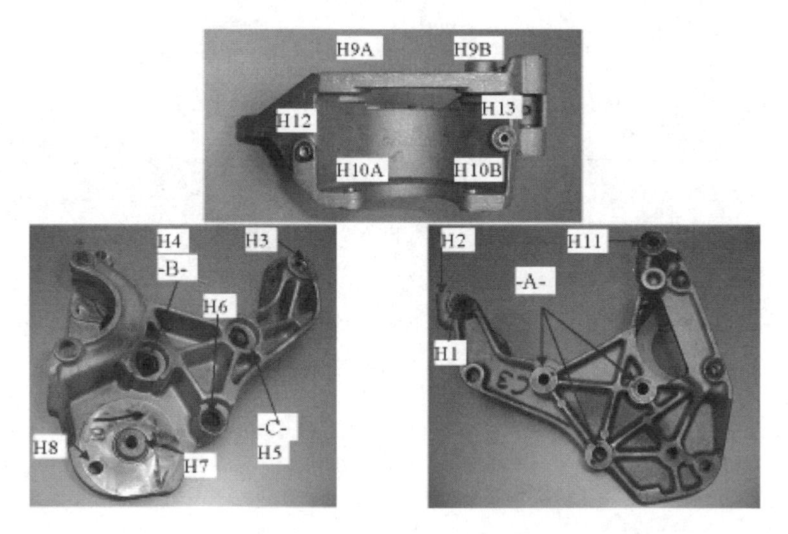

Figure 35.3 Power steering bracket

This part is a mounting bracket; therefore, the information with respect to the tolerances and geometric relationships was considered critical. The total and unique number of product feature callouts are N = 80 and n = 47 respectively. Substituting into Equations (35.1) and (35.2): $H = log_2(80+1) = 6.34$; $D_R = 47/80$

= *0.59*. The product relative complexity coefficient $c_{product}$ is 0.105. For the sake of brevity, a detailed analysis for the product relative complexity coefficient $c_{product}$ is not presented here. This value was determined by taking into account the relative effort to produce the features by considering the number of features, the part material, shape, geometry and tolerances. By substituting these values into Equation (35.3), the product complexity index *CI* is: $CI = (0.59 + 0.105)*6.34 = 4.406$.

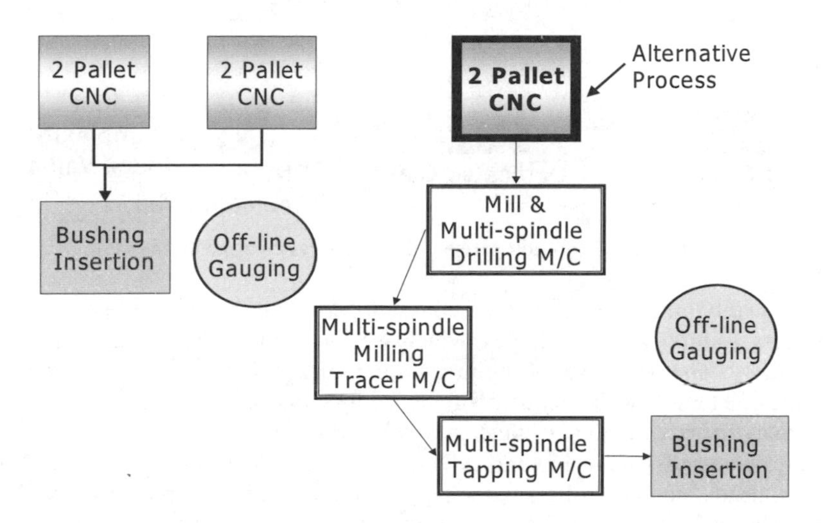

Figure 35.4 CNC and dedicated manufacturing cell layouts

The process complexity elements and analysis are presented in Table 35.1. The factors being assessed are: (i) mechanical (machines, fixtures, tools and spindles); (ii) controls (both flexible and dedicated); (iii) production control (material handling and flow paths); and (iv) quality assurance aspects (gauging and in-process features).

Recall from Equation (35.4) that $PI = \Sigma pc_X + CI$. For scenario (A) summarized in Table 35.1, the detailed calculation for the process complexity index PI_A =. $(0.667*2.00 + 0.154*3.81 +0.950*4.39 +0.500*1.58 +0.500*1.58 +0.500*2.32 + 1.000*1.58\ 1.000*3.70 +0.643*5.43\ 1.000*4.95 +1.000*1.58) + 4.406$, which results in $PI_A = 26.657$.

The overall process complexity indices, and the contribution of the product complexity index is shown in Figure 35.5. The dedicated machine work cell process is approximately 20% more complex than the CNC process, and the modified dedicated process is approximately 30% more complex.

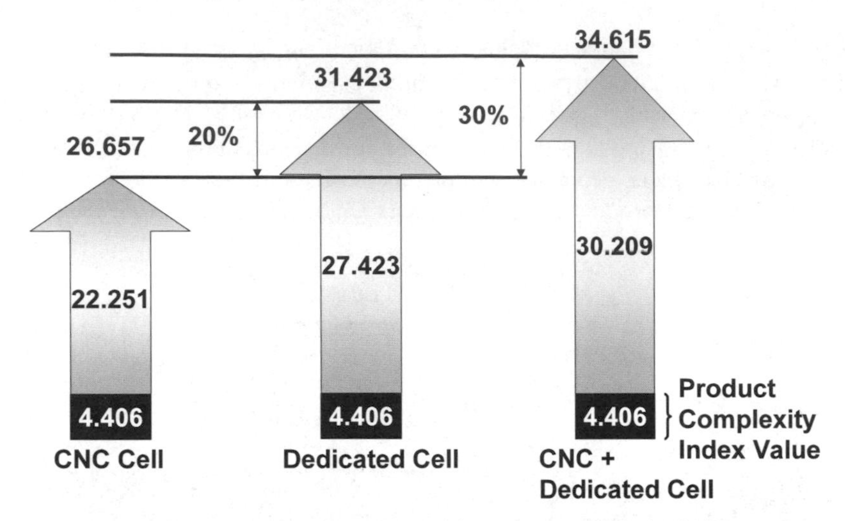

Figure 35.5 A comparison of the process complexity indices

The quality assurance aspects are common across all the process scenarios. The in-process features and the gauges, which were designed for inspecting this part, are identical. The variations of the process complexity indices are associated with the other factors. The breakdown of the factors that contributed to the process complexity index variations is illustrated in Figure 35.6.

The dedicated machine work cell had several unique elements – including fixtures, heads, and controllers, which contributed to a higher process complexity index. The influence of quality issues from the foundry or machine breakdowns within the dedicated work cell impacted the process complexity significantly as a CNC machine and the necessary accoutrements were constantly needed as a stand-by. This had several repercussions with respect to production control, quality, reliability and spare part procurement.

Upon comparing the individual factors (Figure 35.7), it can be seen that the tools and the hand gauge components have significantly higher complexity measures than the machine component. This means that it will tend to take longer for people to learn about the gauges and tools as compared to familiarizing themselves with the machines compared to the CNC cell. Typical for dedicated equipment, the machines, fixtures and spindles/heads were designed for a specific application; hence, these results are expected. Conversely there is less diversity in the tooling, as dedicated machines generate several identical features simultaneously, whereas flexible machines manufacture features serially.

Table 35.1 Process Complexity Comparison for Different Steering Pump Bracket Manufacturing Methods

A)

CNC process + Bushing Machine	**Total, N**	H	**Distinct, n**	D_{Rx}	pc_x
Machines	3	2.00	2	0.667	1.333
Fixtures	13	3.81	2	0.154	0.586
Tools	20	4.39	19	0.950	4.173
Spindles/ Head/ Lead screws	2	1.58	1	0.500	0.315
Flexible Controllers	2	1.58	1	0.500	0.315
Dedicated Controls	4	2.32	2	0.500	0.215
Material Handling Operations	2	1.58	2	1.000	1.585
Flow Paths	12	3.70	12	1.000	3.700
In-process Features	42	5.43	27	0.643	3.488
Gauges - hand	30	4.95	30	1.000	4.954
Gauges - relation	2	1.58	2	1.000	1.585
SUM - **A**					**22.251**
Process Complexity Index - **A**					**26.657**

B)

4 Dedicated Machines Work Cell	**Total, N**	H	**Distinct, n**	D_{Rx}	pc_x
Machines	4	2.32	4	1.000	2.322
Fixtures	4	2.32	4	1.000	2.322
Tools	24	4.64	17	0.708	3.289
Spindles/ Head/ Lead screws	13	3.81	13	1.000	3.807
Flexible Controllers	3	2.00	2	0.667	1.333
Dedicated Controls	1	1.00	1	1.000	1.000
Material Handling Operations	4	2.32	4	1.000	2.322
Flow Paths	1	1.00	1	1.000	1.000
In-process Features	42	5.43	27	0.643	3.488
Gauges - hand	30	4.95	30	1.000	4.954
Gauges - relation	2	1.58	2	1.000	1.585
SUM - **B**					**27.423**
Process Complexity Index - **B**					**31.829**

C)

1 CNC+4 Dedicated Machines Cell	**Total, N**	H	**Distinct, n**	D_{Rx}	pc_x
Machines	5	2.58	5	1.000	2.585
Fixtures	8	3.17	5	0.625	1.981
Tools	26	4.75	19	0.731	3.475
Spindles/ Head/ Lead screws	14	3.91	14	1.000	3.907
Flexible Controllers	4	2.32	3	0.750	1.741
Dedicated Controls	2	1.58	2	1.000	1.585
Material Handling Operations	5	2.58	5	1.000	2.585
Flow Paths	4	2.32	4	1.000	2.322
In-process Features	42	5.43	27	0.643	3.488
Gauges - hand	30	4.95	30	1.000	4.954
Gauges - relation	2	1.58	2	1.000	1.585
SUM - **C**					**30.209**
Process Complexity Index - **C**					**34.615**

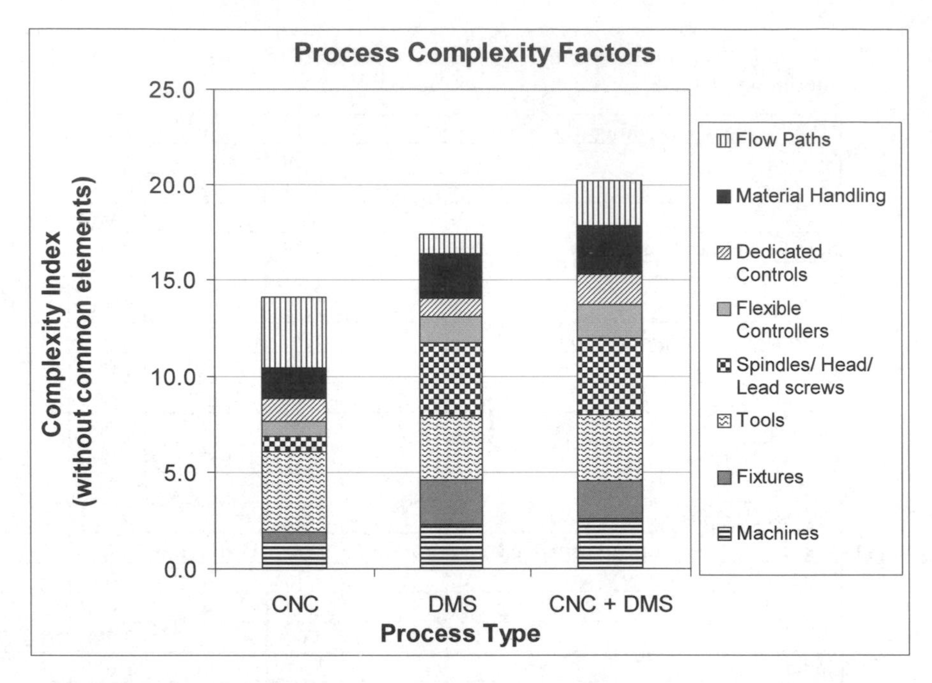

Figure 35.6 A detailed comparison of the process complexity indices' factors

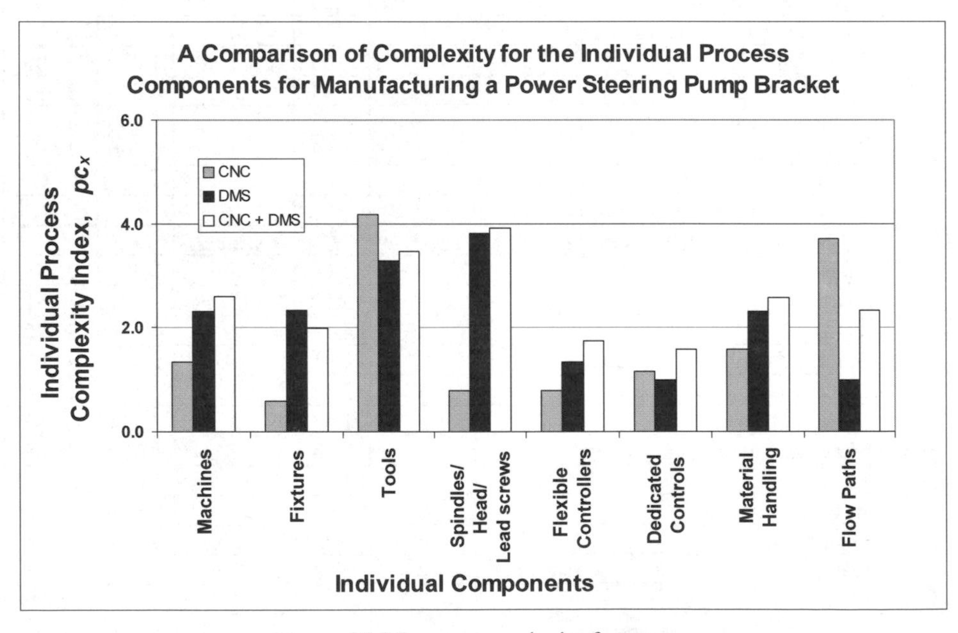

Figure 35.7 Process complexity factors

In general, the following should hold true when comparing dedicated versus flexible equipment:

$$D_{Rfixture}\text{ DMS} > D_{Rfixture}\text{ CNC cell} \quad (35.6)$$

$$D_{Rmachine}\text{ DMS} > D_{Rmachine}\text{ CNC cell} \quad (35.7)$$

$$D_{Rspindle}\text{ DMS} > D_{Rspindle}\text{ CNC cell} \quad (35.8)$$

$$D_{Rtool}\text{ DMS} < D_{Rtool}\text{ CNC cell} \quad (35.9)$$

More constituents can be added (software interfaces, tool setups, *etc.*.), and a more in-depth information quantity and diversity analysis could be performed for each constituent (*i.e.* the fixture information should be extended to include the locating datum, the type and quantity of part touching details and supports, and so forth). However upon performing a first level analysis, the general trends can be established.

35.3 Summary and Conclusion

A model of complexity was created that is a function of the quantity of information, diversity of information and the information content. This model was applied to determine the process complexity for a dedicated and CNC machining line that was used to manufacture a power steering pump bracket. The areas of complexity were easily identified for the various processes. These are the areas that could potentially introduce risk into the system; hence, this simple complexity model can be used as an assessment and decision making tool to proactively address implementation and operational business issues.

Wiendahl and Scholtissek [12] have indicated the need for research efforts to support modularization, simplicity and segmentation. This index clearly highlights the influence of common components. As the diversity ratio, $D_{R\ process,\ x}$, and the relative complexity coefficient, $c_{process,\ x} \to 1$ and the information entropy $H_{process,\ x} \to \infty$ for any factor x, the difficulty to manage that factor also increases, such as the ability to train, troubleshoot and performance maintenance.

This framework to assess process related complexity can be used in any design and manufacturing environment. The level of analysis can be as superficial or detailed as required, and yields results that are readily understood by all players. It identifies the areas of process complexity so they can be managed based on the capabilities of the human players within a particular environment.

35.4 References

[1] Tichy, N.M., Sherman, S., 1993, "Control Your Destiny or Someone Else Will," *HarperBusiness*.

[2] Feldman, K., Slama, S., 2001, "Highly Flexible Assembly – Scope and Justification," *Annals of the CIRP*, Vol. 50, No. 2, pp.489-498.
[3] Skinner, W., 1978, *Manufacturing in the Corporate Strategy*, John Wiley and Sons.
[4] Wiendahl, H.P, Hernández, R., 2001, "The Transformable Factory Strategies, Methods and Examples," *CIRP Int. Conf. on Agile, Reconfigurable Manuf. Proc.*, CD ROM.
[5] Kjellberg, A., 1999, "Teams – What's Next? From Fragmentation and Consciousness to Responsiveness by Competence Management for Modular Manufacturing Learning," *Annals of the CIRP*, Vol. 48, No. 2, pp. 599-609.
[6] Cooper, W.W., Sinha, K.K., Sullivan, R.S., 1992, "Measuring Complexity in High Technology Manufacturing: Indexes for Evaluation," *Interfaces*, pp. 38-48.
[7] Guenov, M.D., 2002, "Complexity and Cost Effectiveness Measures for Systems Design," *2nd Int. Conf. of the Manuf. Complexity Network*, pp. 455–466.
[8] Kim, Y., 1999, "A System Complexity Approach for the Integration of Product Development and Production System Design," *M.A. Sc. Thesis, MIT.*
[9] Suh, N.P., 2001, *Axiomatix Design: Advances and Applications*, Oxford Univ. Press.
[10] Urbanic, R.J., ElMaraghy W.H., 2003, "Modelling of Participatory Manufacturing Processes," *Proceedings of the CIRP 2003 Design Seminar*, pp. 58-70.
[11] ElMaraghy, W.H., Urbanic, R.J., 2003, "Modelling of Manufacturing Systems Complexity," *The Annals of CIRP*, Vol. 53/1, pp. 363-366.
[12] Wiendahl, H.P., and Scholtissek, P., 1994, "Management and Control of Complexity in Manufacturing," *Annals of the CIRP*, Vol. 43/2, pp.533-540.

36

Human Modeling in Industrial Design

Mahmoud Shahrokhi, Mamy Pouliquen, and Alain Bernard

Abstract: The great importance of human aspects in industrial environments have changed the viewpoints of designers and developed Human-Centered design approaches. One of the fundamentals of this approach is to consider human factors at all stages of the design process. The integration of human factors in the design process phases requires effective use of the appropriate human models. This paper presents definitions of human models and their classification in industrial applications with emphasis on industrial design processes. We also focus on the application of the human models in a human-centered industrial system approach. Specifically, we discuss future approaches relevant to the use of human models in the virtual environment.

Keywords: ergonomics, human model, human engineering, virtual reality, industrial systems design, and human-centered design

36.1 Introduction

Human models are used to study and solve problems across a very broad spectrum of domains. These problems combine aspects of the material world (such as the workplace, the economy, medicine), and aspects of the conceptual world that people create (such as art, advertising, organization). New moral, social and commercial criteria and standards increase the role and importance of security and comfort of end users. These changes lead designers to use human models through the design process to provide more safe and applicable productions.

The increased need for more rapid design and redesign processes in a competitive environment has resulted in the appearance of a new generation of human models. The various human modeling techniques and their applications will be presented.

36.2 Objective

The main objective of this paper is to offer a definition of human models and to present the application of human models through the industrial systems design process.

36.3 Essential Concepts

"The worker should be given the possibility to participate in the design of his/her own working conditions and in development work that concerns his/her work" The Swedish Work Environment Law [1].

To assess the importance of man in the industrial environment, we should not only compare the role of man and machine in all intervention modes (normal operating mode, maintenance mode, degraded operating mode, *etc.*.) but also the role of man in present modern industries, and the role of man in the past. The development of modern technologies especially in control aspects, and increasing automation, decreases the amount of direct human intervention. However moral aspects, the professionalization of operations and repairs, human quantity limitations, and finally human (direct & indirect) costs put man in the driving seat of the design process. In this manner Human-Centered design process are presented. Undoubtedly human models are among the most important requirements of this approach. Many modern industrial design approaches use human models to minimize accidents and illnesses due to chronic physical and psychological stress, while the goal is to maximize productivity and efficiency.

There are a variety of classical human models. These models were developed to respond to different needs and thus use different techniques. One problem in these approaches with current human models is the vast domain of human characteristics that require hyper-disciplinary knowledge-based models. In reality the definition of this sophisticated model and its justification is a new challenge that perhaps will never be solved. The main purpose of this study is to present human models and to offer a classification for industrial human models. Regarding the survey, the intention is to identify new integrated human models dependent on the future needs of industrial system designers. Product, service or system of organization involves more than just its technical characteristics.

First, we describe the most important concepts that have been used during the development of human models and the work carried out with them. The concepts related to the industrial workplace are also described.

The workplace designing process is directly related to human satisfaction, and the probable benefits of well-designed jobs, equipment, and workplaces are improved productivity, safety, health, and increased satisfaction for the employees [2]. It is a collection of multidisciplinary efforts that include designing physical spaces and machines, materials, processes and organisations to produce objects. It consists of organised activities with the aim of safe and effective designing of production systems in minimum time. This process needs study to evaluate the degrees of human's adaptation in their environment. The term 'environment' is taken to cover the ambient human working environment but also his tools and materials, his methods of work, the organization of his work, and the psychological and sociological interactions, either as an individual or within a working group [2].

In this way design at highest level also consists of developing technical (logical and physical) solutions related human aspects (policy, organization, planning & implementation, evaluation, and action for improvement [3]). In recent years new design approaches have developed and now we emphasis on the participation of the users through all the design process. Nowadays users actually cooperate in the making of the design decisions. They define their criteria, participate in alternative identification, and evaluate their criteria satisfaction.

Ergonomics is a relatively new branch of science. It may be defined as [4]: "study of the interaction between workers and their tools, machines tools, and work processes". It serves as a repository and source of data and principles that can be validly applied to the specification, design, evaluation, operation, and maintenance of products systems that are intended for safe, effective, satisfying use by individuals, groups, and organisations [5]. This knowledge is applied to design of complex technical systems or work tasks, equipment, workstations, or tools and utensils used at work, at home, or during leisure time [6]. Human engineers use ergonomics to [2]:

- study human-machine interactions to insure that the equipment operational requirements do not exceed human abilities
- consider human performance tolerance, thereby insuring optimal speed, accuracy, and quality of performance; eliminating hazards; and maximising the comfort of the operator.

There are three basic stages [7], in general system design approaches: (1) defining the problem; (2) developing a solution; and (3) evaluating the solution. However in a Human-Centered Design (HCD) process, we must not only show that the hardware and software under design meet specifications, but we must also demonstrate that the resulting hardware and software will provide the necessary support for the human team trying to meet overall task objectives and that the capabilities and limitations of the human operators have been taken into consideration in all stages of design [8]. Therefore the evaluation procedure in the same way must verify that the system satisfies the required functionality and also the user wants and desires [7].

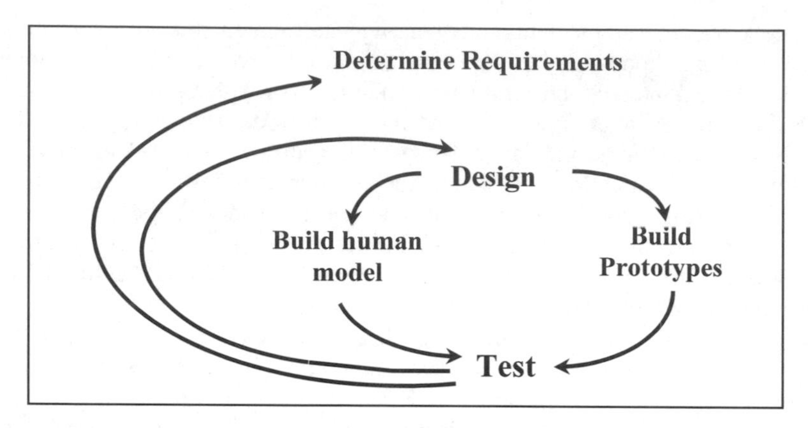

Figure 36.1 Uses of human performance models in system validation

Constraints identification is one of the other essential aspects in this approach. These comprise manpower constraints (such as, how many people), personnel characteristic constraints (such as, what kind of people), and training constraints relevant human resources. Also they comprise performance requirements (such as time, accuracy, risk), function allocations constraints, and communication requirements, relevant system planning and organization [9].

Human modelling is actually the building of models of how people perform or accomplish something. Generally a human model is a sophisticated person (or persons) who does something in a particular condition/way. In order to introduce human models, first we must have a conceptual classification of them. Up to now human beings have had a basic role in the most important and critical tasks in industrial workplaces. These tasks are concentrated in control panel operation, planning, repairs and emergency activities. In contrast, highly individual and sophisticated models of how humans function and behave are almost nonexistent. In this context, humans are the "weakest link" in the system [10]. This is because a human has various and complicated behaviours and he adapt his mental and physical characteristics with the environmental limits, and therefore he is relatively not a stable and predicable object. In this environment human models attempt to represent and predict human behaviours.

Industrial human modeling techniques include computation, mathematical, experimental or methodological formulations that have been used to build models of human competence/performance and they are applied to system design, operation, or evaluation. The term 'industrial human model' (IHM) is typically used to refer to a model that simulates some aspect of human performance/presence in an industrial workplace. Models include complete formulations (or families of them) that attempt to describe, predict, or prescribe aspects of human constraint, competence or performance [9]. IHMs are utilised to evaluate different design alternatives and predict user requirements. They play various roles during system design, ranging from generating design concepts to affording simulation-based design evaluation and training.

A human model may simulate the reach distance of the operator, the amount of time it will take for a team to complete a series of routine procedures, or the reasoning that goes on in a person's head as he tries to identify a new radar track [11].

Ergonomic human model tools are some of the most useful human engineering tools in industrial system design processes but there are also many other human modeling tools that provide widespread design applications. The most essential goal of defining human models, is estimating human acceptability through quantitative-qualitative studies to provide a better response to the consumer's expectations by reintroducing the human factor at the earliest stage of technological and commercial decisions, and from the outset of the initial design phases which covers areas such as ergonomic design, sensorial design, cognitive ergonomics, sociology, *etc.*. [12].

Given these different types of goals, there are different ways to build human models, and there is no single "best" and ideal approach for all applications [8].Human models must represent highly complex and adaptable systems [13], achieving this goal requires a wide spectrum of models, that each of those attempt to model human behaviour from a unique viewpoint. According to [9], human models are classified into 18 categories. This classification covers all the human models in the different approaches, especially in mental effort envisages. In ordinary industrial design, specially physical and ergonomical aspects of work team and their interactions with other parts of workplace system is interested.

36.3 Discussion

Today human modelling technology has become a proven contributor to the systems engineering process [9]. In design process, modeling is an organised activity consisting of a series of steps, which may be repeated a number of times to develop an appropriate model, they are:

1. **System identification:** defining the boundary of system (and aspect of interest), and the environment that influences it,
2. **Component definition:** the identification of the major subsystems, or important entities,
3. **Relationships identification:** specifying the interlinks between those subsystems, entities, and environment,
4. **Knowledge organization:** Recalling relevant domain knowledge and facts pertinent to the situation of subsystems,
5. **Model type selection:** recalling appropriate modeling archetypes that may be applicable to the situation,
6. **Model creating:** creating a particular model by defining all its elements,
7. **Model using:** applying and evaluating the model, testing it against the problem situation and repeating the modeling process if necessary.

It is clear that developing human models is not simply a mechanical application of a set of rules, it involves creative intelligence, and the ability to discriminate among all the available facts, knowledge, and archetypes to select those most appropriate to the situations and questions. By reducing cost and risk [9], human performance modeling technology will play a significant role in both the design and operation of future complex systems.

The models are interconnection tools between real world and human cognition. Future human models will apply in sophisticated human workplaces; therefore these models will change respectively with future workplace progression. They must support industrial design processes with future technology and approaches.

We can classify contemporary human models related to technical concepts as:

Descriptive Models: Descriptive human models are models that present classified knowledge about human behavior, and almost exclusively rely on our visual modality. These models are not scientific tools and to understand their subjects only basic information, related workplace nature and human task, is necessary. In these models information is presented in the form of texts, graphical diagrams, plots, films or charts. The purpose of these representations is to structure the underlying data and make it understandable [14]. Due to their simplicity they can be reviewed by users at various levels and can be effective models for education and co-operation work. As such, they have a special role to play in all kinds of modelling and non-modeling work. Procedural flowcharts relating human tasks, and the various diagrams applied in work and time studies are some examples of these models. Figure 36.2 demonstrates an example of a descriptive model.

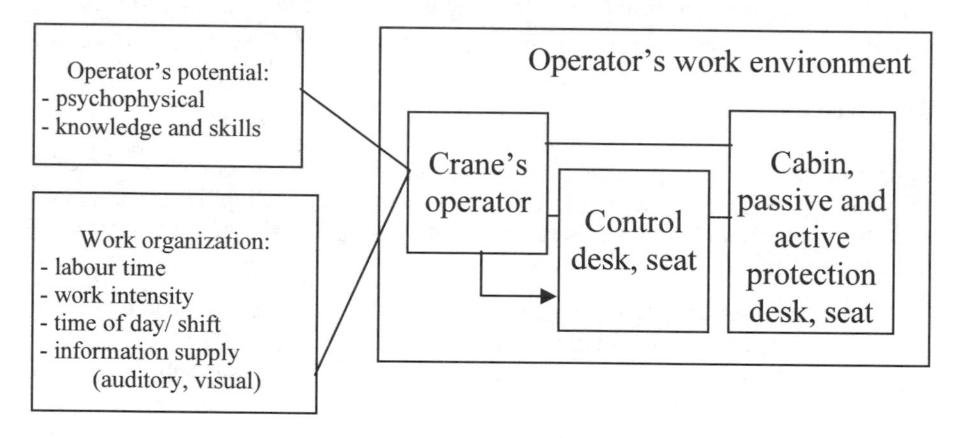

Figure 36.2 A descriptive model of operator-crane interaction

Scale Models (Manikins): A scale model has the same appearance as the original system/object for proportional size and detail [16], and Manikin is a scaled anatomical model of the body or a part of the body, especially for use in medical or art instruction. Manikin postural analysis permits users to quantitatively and qualitatively analyse whole body and segments localised postures to determine operator comfort and performance in accordance with comfort and safety standards. The manikin measures the local parameters that might not be correctly numerically modelled. The manikins also allow three-dimensional visualisation of the reaction responses and can be used to study human interaction in many other environments, such as buildings, aircraft, and spacecraft [17]. A thermal human scale model is developed for thermodynamic studies [17]. This model demonstrates a manikin model that is integrated with a computer control and measurements system. A communication system is constructed to control manikin inter-subsystems, and it receives the transmitted data from the manikin.

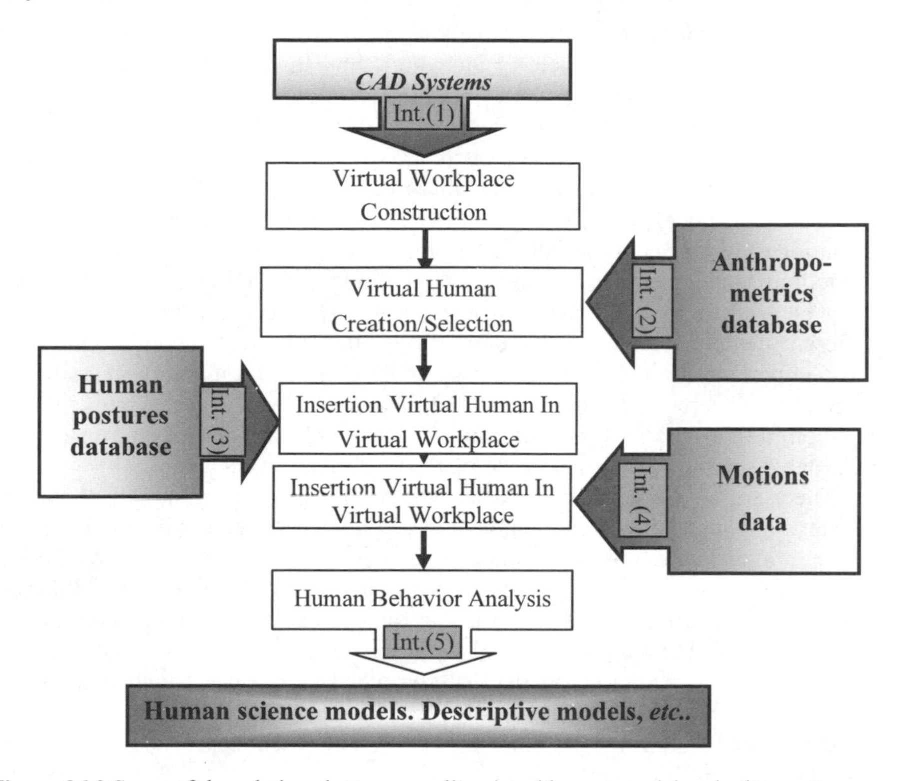

Figure 36.3 Some of the relations between reality virtual human model and other systems

Mathematical Models: A mathematical model is a representation comprised of procedures algorithms and mathematical equations or relationships that can be discreetly solved.

Sometimes the models employ techniques of numerical approximation to solve complex mathematical functions for which specific values cannot be derived [18]. The mathematical human models utilise widespread domain of mathematical techniques that are used for human behavior modelling and optimisation. In this way, they can be divided as: (a) Operational research models [see 22], (b) Probability and statistical models, (c) Fuzzy logic models, (d) Neural network models, and (e) Task network models, or a combination of them.

Human Science Models: Human sciences are concerned with the systematic study of nature and the real characteristics of a human being. Human science represents all aspects of the human existence, including mental, social and physical specifications. A human science model is a presentation that interprets the nature of the human with emphasize experimental researches.

The human science models like other natural sciences are based on theory examination, and they are dependent on their assumptions and examination quality. Therefore their results are not absolutely correct. These kinds of models are divided into three grand categories: (a) physiological models, (b) psychological models, and (c) anatomical (biomechanical) models.

Computer-based Models: Recent technological progresses in computer sciences affect all scientific approaches, including human modelling domain. They contribute to all described models, but here we address the modelling techniques that essentially are based on computer processing, and without it are not discussible. We categorize these models as: (a) Cognitive simulation models, (b) Numeric simulation, and (c) Virtual reality models.

A human model may focus on individual aspects of human capabilities and behavior (physical body or internal information processing). Each kind of these modelling techniques has advantages and drawbacks. Different human models must be used to provide necessary design data. This data can be shared between different modeling tools and therefore model integrity will increase. One of the most important specifications of human models is their capability of integrating with other different human models, and also other work place system technical solution design tools. This integration can be electronic import, export, and automated sharing of information, or manual information exchange.

There are also integrative models, which attempt to integrate all of multiple human components into a single model to provide an integrated human model [9]. There are different opinions concerning these models, according to [19] an industrial human model is selected and developed in a careful and limited domain, and accurately simulates some aspects of human performance, and are not concerned with building a complete replica of a human. He emphasis that "we won't ever have one perfect model that completely replicates a human being. The complexity of human nature is one obvious reason. But another reason is that there are many different ways to build a model, and each approach helps achieve different goals".

In reality there are various scientific attempts to create 'total human model', and there are many integrated human models that consist of different techniques and are concerned with various human aspects [4, 12]. Changes of workplace demands influence industrial design, and cause more necessity for whole and precise modeling techniques at the same time. Integrated models in different sciences (such as mathematics, human sciences, information technology), in the form of flexible and rapid applications in uncertain environments will be needed.

Satisfaction of these requirements implies development of new human modeling approaches with intelligent instruments and interfaces in information gathering and transforming. On the other hand, human modeling systems must be compatible and integrate-able with other design systems, and with the whole design system.

Future human models will integrate a very broad spectrum of science and experience domains, to create multi-aspect models, interlinked with other design models. These models will have some aspects of ergonomics, psychology, information technology, and the various domains of application, such as health systems, production systems and logistics.

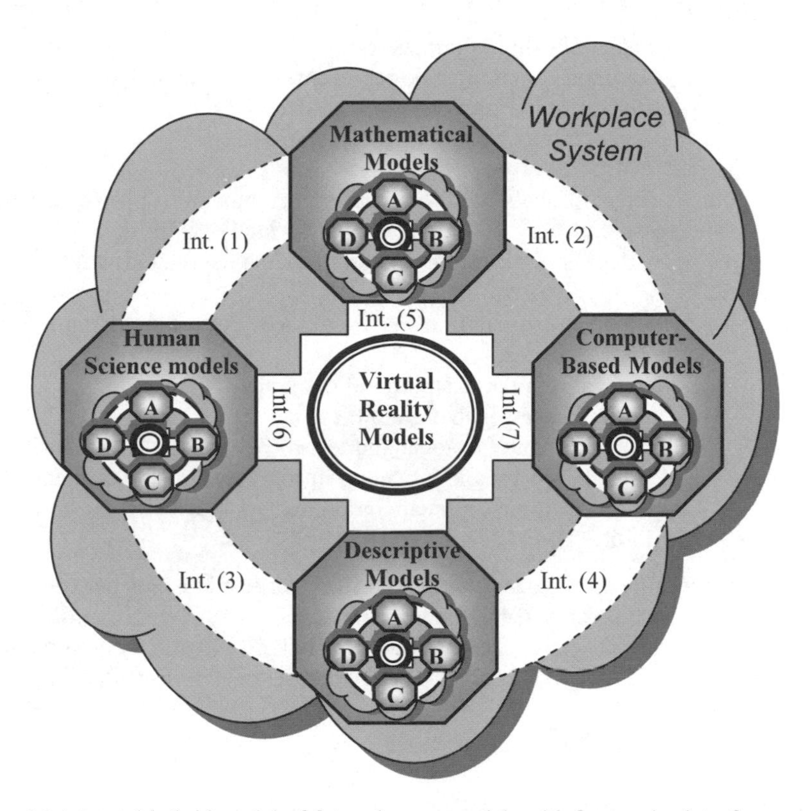

Figure 36.4 A total-hybrid model of future human models with fuzzy-crisp interfaces

They will require large libraries of knowledge and information, effective systems for interrelation between them and various external inputs and the adaptation of most appropriate model(s) and techniques. They also require interlinked artificial systems that will have elements of creativity that will think and learn.

This system integration will be divided into three phases [20]:

- decomposition of all systems to be integrated,
- singling out of the common elements of integration, and
- integration of the decomposed systems.

The common elements of the mentioned models are:

- the same interest subjects,
- similar resource requirements,
- system complementary and data interconnection,
- the same measurement, analysis and improvement concepts, and
- the same organization and environment space.

The degree of model integration is dependant on the human engineering approaches and for this the designer should address a few basic aspects before defining IHMs:

- the object of using of human model;
- the parameters that must be represented in the model;
- interrelationships between product, user(s), and environment;
- the design in the product usage scenario [21]

Future human models will be applied in human sophisticated workplace; therefore these models will change appropriately with the future workplace progresses. Discussion of future human models is related with the following factors:

- technological revolutions that reduce human intervention modes in industrial systems,
- future legal, standards and moral demands,
- scientific progresses in ergonomics,
- human work nature (logical and technical supports, work conditions),
- software, hardware and procedural progresses in human modeling, and
- appearance of new human modeling approaches (for more integration, flexibility).

In these models the data interfaces will make relationships between mathematical, human sciences and computer based subsystems, and system created data, which can be shared freely. They will include a hierarchy of software agents to facilitate data integration and co-ordination in a network-centric multi-sensor environment.

36.4 Conclusion

Future human models must support industrial design processes by applying future technology and approaches. Human model's approaches will be increasingly widely integrated in virtual engineering systems. This is because virtual reality has flexible tools to combine comprehensible presentation with the technological capability of data importing/exporting. These models will use static and mobile agents to collect data from dispersed, heterogeneous data sources, process and fuse the data, and present the resultant information to the user in a virtual environment.

36.5 References

[1] Gulliksen, J., *et al.*, 1998, "User Centered Design – Problems and Possibilities," http://www.acm.org/sigchi/bulletin/1999.2/gulliksen.pdf

[2] Licht, D.M., and Polzella, D.J., "Human factors, ergonomics, and human factors engineering: an analysis of Definitions," *Harry G. Armstrong Aerospace Medical Research Laboratory*, U.S.A. http://iac.dtic.mil/hsiac/docs/Human_Factors_Definitions.pdf

[3] SafeWork, International Labour Office (ILO), 2001, "Guidelines on occupational safety and health management systems (ILO-OSH 2001)," Geneva Switzerland.

[4] School of Human Kinetics - Laurentian University, 2003, "Introduction to Economics" http://209.91.162.35/FeedStream/Content/Lecture%201-introduction%20to%20ergonomics.pdf

[5] Morino, E., *et al.*, 1998, "Working postures analysis using virtual reality - a tool to support ergonomic analysis of the work place," *Proc. of the 5th Pan-Pacific Conference on Occupational Ergonomics,* Federal University of Santa Catarina.

[6] Kroemer, K.H.E., 2002, "Ergonomics: Definition of Ergonomics," *National Safety Council*, http://www.nsc.org/issues/ergo/define.htm

[7] Keates, S. *et al.*, "Combining utility, usability and accessibility methods for Universal Access: the Information Point case study," *University of Cambridge*, UK, http://rehab-www.eng.cam.ac.uk/papers/lsk12/uahci01/ip/

[8] Campbell, G.E., and Cannon-Bowers, J.A., 2002, "Human Performance Model Support for a Human-Centric Design Process," *Naval Air Warfare Centre Training Systems Division- SC-21/ONR S&T Manning Affordability Initiative*

[9] Zachary, W. *et al.*, 2001, "The application of human modelling technology to the design, evaluation and operation of complex systems," *Advances in Human Performance and Cognitive Engineering Research*, Vol. 1, pp. 199–247.

[10] Kanade, T., 2003, "Message from Director," *Digital Human Research Center*, Japan http://www.dh.aist.go.jp/message-e.htm

[11] Campbell, G.E., and Cannon-Bowers, J.A., "The Application of Human Performance Models in the Design and Operation of Complex Systems," *Naval Air Warfare Center Training Systems Division,* Orlando, Florida, U.S.A.

[12] Maxant, O., and Piat, G., "Thinking, elaborating and evaluating future offers in user- oriented design process : to produce perceptive innovations from conceptual ideas," *CREATEAM® Research & Development EDF (Electricité de France)*, France http://www.grenoble-soc.com/proceedings03/Pdf/26-MAXANT.pdf

[13] George, P.E.G.R., and Cardullo, F., "Application of Neuro-Fuzzy Systems to Behavioral Representation in Computer Generated Forces," U.S.A. http://www.link.com/pdfs/neuro-fuzzy.pdf

[14] Häggqvist, M., and Lundqvist, A., "Mental grasp the development of a multi-sensory Mental Rotation Test platform," *European Union Structural Funds* - Umeå University http://www.cs.umu.se/education/examina/Rapporter/441.pdf

[15] Szpytko, J., *et al.*, 2002, "Operator – specialist crane system model," *International Carpathian Control Conference (ICCC)*, Malenovice, Republic CZECH.

[16] Deliang, Chen, 2003, "Systems Approach and Models," Göteborg University. http://www.gvc.gu.se/ngeo/deliang/lec-2.pdf

[17] Office of Transportation Technologies (OTT), 2003, "Thermal Comfort Manikin." http://www.ott.doe.gov/coolcar/manikin.html

[18] DoD Training With Simulations Handbook Strategypage http://www.strategypage.com/prowg/simulationshandbook/chp_1.doc

[19] SC21/ONR S&T Manning Affordability, 1998, "Human Performance Models & Modeling Tools Enabling Technology," http://www.manningaffordability.com/S&tweb/t1hpm/sld001.htm

[20] Savic, S., 2001, "Integration of management systems in terms of optimisation of workplace human performance," Faculty of Occupational Safety, University of Nis, Series: *Working and Living Environmental Protection,* Vol. 2, N° 1. pp. 27-38

[21] Laurenceau, T., 2001, "Benchmarking Digital Human Modeling Tools for Industrial Designers," *IDSA Design Education* Conference Papers, Industrial Designers Society of America, U.S.A.

Part VIII

Design Intent and Tolerancing

37

On the Merging of Geometric Models Based on Hierarchical Context

J.A. Knowlton, and Michael J. Wozny

Abstract: Recently, efforts have been made to define the role of "context" in the Product Realization Process. This paper treats only a small slice of this problem and applies the notion of context to the automatic merging of geometric models created with the computer graphics language OpenGL. Context can be thought of as a set of properties or environmental variables of some entity that constrains or governs the behavior of that entity. The entity in this case is a three-dimensional geometric model and its context is the set of properties for viewing that model: lighting, viewing parameters, material reflective properties, colour. Now suppose a geometric model is made up of a collection of sub-models, each within its own context. If each context is associated with an integer, then the contexts can be ordered hierarchically. Thus the topmost context in the hierarchy becomes the global context for all the sub-models in the collection. Stated in another way, once a contextual hierarchy is defined, then the structure for combining these sub-models is established independently of when and in what individual contexts the sub-models are created. Consequently, context allows a concurrent generation of models within a formalized structure that automatically deals with conflict resolution – albeit in a limited way in this work. This paper describes a compiler in XML that will merge the OpenGL files automatically.

Keywords: Geometric Modeling, Visualization, Context

37.1 Introduction

The present work is motivated in part by the presentation of John Mills at the 2001 CIRP Design Seminar in Stockholm, which dealt with "The Role of Context in the Product Realization Process" [1]. His paper proposed a definition of "context" applied to product realization, the goal being a "... product representation which changes, eventually automatically, as the context within which a worker performs their task changes." To effectively manage knowledge within the product realization process, one must understand what the context is and how it influences the generation, capture and use of knowledge.

Context

The notion of context is not new. The artificial intelligence community has intensively studied it for decades in sub-areas such as natural language processing, knowledge representation and reasoning, and intelligent information retrieval, among others. See Akman [2] for a review of the literature. Much progress has been made in connection with natural language. However, McCarthy [3] believes, "... the main AI uses of formalized context will not be in connection with communication, but in connection with reasoning about the effects of actions directed to achieving goals." Product realization certainly falls within this ambitious scope.

There is a growing literature on abstract formal theories of context and the use of quantificational logic that enables the representation of relations between contexts, operation on contexts, and lifting rules of facts in different contexts; for example, the *ist* (p, c) predicate that asserts proposition p is true in the context c [4, 5]. There is also ongoing work to use context logic to facilitate the integration of information sources and perhaps extend the knowledge interchange format, KIF [6].

Such progress portends an area rich enough to ultimately impact product modelling, as the Mills, *et al.*, paper suggests. Replacing *ad hoc* structures with those based on a more rigorous logical framework will allow the product realization process to be more effective in meeting competitiveness, flexibility and time-to-market goals.

Scope of this Work

The present work deals with only one small portion of this topic: to apply the notion of "context" to the problem of merging three-dimensional graphic models.

Context can be thought of as a set of properties or environmental variables of some entity that may constrain or govern the behaviour of that entity. The entity in this case is a three-dimensional geometric model, comprised of sub-models. Its environmental variables are lighting, material properties, colour, *etc.*. In this problem, contexts and entities are assumed to be nested, so that the context of a larger entity will influence the contexts of those inside it. Consequently, context ordering can be handled in a simplistic way, namely, association with integers.

Context provides a useful means for organizing the implementation of a geometric model comprised of multiple sub-models; a hierarchy of context levels can be employed to determine how each sub-model relates to the whole.

The software implemented in this project, embodied in the Context Manager, takes as input separate sub-models created as OpenGL files and merges the files automatically, according to their associated contextual level. Context is the basis for determining how the separate pieces are linked together. The graphical properties of those components with a higher-level context constrain or influence those components with lower-level contexts. Consequently, the sub-models are ranked according to the hierarchy of context levels, the lower level components exerting less overall influence than the higher level components. Results demonstrating this capability are given in Section 37.4.

OpenGL is a portable (actually a defacto standard) application programmers' interface (API) that allows, in part, programmers to specify the objects and operations needed to produce colour images of three-dimensional objects. It has a state architecture so that the created geometry is rendered with whatever colour, orientation, texture, *etc..*, have been previously defined as the state. OpenGL is high level and has extensive libraries for the creation of three-dimensional geometries. There is also available a comprehensive library of geometric constructions (Open Geometry GL) based on OpenGL routines [7]. OpenGL is a registered trademark of Silicon Graphics, Inc.

37.2 Context Manager

The Context Manager is a stand-alone Java application that performs three main operations: Compile, Link and Reverse Compile. Its user interface, shown in Figure 37.1, denotes these functional areas.

37.3 Software Organization

The functional areas of the application in Figure 37.1, the graphical user interface (GUI) and the utility Matcher.java are all implemented as separate java classes.

ContextManager.java

This file implements the ContextManager class, which handles all tasks related to the user-interface, including painting the GUI on the screen and handling all events triggered by the user.

When the user chooses a file for compilation from OpenGL to XML, several characters are appended to the filename to encode the context level and change the extension. The new filename is used to create the appropriate FileInputStream and FileOutputStream, which are in turn wrapped in a DataInputStream and a DataOutputStream, respectively. The latter are required arguments to the XMLCompiler class.
Since the Linker accepts an arbitrary number of files a Java Vector (dynamically sized array) is used to hold the filenames. As the filenames are chosen, they are added to the Vector in order of context level.

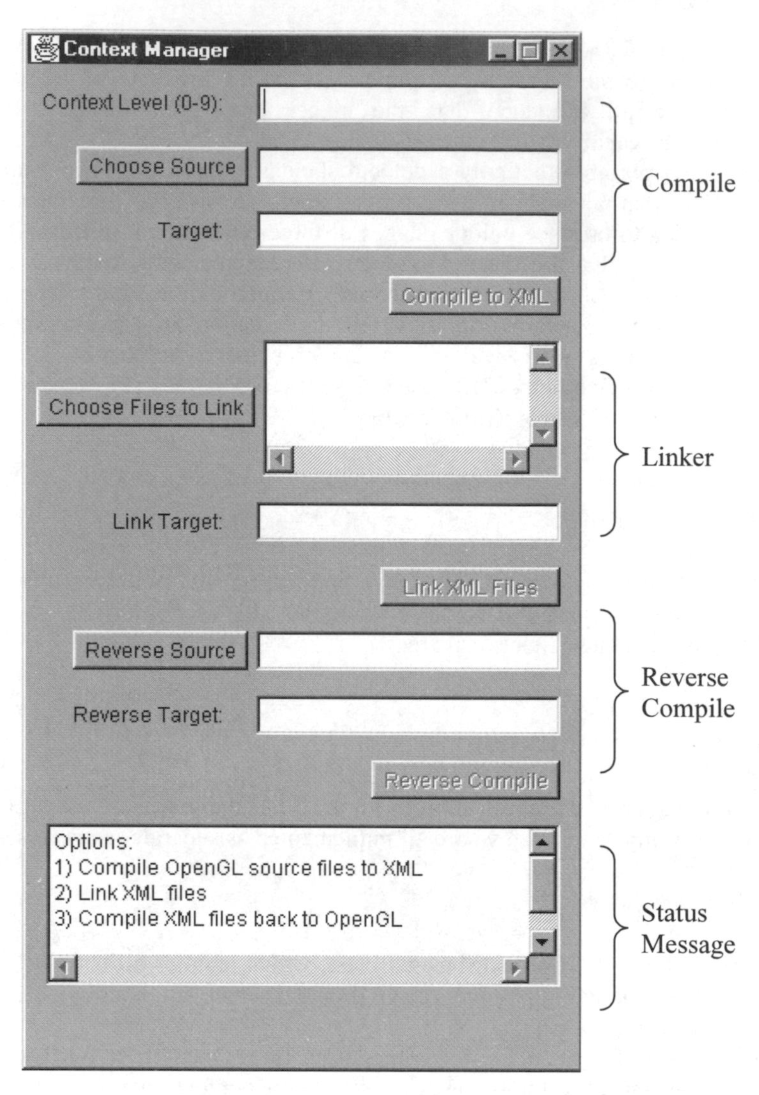

Figure 37.1 Context manager interface and functional areas

XMLCompiler.java

The XMLCompiler class takes three arguments: the calling ContextManager instance, a DataInputStream connected to the OpenGL file to be compiled, and a DataOutputStream connected to the XML file to be created.

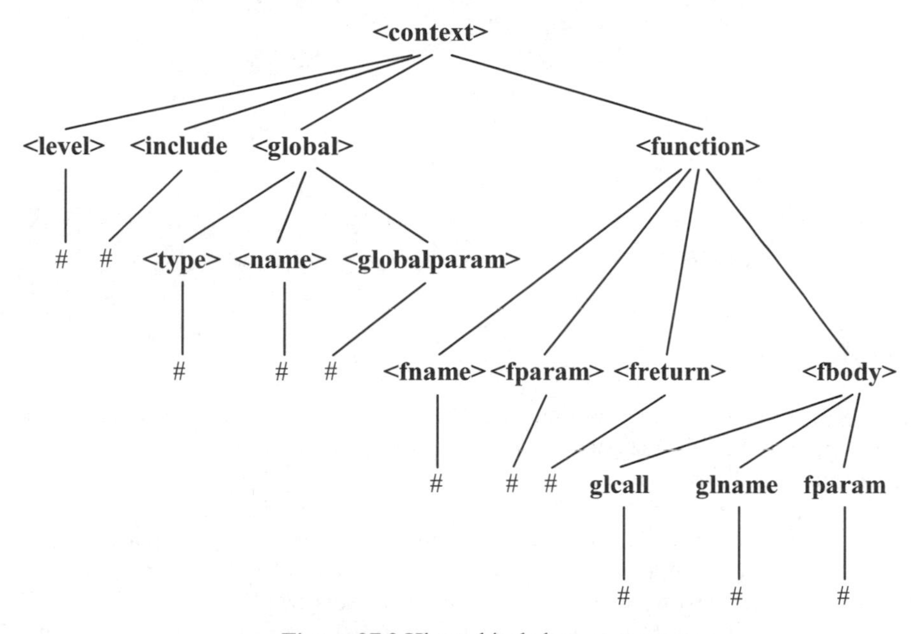

Figure 37.2 Hierarchical element structure

The compilation itself uses a top-down parsing scheme. At the root level, the XML processing instructions, as well as start- and end- tags, are written to the target file. After the start-tag is written, the input file is parsed, starting with C++ include statements and global variables. Whenever a function definition is encountered, a method is invoked specifically to handle the function. Similarly, function calls and arguments are handled by separate methods.

The target XML file is constructed according to the element structure shown in Figure 37.2.

The <context> tag, at the root level of the XML document, encapsulates the entire contents of the OpenGL file. The pound sign (#) indicates a child leaf that contains PCDATA (parsable character data), such as text or numerical data. Underneath <context> are: the <level> tag, containing the integer representing a user-provided context level; the <include> tag(s), containing the C++ include statements; the <global> tag(s), containing the globally declared variables; and the <function> tag(s), representing the top level of a C++ function.

A function, in turn, contains a function name <fname>, parameters <fparam>, a return type <return>, and a function body <fbody>. Within the function body are the individual function calls <glcall>, which have function names <glname> and parameters <fparam>. Typical C++/OpenGL constructs are shown in Figure 37.3 and the corresponding XML elements in Figure 37.4.

```
#include <GL/glut.h>
...
Glfloat low_brick_mat_diffuse0[]={0.8, 0.4 0.4 1.0};
...
int main(int argc, char** argv)
{
        ...
    init();
    glutDisplayFunc(display);
        ...
        return 0
}
```

Figure 37.3 Typical constructs in C++/OpenGL code

XMLLinker.java

The XMLLinker accepts multiple XML files, which have previously been generated by the XMLCompiler, and outputs a single file representing a re-organization of all the input data.

The constructor of the XMLLinker takes three arguments: the calling ContextManager class, a Java Vector containing the names of all input files, and the filename of the file to be output. The Vector of input files built by the ContextManager is sorted in ascending order by context level.

This makes it easy for the Linker to loop through the Vector and parse, using the Xerces1.4.4 XML parser, each input XML document in order of context.

The first two files in this Vector, *viz.* file (0) and file (1), respectively, are parsed simultaneously. File (0) represents context level 0 and is designated the "rootDoc". This designation remains constant throughout the linking process. Every other file to be linked is referred to by the variable name "subDoc". The XML elements contained in file (0), the rootDoc, are compared systematically with the XML elements found in file (1), the first subDoc.

For example, both files will have several <function> elements, containing the data needed to reconstruct C++ functions. Each of the files' <function> elements are parsed and examined for differences. This comparison logic is implemented through the use of the Matcher class (Section 37.5). The Matcher object serves as a container with get and set methods, allowing a particular XML element to be flagged for later processing. A Matcher object is created for each <function> element, and all of the Matcher objects are stored in Vectors.

The Matcher object is only required to represent the elements contained in a subDoc. The rootDoc is parsed separately for each individual subDoc. Comparisons between the elements of two different files are performed by iterating through the Matcher Vector and setting a Boolean flag (isMatch) to true or false for each element. If the two files are found to contain an element with identical text or data, then the value of the corresponding Matcher element is set to true; otherwise, it is set to false.

```
  <?xml version="1.0" ?>
- <context>
    <level>2</level>
    <include>"GL/glut.h"</include>
    <include>"stdlib.h"</include>
  - <global>
      <type>"GLfloat"</type>
      <globalname>"low_brick_mat_diffuse0[]"</globalname>
      <globalparam>"0.8"</globalparam>
      <globalparam>"0.4"</globalparam>
      <globalparam>"0.40"</globalparam>
      <globalparam>"1.0"</globalparam>
    </global>
     ...
  - <function>
      <fname>"main"</fname>
      <fparam>"int argc"</fparam>
      <fparam>"char** argv"</fparam>
      <freturn>"int"</freturn>
    - <fbody>
       ...
      - <fcall>
          <fcallname>"init"</fcallname>
        </fcall>
      - <glcall>
          <glname>"glutDisplayFunc"</glname>
          <fparam>"display"</fparam>
        </glcall>
         ...
      - <fcall>
          <fcallname>"return"</fcallname>
          <fparam>"0"</fparam>
        </fcall>
      </fbody>
    </function>
  </context>
```

Figure 37.4 Corresponding XML elements

Once parsing is complete for a particular set of document elements, for example, the <function> element, the Matcher Vector will contain a complete list of these elements with flags to denote whether or not they have been matched. Beginning with the <function> element: if a match has been found, *i.e.*, an element of the Matcher Vector contains a Boolean value of true for is Match () then the pair of matching <function> elements is passed on to another method, to be compared at a finer granularity. If a mismatch has been discovered, then the mismatched element is saved in a Vector to be added to the rootDoc at a later point.

The Linker continues in this manner, having started at the top of the document hierarchy, comparing elements for matches and, when matching elements are found, passing them on for further processing. All mismatched elements are stored in a Vector and added to the rootDoc at the end of the linking operation.

Each input file is parsed individually against the rootDoc. Once all files have been parsed and compared for matches/mismatches, the rootDoc will contain a copy of every unique element found in all files.

Certain subDoc elements are automatically marked for addition to the rootDoc. Among these are <glcall> elements containing calls to PushMatrix () and PopMatrix (), glut function calls, and other function calls such as glTranslate (), glScale () and glRotate (). Similarly, certain subDoc elements are automatically excluded from the rootDoc, such as lighting and ClearColor () calls.

ReverseCompiler.java

The Reverse Compiler is concerned mainly with parsing a single XML document and re-formatting the data so that it can be compiled and run as a C++ file. The input file is expected to have been generated by the XMLLinker, and to be a valid and well-formed XML document adhering to the hierarchical structure diagrammed in Figure 37.2. The input document is simply traversed, using the Xerces1.4.4 XML parser, in a recursive fashion, its elements processed and re-formatted as C++/OpenGL code.

Matcher.java

The Matcher class is a utility containing: (1) a single XML Node object from some XML subDoc, which may or may not already be contained in the rootDoc; (2) character data (PCDATA) from that Node's child element, representing the actual data inside the XML document; (3) a Boolean flag, which is set to true only if the Node in question has an exact duplicate in the rootDoc.

Matcher objects are created for all instances of a particular XML element, for example, the <function> element. These are all placed in a single Vector, so that comparisons with another document can be made by simply iterating through the Vector. Generally, all of the Matcher objects that have been flagged as having a match in the rootDoc are excluded from further processing. The Matcher objects that have no corresponding match in the rootDoc are, in general, added to the rootDoc at the end of the Linking operation.

37.4 Results and Discussion

The screen shots in Figure 37.5(a), (b), (c) depict examples of separately rendered OpenGL files that are to be processed by the Context Manager.

In Figure 37.5(a), OpenGL file, ground.cpp, is given context level 0. In addition to drawing a geometric plane representing ground in the image, this file also establishes the background colour as well as the lighting and view parameters for all subsequent merged files.

In Figures 37.5(b) and 37.5(c), building1.cpp and building2.cpp, respectively, represent the contents of separate files, each containing a model of a building. These two files are arbitrarily assigned context levels greater than 0. These files are then automatically merged into the single environment shown in Figure 37.5(d).

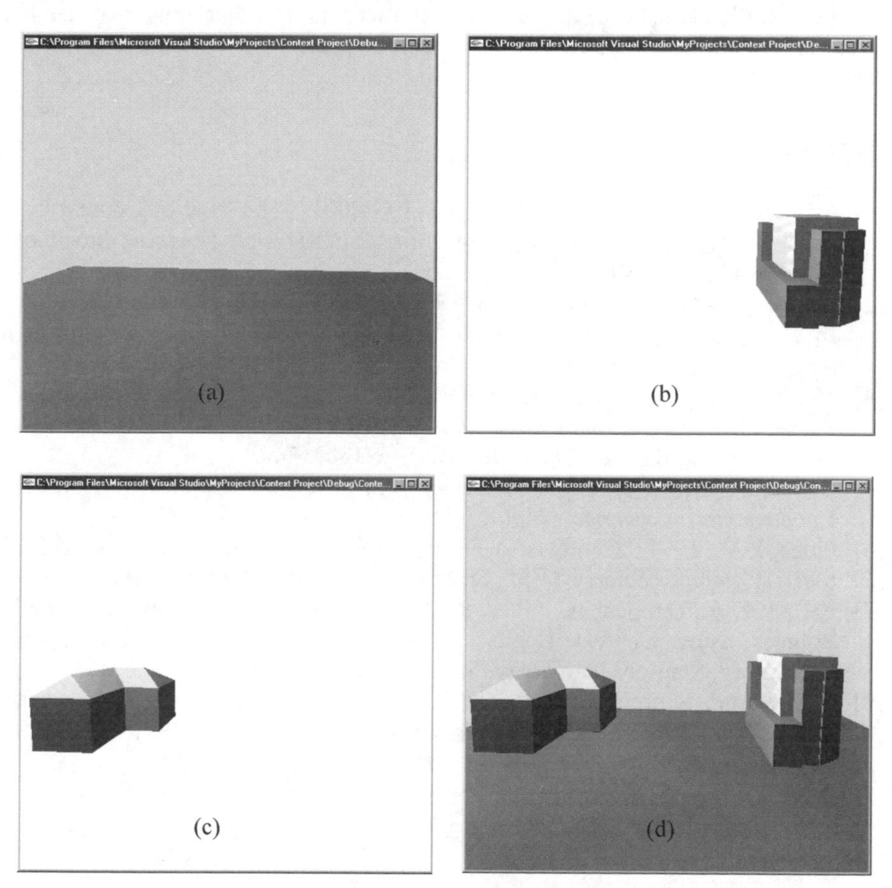

Figure 37.5 Screen shots of example

This paper demonstrated a means for incorporating context into the realm of geometric modelling and computer graphics. The Context Manager allowed multiple geometric models, represented as OpenGL files, to be automatically merged into a single composite model (image) according to pre-established rules. In this case, context level 0 established the global lighting, viewing, buffering and Clear Colour (background) settings.

Future work includes expanding the current subset of the OpenGL API functionality to include: NURBS surfaces, texture mapping, menu-driven user interfacing, as well as support for C++ loop constructs (for-loops, while-loops) and declaration of local variables inside functions. Additional work is also needed to support the nesting of more than two distinct context levels. For more complicated models, it is desirable to have finer control over the individual context levels, allowing greater distinction among them. Also a richer context structure could be investigated.

Looking at the bigger picture, geometric modelling allows for a hierarchical structure of contexts. But this may not be the case for other elements of the Product Realization Process. In reality there is no one universal context. Eventually some assumption breaks down somewhere!

37.5 References

[1] Mills, J.J, Goossenaerts, J.B.M., Pels, H.J., 2001, "The Role of Context in the Product Realization Process," *Proc. Intl. CIRP Design Seminar*, Stockholm, Sweden, pp. 175-180.

[2] Akman, V., 2002, "Context in Artificial Intelligence: A Fleeting Overview," In: *La Svolta Contestuale*, C. Penco, ed., McGraw-Hill, Milano

[3] (see http://www.cs.bilkent.edu.tr/~akman/papers.html).

[4] McCarthy, J., 1989, "Artificial Intelligence, Logic, and Formalizing Common Sense," In: *Philosophical Logic and Artificial Intelligence*, R.H. Thomason, ed., Dordrecht, Kluwer, The Netherlands, p.180.

[5] Buvac, S., Buvac, V., Mason, I.A., 1995, "Metamathematics of Contexts," *Fundamenta Informaticae*, Vol. 23(3).

[6] Guha, R.V., 1991, "Contexts: A Formalization and some Applications," *Ph.D. thesis, Computer Science Dept., Stanford University*, Stanford, CA.

[7] Farquhar, A., Dappert, A., Fikes, R., Pratt, W., 1995, "Integrating Information Sources using Context Logic," *Report KSL-95-12*, Knowledge Systems Laboratory, Stanford University, Stanford, CA.

[8] Glaeser, G., Schrocker, H.-P., 2002, *Handbook of Geometric Programming using Open Geometry GL*, Springer-Verlag, New York.

38

Haptic Virtual Prototyping for Design and Assessment of Gear-shifts

Martijn Tideman, M.C. van der Voort, and Fred J.A.M. van Houten

Abstract: Traditionally, a designer forms the link between the customer and the final product by interpreting customer demands and desires and translating them into geometry. By combining 3D CAD systems and software tools for analysis, a designer is able to examine whether the created geometry complies with these customer demands and desires. However, in the process of translation and examination, a measure of subjectivity is added to the design. A virtual prototyping environment (VPE) can be created by utilizing Virtual Reality technology, in which the customer is able to specify the product's behavior in a direct way, *i.e.* without designer interference. In this way, not only is the design process is made more objective, but also significant amounts of time and money are saved since less physical prototypes are required. This paper describes the design and evaluation of a VPE for manually operated gearboxes in passenger cars. Based on measurements taken of the gearlever on a test vehicle, an application is designed that simulates its gearshift feel. This application incorporates a commercially available haptic device. In order to determine whether the virtual gearshift feel conforms with the real gearshift feel, a usability test is performed. The test group considered the feel of the simulated "virtual" gearshift to be quite similar to the "real" gearshift feel of a test vehicle. By further developing this VPE, it should become possible to define gearshift feel by customer assessment through haptic simulation, after which the physical gearbox is designed in such a way that it matches the preferred shifting behavior.

Keywords: Behavior Based Design, Virtual Reality, Haptic Interfaces, Gear-shift Feel

38.1 Introduction

In present-day product design practice, many software tools are available to support the designer in his/her creative process. First of all, geometry can be created and evaluated by using a solid based 3D CAD system. Next, various kinds of analysis tools are available for examining whether the created geometry complies with the designer's demands and desires. As these desires usually relate to the behaviour of the forthcoming product, so do the analysis tools. Tools that exploit dynamic and kinematic simulation, the Finite Element Method, and Computational Fluid Dynamics are mature and affordable these days. Moreover, the interfacing —on the one side with the CAD system and on the other side with the designer— has very much improved. In this way integrated design environments are born, which have positive effects on the overall performance of the final product, and development time and costs are reduced.

Within the development of integrated design environments, there are still some gaps that need to be filled. One of them is the direct integration of customer wishes into the design. Traditionally, a designer forms the link between the customer and the final product by interpreting customer demands and wishes and translating them into geometry. However, the designer has his/her own unique way of interpreting and translating. Consequently, the design will inevitably contain a degree of subjectivity with regard to its geometry and its behaviour. Ideally, this process should be made more objective. That is, the customer should be able to specify the product behaviour in a direct way, *i.e.* without designer interference.

This gap can be filled by the application of Virtual Reality (VR) technology.) A virtual environment can be created by utilizing Virtual Reality interfaces (*e.g.* haptic devices, stereoscopic displays, 3D sound devices. VR interfaces record the user's actions and —in turn— stimulate his/her senses. In this way, the user experiences the illusion of having some kind of interaction. A virtual environment especially suited for product design applications is called a Virtual Prototyping Environment (VPE). It enables the evaluation of specific characteristics of a candidate design without having a physical prototype. When coupled with a CAD system, a VPE enables an inversion of the design process. Instead of creating geometry and evaluating the behaviour afterwards, the product behaviour can be defined first, and then the corresponding geometry is created [1]. This so-called behaviour based design provides a tool for the direct integration of customer wishes into the design, so they will be able to adjust the behaviour of a future product to exactly meet their requirements. In this way, the design process is no longer the exclusive domain of the designer and as a result, his/her subjective way of interpretation and translation of customer wishes is prevented. Moreover, since less physical prototypes are required, significant amounts of time and money are saved within the design process.

Research is performed on the creation of environments for virtual prototyping applications at the Laboratory of Design, Production and Management of the University of Twente. One of the projects is the development of a VPE for manual transmissions in passenger cars. This paper describes the design and evaluation of a Virtual Gear-shift Application: an application that provides the illusion of manual shifting gears in a passenger car. By further developing this application, it should become possible to define the desired gearshift feel, after which the physical gearbox can be designed to match the customer's requirements. In this way, less physical prototypes are required and gearboxes can be adjusted for specific groups of drivers (*e.g.* old/young, sporty/comfy).

38.2 Approach

To date, there are no findings on whether or not, and how, the gearshift feel of a manual transmission in a passenger car can be simulated. In order to obtain knowledge and gain experience of simulating gearshift feel, this research is limited to the development and evaluation of an application that simulates the feel of one single existing gearbox.

Therefore, the gearlever of a passenger car is equipped with a system that measures the gearlever's motion related to the forces induced on the operator's hand. Based on measurements performed with this system, an application is designed that simulates this particular gearshift feel, *i.e.* this gearbox-specific relation between motion and forces. This application incorporates a commercially available haptic device. In order to determine whether the virtual gearshift feel conforms with the real gearshift feel, a usability test is performed. Within this test, a group of ten participants are asked to compare the "gearboxes" by means of a questionnaire.

38.3 Haptic Interfaces

Using a haptic interface can simulate the gearshift feel of a manual transmission. A haptic interface is a device configured to provide haptic information for a human. Just as a video interface allows the user to see a computer-generated scene, a haptic interface permits the user to "feel" it [2].

Haptic interfaces have two basic functions [3]. The first is to measure forces, positions and their time-derivatives at the operator's hand (or other body locations). The second is to display forces and positions for the operator under the control of the computer running the VR simulation.

Two fundamental methods for controlling haptic interfaces exist [4]. When a position is input to the control loop and forces are fed back to the operator, we speak of impedance control. Alternatively, the simulation can use admittance control, in which forces applied to the end effector are sensed and positions are fed back through the haptic device.

Impedance control and admittance control are dual, not only in their cause-and–effect structure, but also in their performance. The impedance-controlled device is typically lightweight, backlash free, and it renders low mass [5]. Consequently, performance is lacking in the region of higher forces, high mass, and high stiffness. Adding complex end effectors magnifies the problem. Admittance controlled interfaces, on the other hand, are capable of rendering very high stiffness and minimal friction. However, in order to prevent the device from becoming unstable, a small mass needs to be rendered constantly. Admittance controlled devices are very suitable for larger workspaces and for carrying complex end effectors with many degrees of freedom. Moreover, because forces are sensed rather than computed in real time, admittance control has the advantage of reduced modelling computation.

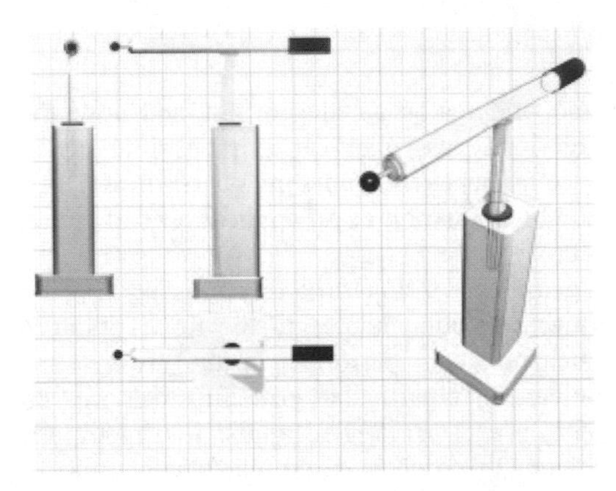

Figure 38.1 The haptic interface selected for the Virtual Gear-shift Application

The haptic interface selected to develop the Virtual Gear-shift Application is a commercially available admittance controlled device [6] (Figure 38.1). The reason for this choice is that in order to simulate the gearshift feel of a manually operated transmission, it is expected that high force and high stiffness need to be rendered, of which an impedance device is not capable. Moreover, during gear shifting, the operator performs quick movements, which result in great accelerations of the end effectors, for which an admittance-controlled interface offers a more accurate and more stable simulation. The end effector is chosen to be the upper part of a gearlever.

38.4 Gear-shift Feel

Operation of a manual transmission in a passenger car takes place by means of interaction between the operator's hand and the user-interface of the transmission, *i.e.* the gear knob. Moving the gear knob is done by muscle power, applied through the skeletal system and the contact area between the hand and the gear knob. Depending on the resistance force induced by the complete transmission system and applied on the hand through the contact area, the totality of gear knob and hand executes a certain movement. Therefore, the interaction is nothing but a relation between motion and forces.

Although the interaction between operator and gear knob and the so-called "gear-shift feel" are closely connected, there is a subtle difference. As described above, the interaction can be expressed by an objective set of discrete motion and force data. However, when the operator experiences this interaction, a subjective layer is added. The operator will interpret the relation between motion and forces, after which he judges the gearbox to be stiff, loose, smooth, silky, clunky, or rubbery, *etc.*. The actual gearshift feel consists of both the objective and subjective layer.

This leads us to an important issue in designing and evaluating virtual environments. During the design phase, there is an attempt to duplicate precisely, the interaction between the user and the real world. However, the quality of a virtual environment is not determined by the degree of perfection to which this duplication has been achieved. It is determined by the degree to which the user interprets the copied interaction correctly, *i.e.* the degree to which the user gets the same experience, as he would have in the real world. Therefore, in order to evaluate a virtual environment, the real and virtual experiences should be measured and compared rather than the real and virtual interactions.

In connection with this issue, the design of the Virtual Gear-shift Application will be based on the measured *interaction*, *i.e.* the relation between motion and forces, whereas the evaluation will be done by measuring and comparing the user's *experience*. An adequate way to measure experiences is by exploiting a questionnaire.

38.5 Design of the Virtual Gear-shift Application

The Measurement System

For simulation of the interaction between operator and manual transmission, the relation between forces practiced on the gear knob and its resulting movement has to be determined. The relation between forces and movements in a mechanical system can be represented by:

$$\boxed{\sum F_x(x,t) = m_x(x,t)\cdot\ddot{x}(t) + d_x(x,t)\cdot\dot{x}(t) + k_x(x,t)\cdot x(t)} \tag{38.1}$$

Table 38.1 Explanation of the symbols used in Equation (38.1)

Symbol	Description	Unit
$\sum F_x(x,t)$	The sum of all forces in x-direction	N
$m_x(x,t)$	Mass in x-direction	kg
$d_x(x,t)$	Viscous damping coefficient in x-direction	Ns/m
$k_x(x,t)$	Spring constant in x-direction	N/m
$\ddot{x}(t)$	Acceleration in x-direction	m/s^2
$\dot{x}(t)$	Velocity in x-direction	m/s
$x(t)$	Position in x-direction	m

The functioning of the haptic interface used for the Virtual Gear-shift Application is also based on Equation (38.1):

1. $\Sigma F_x(x,t)$ is measured by the force sensor;
2. The values of $m_x(x,t)$, $d_x(x,t)$ and $k_x(x,t)$ are specified by the virtual model;
3. The values for $\ddot{x}(t)$, $\dot{x}(t)$, and $x(t)$ are calculated;
4. The calculated values are translated into a DC current that drives an electrical motor and thereby moves the robot arm in a certain direction.

There are three design parameters at one's disposal: mass (inertia), viscous damping, and stiffness (compliance). The measurement system in the test vehicle is designed in such a way that these three parameters can be determined for every position (x) at any point of time (t). It therefore consists of a force sensor and a motion tracker mounted onto the test vehicle's gearlever (Figure 38.2). The system logs the gearlever's spatial orientation as well as the forces induced on the operator's hand. By taking the first and second time derivative of the gear knob's orientation, motion data (*i.e.* velocity and acceleration) can be extracted.

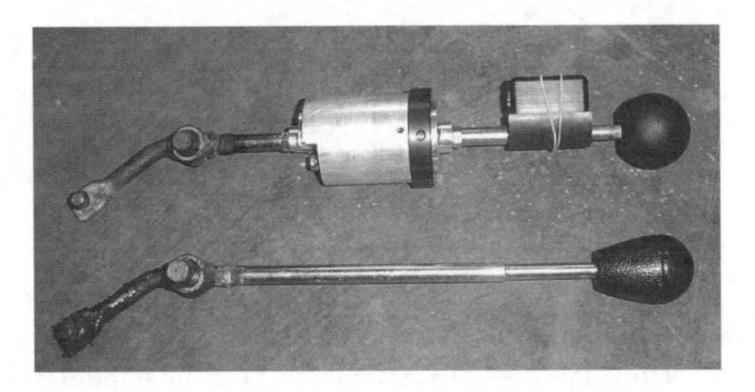

Figure 38.2 The gearlever equipped with the measurement system compared to the original gearlever

Measuring the Interaction

Measurements are performed on the test vehicle's gearlever using the measurement system. Due to practical limitations, these measurements take place under "showroom-shifting" circumstances, *i.e.* while the vehicle is at rest, the engine is not running, and the clutch is disengaged. A group of ten participants is asked to operate the gearbox according to a number of prescribed patterns. The main results of this test are:

- During gear shifting, the forces induced on the operator's hand do not show a significant variance for different operators;
- During gear shifting, the forces induced on the operator's hand do not show any variance for different operating speeds;
- During gear shifting, the forces induced on the operator's hand are mainly induced by the locking mechanism and the synchronization unit;

- As the locking mechanism of the gearbox consists of a spring-loaded ball that moves into a groove when a particular gear is engaged and out of it when a gear is disengaged, this effect may be modeled as compliance;
- As the synchronization unit of the gearbox consists of two conical faces that make frictional contact just before a gear is engaged, this effect may be modeled as Coulomb friction.

Design of the Virtual Environment

Programming the haptic device simulates the interaction between operator and gear knob. Therefore, out of the measurement results, a mechanical model of the gearbox is extracted. This model can be split into two sub models: one for movements used for *selecting* a particular gear ratio (horizontal direction) and one for movements used for *engaging* the selected gear ratio (vertical direction). The model only contains those mechanical parts that mainly cause the forces induced on the operator's hand. These are the locking mechanism and the synchronization unit of the gearbox, which are modelled as compliance and Coulomb friction respectively. Both sub models are shown in Figure 38.3.

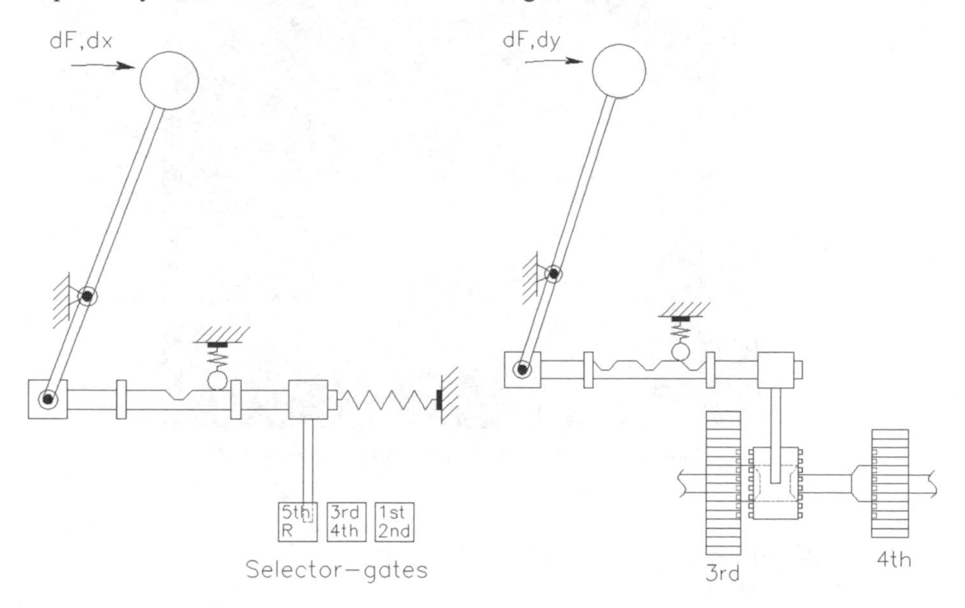

a. Movements in horizontal direction *b. Movements in vertical direction*

Figure 38.3 Mechanical model of the gearbox

Measuring the interactions at the end effector of the haptic device and comparing them to the interaction measured in the test vehicle results in several iterations in design. This way the mechanical model is fine-tuned so that the measured simulated interaction resembles the measured real interaction as closely as possible.

In order to provide a true illusion of gear shifting, not only the interaction between operator and gear knob should be simulated, but this interaction should also be embedded in a realistic context. That is why the haptic device is placed inside a mock-up of the test vehicle. This is done in such a way that the location of its end effector exactly coincides with the usual location of the gearshift knob.

In order to evaluate the gearshift feel, the influence of visual differences between the real gearlever and the haptic device on the evaluation of the virtual gearbox should be avoided. In normal driving situations, drivers do not look at the gearlever while changing gears. Messages are shown on a big screen in front of the mock-up to distract drivers' visual attention from the gearlever in order to force them to change gears based on feel.

The implementation of the Virtual Gear-shift Application is shown in Figure 38.4.

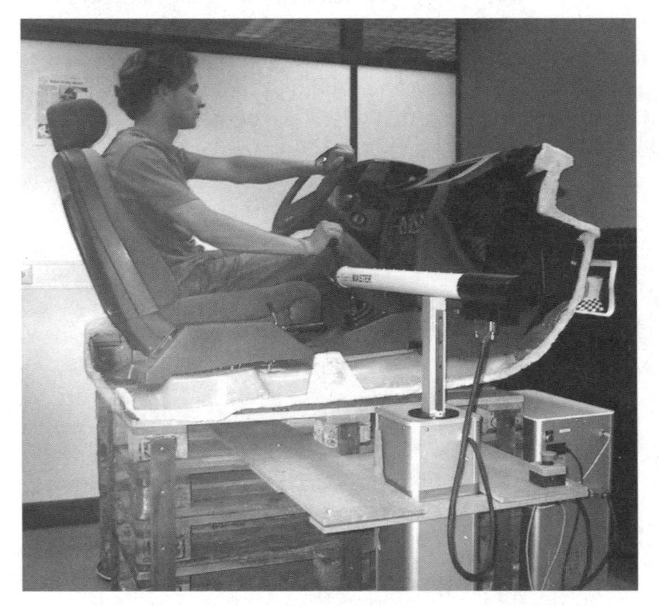

Figure 38.4 The Virtual Gear-shift Application

38.6 Evaluation of the Virtual Gear-shift Application

Evaluation Setup

The objective of the evaluation is to determine the quality of the designed Virtual Gear-shift Application. As the quality of a virtual environment is determined by the degree to which the user gets the same experience, as he would have had in the real world, the evaluation should consist of comparing the virtual gear shifting experience to the real experience. Ideally, this would be done by blindfolding a participant and placing him inside the driver's seat of either the test vehicle or the

Virtual Gear-shift Application. If participants were not able to discriminate between the gearshift feel of the real and the virtual gearboxes, the quality of the simulation would be considered optimal. Unfortunately, due to the fact that other clues (*e.g.* entrance to the mock-up and the test vehicle respectively) could play a part in the discrimination between both "gearboxes", this method can't be used. Therefore, the user experiences are measured and compared by exploiting a questionnaire. A completed questionnaire should give a personal opinion on the gearshift feel. By comparing a participant's opinion on the real and the virtual gearbox, the quality of the application can be determined.

The questionnaire is compromised of four types of questions:

1. Questions concerning the general feel of the gearbox;
2. Questions concerning the individual forward gears;
3. Questions concerning the reverse gear compared to the forward gears;
4. Questions concerning movements to the right and to the left of neutral position.

Evaluation Results

After evaluation of the questionnaires, it turned out that participants often gave a similar or equal judgment. Apart from some minor discrepancies that should be corrected in a future redesign, generally, the simulated gearshift feel was judged to be similar to the original gearshift feel. The most convincing indication that the Virtual Gear-shift Application simulates the gear-shift in a natural way is the observation that during the usability testing, participants were able to look upon the Virtual Gear-shift Application as if it were a real gearbox: they treated it the way a gearbox is naturally treated and they spoke about it as if it were real, *e.g.* "I feel more play than in my own car" or "Switching from 3rd to 4th gear goes smoother than from 1st to 2nd gear". It is therefore concluded that it is indeed possible to create an application that realistically simulates the gearshift feel of a manual transmission in a passenger car.

The two main discrepancies found between the real and the virtual gearbox are:

- Operation of the Virtual Gear-shift Application requires more effort than operation of the test vehicle's gearbox;
- The gearlever of the Virtual Gear-shift Application feels stiffer than the gearlever of the test vehicle.

These findings provide input for a future redesign of the Virtual Gear-shift Application. Further analysis of these outcomes revealed that:

- The effort required for engagement of a gear is judged on the basis of the total work (force times displacement) performed rather than on the basis of maximum required forces;
- While operating a transmission in a passenger car, actions far from the body are judged to require more effort than actions close to the body.

38.7 Conclusions

Apart from some minor discrepancies that should be corrected in a future redesign, generally, the simulated gearshift feel was judged to be similar to the original gearshift feel. Moreover, during the usability testing, participants were able to look upon the Virtual Gear-shift Application as if it were a real gearbox. So, the conclusion is that it is indeed possible to create an application that realistically simulates the gearshift feel of a manual transmission in a passenger car.

By further developing this application, it should become possible to define the desired gearshift feel, after which the physical gearbox is designed in such a way that it matches the customer's desires. In this way, the customer will be able to specify product behaviour in a direct and clear way. As a result, gearboxes can be designed or adjusted for groups of drivers (*e.g.* old/young, sporty/comfy). Moreover, since less physical prototypes are required, significant amounts of time and money are saved within the design process.

In order to do this, the Virtual Gear-shift Application needs to be linked to a CAD system. From the CAD model of a gearbox, a model needs to be extracted that can be loaded into the haptic device. This model should specify the mass, viscous damping and stiffness parameters for every position of the gear knob at any point of time. Future research will be concerned with this issue.

Besides the implementation of linking the Virtual Gear-shift Application to a CAD system, other components could be added to the designed Virtual Prototyping Environment as well. In the future, haptics for simulation of the clutch feel, the steering feel or the suspension feel could be implemented. It then grows from a VPE for gear shifting only to a VPE for a complete experience of driving a passenger car. In this way, all kinds of personal preferences with respect to driving behaviour could be identified and as the next step be physically realized.

38.8 References

[1] van Houten, F.J.A.M., 2001, "The Use and Development of Haptic Devices and Virtual Reality as Engineering Tools," *Proceedings of the 34th CIRP International Seminar on Manufacturing Systems*, pp. 275-283.

[2] Adams, R.J., 1999, *Stable Haptic Interaction with Virtual Environments*, University of Washington.

[3] Salisbury, J., & Srinivasan, M., 1996, "Virtual Environment Technology for Training (VETT), 1992," In: G. Burdea, *Force and Touch Feedback for Virtual Reality*, John Wiley & Sons, ISBN 0-471-02141-5.

[4] Burdea, G., 1996, *Force and Touch Feedback for Virtual Reality*, John Wiley & Sons, ISBN 0-471-02141-5.

[5] Adams, R.J., & Hannaford, B., 2002, "Control Law Design for Haptic Interfaces to Virtual Reality, 2001," In: R. Q. van der Linde, P. Lammertse, E. Frederiksen, & B. Ruiter, *The HapticMaster: a New High-Performance Haptic Interface, Proceedings of Eurohaptics*, pp. 1-5.

[6] van der Linde, R.Q., Lammertse, P., Frederiksen, E., & Ruiter, B., 2002, "The HapticMaster: a New High-Performance Haptic Interface," *Proceedings of Eurohaptics*, pp. 1-5.

39

Predicting Design Quality through Sensitivity Modeling

Luc Laperrière, Walid Ghie, and Alain Desrochers

Abstract: This paper presents a modeling approach that can be used as an engineering design tool to predict the effects of various design choices on product quality. The mathematical model provides a rigorous functional relationship between dependent and independent variables. The dependent variables quantify product quality in the physical domain in terms of design functional requirements the product must possess. The independent variables quantify the design choices in terms of nominal dimensions, degrees of freedom and tolerances. An example application is presented to illustrate how product quality can be achieved by appropriately tuning the design parameters in a constrained design context.

Keywords: design, tolerances, quality, sensitivity, modeling

39.1 Introduction

The design of a product starts with customer needs that are transformed into Functional Requirements (FRs). These, in turn, are transformed into the physical characteristics of product parts (material, shape, dimensions, tolerances, etc). Focusing on FRs, these are known to be represented at various levels of abstraction (FR1, FR1.1, FR1.1.1), known as function decomposition. Systematic approaches or methodologies for function design (*e.g.* where FRs are used) have therefore been proposed [1-2]. Other researchers suggest that the knowledge structure, which feeds the various abstraction levels in the functional design task, is also of prime importance. To date, functional design methodologies and the role that knowledge structure plays in their application is still the subject of active research [3].

In this paper, we are not as much interested in the pure functional domain as we are in the more detailed physical domain. The motivation is best illustrated by the following dilemma: the more detailed the design becomes, the more knowledge we have about it, but the more difficult and expensive it becomes to modify it. In other words, as the design task progresses from the functional domain to the physical domain, design changes become more critical.

Although the above dilemma clearly suggests the importance of making the right choices earlier in the design cycle, our industrial experience has shown that many important design changes still happen very late in the design phase for all sorts of reasons, for example, a part radius that was changed after a very expensive steel block, used to make the required plastic injection mold, had been CNC machined. It is believed that among the possible reasons for such late changes is the lack of appropriate design tools that can help predict how quality is influenced when all parts are brought together during assembly. Designers sometimes make blind choices on individual parts the consequences of which on quality appear only at assembly time, or even during product servicing. This raises the question of how the design task is performed in the physical domain, where FRs cannot be decomposed further. It seems that there are generally four approaches:

1. The FR is calculated using pre-defined techniques, for example, floating of fixed fastener principles [4];
2. The FR is calculated using computer-aided tolerancing tools implemented in commercial CAD software (AnaTol, Tasysworks, Tolsys, Visvsa, Cetol 6σ, *etc..*);
3. The FR is selected from experiments on different physical or virtual prototypes, each with different FR values;
4. The FR is determined by the designer from standard conditions or past experience.

The first approach above can only handle simple cases. The second approach mainly handles statistical tolerancing through Monte Carlo simulations. Recently, the third method has found a virtual counterpart which enables the designer to "feel" the effects of various FR choices using immersive virtual environments that recreate reality through perceptions with 3-D imagery involving sight, sound and touch [5]. These so called "haptic" technologies are rather new and their degree of acceptance and implementation remains to be confirmed. This leaves us with the last method of "past experience" which is still widely used. Consider that we now live in a world where:

1. Development cycles are shorter;
2. Emerging concerns like recycling and sustainable development constrain design choices;
3. Quality is the rule.

It is clear that iterative approaches and past experiences become even more limited. What is needed is a tool that captures the *sensitivity* of design choices that occur late in the design process, in order to *predict* their effects with respect to important concerns like the ones above.

In this paper we will therefore look at FRs that have been decomposed all the way down to the physical domain. We will not be concerned about the function decomposition path, which led to a particular value of some FR in the physical domain, nor about the knowledge structure used to discover such a path. We will simply assume that the designer did a good job at identifying key FRs that are known to be indispensable for proper product assembly, function and quality. The paper will present a sensitivity-modeling tool that can be used to predict the effects of design choices on such product quality. The design choices are represented in terms of the design nominal dimensions, degrees of freedom (DOFs), and tolerances. The product quality is defined in terms of physical domain FRs to be satisfied by such design choices.

The next section of the paper presents some background work that led to the formulation of the sensitivity model. Section 39.3 presents an example application. The last section concludes the paper.

39.2 Sensitivity Modeling

Prior Work

In previous papers we havc presented a tool for tolerance analysis [6-7], which uses an interval arithmetic formulation:

$$\begin{bmatrix} [\underline{u} & \overline{u}] \\ [\underline{v} & \overline{v}] \\ [\underline{w} & \overline{w}] \\ [\underline{\alpha} & \overline{\alpha}] \\ [\underline{\beta} & \overline{\beta}] \\ [\underline{\delta} & \overline{\delta}] \end{bmatrix}_{FR} = \left[[J_1 J_2 J_3 J_4 J_5 J_6]_{FE_1} \; [\ldots\ldots]_{FE_{n-1}} \; [J_{6n-1} J_{6n-2} J_{6n-3} J_{6n-4} J_{6n-5} J_{6n}]_{FE_n} \right] \begin{bmatrix} \begin{bmatrix} [\underline{u} & \overline{u}] \\ [\underline{v} & \overline{v}] \\ [\underline{w} & \overline{w}] \\ [\underline{\alpha} & \overline{\alpha}] \\ [\underline{\beta} & \overline{\beta}] \\ [\underline{\delta} & \overline{\delta}] \end{bmatrix}_{FE_1} \\ [\ldots \quad \ldots]_{FE_{n-1}} \\ \begin{bmatrix} [\underline{u} & \overline{u}] \\ [\underline{v} & \overline{v}] \\ [\underline{w} & \overline{w}] \\ [\underline{\alpha} & \overline{\alpha}] \\ [\underline{\beta} & \overline{\beta}] \\ [\underline{\delta} & \overline{\delta}] \end{bmatrix}_{FE_n} \end{bmatrix} \quad (39.1)$$

where:

FE_i: Functional Element (part feature) is applied with a small displacement;

$\underline{u}, \underline{v}, \underline{w}, \underline{\alpha}, \underline{\beta}, \underline{\delta}$: lower limits of the 6 small displacements that define the tolerance (uncertainty) zone for an internal or kinematic pair [6];

$\overline{u}, \overline{v}, \overline{w}, \overline{\alpha}, \overline{\beta}, \overline{\delta}$: higher limits of the 6 small displacements that define the tolerance (uncertainty) zone;

$[J_1 J_2 J_3 J_4 J_5 J_6]_{FE_i}$: 6x6 Jacobian matrix maps the contribution of the ith tolerance (uncertainty) zone into FR space.

On the right hand side of Equation (39.1), we have several sets of 6 small displacement intervals, each mapping the 3-D tolerance zone on a particular Functional Element (FE) into 6-D small displacements space (3 translations and 3 rotations). Figure 39.1 presents the mapping for a plane FE from the 3-D tolerance "t", representing the variable distance between the two planes, to the 6-D small displacements, representing the DOFs of the nominal plane within the toleranced region. The figure also shows that only intervals on "w", "α" and "β" are relevant since the three remaining small displacements are invariant degrees for a plane and are therefore assumed to be null intervals [0,0]. Such a mapping yields a Small Displacement Torsor with Interval (SDTI), and for the reference frame in Figure 39.1, it is expressed as follows:

$$STDI_{plane} = \begin{bmatrix} [\underline{u} & \overline{u}] \\ [\underline{v} & \overline{v}] \\ [\underline{w} & \overline{w}] \\ [\underline{\alpha} & \overline{\alpha}] \\ [\underline{\beta} & \overline{\beta}] \\ [\underline{\delta} & \overline{\delta}] \end{bmatrix}_{Plane} = \begin{bmatrix} [0,0] \\ [0,0] \\ \left[-\frac{t}{2}, +\frac{t}{2}\right] \\ \left[-\frac{t}{L_1}, +\frac{t}{L_1}\right] \\ \left[-\frac{t}{L_2}, +\frac{t}{L_2}\right] \\ [0,0] \end{bmatrix}_{Plane} \tag{39.2}$$

On the left hand side of Equation (39.1), we have a set of 6 small displacement intervals that map the 3-D space that the FR under study must occupy for maintaining quality in a 6-D small displacements space. In the middle of Equation (39.1), the Jacobian performs the necessary transformations to map the cumulative effect of local tolerances expressed on the right hand side to the global FR space on the left hand side.

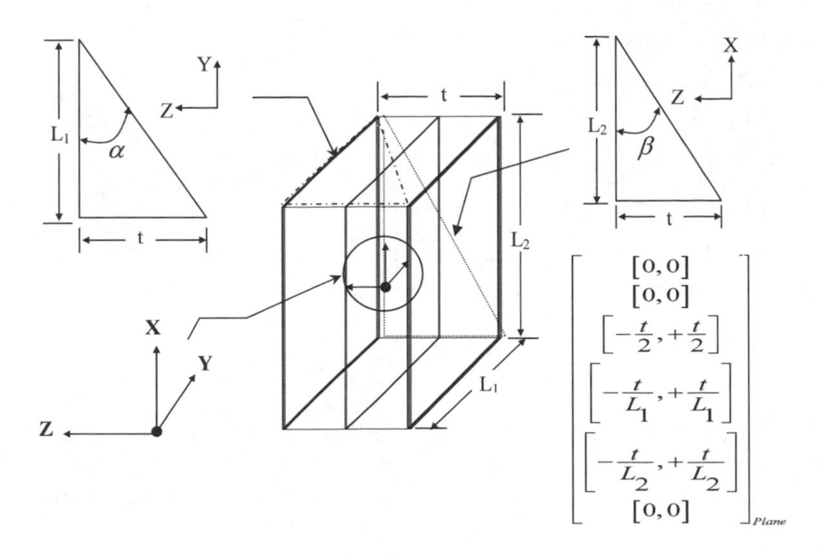

Figure 39.1 Mapping of a 3-D tolerance "t" into a 6-D SDTI, case of a plane

Jacobian Dissected

In this paper, we extend the functionality of this tolerancing tool and provide an interval arithmetic formulation for the terms in each 6x6 Jacobian as well:

$$[J]_{FE_i} = \begin{bmatrix} \begin{bmatrix} [\underline{a}_{11},\overline{a}_{11}] \\ [\underline{a}_{21},\overline{a}_{21}] \\ [\underline{a}_{31},\overline{a}_{31}] \\ [\underline{a}_{41},\overline{a}_{41}] \\ [\underline{a}_{51},\overline{a}_{51}] \\ [\underline{a}_{61},\overline{a}_{61}] \end{bmatrix}_{J1} \begin{bmatrix} [\underline{a}_{12},\overline{a}_{12}] \\ [\underline{a}_{22},\overline{a}_{22}] \\ [\underline{a}_{32},\overline{a}_{32}] \\ [\underline{a}_{42},\overline{a}_{42}] \\ [\underline{a}_{52},\overline{a}_{52}] \\ [\underline{a}_{62},\overline{a}_{62}] \end{bmatrix}_{J2} \begin{bmatrix} [\underline{a}_{13},\overline{a}_{13}] \\ [\underline{a}_{23},\overline{a}_{23}] \\ [\underline{a}_{33},\overline{a}_{33}] \\ [\underline{a}_{43},\overline{a}_{43}] \\ [\underline{a}_{53},\overline{a}_{53}] \\ [\underline{a}_{63},\overline{a}_{63}] \end{bmatrix}_{J3} \begin{bmatrix} [\underline{a}_{14},\overline{a}_{14}] \\ [\underline{a}_{24},\overline{a}_{24}] \\ [\underline{a}_{34},\overline{a}_{34}] \\ [\underline{a}_{44},\overline{a}_{44}] \\ [\underline{a}_{54},\overline{a}_{54}] \\ [\underline{a}_{64},\overline{a}_{64}] \end{bmatrix}_{J4} \begin{bmatrix} [\underline{a}_{15},\overline{a}_{15}] \\ [\underline{a}_{25},\overline{a}_{25}] \\ [\underline{a}_{35},\overline{a}_{35}] \\ [\underline{a}_{45},\overline{a}_{45}] \\ [\underline{a}_{55},\overline{a}_{55}] \\ [\underline{a}_{65},\overline{a}_{65}] \end{bmatrix}_{J5} \begin{bmatrix} [\underline{a}_{16},\overline{a}_{16}] \\ [\underline{a}_{26},\overline{a}_{26}] \\ [\underline{a}_{36},\overline{a}_{36}] \\ [\underline{a}_{46},\overline{a}_{46}] \\ [\underline{a}_{56},\overline{a}_{56}] \\ [\underline{a}_{66},\overline{a}_{66}] \end{bmatrix}_{J6} \end{bmatrix}_{FE_i} \tag{39.3}$$

It can be shown that the terms in each 6x6 Jacobian have the following general form:

$$[J]_{FE_i} = \begin{bmatrix} R & D \\ 0 & R \end{bmatrix}_{FE_i} \tag{39.4}$$

The following is an interpretation of each term:

$$\begin{bmatrix} [\underline{u} & \overline{u}] \\ [\underline{v} & \overline{v}] \\ [\underline{w} & \overline{w}] \\ [\underline{\alpha} & \overline{\alpha}] \\ [\underline{\beta} & \overline{\beta}] \\ [\underline{\delta} & \overline{\delta}] \end{bmatrix}_{FR} = \begin{bmatrix} \boldsymbol{R} & D \\ 0 & R \end{bmatrix}_{FE_i} \cdot \begin{bmatrix} [\underline{u} & \overline{u}] \\ [\underline{v} & \overline{v}] \\ [\underline{w} & \overline{w}] \\ [\underline{\alpha} & \overline{\alpha}] \\ [\underline{\beta} & \overline{\beta}] \\ [\underline{\delta} & \overline{\delta}] \end{bmatrix}_{FE_i} \qquad (39.5)$$

The top left 3x3 transformation matrix “R” serves the purpose of mapping the locally expressed translational variables “u”, “v” and “w” of the corresponding SDTI. This region of the Jacobian represents the contribution that the translational part of the SDTI has to the translational part of the FR space.

The same matrix “R” is repeated at the bottom right of Equation (39.4) to transform the locally expressed rotational variables “α”, “β” and “δ” of the corresponding SDTI. Similarly, this region of the Jacobian represents the contribution that the rotational part of the SDTI has on the rotational part of the FR space.

Since local rotations of an SDTI can contribute to remote translations in FR space, the top right 3x3 matrix “D” typically contains nominal dimensions and translational DOFs on the mechanism that act as lever arms. These multiply the rotational terms of the local STDI to provide the small translational effects in FR space. Since DOFs variables of the mechanism appear in transformation “D”, this matrix will be different depending on the mechanism’s possible geometrical configurations (extended, retracted, etc…).

Finally, since local translations of an SDTI do not contribute to remote rotations in FR space, the bottom left 3x3 null matrix (noted “0” in Equation 39.4) also appears.

The idea expressed by Equation (39.3) is to provide bounds on nominal variables appearing in term “D” above to reflect physical constraints that a designer has to work with during design, for example, geometric interference between parts, servicing space available, limiting certain dimensions for weight consideration, etc, which can all be expressed as intervals. The multiplication of the resulting Jacobian with the various SDTIs using interval arithmetic rules will help discover how the FR is influenced within such bounds of the nominal variables. Looking back at Equation (39.1), we see that the tool provides a functional relationship between design choices, namely; nominal dimensions and DOFs intervals in the Jacobian along with tolerances in the SDTI, and product quality, namely; physical domain FR space. The process of studying how quality is affected for various instances of the mechanism as reflected by the defined intervals is called *sensitivity modeling.* For the purpose of this paper, we will therefore call the unified Jacobian-torsor model of Equations (39.1) and (39.3) a *sensitivity model*.

$$\begin{bmatrix} [\underline{u} & \overline{u}] \\ [\underline{v} & \overline{v}] \\ [\underline{w} & \overline{w}] \\ [\underline{\alpha} & \overline{\alpha}] \\ [\underline{\beta} & \overline{\beta}] \\ [\underline{\delta} & \overline{\delta}] \end{bmatrix}_{FR} = \begin{bmatrix} R & D \\ 0 & \boldsymbol{R} \end{bmatrix}_{FE_i} \cdot \begin{bmatrix} [\underline{u} & \overline{u}] \\ [\underline{v} & \overline{v}] \\ [\underline{w} & \overline{w}] \\ [\underline{\alpha} & \overline{\alpha}] \\ [\underline{\beta} & \overline{\beta}] \\ [\underline{\delta} & \overline{\delta}] \end{bmatrix}_{FE_i} \tag{39.6}$$

$$\begin{bmatrix} [\underline{\boldsymbol{u}} & \overline{\boldsymbol{u}}] \\ [\underline{\boldsymbol{v}} & \overline{\boldsymbol{v}}] \\ [\underline{\boldsymbol{w}} & \overline{\boldsymbol{w}}] \\ [\underline{\alpha} & \overline{\alpha}] \\ [\underline{\beta} & \overline{\beta}] \\ [\underline{\delta} & \overline{\delta}] \end{bmatrix}_{FR} = \begin{bmatrix} R & \boldsymbol{D} \\ 0 & R \end{bmatrix}_{FE_i} \cdot \begin{bmatrix} [\underline{u} & \overline{u}] \\ [\underline{v} & \overline{v}] \\ [\underline{w} & \overline{w}] \\ [\underline{\alpha} & \overline{\alpha}] \\ [\underline{\beta} & \overline{\beta}] \\ [\underline{\delta} & \overline{\delta}] \end{bmatrix}_{FE_i} \tag{39.7}$$

$$\begin{bmatrix} [\underline{u} & \overline{u}] \\ [\underline{v} & \overline{v}] \\ [\underline{w} & \overline{w}] \\ [\underline{\alpha} & \overline{\alpha}] \\ [\underline{\beta} & \overline{\beta}] \\ [\underline{\delta} & \overline{\delta}] \end{bmatrix} = \begin{bmatrix} R & D \\ \boldsymbol{0} & R \end{bmatrix}_{FE_i} \cdot \begin{bmatrix} [\underline{\boldsymbol{u}} & \overline{\boldsymbol{u}}] \\ [\underline{\boldsymbol{v}} & \overline{\boldsymbol{v}}] \\ [\underline{\boldsymbol{w}} & \overline{\boldsymbol{w}}] \\ [\underline{\alpha} & \overline{\alpha}] \\ [\underline{\beta} & \overline{\beta}] \\ [\underline{\delta} & \overline{\delta}] \end{bmatrix}_{FE_i} \tag{39.8}$$

39.3 Example Application

The centering pin mechanism in Figure 39.2 will be used to demonstrate the use of the tool. Let's begin by stating a few general FRs regarding this mechanism:

1. FR1: swing between 40 and 50 mm (centering parts of different diameters);
2. FR2: adaptable capacity of 50 mm (centering parts of different lengths);
3. FR3: vertical centering precision of ±1mm.

Out of these three FRs, FR3 is clearly of a varying nature and require the identification of a tolerance chain to pinpoint dimensions/tolerances affecting it. Therefore, in Figure 39.2, the variables pertaining to this chain are labelled: nominal dimensions A, B, C, D, F; translational DOF E (see below); and dimensional and geometric tolerances on some features. It can be shown that all these appear in the mathematical expression that transforms the tolerance chain into its corresponding sensitivity model (*e.g.* Equations (39.1) and (39.3) in symbolic form).

The label "E", which does not appear in Figure 39.2, is used to take into consideration the interaction between FR2 and FR3; that is, "E" will be the variable that represents the mechanism in various configurations (retracted or expanded, see Figure 39.3). Note that this variable not only directly affects capacity (FR2) but also precision (FR3) due to the lever arm effect.

In this first draft design, we assume that this DOF is provided by a cylindrical fit H8/g6 between the horizontal pin and hole. We also assume that there exists an identical mechanism to that in Figure 39.3, which will support the other end of the cylindrical part to be centered, such that FR2 is in fact 25 mm at each end. Therefore, we will model "E" as the interval [0, 25] with 0 = retracted, and 25 = expanded.

Table 39.1 presents some initial, manually generated values for the variables "A" to "F". These initial values are all expressed as intervals in the table, except for the chosen clearance fit H8/g6, which remains expressed as such. The first column "initial guess" shows a physical instantiation of the mechanism, where the intervals on the nominal dimensions have been assigned the same bounds. The first column also presents manually generated initial values for the relevant dimensional and geometric tolerances on the relevant parts.

The last line in the table computes the resulting translational vertical effect at the tip of the pin, FR3, which is obtained by inserting the above interval values into the sensitivity model pertaining to FR3 (*e.g.* Equations (39.1) and (39.3) in numerical interval form). Note that the parallel tolerance values, which were left blank in the drawings, are labelled "n/a" in the first column to reflect that initially we would not like to make use of them. Finally, we note that the interval B=[55,55] directly satisfies FR1 and that the interval E=[0, 25] directly satisfies FR2, although we will see that "E" also influences FR3.

We will now go through an iterative manual procedure and try to converge towards a design that meets FR3 since in Table 39.1, we see that the initial guess values generated manually lead to a larger interval than the one that is acceptable for FR3, *i.e.* [-2.10, 2.10] instead of [-1, +1]. The result of the changes at each step (*e.g.* different columns in Table 39.1) is obtained by inserting the modified intervals in each trial back into the sensitivity model (Equations (39.1) and (39.3)) pertaining to FR3.

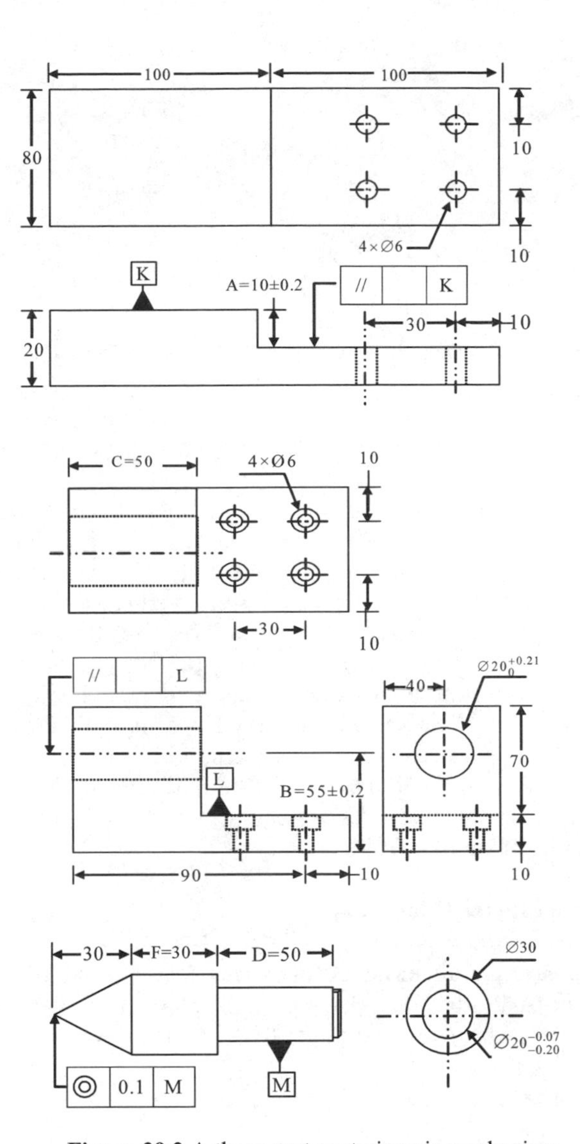

Figure 39.2 A three-part centering pin mechanism

Figure 39.3 Definition of interval E=[0 (right), 25 (left)]

Investigating Nominal Dimensions with No Intervals

We first note that if C=D=50 in the initial draft, then when the mechanism is retracted (E=0), the mating cylindrical surfaces overlap by their full length of 50 mm. However, when the mechanism is extended (E=25), the mating cylindrical surfaces overlap by only 25 mm. The possible small vertical movement of the pin, permitted by the clearance fit H8/g6, is larger as the size of the overlap (mating surfaces) decreases. This suggests increasing "D". If "D" is set to 75 mm, then we always have a 50 mm overlap in any configuration (retracted and extended). By inserting this new value in the sensitivity model, we now obtain FR3 = [-2.01, 2.01] (see Table 39.1). This represents a very small improvement. We conclude that for the chosen clearance fit H8/g6, and the desired capacity FR2 = 50 mm, the sensitivity of FR3 to variations in "D" is negligible (4.2% increase in precision from D=50 mm to D=75 mm).

Investigating Dimensional Tolerances

We will now decrease the dimensional tolerances for "A" and "B" by half their initial values, from [-0.2, +0.2] to [-0.1, +0.1]. We therefore obtain a significant improvement for FR3, which now stands at [-1.10, +1.10].

Investigating Geometric Tolerances

We continue by adding two parallelism tolerances of 0.1 each for "A" and "B". We get [-0.75, +0.75] for FR3, which now represents excess in precision. At this point, we can therefore analyze where we could be less stringent for this design. We will assume the chosen clearance fit H8/g6 is a good candidate for decreasing its precision.

Investigating Gaps and Fits

We now choose the clearance fit H10/g9, which results in the interval [-0.99, +0.99]; this is very close to the desired interval.

Table 39.1 Iterative design procedure using intervals in the sensitivity model

Initial Guess	Change Nom. Dim. D	Change Dim. Tol. on A, B	Add Geom. Tol. on A, B	Change Clearance Fit	Sensitivity to D	Sensitivity to E	
[10,10]	[10,10]	[10,10]	[10,10]	[10,10]	[10,10]	[10,10]	**A**
[55,55]	[55,55]	[55,55]	[55,55]	[55,55]	[55,55]	[55,55]	**B**
[50,50]	[50,50]	[50,50]	[50,50]	[50,50]	[50,50]	[50,50]	**C**
[50,50]	**[75,75]**	[75,75]	[75,75]	[75,75]	**[50,75]**	[50,75]	**D**
[0,25]	[0,25]	[0,25]	[0,25]	[0,25]	[0,25]	**[0, 40]**	**E**
[30,30]	[30,30]	[30,30]	[30,30]	[30,30]	[30,30]	[30,30]	**F**
[-0.2, +0.2]	[-0.2, +0.2]	**[-0.1,+0.1]**	[-0.1, +0.1]	[-0.1, +0.1]	[-0.1, +0.1]	[-0.1, +0.1]	**Dim. Tol. on A, B**
N/A	n/a	n/a	**// 0,1**	// 0,1	// 0,1	// 0,1	**Geom. Tol. on A, B**
H8/g6	H8/g6	H8/g6	H8/g6	**H10/g9**	H10/g9	H10/g9	**Clearance Fit**
[-2.10, +2.10]	**[-2.01, +2,01]**	**[-1.10, +1.10]**	**[-0.75, +0.75]**	**[-0.99, +0.99]**	**[-1.23, +1.23]**	**[-1.96, +1.96]**	**FR3**

Investigating Nominal Dimensions with Intervals

Now that we have a satisfying design, we can go back and check the sensitivity of the final design on dimension "D", by letting D=[50, 75] in the computations. As can be seen, we now get w = [-1.23, +1.23]. This means that the new looser fit H10/g9 drastically increases the sensitivity of FR3 to "D" (a 24.2% decrease in precision from D=75 mm to D = 50 mm).

Investigating Degrees of Freedom with Intervals

We will finally investigate the sensitivity of this design with respect to FR2, *i.e.* we will test the potential for this design to serve as a centering device with larger capacity, say from FR2 = 50 mm to 80 mm. We therefore define E=[0, 40] and get w= [-1.96, +1.96]. This DOF therefore has a major impact on FR3 and we discover the interdependence between FR2 and FR3: the larger FR2 is the larger FR3 is when all other variables are fixed. Further investigation shows that one possible configuration of the assembly is when E=40 mm and D=50 mm, for which the mating surface is only 10 mm, which decreases the overall precision for FR3.

39.4 Conclusion

A unified Jacobian-torsor model with intervals has been used to detect the sensitivity of some detailed level FRs to nominal dimensions, DOFs, dimensional tolerances and geometric tolerances. It is a very powerful design tool that can help understand the fundamental relationships between important design variables. This understanding translates into a decision making tool regarding final design values.

The tool is also very useful to study how single mechanisms could be customized in different variants for different markets. Each market would likely give rise to different FRs (one instance of a product variant), so that the sensitivity of nominal dimensions, tolerances and DOFs for all the variants in a family could be simultaneously investigated. The parts where detail design could remain the same would be easily identified and exploited and potentially be fabricated in mass production.

In the example that was presented, changes in the various intervals were manually input separately and the sensitivity model was run for each case, leading to an iterative manual procedure (Table 39.1). We are now working at integrating the model to an optimizer that can perform the search for optimal values automatically using *data reconciliation* techniques.

39.5 Acknowledgements

The authors would like to thank the Natural Science and Engineering Research Council (NSERC) of Canada for financial support through the AUTO21 Network of Centres of Excellence and Discovery Research programs.

39.6 References

[1] Pahl, G., and Beitz, W., 1996, *Engineering Design: Systematic Approach,* Springer-Verlag, Berlin, 2nd Edition.

[2] Suh, N.P., 1990, *The Principles of Design*, Oxford University Press, Oxford, New York.

[3] Meijer, B.R., Tomiyama, T., van der Holst, B.H.A. and van der Werff, K., 2003, "Knowledge Structuring for Function Design," *Annals of CIRP*, Vol. 52/1, pp. 89-92.

[4] Foster, L.W., 1994, *Geo-Metrics 3M-The metric Application of Geometric Dimensioning and Tolerancing Techniques*, Addison Wesley Pub. Co.

[5] Burdea, G., 1996, *Force and Touch Feedback for Virtual Reality*, John Wiley & Sons, New York.

[6] Desrochers, A., Ghie, W. and Laperrière, L., 2003, "Application of a Unified Jacobian - Torsor Model for Tolerance Analysis," *ASME Journal of Computing and Information Systems in Engineering: Special Issue on Computing Technologies for GD&T*, Vol. 3, No. 1, pp. 2-14.

[7] Laperrière, L., 2002, « Modèle de sensitivité global en tolérancement assisté par ordinateur, » *Revue de CFAO et d'informatique graphique, De la CAO géométrique vers une CAO fonctionnelle, Spécial CFAO au Québec*, Hermes Science, Vol. 17, No. 1-2, pp. 61-78.

40

Computer Aided Tolerancing - Solver and Post Processor Analysis

Serge Samper, Jean-Philippe Petit, and Max Giordano

Abstract: The world of the designer is three dimensional, and the language of tolerancing is a set of ISO specifications. We have built a methodology in order to compute geometric specifications on parts and clearances in joints through a mathematical model based on the small displacement torsors. A tolerancing object becomes a 6D object thanks to the developed solver. One objective is to represent 6D polytopes in the 3D world of the designer in order to inform him of the results for his tolerancing choices: assemblability performance, best and worst precision zones, and functional requirements. Therefore, it is necessary to indicate, the results to the designer graphically. This representation will be done in a CAD application by means of zones (3D volumes), which will be associated with functional features of the mechanism. An assembly example is presented to illustrate this method.

Keywords: Tolerancing analysis, CAD, 6-polytopes

40.1 Tolerancing and Functional Requirement

Tolerancing is a standardized language; it follows the designed product in design, manufacturing and control processes. Its goal is to fix maximum deviations on geometric parts: those parts are manufactured imperfectly. It is an important operation, which influences the functional requirements and the cost of the final product. In spite of the continual development of Computer–Aided Design software, it is amazing to note that none offers an integrated, systematic and automatic tool able to find optimal tolerances (from a qualitative and/or quantitative point of view). In the best of cases, computer-aided tools can only check the coherence of a geometric specification type chosen according to the toleranced feature.

Since the beginning of the 80's and in particular with Requicha's works [1], many research groups have worked in this subject area, based on various mathematical models. One can hold up as examples the PACV model (Proportioned Assembly Clearance Volume) used by D. Teissandier [2] or the Tolerance-Map model of J.K. Davidson [3]. These models enable translation of the standards in a formal way (that is passage of the standard towards the model). This data is then used to carry out the tolerancing analysis of the mechanism. The following step consists of carrying out the reverse passages *i.e.* a mathematical model towards the standard. However, the methods developed to date do not integrate this aspect of reverse passages and the results of analysis remain most of the time incomprehensible for the designer who awaits readable and concrete answers for his tolerancing choices.

We propose in the following to present the analysis method developed in LMécA [4], the passage "model towards standard". This method is based on the model of clearance and deviation domains [5], which translates geometric specifications on parts and clearances in joints into 6 dimension domains. Various geometrical operations on these 6-polytopes [6] allow one to determine if the tolerances chosen by the designer satisfy the specifications. However, these representations still belong to the mathematical model and are not readable for the designer. We will thus describe a procedure of post processing which is used to carry out the passage of 6D towards 3D (model towards standard). The designer will have the possibility of visualising the consequences of his choices via a graphic representation through 3D zone(s) displayed on the assembly definition drawing of. The designer then will be able to check if the chosen tolerancing meets the functional requirements expressed by the specifications: assemblability, accuracy requirement, non-contact conditions, *etc.*.

40.2 Model of the Clearance and Deviation Domains

This section deals with the first step of our method of geometric tolerance analysis. It consists of translating standard geometric specifications on parts and joints constituting the assembly through a mathematical formalism.

Tolerance Zone and Associated Feature

International standards [7] allow representation of every geometric specification by a tolerance zone (2D or 3D), which is built on the nominal geometry of the toleranced feature. The tolerance will be valid if the real feature (that is the theoretical geometric element) associated with the nominal feature lies inside the tolerance zone (Figure 40.1).

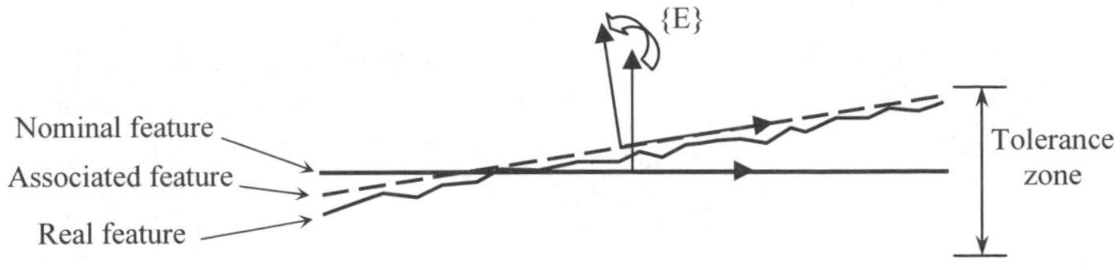

Figure 40.1 Tolerance zone

Tolerance Zone to Deviation Domain

In our model, a datum frame is attached to each functionally associated feature and another frame is attached to each part. The displacements of the associated feature inside the tolerance zone are assumed small enough. It is then possible to express the positions of the associated frame relative to the general frame in the form of a small displacement torsor [8] (6 components *i.e.* 3 translations and 3 rotations) called a deviation torsor. When the associated and nominal features are merged, the corresponding small displacement torsor is equal to the null torsor. The general form of a deviation torsor is noted:

$$\{E\} = \begin{Bmatrix} Tx & Rx \\ Ty & Ry \\ Tz & Rz \end{Bmatrix} \tag{40.1}$$

The set of values of all the deviation torsors define a domain in the 6D configuration space called the deviation domain, noted as [E]. The associated feature can be a point, a segment or a polygon (in the opposite case, it can be polygonized). The deviation domain is reduced to consider each maximum displacement of characteristic vertices of the associated feature inside the tolerance zone. This observation allows the representation of those displacements by a set of in-equalities. With a polyhedral computation code [9], it is possible to generate all the vertices of the convex 6-polytope from the set of in-equalities in $\mathbb{R}^6$. This double definition (vertices and in-equalities) is necessary for several geometric operations inn different domains in the model.

Example:
In Figure 40.2, we show how a torsor is built for a chosen geometric specification (that is tolerance symbol and type of feature). The corresponding zone limits the displacement of the associated feature. Those limits can be computed only from the displacement of the 4 vertices of the rectangle (so 8 in-equations).

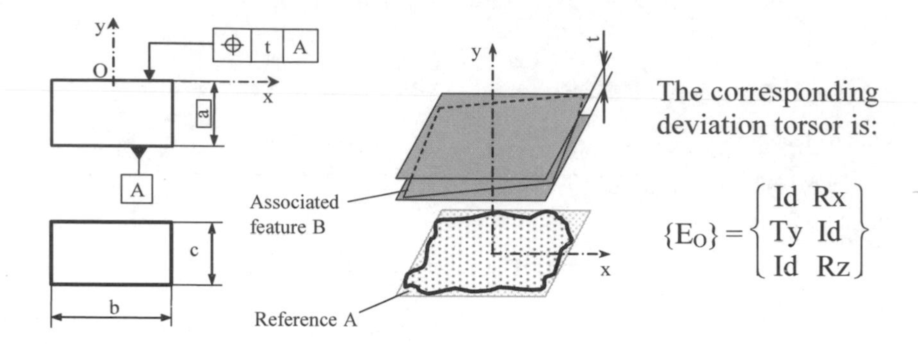

Figure 40.2 Tolerance zone associated with a specification

The deviation domain corresponding to the zone of Figure 40.2 is computed by the following set of in-equalities:

$$\begin{cases} -t \leq b\,R_z + c\,R_x + 2\,T_y \leq t \\ -t \leq b\,R_z + c\,R_x - 2\,T_y \leq t \\ -t \leq b\,R_z - c\,R_x + 2\,T_y \leq t \\ -t \leq b\,R_z - c\,R_x - 2\,T_y \leq t \end{cases} \tag{40.2}$$

And the 6 following vertices are obtained with $b = 5$, $c = 3$ and $t = 0,1$:

$$V_1\begin{pmatrix}0\\0\\0.02\end{pmatrix}, V_2\begin{pmatrix}0.1\\0\\0\end{pmatrix}, V_3\begin{pmatrix}0\\0.0333\\0\end{pmatrix}, V_4\begin{pmatrix}0\\0\\-0.02\end{pmatrix}, V_5\begin{pmatrix}0\\-0.0333\\0\end{pmatrix}, V_6\begin{pmatrix}-0.1\\0\\0\end{pmatrix} \tag{40.3}$$

The 8 in-equalities (8 facets) of (40.2) and the 6 vertices of (40.3) allow computing the deviation domain through its 3-polytope form shown in Figure 40.3.

To sum up this section, we can say that in the proposed model each standard geometric specification can be translated by a deviation domain (a set of in-equalities and the list of its vertices coordinates in a 6D space).

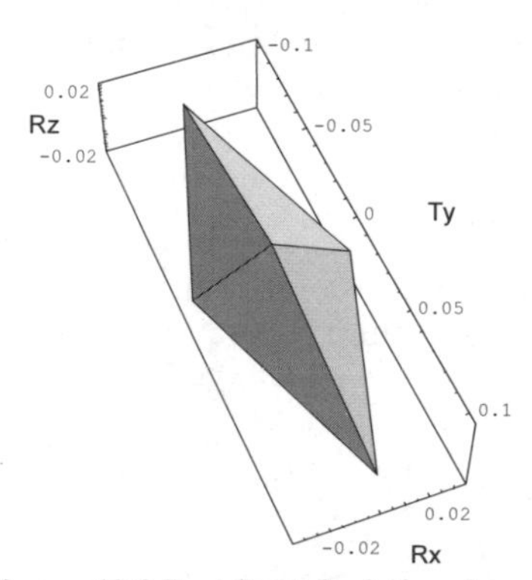

Figure 40.3 Resultant deviation domain

Clearance Domain

A joint is constituted of two parts. The clearance inside the joint allows for the writing of a clearance torsor, noted as *{J}*, which represents the small displacements of a part (of an associated datum frame actually) in relation to another. The values of the torsor depend on the contact conditions (clearance value and topology of the contact surfaces).The clearance domain is computed in the same manner as defined for the deviation domain.

We can associate with each joint a clearance domain (6-polytope) which will be defined by a system of linear inequalities and a set of coordinate vertices in a 6D configuration space.

Example for a Cylindrical Joint:

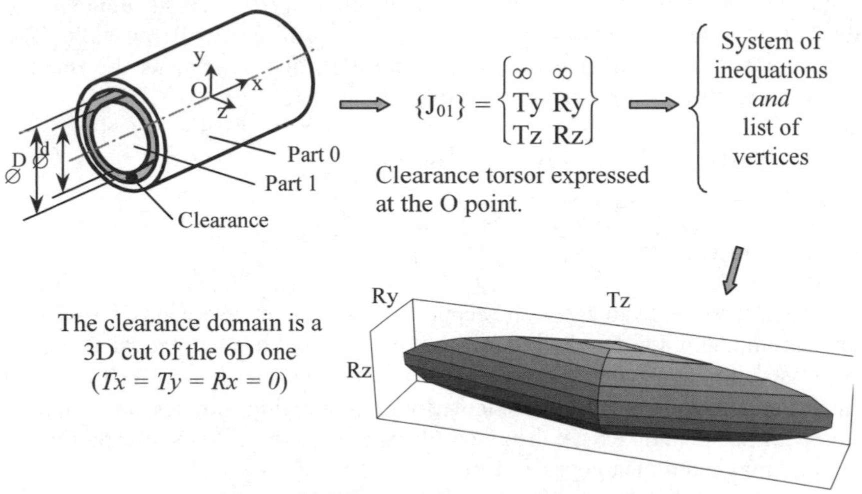

Figure 40.4 From a joint to its clearance domain

Note: Directions of the domain corresponding to the degrees of freedom of the joint (*Tx* and *Rx* in the example) are unbounded: displacements are infinite in those directions.

40.3 Analysis Treatment

Once every geometric specification is translated by a deviation domain and every joint by a clearance domain, diverse geometric operations on 6-polytopes enable the tolerance analysis. The main operations are:

- the Minkowski addition noted ⊕,
- the intersection ∩,
- the Sweeping-Intersection noted ⊖ which consists "graphically" in sweeping a (usually clearance) domain on the boundaries of a (usually deviation) domain

and keeping the intersection of all the shifted (clearance) domains. The resulting domain of such an operation is called residual clearance domain and is noted as [R].

To illustrate the treatment method, we will take the example of an assembly of two parts (*0* and *1*) linked by three parallel branches *A*, *B* and *C*. $[R_{0A1}]$, $[R_{0B1}]$ and $[R_{0C1}]$ are calculated (see Equation 40.4) at the same point.

$$[R_{0X1}] = [J_{0X1}] \ominus [[E_{0X}] \oplus [E_{X1}]] \quad (40.4)$$

Assemblability

If the intersection of the three residual clearance domains $[R_{0\_1}]$ exists, then we can ensure that the assembly between part 0 and part 1 will always be possible. The resulting domain of this intersection (Equation 40.5) also represents the relative position between the frames associated with each part.

$$[R_{0\_1}] = \cup_X [R_{0X1}] = [R_{0A1}] \cup [R_{0B1}] \cup [R_{0C1}] \quad (40.5)$$

Accuracy Requirement

These tools also allow us to check the accuracy requirement between two surfaces. The maximum and minimum deviations on these surfaces are known because they are expressed through geometric specifications. On the one hand, the minimum residual clearance domain $[R_{0\_1}]$ is obtained with minimal clearances and maximal deviations. If this domain exists, the assemblability requirement is satisfied. On the other hand, maximum clearances and minimum deviations provide the maximum residual clearance which is significant for the bigger deviations between surfaces.

Cost

The decrease of the product cost can be obtained by affecting the magnitudes of tolerances. We saw that the assembly requirement depends on the residual clearance domain, which is calculated using several parameters (tolerance values, joint dimensions…). By decreasing the values of tolerances and hence decreasing the cost, the method consists of reaching the smallest residual clearance domain as possible. If the resulting domain is reduced to a point, the tolerance value will be considered as optimal for this functional requirement.

40.4 Post Processing

Once the solving is complete, objects are 6D domains. In order to see them, we can project them in the 3D space as zones. For example, we can say the assembly is possible if the global residual clearance exists (6D result).

The designer knows if the tolerancing is good or not, based on the solution, but he would also know how to modify tolerances. A Boolean result is not sufficient. We propose here to show results in the form of zones. They can be:

- Best precision zone
- Worst precision zone
- Residual clearances
- Other functional requirements can also be computed.

The designer can then modify his choices. For example, if residual clearances are sufficient, corresponding tolerances can be increased so that the cost will be lower or manufacturing is more feasible. Then the designer can solve again and see the results. Tolerancing analysis can be used in the same way as a finite element analysis in an iterative design process.

Method

The results of our computing are 6D domains as follows:

- Set of in-equalities (constrains in 6D space)
- Set of vertices (which describe the corresponding 6D domain).

We make a projection of this 6D domain into a 3D zone, attached to a feature as a 3D view of the result. For each point of the feature, we make a displacement (3D) about a torsor with its coordinates taken in the 6D domain.

This could represent a lot of computations but we can be faster by taking only the vertices of the 6D domain. Then we compute only the vertices of the polygon containing the feature. In this way, we have obtained the largest convex zone corresponding to the domain.

Different Projections

It would be interesting to have a real correspondence between a 3D zone and the 6D domain. Unfortunately, this is not always possible.

We can build the 6D domain for any zone, but any domain is not a zone. We can therefore build the following three zones:

a) the smallest outer convex zone for any 6D domain.
b) a non convex zone for any 6D domain.
c) the biggest inner convex zone for any 6D domain.

In fact, the zone of case *b* is outside of *c* zone and inside *a* zone. Figure 40.6 below shows the three zones for a given domain.

Next, for each point of the feature, we should apply all the possible displacements of the associated feature according to the 6D domain. Then we should obtain the corresponding zone (case *c*). The simplest 3D object is the convex one. In order to show simple but accurate information, we propose to show the two convex (outer and inner) domain to the designer.

Examples

A Simple Axis

Here, we explain the concept with a very simple example: an axis. We have obtained a result in the form of a domain. This domain is then translated into a zone.

From the Zone to the Domain

A zone is computed to obtain the domain corresponding to all the possible displacements. Thus we can know all the in-equalities and the corresponding vertices of the 6D domain. This domain can be cut to see its representation in a 2D space, as shown below.

Let us consider a single axis with its zone as shown in the Figure 40.5. We first show the domain in 2D corresponding to this zone. We can observe that each vertex of the domain corresponds to an extreme position (four, in 2D) of the AB axis in the zone.

L is the distance between A and B; h is the height of the zone. Then, $Ty_{MAX}=h/2$ and $Rz_{MAX} \approx h/L$.

The 2D domain is a cut of the 6D domain.

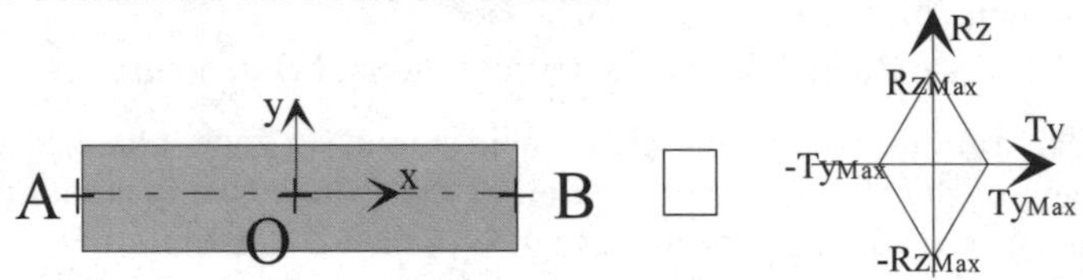

Figure 40.5 A-B axis zone and its domain

From the Domain to Zones

We can illustrate the cases presented in 40.2 with the domain presented in the Figure 40.6.

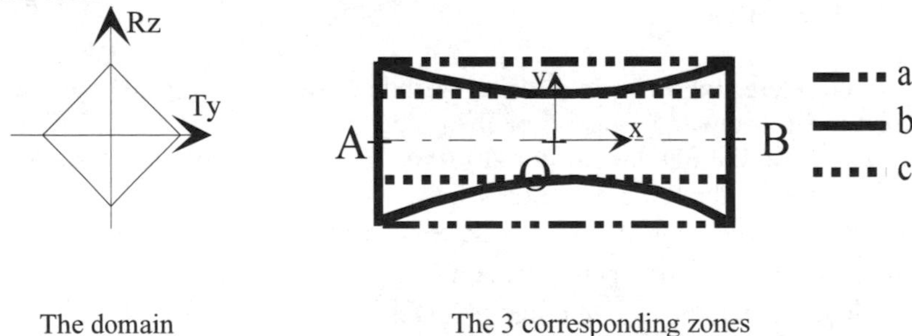

Figure 40.6 Domain and zones

In the resulting domain, there are no portions that correspond exactly to a convex zone ($Ty_{MAX}=h/2$ and $Rz_{MAX} = h/L$); the computation of the corresponding zone gives (point by point) case *b*. We can also compute the *a* outer zone and the *c* inner zone by testing only the A and B points.

A Spherical Drilling Tool

Let us consider the analysis of a drilling tool. In Figure 40.7, we have two parts linked by four joints (two plane joints and two cylindrical joints).

We have specified tolerances (not detailed here) for each surface, *Ai*, *Bi*, *Ci* and *Di,* belonging to each part *i* (*0* and *1*). We have chosen clearances in joints (*A, B, C,* and *D*) and solved the assembly. The assembly is possible because *[R0_1]* (minimum clearance domain, computed as shown in Equation 40.5) exists. *[R0_1]* is also the precision domain (3D view) presented in Figure 40.8

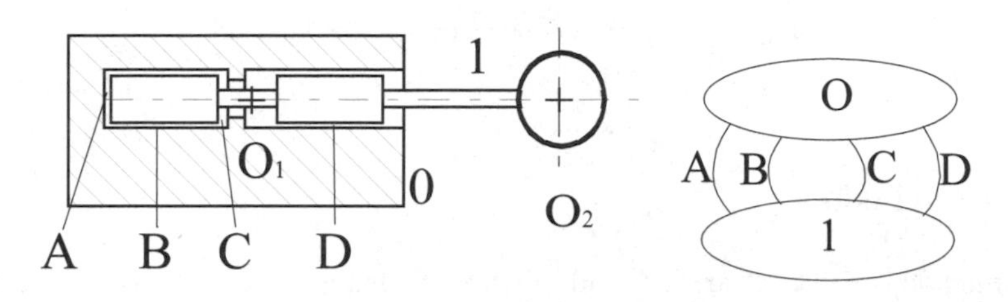

Figure 40.7 Simplified drilling tool

The designer would know where the tool could be according to this domain. Thus we project this domain and obtain the precision zone shown in Figure 40.9. The precision zone is the set of possible positions of the points of the sphere according to the geometric specifications.

It is possible to see the results and verify their suitability, and then to change the geometric specifications.

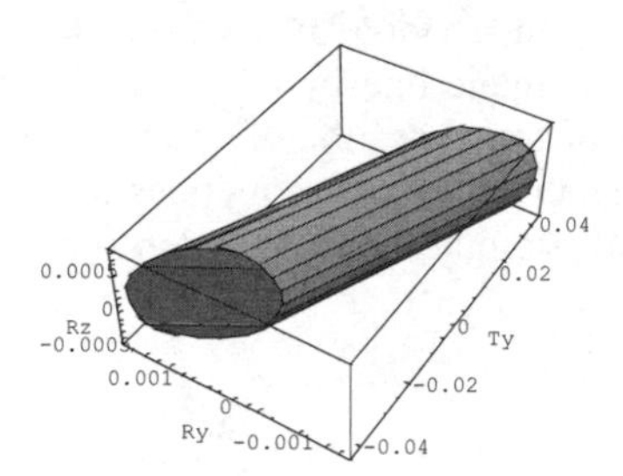

Figure 40.8 Precision domain at the point O_2

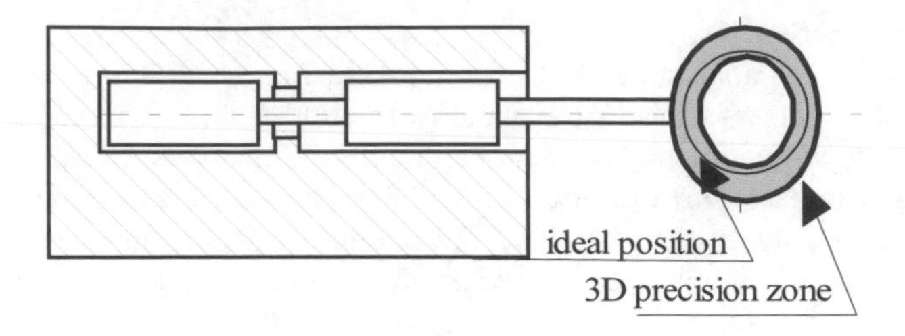

Figure 40.9 Precision zone (magnified volume)

40.5 Conclusion

The method of the clearance and deviation domains allows us to compute tolerancing in the 6D space of the torsor components. This computation allows testing assemblability and functional requirements. The solving is made and the results are 6D domains, such as minimal clearances or precision domains.

The designer usually needs to modify choices and test a new solution, thus we show results in the 3D space in the form of zones around associated features. The post processor proposed here makes it possible to see the results in 3D, which can be projected on a feature, as in the example.

The aim of our work is to build a method of Computer–Aided Tolerancing for the designer, and thus to build a CAT pre-processor, solvers and post processors (like the FEM for structural analysis). The features, the assembly, the functional requirements, and the tolerancing are input to the pre-processor. The solving of the assembly is made, and then a Boolean gives the assembly test. The solving of functional requirements is made one by one and the residual clearance domain is output. The post processor translates the 6D domain into a 3D zone on a feature in order to inform the designer of the consequences of his choices. Then the designer can modify the input and compute again in order to reach optimal tolerances.

40.6 References

[1] Requicha, A.A.G., 1983, "Toward a theory of geometric tolerancing," *The International Journal of Robotics Research*, Vol. 2, N° 4, pp. 45-60.

[2] Teissandier, D., Couétard, Y., Gérard, A., 1999, "A computer aided tolerancing model: proportioned assembly clearance volume," *Computer-Aided Design 31*, pp 805-817.

[3] Davidson, J.K., Mujezinović, A., Shah, J.J., 2002, "A new mathematical model for geometric tolerances as applied to round surfaces," *Journal of Mechanical Design*, Vol. 124, pp 609-622.

[4] Giordano, M., Duret, D., "Clearance space and deviation space. Application to three dimensional chain of dimensions and positions," *3rd CIRP International Seminar on Computer-Aided Tolerancing*, pp. 179-196.

[5] Petit, J-Ph., Samper, S., Giordano, M., 2003, "Minimum clearance for tolerancing analysis of a vacuum pump," *Proceedings of the 8th CIRP International Seminar on Computer-Aided Tolerancing*, pp. 43-51.

[6] Fukuda, K., Petit, J-Ph., 2003, "Optimal tolerancing in mechanical design using polyhedral computation tools," *Proceedings of the 19th European Workshop on Computational Geometry*, pp. 117-120.

[7] *ISO 1101* "Technical drawings. Geometrical tolerancing. Tolerancing of form, orientation, location and run-out. Generalities, definitions, symbols, indications on drawings," 1983.

[8] Bourdet, P., Mathieu, L., Lartigue, C., Ballu, A., 1995, "The concept of the small displacements torsor in metrology," *Advanced mathematical tools in metrology*, Oxford.

[9] Fukuda, K., 2001, *cddlib reference manual, cddlib Version 092a,* McGill University, Montreal, Canada.

41

A New Method for Integrated Design and Tolerancing

Pascal Hernandez, Max Giordano, and Gaétan Legrais

Abstract: The dimensional and geometrical tolerancing of machine elements is an important step in the design and manufacturing of a product. Unfortunately, tolerancing takes place late in the current design processes. Generally, it is only in the detail drawings of the parts that the tolerances are determined qualitatively and quantitatively. Some design problems appear which could have been detected upstream if the tolerances had been introduced from the very start. In the proposed design process, the mechanism is defined from a minimal kinematics structure to a detailed geometry. The tolerancing method is directly integrated into this design process. There is an inevitable growing complexity of the mechanical structure. Some technical choices are carried out at each level and it would be interesting to evaluate their geometrical influence on the expressed conditions. Therefore, we propose to deal with the problem of tolerance in an integrated manner with the process of design. The recursive top-down design and tolerancing process is general. The different design solutions, and technological choices, directly influence the dimensional and geometrical tolerances. We present a graph tool, which allows definition of the topology of the mechanism, during all phases of its design. The tolerancing graph is translated into ISO standards conforming tolerances. Different views are possible depending on the detail level needed by the designer. During the design process, the graph is simultaneously updated. An example is studied with the different steps to illustrate this integrated method. The influence of different possible design solutions on the tolerances is compared in order to validate these choices.

Keywords: tolerancing, graph, mechanical system, top-down design, tolerance synthesis

41.1 Introduction

Design and Tolerancing

The dimensional and geometrical tolerancing of machine elements is an important step in the design and manufacturing of a product. It ensures the satisfaction of functional conditions expressed in a geometrical form. Moreover, the tolerancing step allows choice of the different processes necessary for manufacturing the parts. One can thus have an idea of the various manufacturing costs and validate the design or modify it if necessary.

Tolerancing takes place late in the current design processes. It is only in the detail drawings of the parts that the tolerances are determined in qualitative and then quantitative form. Some design problems appear which could have been detected upstream if the tolerances had been introduced from the very start. The use of CAD has increased recently. Several methods of design are generally used. Either the parts are designed separately and then they are assembled, or all the parts are designed directly in a model of the assembly, or both methods are used. This last method seems inevitable since standard components or already existing sub-assemblies are employed. One is also forced to coordinate the projects corresponding to complex products, so that several designers can work simultaneously.

In the proposed design process, first, the mechanism is defined as a minimal kinematic structure. Then, one substitutes for the kinematic connections of the assemblies in series, or in parallel, because of convenience, manufacturing or distribution of efforts. The parts are also divided into several parts for the following step, for reasons of assembly, manufacture or economics. There is then an inevitable growing complexity of the mechanical structure. Some technical choices are carried out at each level and it would be interesting to evaluate their geometrical influence on the expressed conditions. Therefore, we propose to deal with the problem of tolerance in a way integrated with the process of design [1, 2].

The proposed tolerancing method is directly integrated into the design process. The first step, for the designer, consists of defining, in a non-ambiguous form, the external requirements and the features referenced by them. The first step of tolerancing is to satisfy, on the one hand, the external conditions and on the other hand, the internal conditions depending on the technical choices of the designer. These conditions are translated into tolerances between the features of similar parts and clearances at the joints.

In the second step, the designer carries out some choices by breaking down the structure, with more details and more precise technological data. For example, a joint is made of different connected faces. New joints and new parts appear. Some parts in the first step become sub-assemblies with their own joints and parts.

More generally, some joints and some parts for a step will be, in fact, a sub-mechanism at the following step of the design process. Tolerance synthesis follows this process. The tolerances defined at one step become external tolerances at the next step. They are transferred and distributed between the parts, while new assembly requirements appear due to the new technical choices.

The recursive top-down process can be generalised. The different design solutions in terms of technological choices influence directly the dimensional and geometrical tolerances.

Literature Synopsis

Many authors considered the problem of the integration of tolerancing in CAD systems. Almost all current CAD systems are history-based and use feature modelling, parametric and boundary representation as the main geometric model, but a precise and robust tolerance model associated with the geometric model is more difficult to define. The model of the TTRS was planned to connect the tolerances to the geometry by a binary tree of functional features [3, 4]. This method is applied with difficulty, to complex mechanisms and assumes that the parts are well defined. The designer determines the features associations.

Bradley and Moropoulos [5] proposed a unified approach to improve assistance with the design at the early stages. They imagine a model of relations between geometric features enabling tolerances to be allotted according to the parameters of the geometric model. But application to the geometrical tolerances in conformity with the ISO seems more difficult to integrate in this model. Hoffman and Joan-Arinyo [6] proposed a master model to established coherence between tolerance and geometry. The geometry and the tolerances are regarded as two particular views downstream from the master model. But no tool helps the designer to choose the functional tolerances during the design process. We propose a tolerancing method directly integrated into this design process.

41.2 Top-down Design Method for Tolerancing

Top-down Designing

The design step of a mechanical system begins with drafting and validation from the requirement list. Though it can be modified thereafter, it contains for the time being, the description of all the functions to be fulfilled. It comprises information about satisfaction conditions for certain criteria. The designer must imagine the adequate product, taking into account his know-how or knowledge. He must then define this product entirely so that its manufacture is possible.

Kinematics is defined from the very start. It must correspond to various functions. For the mechanical systems design, which we are particularly interested in, it is a key stage. At this stage, the product is described as parts whose relative movements are limited by kinematic joints, there should be only one joint between two parts.

Then, the designer proceeds to two types of modifications. In order to technically construct the joints, the designer can seek a series association of several complementary joints having fewer degrees of freedom. For example, a point contact can be carried out by the series association of a spherical joint and a planar joint. It can also break up a joint into several parallel joints. A revolute joint can be ensured by two bearings in parallel, each one using a standard ball bearing. Intermediate parts then appear.

In order to ease the assembly, maintenance or manufacture, the parts can be divided into several parts which will be embedded one inside the other (Figure 41.2). It is understood that the process of design, kinematics in particular, develops from the basic concept to the local technical solutions. The various structural elements are defined all together and then one part at a time, as in the case of a welded casing, for example. We propose to treat the tolerancing process in a similar way. We will seek to start with a global overview, and then progress towards the individual tolerances, while preserving the links between the two.

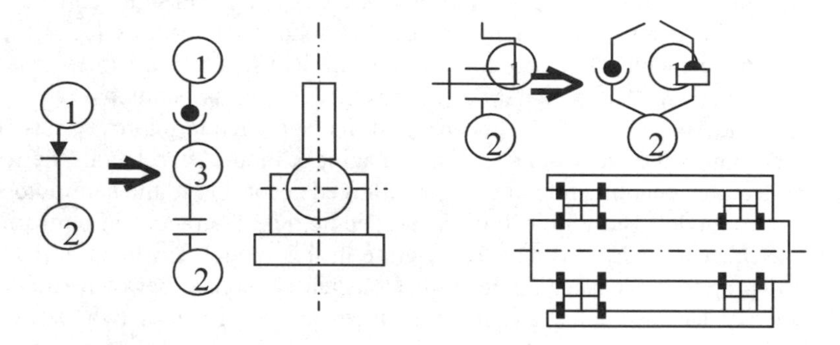

Figure 41.1 Serial and parallel breaking

Figure 41.2 Structure divided in to two sub structures

Tolerancing Process

From the beginning of the design, the kinematics of the machine is defined. We then know the joints and the parts, which they connect. The topology of such a system is usually represented by a kinematic graph; the vertices are the parts connected by arcs representing the joints. It is possible to locate on this graph two types of geometrical variations: defects of the parts, and defects of connection. They are both present between various reference frames placed on the parts.

The defects of the parts are constant, *i.e.* their value is contained in a field depending on the precision of the manufacture. They exist between the reference frames of the joints of the same part. The variations of the joint correspond to constant deviation and clearance, *i.e.* variations occurring during operation. They are present between the reference frames of the joint two different parts.

The tolerable amount of variation can be defined at the beginning of the design. The defects of the parts and the defects and clearances of the joints must be compatible with the expressed functional requirements. The functional conditions must be expressed in a non-ambiguous manner. Generally, it is necessary to define the tolerable limits for the defects, clearances, or compositions between the two. The evolutions during kinematic operation must also be defined. The specificity related to the efforts can be considered if necessary.

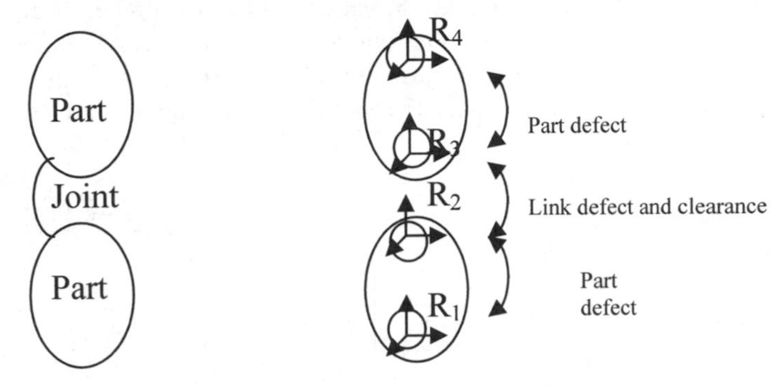

Figure 41.3 Variations of position between reference frames

On the basis of this initial attribution, the evolution of the design must be consistent. At the time of the decomposition of a joint, the allotted defects and variations become conditions to be fulfilled by the various elements. At the time of the decomposition of a part, the geometrical defects between the reference frames of the joints become external conditions, which must be met and ensured by the embedding process. The process is repeated as one progresses in the design and as the solutions are defined.

This tolerancing process is thus closely related to the selected design process. A geometrical data exchange takes place naturally, in particular with regard to the relative positions of the various reference frames. Some choices may need to be discussed and, if necessary, we may have to go up a level in order to reconsider them.

Implementation Example

Let us take the example of a machine tool with three moving axes that is made up of a structure, two positioning tables (X and Y) and a spindle with Z movement. The structure of such a machine is in the first stage made up of a single loop: the part is linked to the table Y, linked to table X, and linked to the structure of the machine, where the spindle and then the tool are attached. We identify six embedding joints and four kinematic joints based on the condition expressed between the part and the tool. There are eleven main parts. The functional condition is thus the result of the twenty-one geometrical variations, which can be placed in the loop. Six of them are position defects due to embeddings. Eleven are geometrical defects of the parts. Four are operating defects and clearances present due to the degrees of freedom.

Let us imagine that we rationally allotted the tolerable quantities of variations during this stage. For example, in the structure of the machine, a variation of position takes place between the reference frame of the embedding joint (X-table) and the reference frame of the spindle. Then the designer can break up the structure into several parts. Let us imagine that he decides to design a higher arm related to a lower part of the structure, both bound by an embedding. Before technically specifying the chosen solution, to realize this embedding, one qualitatively understands the resulting modifications. On the higher arm, there will be a geometrical defect between the reference frames of its two embeddings, similar to that on the lower part of the structure. Then there will be a joint defect at the place of embedding. It is necessary that the three cumulative defects remain lower than the variation.

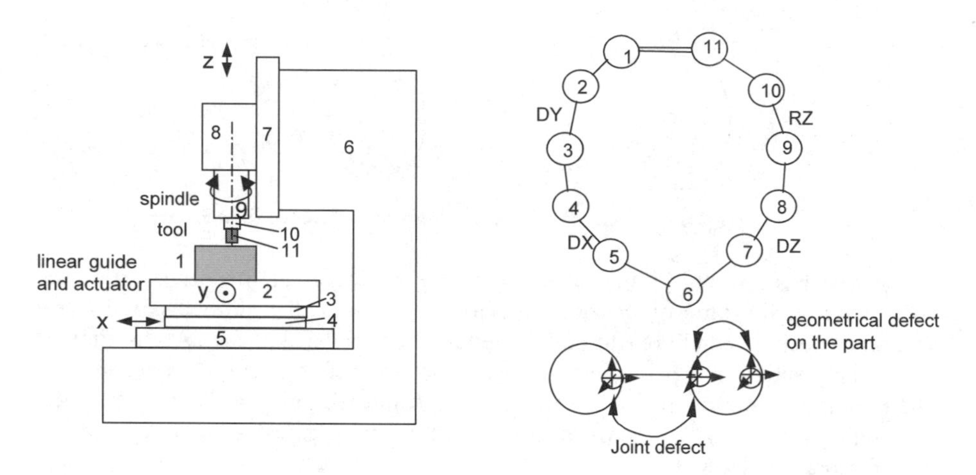

Figure 41.4 Geometrical variations in loop for testing their influence on requirements

41.3 Nested Tolerancing Graph

Before the presentation of the tolerancing graph, some definitions of the main concepts are listed below to provide insights into their meaning.

A feature can be defined here as an elementary face of a part or a set of faces, or a part of a face, or an extension of a face (used for a projective tolerance zone, for example). An elementary feature is a single face of a part among the seven classes: planar, spherical, cylindrical, prismatic, helical, revolute and complex.

A geometric functional requirement is a set of allowed positions and or orientations for a feature with regard to a datum reference frame. This frame is built from features and may possess some degrees of freedom. For example, if the datum is a simple plane face, the datum reference frame also has the three degrees of freedom of the plane. Two or more datum features with some specific priority may be used to establish the reference frame.

The functional features are the different features that occur in the joints between the parts and those concerned with functional requirements.

Three Graph Types

The graph represents the topological structure of a mechanism. It complements the geometric representation. There are three types of graphs: kinematic, contact and tolerancing graphs.

In a kinematic graph, the vertices are the parts and the edges are the joints between two parts. Usually, nine types of joints occur: embedding planar, spherical, cylindrical, revolute, prismatic, and screw joints, and the higher kinematic pairs that can be achieved by linear, circular or point contact. A joint often represents a sub-mechanism including different parts and joints, but the graph is generally simplified as much as possible and only the equivalent joint is represented. For example, a revolute joint between a shaft and a support can include ball bearings, a screw nut, clips, ring, etc, but these auxiliary parts do not appear in the kinematic graph. Different parts embedding each other are generally regarded as a single part. This type of graph is used for kinematic or dynamic analysis.

In a graph of contacts, the vertices are elementary functional surfaces among the 7 classes. The edges are of two types: those that connect the faces of the same part and those that define the contacts between the faces of different parts. This type of graph provides richer information than the kinematic graph. However, one difficulty is the choice of the arcs, which connect the faces of the same part. Generally, the vertices are surrounded rather than linked with one another to form a complete graph. The graph of contact is often used for the synthesis of tolerances [7, 8].

The graph of tolerance, or tolerancing graph, is defined by the following axioms:

- The vertices are the functional features and the tolerances. The arcs specify the toleranced feature and the datum.
- Two types of tolerances are distinguished. The first is intrinsic tolerance, which only relates to one feature. It is a form or size tolerance, for example, a tolerance on the diameter of a cylinder or a sphere, the angle of a cone, or the two intrinsic parameters of a torus. The second type of tolerance is a relative tolerance, which relates at least two features. For example, a dimension tolerance between two parallel opposite planes, a geometrical tolerance with reference, a form tolerance in a common zone, *etc.*.
- Any tolerance enables us to define a frame that can be a datum for another tolerance. The reference frame can have degrees of freedom. For example, a tolerance on the diameter of a cylinder returns the axis of this cylinder. The reference frame has two degrees of freedom (rotation and translation along its axis). A tolerance between two parallel opposite planes returns the median plane.
- When there are several toleranced elements for the same tolerance, these elements are used as the reference – this is called self-referencing.

The graph is a simplified representation. It can thus happen that several solutions of tolerances for one part, given in terms of standardized specifications, correspond to the same graph.

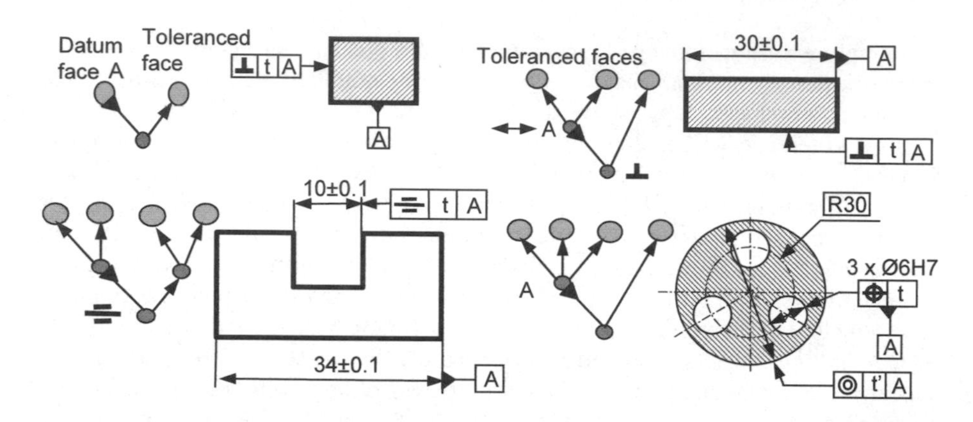

Figure 41.5 Example of current standardized tolerances and the associated graph

Nested Graph

The three types of graphs can be integrated in only one diagrammatic representation. A vertex will be a functional feature or a group of features or a tolerance or a functional requirement. The edges of the graph are either the joints, or the contacts, or the links between tolerances.

Moreover, for the same mechanism there will be several graphs, which correspond each to a more or less important level of simplification. These various graphs will be structured so that an edge or a vertex can be broken up into a subset represented by another graph. At the early design stage, the graph is only constituted by the main parts and joints. On this graph, it will first be possible to make the geometrical functional requirements appear. The mechanical topology and its evolution are represented by this evolving graph. During the design process, the graph is simultaneously updated. Thanks to an appropriate data structure, different views are possible depending on the detail level needed by the designer.

The geometrically allowed variations are installed corresponding to defects or clearances. The description made on the graph will enable us to identify the influences of these variations on the functional conditions. Then the tolerancing graph is translated into ISO tolerances and allows a quantitative evaluation of the intervals of tolerances. The computation of tolerance values can be done according to the model presented in [9].

Example

An example is studied with the different steps to illustrate this integrated method. At each step, a partial parametric geometry is obtained in parallel with the topology graph and a partial associated tolerancing. The influence of the different possible design solutions on the tolerances can be compared in order to validate these choices.

For example, one must design a roller intended to support and guide fine and fragile materials in bands. The geometrical functional requirements are established between the cylindrical roller and the plane support. The position and orientation of the cylinder relative to the bored plan of four tapped holes used for fixing by screws, must be sufficiently precise. The graph and the diagram Figure 41.6 specify these functional conditions.

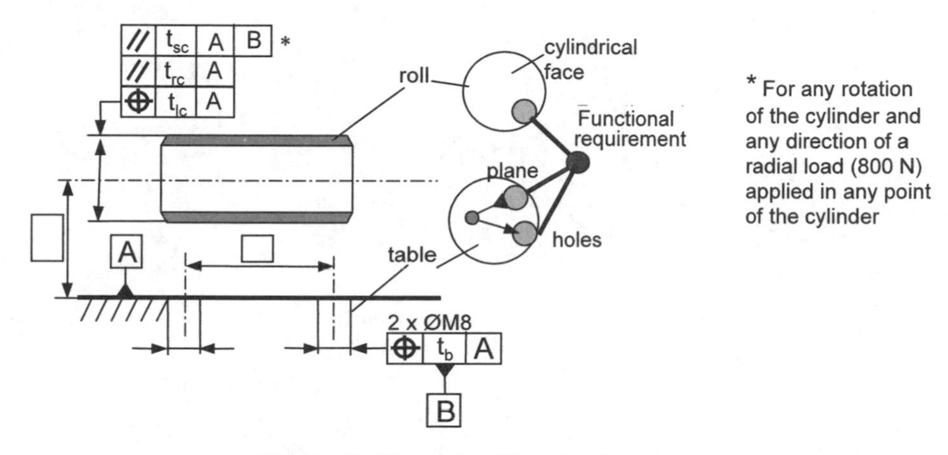

Figure 41.6 Example of functional requirements

The solution can be seen as a simple revolute joint. The mechanism considered is formed of three parts. The external functional condition will result in two functional conditions, which are translated into terms of ISO tolerances. Models of calculation enable us to determine the relation between the values of the tolerances assigned to the functional requirement and those assigned to each part. However, this type of calculation will not be developed here [9].

In the following stage, two joints will define the revolute joint. The embedding joint will also be replaced by joints in parallel, while a new part (the axis) will appear in the loop. The tolerances will become the external conditions for the subsets and will lead to the new tolerances represented in Figure 41.7. The process will continue for the detailed design. For each part, the functional elementary faces will be defined and toleranced. We will then have to build only the parts as manifold solid, which allows the realization of the functional faces. It is possible to imagine a data structure and links with the geometrical model in order to support this design process.

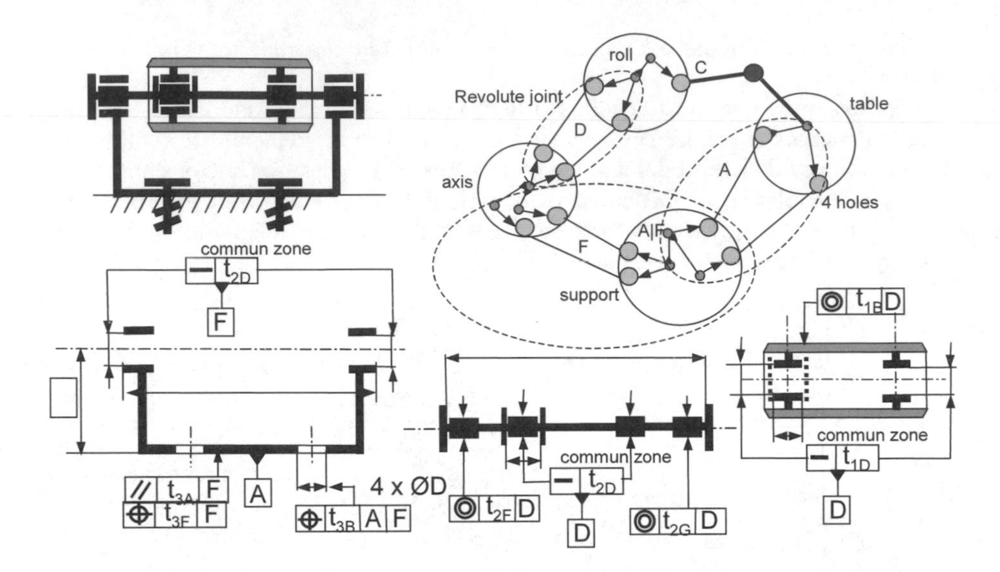

Figure 41.7 A step in the design and tolerancing process

41.4 Conclusion

The proposed design methodology may appear ambitious but it seems necessary to validate the various stages of design or to divide the total project into sub-projects. It must be done in an integrated manner within CAD in particular. The technical choices of joints and characteristics of the parts then become parameters on which the designer will be able to act with a pragmatic aim of designing a functional product.

Our present and future work targets the integration of tolerance synthesis simultaneously not only with the design process but also with the assembly process.

41.5 References

[1] Giordano, M., Hernadez, P., Samper, S., Pairel, E., 2003, "Expression of geometric functional requirements dedicated to concurrent engineering," *International CIRP Design Seminar, Methods and tools for Co-operative and Integrated Design*, 12-14 May 2003, Grenoble, France.

[2] Hernandez, P., Legrais, G., Giordano, M., 2003, "Process definition for computer aided tolerancing," *The 8th CIRP Int. Seminar on Computer Aided Tolerancing*, April 28-29, 2003, Charlotte, North Carolina, USA.

[3] Salomons, O.W., *et al.*, 1996, "A computer aided tolerancing tool I: tolerance specification and II: Tolerance analysis," In: *Computers in Industry*, Vol. 31, pp.161-186.

[4] Desroschers, A., 2003, "A CAD/CAM Representation Model Applied to Tolerance Transfer Method," *Journal of Mechanical design, Transaction of the ASME*, Vol. 125.

[5] Bradley, H.D., Maropoulos, P.G., 1998, "A relation-based product model for computer-supported early design assessment," *Journal of Materials Processing Technology*, N°76, pp. 88-95.

[6] Hoffman, C.M., Joan-Arinyo, R., 1998, "CAD and the product master model," *Computer-Aided Design*, Vol. 30, N°11, pp. 905-918.

[7] Kandikjan, T., Shah, J.J., Davidson, J.K., 2001, "A mechanism for validating dimensioning and tolerancing schemes in CAD systems," *Computer-Aided Design*, Vol. 33, pp.721-737.

[8] Dufaure, J., Tessandier, D., 2003, "Geometric tolerancing from conceptual to detail design," *The 8th CIRP International Seminar on Computer Aided Tolerancing*, April 28-29, 2003, Charlotte, North Carolina, USA.

[9] Giordano, M., Kataya, B., Pairel, E., 2003, "Tolerance analysis and synthesis by means of clearance and deviation spaces," *Geometric product Specification and Verification: Integration of Functionality, selected conference paper of the 7th CIRP*, Kluwer Academic Publishers.

42

Contact and Channel Model for Pairs of Working Surfaces

Albert Albers, Norbert Burkardt, and Manfred Ohmer

Abstract: The elementary design model “Contact and Channel Model” (C&CM) is a new approach to the treatment of technical systems. It connects the abstract level of the function of a technical system with the detailed level of the system’s real shape. This connection is generated by the description of the areas relevant to the function of the system: the Working Surface Pairs and the Channel and Support Structures linking them. Basic hypotheses concerning C&CM are described in [9]. They define the connections between a pair of working surfaces and their linking channel and support structure. Two important hypotheses are that functions can only be generated in Working Surface Pairs and that the fulfilment of any technical function needs at least two Working Surface Pairs and a connecting Channel and Support Structure. In this paper, these two basic hypotheses are validated and explained by means of technical examples.

Keywords: design theory, element model, analysis, synthesis, thinking process

42.1 Introduction

The design of technical systems includes many subconscious processes, which are difficult to determine, and which require a designer with a great deal of experience and expertise.

Although the design process cannot be automated, it is still possible to support the designer in his/her thinking process. In order to be helpful to the designer in everyday work, the support is to precisely describe all technical processes. These should be easily applicable and the designer should be aware, at every stage of the design process, of the consequences his/her changes may cause concerning the entire system.

At the Institute of Product Development at the University of Karlsruhe (TH) the elementary thinking model “Contact and Channel Model – C&CM“ has been used successfully in research and product development for several years. One advantage of C&CM is the abstraction of a technical system, which is very clear and easily applicable to the properties relevant to its function.

The central message, functions are always fulfilled by means of the Working Surface Pairs of the technical system and their connecting Channel and Support Structures, will be further explained and verified in this paper. The objective of this paper is to demonstrate the character of Working Surface Pairs as a condition to fulfil a predefined function. It is referenced to the basic theories of C&CM [3, 9].

42.2 Elementary Model C&CM

The elementary design model C&CM – Contact & Channel Model – was developed at the Institute of Product Development of the University of Karlsruhe (TH) in 1997 [3, 9]. The successful application of this model to several design problems in research, and increasingly to engineering practices, shows that this model is a great help for the designer.

C&CM describes the correlation between the design and the function of technical systems. This correlation exists in the form of the Working Surface Pairs of the system and the Channel and Support Structures linking them. They are also geometrical characteristics of the system as they are the areas where the functions are fulfilled.

One reason for the success of this model is that it does not reduce the system to formulas and matrices, like Roth [12] and Hubka [7]. C&CM is a method that supports the designer in his/her “normal” thinking process. It makes it easier to switch between the abstraction levels of function and design.

There are several basic definitions and propositions that help the designer keep the whole technical system in mind even when he is working on a very special detail of the system.

42.3 Importance of the Pair Character in C&CM

The aim of this paper is to show the relevance of the hypothesis that functions cannot be fulfilled in unique working surfaces and that Working Surface *Pairs* are always necessary to realize a function in a technical system. Moreover, the message that at least *two* Working Surface Pairs and one Channel and Support Structure linking them are required for fulfilling a technical function will be validated.

Basic Definitions of C&CM

The basic hypothesis I [3, 9] reads as follows:
"Every basic element of a technical system fulfils its function by interacting with at least one other basic element.

The actual function – and thus the desired effect – is only possible by means of the contact of one surface with another. These surfaces are working surfaces and together form a Working Surface Pair.

"This basic hypothesis clearly states that one single working surface cannot fulfil a function. It is also not possible to draw conclusions concerning the function of a technical system from the properties of a single working surface when the properties of the other working surface, which forms a Working Surface Pair with the first one, is unknown."

Furthermore, the theory of basic hypothesis II says: "The function of a technical system or a technical subsystem is basically realised by at least two Working Surface Pairs and one Channel and Support Structure connecting them.

In this context only the properties and the interactions of the two Working Surface Pairs and the Channel and Support Structure connecting them determine the function" [3].

This means that a technical system can only fulfil a function when it provides at least two Working Surface Pairs as well as one Channel and Support Structure linking them.

All system quantities – material, energy and information – are conducted via Working Surface Pairs into the technical system and out of it. Inside the technical system the conducting is realized via Working Surface Pairs and Channel and Support Structures. Apart from that, if necessary, the system quantities are also stored in the Channel and Support Structures.

Evidently, it can be concluded that a single surface can never have a function.

The function can only be fulfilled if there are two working surfaces that form a Working Surface Pair. And this Working Surface Pair can only fulfil a function if there is a further Working Surface Pair connected to it by a Channel and Support Structure.

If one of these elements is missing, the considered technical system cannot fulfil a function – neither a desired nor an undesired one.

Example 1: Tolerances and Fittings

The function of fittings and tolerances cannot be explained with the aid of single surfaces. Fittings and position tolerances cannot even be defined by means of a single working surface. The fact that fits and tolerances always ensure the fulfilling of a technical function also confirms the theory, explained above and defined in [9] by the basic hypotheses I and II, that individual working surfaces cannot fulfil functions.

In the following, this message is explained by means of dimensional, form and position tolerances as well as fittings.

Tolerances generally describe the admissible deviations of a component from the theoretically exact nominal value. The indication of tolerances for manufacturing is necessary, as the components cannot be manufactured with only theoretical measures. The magnitude of tolerances is to be selected in such a way that the technical function of the component or the component system can be easily fulfilled while its manufacturing is as economical as possible. Therefore, one basic rule can be directly derived: the tolerances are to be selected as large as possible and as small as necessary.

The designer can only usefully carry out this definition when he knows the function of thc component for which he determines the tolerances. The function of a component is fulfilled in its Working Surface Pairs. Therefore, only the Working Surface Pairs are to be tolerated. Tolerating of the boundary surfaces is generally determined with the aid of general tolerances. Boundary surfaces are only tolerated in exceptional circumstances. This generally prevents the tolerated boundary surface colliding with another surface of the machine system due to an imprecision, which is geometrically very large, and thus fulfils an undesired function by becoming a working surface.

Dimensional Tolerances

Dimensional tolerances describe the admissible deviations of individual dimensions of a component from the theoretically exact nominal value. This means that the property of a Working Surface Pair will not be determined; only the property of a single working surface will be determined. Since according to [9] a single working surface does not fulfill a function, it cannot be determined by means of a single dimensional tolerance whether and how the technical function of a component is fulfilled in its system connection.

If, as in Figure 42.1, the diameter of a shaft end is described with

$$\varnothing\ 50\ \mathrm{h6}, \tag{42.1}$$

The shaft has a diameter of at least 49.84 mm and 50.00 mm at the most. [5]

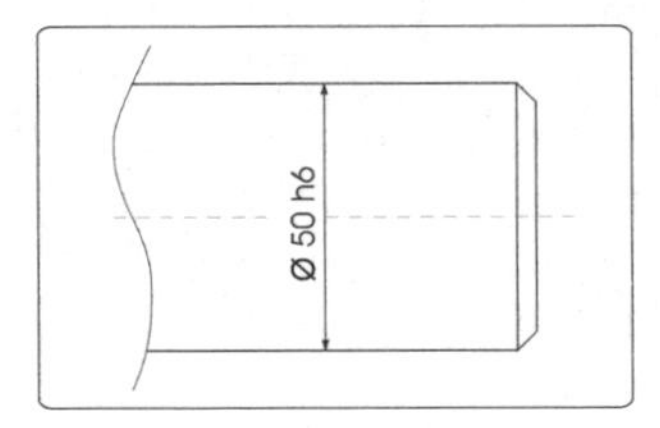

Figure 42.1 Dimensional tolerance of a working surface

From this information the function of this shaft cannot yet be derived. This is only possible with the aid of the tolerance of the counter-working surface as described in the section on "Fittings". The measurement Ø 50 h6 alone does not determine if the shaft can transmit a torque, for example.

The function of the surface considered can thus not be derived from the tolerance of the individual working surface but only from the tolerance of the Working Surface Pair.

Fittings

Fittings result from the tolerances of two paired components. As explained in the section on "Dimensional Tolerances" it is not possible to draw any conclusions concerning the function of a single working surface from its dimensional tolerance as it does not describe a Working Surface Pair and functions always require Working Surface Pairs.

The fitting of a Working Surface Pair defines its function in the system connection. The shaft presented in Figure 42.1 can form a clearance fit with a hub with a diameter of Ø 50 H7 or an interference fit with a hub with a diameter of Ø 50 R7.

Figure 42.2 shows a combination of the shaft with a hub where the inner working surface is tolerated with a dimensional tolerance of Ø 50 H7. The combination of the two working surfaces forms a Working Surface Pair with the fitting Ø 50 H7/h6. This is a clearance fit which means that by this fitting, no normal forces F_N perpendicular to the working surface pair in a radial direction are generated and thus according to the Coulomb law,

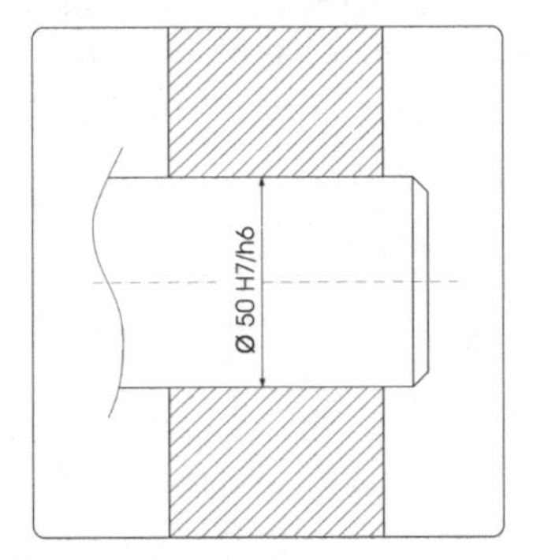

Figure 42.2 Fitting of a Working Surface Pair

$$F_R = \mu \cdot F_N, \tag{42.2}$$

Note that no friction forces F_R can be transmitted in a tangential direction. Since the torque transmitted by this joint can be calculated as the product of the tangential friction force F_R and the radius of the Working Surface Pair r_{WSP},

$$T = r_{WSP} \cdot F_R, \tag{42.3}$$

it is evident that the working surface pair can't transmit a torque due to its properties. Therefore, its function is limited to the transmission of radial forces.

A Working Surface Pair that forms a fitting of Ø 50 R7/h6 for example can additionally transmit tangential forces and hence a torque such as the elastic deformation of the channel and support structures, due to the fitting, generates a normal force in the working surface pair. According to Equations (42.2) and (42.3), the transmittable torque can be calculated.

As a result of this consideration and basic hypothesis II (see "Basic Definitions of C&CM"), the function "transmit torque" cannot be fulfilled by the considered Working Surface Pair alone. Even the third Newton's axiom states that action equals reaction [10]. In order to fulfil the function "transmit torque", the shaft thus needs at least one more Working Surface Pair in which the torque can be passed into another component. The same applies to the hub, which may be a gearwheel, for example. In this case, the second Working Surface Pair is located at the tooth flank that is engaged with the counter gear.

Form Tolerances

Form tolerances, similar to dimensional tolerances; determine the maximum admissible deviations of the form of a component with which it can reliably fulfil its function. Figure 42.3 shows a typical form tolerance – cylindricity. Without knowing anything about the properties of the respective counter working surfaces, this form tolerance of the working surface provides no information about the technical function that can be fulfilled in the formed Working Surface Pair. It does not make any sense to tolerance only one of the two working surfaces in concentricity.

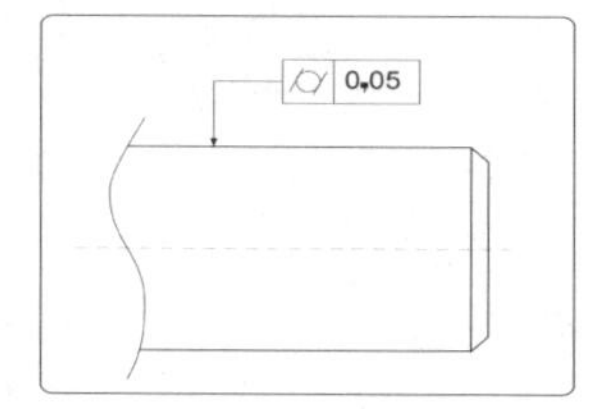

Figure 42.3 Form tolerance of a working surface

Only the additional tolerancing of the counter-working surface determines the property of the Working Surface Pair and therefore ensures the fulfilling of the function. So, we must think in Working Surface Pairs in order to determine a function-oriented form tolerance.

Position Tolerances

All messages concerning the pair character of form tolerances contained in the section on "Dimensional Tolerances" also apply to position tolerances. Regarding position tolerances, it is even easier to realize that technical functions like those explained in basic hypothesis II [9] require at least two Working Surface Pairs in order to be fulfilled.

A positional tolerance defines the property of a working surface with regard to another working surface of the component. Thus, as shown in Figure 42.4, for example, the perpendicularity of the axial working surface to the cylindrical Working Surface of the shaft can be toleranced in order to guarantee an accurate axial force transmission from a roller bearing to the shaft. A functional-oriented tolerancing is only possible regarding both the radial and the cylindrical Working

Surface Pairs of the subsystem and their functional and geometrical interrelations. The connection between these can be easily regarded on the abstract level of C&CM.

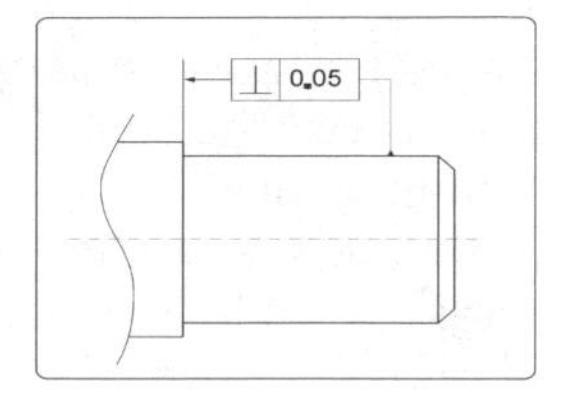

Figure 42.4 Positional tolerance of a working surface

Experiences

In a current exchange of experience with a business partner, it became evident that the abstract thinking of the elementary model is very important for the designer in order to correctly implement these obvious conclusions into the individual design. In the case described above, an inexperienced designer determined form tolerances for a surface, which was always a boundary surface, and never a working surface during the operation of the component. This caused unnecessary high costs.

If the designer in question had used the thinking model C&CM for his considerations, this error would never have occurred as he would have looked for the fitting with the counter working surface of the entire system, as described in the section on "Fittings." It is at this point, he would have noticed the error at the latest.

Example 2: Abrasive Wear

Abrasive wear is a mostly undesired function that occurs in virtually all technical systems. In connection with the basic hypothesis II, "Only the properties and the interactions of the two Working Surface Pairs and the Channel and Support Structure connecting them determine the functions" [9]. It can be directly concluded that this function occurs in a Working Surface Pair.

The message that not only the properties of one of the two working surfaces, but also those of the Working Surface Pair is of considerable importance.

The wear behaviour of a Working Surface Pair is substantially influenced by the geometrical micro- and macro-structure of the two working surfaces as well as by their material properties such as hardness, for example.

If one of the two working surfaces is much harder than the other, the softer one will generally experience more wear. If the properties of the two working surfaces are similar, wear can normally be observed on both surfaces. Especially with metallic materials the wear can be very high when the two working surfaces have almost identical properties. In this case, inter-molecular forces are developed at many places as two similar structures encounter each other. In order to prevent that in special cases, in which one can expect that the lubricating film between two surfaces sliding on each other can break down, metals with completely different structural properties are often employed. Classic examples of this are slide bushings made of brass or bronze, which contain a steel shaft.

Consequently, it is not possible to draw any conclusions concerning the wear behaviour only from the knowledge of one single working surface of the Working Surface Pair. It is always necessary to have information about both partners forming the Working Surface Pair and thus fulfilling the function in the system connection. A brass slide bushing may show excellent wear properties in connection with a steel shaft. If a shaft made of the same brass as the slide bushing is installed into the machine system, both working surfaces may wear out faster than with the original material pairing.

Example 3: Conduction of Electric Energy

The conduction of energy in a technical system is another example which illustrates very clearly the advantage of the pair consideration in the elementary model C&CM. If a form of energy is to be conducted through a system, this energy generally needs to be conducted via several Working Surface Pairs and through several Channel and Support Structures.

Conducting energy in a Working Surface Pair can again only be explained and understood by considering the complete Working Surface Pair. The example of electric current being conducted by means of a switch plug demonstrates why it is not possible to draw a conclusion concerning the change of energy conducted in a Working Surface Pair from changing the property of one of the two working surfaces contained in a Working Surface Pair.

If, due to such a switch plug, the current cannot be completely transmitted, this problem cannot be solved by considering and improving only one of the working surfaces contained in the Working Surface Pair. It is necessary to consider the entire place where the function is fulfilled, *i.e.* the entire Working Surface Pair. Gilding the switch plugs - a very expensive process that is carried out frequently – only makes sense when good conductivity of a gold film on one working surface also shows a better end result in combination with the other working surface. If there is a problem due to corrosion of the counter-working surface, it will not be possible to considerably improve the electrical conductivity by gilding the other working surface. Mechanical problems such as a missing contact pressure in the Working Surface Pair cannot be enhanced this way either.

The basic message of this knowledge can be directly applied to further types of energy conduction such as thermal conduction. By means of the example of conducting electric current the message of the basic hypothesis II (see Section 42.2) [9] can also be clearly understood: The electric current is always conducted in a Working Surface Pair into a part, passes through its Channel and Support Structure and is conducted in the next Working Surface Pair into the next body. If one of the two Working Surface Pairs is changed or even removed, this will affect the entire technical system.

42.4 Conclusions

The examples described above illustrate the importance of the pair character for working surfaces and Working Surface Pairs in the elementary model C&CM as well as the power and the diversity of this instrument.

As the basis for numerous successful research and business projects, the elementary model C&CM has proved to be a precise multi-purpose thinking model with which even the most complex problems from the conception to the validating and manufacturing of technical systems can be reliably solved. The connection of function and shape is a valuable support for the designer whose primary task is to fulfil the required function by defining the embodiment design of a technical system.

Current research projects at the MKL Institute deal with creating rules to support the designer in being able to constantly apply the theory of the elementary model C&CM in everyday work and thus to quickly generate solutions. Apart from that, in the last few years, the elementary model C&CM has also been the basis of the lecture "Mechanical Design" at the University of Karlsruhe (TH). In this lecture, all elements and types of behaviour of a technical system are based on this thinking model. It was proven that this way of thinking makes it much easier for students to understand the complex principle of technical systems [1, 2].

42.5 References

[1] Albers, A., Matthiesen, S. and Ohmer, M., 2003, "Evaluation of the Element Model, Working Surface Pairs & Channel and Support Structures," *Proceeding of International CIRP Design Seminar 2003*, *Methods and Tools for Co-operative and Integrated Design*, May 12-14, 2003, Laboratoire 3S, Grenoble, France.

[2] Albers, A., Matthiesen, S., and Ohmer, M., 2003, "An innovative new basic model in design methodology for analysis and synthesis of technical systems," *Proceeding of 14th International Conference on Engineering Design ICED 03*, August 19-21, 2003, Stockholm, Sweden.

[3] Albers, A., Matthiesen, S., 2002, "Konstruktionsmethodisches Grundmodell zum Zusammenhang von Gestalt und Funktion technischer Systeme - Das Elementmodell" "Wirkflächenpaare & Leitstützstrukturen" zur *Analyse und Synthese technischer Systeme,* Konstruktion, Zeitschrift für Produktentwicklung; Band 54; Heft 7/8 - 2002; Seite 55 bis 60; Springer-VDI-Verlag GmbH & Co. KG; Düsseldorf.

[4] Edited by W. Beitz and K.-H. Küttner, 1994, *DUBBEL – Handbook of Mechanical Engineering*, Springer, Berlin, Heidelberg, New York.

[5] DIN ISO 286: ISO System of limits and fits; Beuth, 1990.

[6] GfT-Arbeitsblatt 7 - Tribologie, Moers, 2002.

[7] Hubka, V., 1984, *Theorie technischer Systme*, Springer, Berlin.

[8] Lindemann, U., and Pulm, U., 2001, "Enhanced Systematics for functional Product structuring," *Proceedings of ICED '01, Design Research –*

Theories, Methodologies, and Product Modelling, Vol. 1, Glasgow, pp 477-484.

[9] Matthiesen, S., 2002, *Ein Beitrag zur Basisdefinition des Elementmodells "Wirkflächenpaare & Leitstützstrukturen" zum Zusammenhang von Funktion und Gestalt technischer Systeme,* ISSN 1615-8113; Hrsg: o. Prof. Dr.- Ing. Dr. h. c. A. Albers; Karlsruhe.

[10] Newton, I., Bernoulli, D., and d'Arcy, P., 1914, *Abhandlungen über jene Grundsätze der Mechanik, die Integrale der Differentialgleichungen liefern,* Ostwald´s Klassiker der exakten Wissenschaften; Leipzig ; Berlin.

[11] Pahl, G., and Beitz, W., 1997, *Konstruktionslehre*, Springer, Berlin.

[12] Roth, K., 1994, *Konstruieren mit Konstruktionskatalogen,* Springer; Berlin.

[13] Tollenaere, M., Belloy, P., and Tichkiewitch, S., 1995, "A part description Model for the preliminary design," *Advanced CAD/CAM Systems – State-of-the-art and future trends in feature technology*; Chapman & Hall, London.

Part IX

Modeling and Design for Manufacturing

43

Manufacturing-driven Design of Sculptured Surfaces

Ahmad Barari, and Hoda ElMaraghy

Abstract: Designers in every industry from automotive, aerospace and telecommunications to medical equipments and biomedical artifacts strive to enhance the product design efficiency in order to cope with changing demands, while delivering higher quality products with shorter lead time and for less cost. Effective integration of the product design and process planning can improve manufacturability and maximize satisfaction of the designer's intent. This paper presents a Design For Machining (DFM) tool that enables designers to the estimate effect of the design decisions on the accuracy of the machined product, particularly those containing sculptured surfaces. Actual variations of machined features can be predicted using the proposed analytical method. The comparison of these variations with the nominal allocated tolerances identifies critical portions of the design where unacceptable deviations may occur after machining. Constraints may be imposed on the design space to take into consideration the manufacturing limitations, increase parts acceptance and reduce scrap and rework. The designers can use these results to guide or drive the product design either by changing the design geometry or by modifying the specified design tolerances. The developed method is applicable to any geometry and is particularly useful and efficient for designing accurate sculptured surfaces. A sculptured surface auto-part is used for illustration.

Keywords: design for manufacturing, tolerance allocation, machining errors

43.1 Introduction

The concept of Design For Manufacturing and Assembly (DFM/A) has become widely used in industry to integrate product design, process planning and manufacture [1]. DFM/A enhances manufacturability, reduces costs and promotes better appreciation of the consequences of design decisions on the product manufacture and assembly. Most currently available DFM/A techniques are focused on the assembly aspects such as reducing the number of parts or product complexity to decrease assembly time and cost. The DFA techniques, in spite of their demonstrated benefits, are still not widely applied. The available Design for Manufacturing (DFM) techniques are even fewer and more underutilized. They are mostly limited to general guidelines for selecting materials and processes or for analyzing manufacturing cost. They also provide design guidelines for parts with standard primitive features.

Machining is important because it plays a role in the manufacture of the majority of parts. The limitations of machine tools accuracy impose new constraints on the design choices of shape, dimensions and tolerances. Machining errors lead to costly rework and scraps. Their effect may be reduced by using highly accurate and very costly machines or by adapting the design specifications to achieve the desired shapes and accuracy.

The specification of the constraints set is one of the major difficulties in managing constrained design. It requires extensive engineering knowledge and experience in the design process. Engineering tools and software have been developed to assist the designers in specifying functionality constraints, while tools for capturing the manufacturing constraints related to geometry are still in their infancy. There is a need for DFM tools that enable the designers to estimate the effect of early design decisions on the resulting product accuracy.

In this paper, we propose an approach that considers the characteristics of the used machine tools and the limitations imposed by their errors and capabilities and their effect on the machined products accuracy. This technique can be used as a "Design For Machining" (DFM) procedure during the design process to enhance the design efficiency.

In this approach, an error model for the machining workspace is developed based on the machine tool capability. The actual machining errors of products are estimated using this error model, which can be used to predict actual variations in the product geometry. Specific areas of the geometry that require special attention can be identified by comparing the actual variations with the nominal tolerances. This enables the designer to consider modifying the design geometry or the specified tolerances early in the design stage.

The proposed DFM technique is useful and effective for designing of individual precise artifacts and components with complex geometries. Products in this group, such as dies and tools, medical instruments and biomedical implants, mostly have critical and important functionality that demands very careful design decisions.

43.2 Currently Available Design for Machining Tools

About 80% of manufacturing decisions are directly related to product design decisions [2]. Design changes at the manufacturing stage are very costly; hence the product design must be optimized for manufacturability at the design stage. Designers must be provided with knowledge of available manufacturing processes and their capabilities, tools and fixtures as well as DFM support tools. Benefits of this approach include improving manufacturability, involving the designer in the downstream processes, better communication between the departments, and shorter time to market of new products [2, 3].

Boothroyd and Dewhurst introduced formal DFM concepts in 1983. These design for manufacturing guidelines were developed based on comprehensive work-studies that related the part characteristics to ease of handling, manipulation, assembly and degree of symmetry, *etc.*. There are many books on DFM and DFA that list good practice rules explicitly or in the form of geometric shapes and their manufacturability attributes [3, 4].

Guidelines and rules presented to design for machining are mostly for identifying undesirable features which can be categorized as: features impossible to machine, features extremely difficult to machine that require the use of special tools or fixtures and features that are expensive to machine even though standard tools can be used [3]. Figure 43.1, shows examples of some DFM considerations for designing of turning and milling parts.

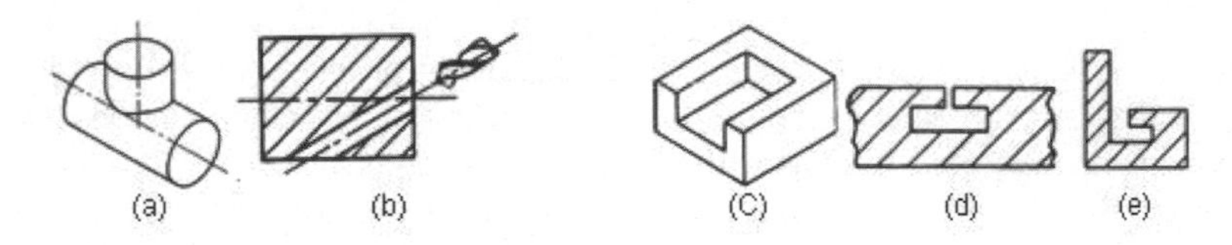

Figure 43.1 Examples of undesirable machining features; (a) impossible to machine, (b) difficult to machine-deflection of drill, (c) expensive to machine, internal sharp corners can not be machined by standard tools, (d) & (e) impossible to machine by standard tool.

In common practices, design for machining refers generally to the ease and cost of material removal operations. Although there are a myriad of research projects around the world in the area of automated DFM/A tools, there are a few commercial solutions available. This is because the DFM solution must be customized for the industry and company in which it is applied. Consequently, there are no universal or systematic DFM evaluation techniques and the implementation of DFM can be seen in design for the available resources and design for a specific process [2].

Most available tools of design for machining are developed for early cost estimation. The machining cost of design is evaluated using a library of machining process. Process variables including speed and feed rate can be varied, as well as some of the design variables such as material type to experiment with the different combinations. The designer can optimize the design parameters with respect to the machining performance and characteristics using the results of these experiments in an interactive environment. Many research attempts focused on machining cost evaluation and machinability of parts with primitive features. The mapping of design features into machining requirements and steps is one of the challenging tasks in these research efforts. Features recognition and feature-based design methods are commonly employed [5, 6, 7].

Another well-developed issue in design for machining research is the decision support tools. Applications of artificial intelligence, expert systems, knowledge engineering, fuzzy logic and neural network were developed to support decision-making considering machining cost and capabilities. Designing for available resources or for a specific process defines the level of design customizing considering parameters such as the product accuracy, quantity and cost. Although cost and accuracy are generally two correlated parameters, but in a customized design space, these two parameters are often treated relatively independently. However, when design is customized for a specific process or an available machine tool, apart from of the cost of production, limitations on achievable accuracy do exist.

43.3 Methodology

Tolerance Allocation in Design

Tolerance allocation is one of the major challenges in product design. It refers to a class of design problems where the tolerance values imposed on dimensions and geometries are selected to satisfy the design constraints [7, 8]. Traditionally, for each part, the design constraints are determined by the interactions between its corresponding mating parts in an assembly. Interactions of all dimensional and geometric tolerances in an assembly create a tolerance chain governing the whole system which should satisfy the assembly functional requirements. Effective DFM supporting tools should ensure that designated tolerances not only satisfy the desired functionality but can also be manufactured by available resources.

Nassef and ElMaraghy developed a method to simulate machining processes of a feature of size to consider machining errors in the process of dimensional and geometric tolerance allocation [9]. They assumed that each geometric attribute is a random variable with a probability distribution. They simulated a manufacturing surface by a set of points with a multinormal distribution. An optimization procedure was used to calculate the probability of violating the assembly functional requirements. This statistical approach has been employed by many investigators to simulate machining errors at the design stage. Other techniques such as artificial intelligence and fuzzy logic were also used to evaluate the machining errors [10].

Many researchers reported that the majority of machining errors are systematic and predictable [11, 12]. Therefore, instead of using statistical random variables to estimate machining errors, more accurate results can be achieved by predicting the errors analytically by modeling the machine structural characteristics.

Systematic Machine Tool Errors

In the machining process, the accuracy of the final product is affected by the error of the relative movement between the cutting tool and the work-piece [11]. The relationship between any point (Figure 43.2), defined by vector $\overline{p}_i$, in the desired geometry and the corresponding point on the actual surface, vector $\overline{p}^*_i$, where index i is the point index and the corresponding machining error vector is designated by $\bar{\varepsilon}_{T_i}$, can be described as:

$$\boxed{\bar{p}^*{}_i = \bar{p}_i + \bar{\varepsilon}_{T_i}} \tag{43.1}$$

Machine tool errors are the major component of the machining errors, which result from all the systematic and non-systematic error sources attributable to the machine tool. Systematic errors are quasi-static in nature; they vary very slowly in time and are related to the structure of the machine itself [11, 12]. They are systematically repeated during the machining process; which creates an interesting compensation potential for this type of errors. Quasi-static errors are reported as the major part of the total machining errors and in general are estimated to account for 70% of total machine tool errors [11, 12, 13].

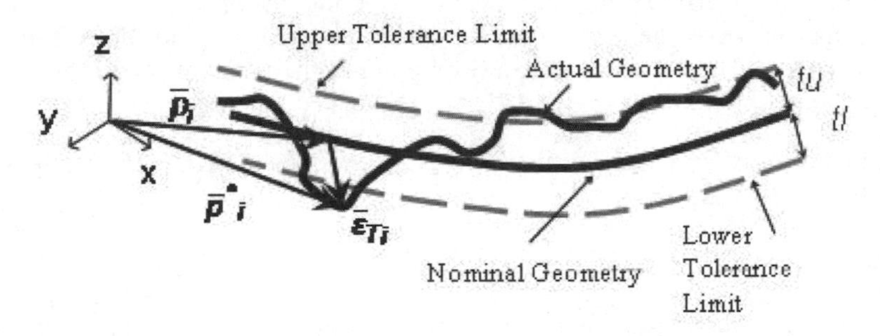

Figure 43.2 Desired and actual geometries

Multi-axis machines are composed of a sequence of elements or links connected by joints to provide motions. The rigid body kinematics of each element and joint can be modeled using homogeneous coordinate transformations to describe the position and orientation of one object with respect to several different coordinate systems. In a typical machine tool with prismatic joints, for each translation axis there exist errors in six degrees of freedom in addition to the intended motion. By assigning a coordinate frame to a slide and using a homogeneous transformation matrix, it is possible to describe the motion of the slide in a reference coordinate system [13]. Therefore, a machine tool error model can be derived for specific machine types. Using the small angle approximation, the desired X slide motion, ***Hx***, with all the unwanted motions can be represented by:

$$\boldsymbol{Hx} = \begin{bmatrix} 1 & -xRz-Sxy & xRy+Szx & xTx+DX \\ xRz+Sxy & 1 & -xRx & xTy \\ -xRy-Szx & xRx & 1 & xTz \\ 0 & 0 & 0 & 1 \end{bmatrix} \tag{43.2}$$

Motions of the other slides, ***Hy*** and ***Hz***, are represented similarly. In the notations used to depict parametric errors, R means rotation, and T means translation. The left hand lowercase letter refers to the moving slide and the right hand lowercase letter refers to the error direction. Three squareness errors are defined as Sxy, Syz and Szx where in these notations S means squareness, and the two following letters indicate that the error is between these two reference axes. Since a machine tool can be considered as a chain of linkages, the spatial relationship between the cutting tool and the work-piece can be easily determined. The transformation matrix of the total system, **H**, for a generic three-dimensional machine tool is as follow:

$$\boxed{H = H_x \cdot H_y \cdot H_z} \tag{43.3}$$

Derivation of matrix ***H*** requires heavy symbolic manipulation. The forth column of matrix, ***H***, is the actual position of the cutter relative to the reference coordinate system. Therefore, for any desired point in the nominal geometry, with coordinates of $\overline{\mathbf{p}} = [DX \quad DY \quad DZ \quad 1]^T$, the actual point due to quasi-static machine tool error is $\overline{p}'_i$ which is calculated as follows where ($\mathbf{j} = [0 \quad 0 \quad 0 \quad 1]^T$):

$$\boxed{\overline{\mathbf{p}}' = \mathbf{H.j}} \tag{43.4}$$

This equation represents a point-to-point relationship between $\overline{p}'_i$ and $\overline{p}_i$. It can be seen that the distribution of machine tool error is a function of nominal geometry and it varies according to the geometrical complexity of the design [13].

Implementation

Using Equation (43.4) the corresponding actual point for any desired point in the nominal geometry can be calculated. These points represent the actual machining surface. The normal distance between any point and the nominal surface specifies the actual machining deviations, $t\overline{m}_i$ s.

$$\boxed{|t\overline{m}_i| = \|\overline{p}_i' - \overline{p}\|} \tag{43.5}$$

By comparing the magnitude of actual variations due to machining, $t\overline{m}_i$, with values of upper and lower limits of design tolerances, tu and tl, the ability to achieve the desired design can be evaluated.

In order to implement the described technique, a NURB (Non-Uniform Rational B-spline) based algorithm is developed. The NURB representation is chosen because it is the most flexible method to model both analytical and/or sculptured surfaces. Therefore, it provides sufficient flexibility to the designer for complex geometries with combination of analytical and sculptured features.

In the developed system, a library of machine tool independent errors is available. Inherent elemental errors in the machine tool workspace are measured and stored in this library during the regular calibration of machine tools. For example, in a typical three axis milling machine 21 independent error parameters, including undesired translations and rotations of each axis and their mutual squareness errors, should be known to apply Equation (43.4). In order to evaluate the designed geometry and assigned tolerances, the designer initially selects an available machine tool from the library and a specific machining set up for the part.

The first step of the algorithm is the equi-parametric meshing of the part's surface. Generated nodes on the surface are sampling locations where machine tool errors at those points will be evaluated. Next, the deviated location of each sampling node is calculated using Equation (43.4).

The resulting set of points represents the actual machined surface. In order to evaluate this surface, using Equation (43.5), a machining deviation vector for each point is calculated. An optimization algorithm is used to find the normal distance between each point and the nominal surface. The evaluated range of variations obtained using this process represents the machining variations zone for the specific design. Simulation and numerical experiments of the described method for variety of geometric primitives and sculptured surface were conducted. The distribution of calculated machining variations for the design being considered is obtained and visualized by assigning a colour map to the range of estimated error values. This feedback to the designer can be used to either adjust the design tolerances or locally modify the geometry and reduce its complexity.

Experimental Results

In order to illustrate the performance of the developed method, an actual part is used as an example in this section. Figure 43.3, shows a section of a car door with typical curvatures of sculptured surfaces found in automotive dies. The initially selected tolerance range for the profile is equal to ±0.1mm.

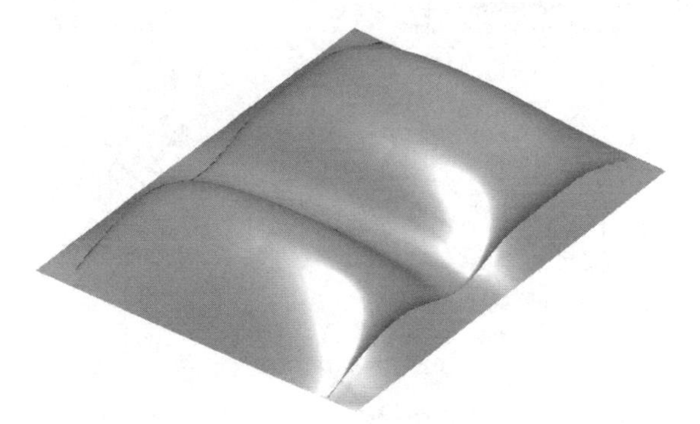

Upper Design Tolerance=0.1 mm
Lower Design Tolerance=-0.1 mm

Figure 43.3 Desired geometry, a forth order NURB using 64 control points

The model's overall dimensions are 350mm×270mm×60mm and it is represented by a fourth order NURB surface with 64 control points. The model analysis starts after selecting the machine tool and part's set up by the designer. The selected machine tool is a vertical CNC milling machine and an actual calibration dataset reported in [12] is used for this simulation.

Next, an equi-parametric meshing of the model for 40×40 sampling nodes is performed. Then for all 1600 specified nodes on the model, the actual machine tool error is calculated. In order to simulate the non-systematic part of machining errors, a normally distribution error is added to the quasi-static errors in all three X, Y and Z directions. The distribution of the applied noises is chosen such that they represent almost 30% of the total machining errors. The distribution of non-systematic errors has a mean equal to zero and standard deviation of 10 μm, which means that the non-systematic errors statistically vary between -30 to 30 μm in any spatial direction. Figure 43.4, represents the machining set up of the meshed part and the actual points obtained using machining simulation. The calculated error vectors are magnified 80 times in Figure 43.4 for better visualization.

In order to quantitatively assess the accuracy of the final machined part, the normal distance between each point and the nominal surface and the machining deviations for this product are evaluated. Graphical representation of machining tolerances is presented in Figure 43.5.

A distribution of errors with maximum of $Emax=0.1441mm$, minimum of $Emin=-0.1384mm$, mean of $Eman=0.0163mm$ and variance of $Evar=0.0036mm$, was obtained. It can be clearly seen that some portions of the surface are not within the designated tolerance range (±0.1mm).

Depending on the other design constraints, the designer may consider modifying the specified tolerance to ensure the machined product quality. Using the information obtained by DFM analysis, assigned tolerances to the critical portions of the part might be increased by adjusting tolerances of their mating parts or other design variables.

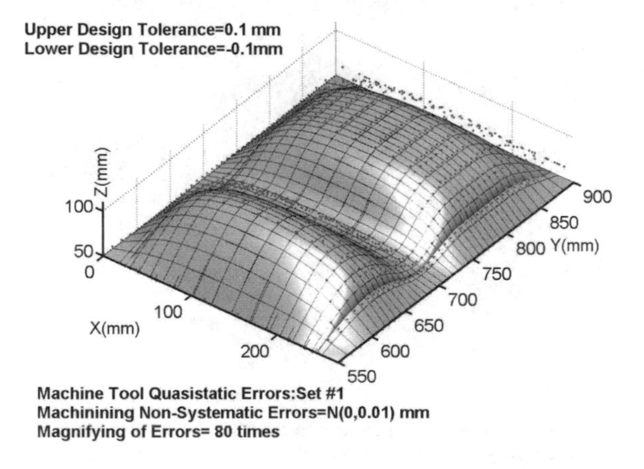

Figure 43.4 Machining set up of meshed part and machining simulation points

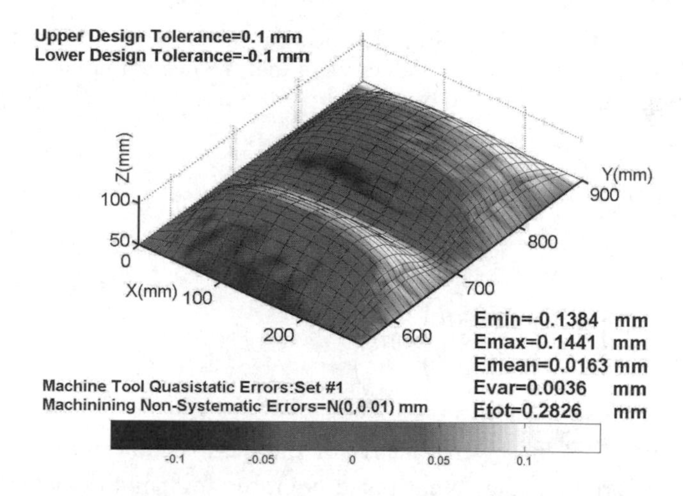

Figure 43.5 Graphical representation of geometric deviations after machining

43.4 Compensation of Machining Errors by Changing the Nominal Geometry

Modification of the allocated tolerance based on the estimated machining errors is a reasonable approach that satisfies the product machinability requirements. However, sometimes other design constraints interfere with adjusting the tolerances sufficiently to satisfy the machining constraints. Minor changes in the nominal geometry, used to generate the machining instructions, to compensate machining errors might be a feasible alternative approach in this situation. To achieve this aim, using Equation (43.4) and an inverse kinematics approach, for any point on the nominal geometry a new corresponding point can be found that in effect maps the machining errors on the nominal geometry. It can be seen that for machine tools with prismatic joints, a linear operator for Equation (43.4) can be found. This operator, $\mathbf{O_Q}$, is a transformation matrix which is called quasi-static errors operator and for a typical three axes machine it is equal to: [13]

$$\mathbf{O_Q} = \begin{bmatrix} 1 & -xRz - Sxy & \Psi_1(xRx,...) & \Psi_4(xRx,...) \\ 0 & 1 & \Psi_2(xRx,...) & \Psi_5(xRx,...) \\ 0 & xRx & \Psi_3(xRx,...)+1 & \Psi_6(xRx,...) \\ 0 & 0 & 0 & 1 \end{bmatrix} \tag{43.6}$$

In this model, Ψ_1 to Ψ_3 and Ψ_4 to Ψ_6 are second and third order polynomial functions of independent machine tool errors. Because of the small magnitude of machine tool errors and consequently all of these functions, the determinant of this matrix is very close to unity which guaranties availability of an inverse matrix for operator $\mathbf{O_Q}$. Therefore, for any point that lies out of the desired tolerance range, the required compensated point, which ensures that the desired point are achieved after machining, can be calculated by:

$$\bar{p}^{Comp.}{}_i = O_Q^{-1} \bullet \left(\bar{p}'_i - \left(\frac{|\overline{tm}_i| - tu}{|\overline{tm}_i|} \right) \times \overline{tm}_i \right) \tag{43.7}$$

Or:

$$\bar{p}^{Comp}{}_i = O_Q^{-1} \bullet \left(\bar{p}'_i + \left(\frac{|\overline{tm}_i| - tl}{|\overline{tm}_i|} \right) \times \overline{tm}_i \right) \tag{43.8}$$

Where the $\mathbf{O_Q}^{-1}$ is the inverse matrix of the quasi-static errors operator and $\bar{p}^{comp}{}_i$ is the corresponding compensated point for any point out of the desired tolerance range. *tu* and *tl* are values of upper and lower tolerance limits. Equations (43.7) and (43.8) are used when the error point is located respectively in the outer side or inner side of nominal surface. Using Equations (43.7) and (43.8), sufficient number of compensated points to fit a new NURB surface, are calculated [9]. By fitting a NURB surface to the obtained compensated points, a new design model is generated, which guaranties obtaining the original desired geometry after machining. As an example, implementation of this method for the part analyzed in the last section is presented. Consider the same nominal geometry with the same machine tool and set up is desired, but, the allocated tolerance range is equal to ±0.05mm that is tighter than required in the previous example. By calculating machining variations, results similar to those presented in Figure 43.4, are obtained. It can be seen that the calculated error for most surface portions is more than ±0.05mm; In this case, adjustment of design tolerance may cause an undesirable change in the allocated tolerances of the whole assembly. In order to solve this problem, for any simulated point whose associated error is more than the defined tolerance range, using Equations (43.7) and (43.8), a compensated point is calculated.

A new nominal geometry is generated by fitting a NURB surface to the obtained set of compensating points, which is slightly different from the original one. The final result is presented in Figure 43.6.

It can be seen that the total deviation of the final product from the original geometry is 0.09998 mm that is within the rage of the specified design tolerance.

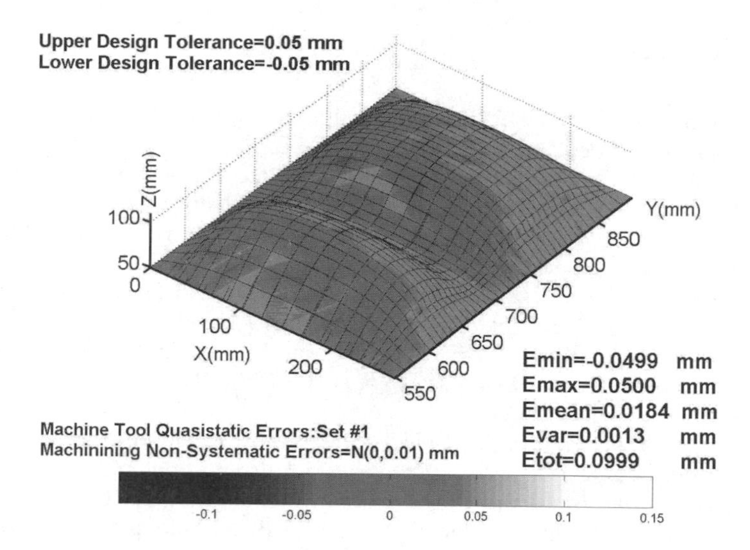

Figure 43.6 Reduction of errors by using a compensating geometry

43.5 Conclusion

In this paper a Design For Machining technique that closely links the manufacturing capabilities to the product design features and reduces manufacturing cost is presented. It allows the designers to evaluate the consequences of design decision on the expected quality of machined products and its manufacturability. These results can be used to modify the nominal geometry and/or re-allocate tolerances in geometry regions whose expected machining errors would exceed the specified tolerance range. This technique can be employed for designing of high precision sculptured surfaces in different industries. Future work includes the optimization of the design variables in view of the results of design for machining analysis.

43.6 Acknowledgements

Guidance provided by Professor G. K. Knopf, Department of Mechanical and Materials Engineering, University of Western Ontario is acknowledged. The authors acknowledge the valuable contributions of Professor Waguih ElMaraghy, Department of Industrial and Manufacturing Systems Engineering, University of Windsor, through the AUTO21 design project. Funding by the Natural Science & Engineering Research Council of Canada (NSERC) & AUTO21 Network of Centres of Research Excellence and collaboration with Canada's National Research Council & its Integrated Manufacturing Technologies Institute (IMTI) are also acknowledged.

43.7 References

[1] Boothroyd, G., 1994, "Product Design for Manufacture and Assembly," *Computer Aided Design*, Vol. 26, pp. 505-515.

[2] Vliet, J.W., and Luttervelt, C.A., 1999, "State-of-the-Art Report on Design for Manufacturing," *4th Design for Manufacturing Conf., ASME Design Engineering Tech. Conf.*, 1999, Las Vegas, Nevada; DETC99-DFM-8970.

[3] Boothroyd, G., Dewhurst, P., and Kinght, W., 2002, *Product Design for Manufacture and Assembly*, Second Edition, N. Y.: Marcel Dekker Inc.

[4] Bralla, J.G., 1986, *Handbook of Prod. Design for Manuf.*, McGraw-Hill.

[5] Gupta, S.K., Regli, W.C., Das, D., and Nau, D.S., 1995, *Automated Manufacturability Analysis: A Survey*, ISR-TR-95-14, NIST, IR 5713 Carnegie Mellon University.

[6] Barari, A., and Arezo, B., 1997, "A Feature Recognition Approach for CNC Milling of 2 1/2 parts," *Proc. of 3rd Iranian Conf. on Manufacture Engineering*, Amirkabir Univ. of Tech., Tehran, Iran, pp. 111-121.

[7] Gadalla, M.A., and ElMaraghy, W.H., 1997, "Tolerancing of Free Form Surfaces," *5th CIRP Int. Seminar on Computer Aided Tolerancing*, 1997, Toronto, ON

[8] Nassef, A.O., and ElMaraghy, H.A., 1997, "Allocation of Geometric Tolerances: New Criterion and Methodology," *CIRP Annals 1997*, Vol. 46/1, pp. 101-106.

[9] Nassef, A.O., and ElMaraghy H.A., 1995, "Statistical Analysis and Optimal Allocation of Geometric Tolerances," *Proc. of the Computers in Engineering Conf. and the Engineering Database Symposium*, ASME 1995, pp. 817-823.

[10] Ji, S., Li, X., Cai, M., and Cai, H., 2000, "Optimal Tolerance Allocation Based on Fuzzy Comprehensive Evaluation and Genetic Algorithm," *Int. Journal of Advanced Manufacturing Technology*, Vol. 16, pp. 461-468.

[11] Barari, A., ElMaraghy, H.A., Knopf, G.K., and Orban, P., 2004, "Integrated Inspection and Machining Approach to Machining Error Compensation; Advantages and Limitations," *Proc. of Flexible Automation & Intelligent Manufacturing (FAIM) 2004*, Toronto, Canada, pp. 563-572.

[12] Hocken, R.J., 1980, "Technology of Machine Tools. Machine Tool Accuracy," *UCRL –52960-5, Lawrence Livermore Laboratory*, University of California, Vol. 5.

[13] ElMaraghy, H.A., Barari, A., and Knopf, G.K., 2004, "Integrated Inspection and Machining for Maximum Conformance to Design Tolerances," *CIRP Annals- Manufacturing Technology 2004*. Vol. 53/1, pp. 411-416.

44

Extended Design for X for Digital Consumer Products

Koichi Ohtomi

Abstract: Worth is assessed throughout the Life Cycle. Then we consider a trade-off between Worth, Cost, and Time. This methodology concerns the selection of design solutions from thousands of combinations of design parameters. The current DfX is considered extended, and so the proposed methodology is called Extended DfX. This methodology is applied to digital consumer product development. The worth of digital consumer products is especially important, but Worth is not always equivalent to performance, whereas it usually is in the case of other products. Therefore, a novel approach is required for assessing Worth in digital consumer products development.

Keywords: DfX, worth, cost, time, customer, function, structure, optimization

44.1 Introduction

The process of product development varies greatly depending on the product field. Figure 44.1 shows an example of classification of the product development pattern. The horizontal axis indicates the size of the product development in proportion to the development cost. The vertical axis indicates whether the objective is mass production for an unspecified client or production ordered by a specific client. Power plants and space equipment correspond to the lower right region. This region is a product field in which the development cost is high and the performance can be investigated thoroughly over a long period of time. The digital consumer product that is the target of this paper is antithetical to the power plant and space equipment.

This region is a product field in which investment in the product development is relatively small and development time is short. The product of this region is customer-driven. Figure 44.2 shows various methods and tools used for product development [1, 2]. They can be extensively used for the above-mentioned power plant and space equipment. On the other hand, it is necessary to apply them selectively and efficiently in the case of digital consumer product development.

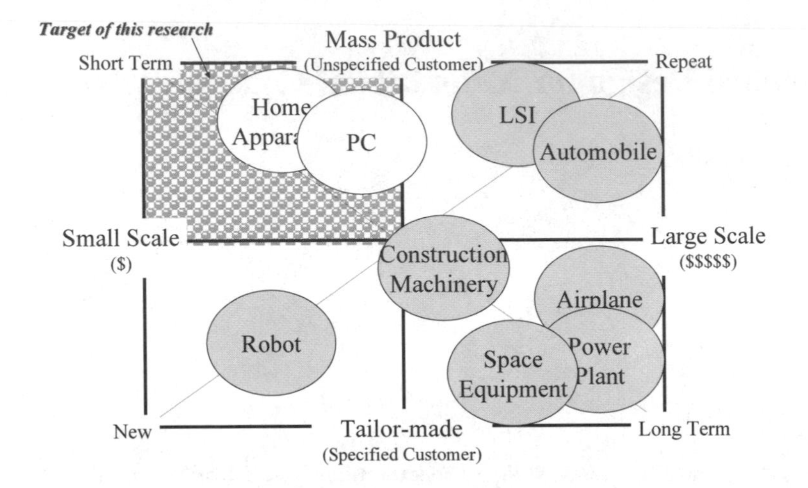

Figure 44.1 Product classification

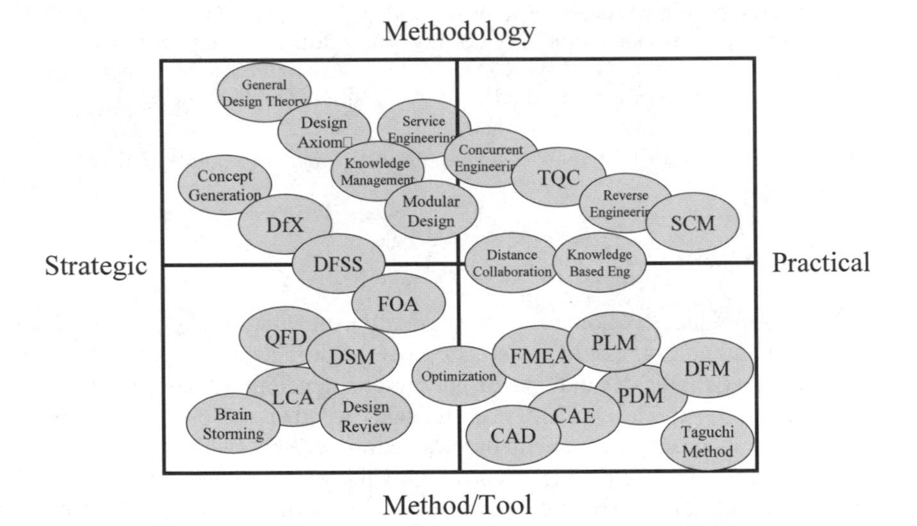

Figure 44.2 Design method/tool

In this paper, we propose a design concept for digital consumer product development, where a customer has the ability to decide the product price in many cases. This causes a manufacturer to make a product that has less variety. As a result, a manufacturer endeavours to reduce costs by improving efficiency and becomes caught up in price-driven competition. In order to break this cycle, it is necessary to assess Worth from the manufacturer's point of view and to reflect the result in product development. Many studies have attempted to evaluate worth/value from the customer's point of view [3, 4]. Where a potential customer requirement is analyzed, and quantified as absolute worth (we define this as Worth) independent of cost. Then, we estimate Cost to realize the above-mentioned Worth by using a Worth/Function/Structure relationship graph [5]. We define this concept as Extended DfX methodology –an extension of the current Design for X (DfX) [6] to Worth-based product development.

44.2 Trade-off between Worth and Cost

We consider, for the sake of simplicity, a product composed of three kinds of parts: a, b, and c. Each of these parts has two kinds of grades: 1 and 2. Then, eight kinds of products can be considered in accordance with the cube of two as shown in Table 44.1. Roughly speaking, the cost is defined as the sum of the cost of each part for eight kinds of products. On the other hand, Worth at the component level increases if the grade is higher. However, unlike the case of a CPU, Worth is not always proportional to price. There are some nonlinear factors. In addition, product Worth itself is not equal to the sum of components Worth. Harmonious balance in a product greatly affects Worth. In addition, Worth strongly depends on the user of the product, when it is used, and where it is used.

Table 44.1 Product Varieties

Product	A	B	C	D	E	F	G	H
Part a	a1	a1	a1	a1	a2	a2	a2	a2
Part b	b1	b1	b2	b2	b1	b1	b2	b2
Part c	c1	c2	c1	c2	c1	c2	c1	c2
Worth	x1	x2	x3	x4	x5	x6	x7	x8
Cost	y1	y2	y3	y4	y5	y6	y7	y8

We assume that Worth of each of the eight kinds of products is obtained by some means. The result is plotted on the Worth/Cost map as shown in Figure 44.3. If the relationship between Cost and Worth is linear, eight kinds of points are plotted on the straight line. However, since Worth is defined through a rather complex process, results will be scattered as shown in Figure 44.3. An actual product consists of dozens of parts and grades. Moreover, the style, the color, weight, size, *etc.*. should be considered for evaluation of the relation between Cost and Worth. Therefore, thousands of product varieties exist. Once Cost and Worth for thousands of product varieties are plotted on the Worth/Cost map, groups of product varieties can be visualized. Then, the boundary of the lowest Cost limit and the highest Worth limit come into view. This boundary is called a Pareto optimal solution. Products B and F in Figure 44.3 correspond to this solution. That is, we can see a group of best solutions by mapping Cost and Worth on the Worth/Cost map. This is why we focus on Worth and compare Worth and Cost on an equal footing.

We consider a trade-off between Worth and Cost, but Time (schedule) may also be included. Product B and Product F are optimal solutions in the current state. The optimal solution does not always satisfy the target solution. In this case, the reduction of Cost and the increase of Worth will be needed in order to approach the target.

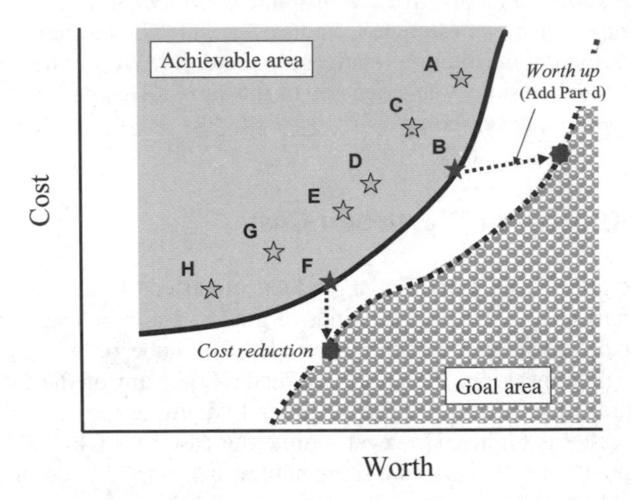

Figure 44.3 Worth/Cost map for eight products (Schematic Graph)

44.3 How to Estimate Worth

Worth is a nonlinear function, and a method to define and quantify it has not been established. Here, we try to consider how to evaluate Worth. The total product Worth is assumed to be W. Worth at the component level is defined as Wi for each part. The additional Worth, generated by the combination of parts and elements into a system, is defined as Wa. In this case, Worth may be defined in two ways as follows:

Case 1. $W = [\Sigma\, Wi \times Wa]^{1/2}$ where there are interactions between parts-level Worth and system-level Worth.

Case 2. $W = \Sigma\, Wi + Wa$, where parts-level Worth is independent of system-level Worth.

We cannot determine which worth formula is correct here, but we think that for the product focused on pleasure of ownership, such as mobile phones in Japan, Case 1, and for products focused on performance, such as notebook PCs, Case 2 is applicable. Moreover, Worth of Wi and Wa is a function of Who, When, and Where as mentioned above. Psychology, ergonomics, design engineering, economics, cultural anthropology, *etc.*. are also relevant to the evaluation of Worth.

44.4 Relation between Worth and Cost

In the design of digital consumer products, the problems concerning the trade-off between Worth and Cost are important. Worth, as perceived by the customer, is Worth can be considered an indicator perceived by the user, whereas Cost can be considered an indicator perceived by the manufacturer. Likewise, in the case of marketing tactics, the market effect leads the enterprise efficiency. Worth changes depending on the targeted customer, and even if it is the same customer, it changes into real time depending on the customer's environment. In general, Worth is predicted on the basis of a questionnaire and an interview with the customers.

Figure 44.4 shows the typical trade-off between Worth and Cost. Category 1 is a linear relationship between Worth and Cost, and this line is considered to be the break-even point. Category 2 is the worst case in that Cost exceeds Worth in all Worth regions. In this case, only Cost reduction and/or Worth decrease is a counter measure. Conversely, Category 3 is the best case in that Cost has fallen below Worth in all Worth regions. This is ideal for profit, but is a rare case for digital consumer products since customers tend to perform severe product assessment.

Category 4 is a case in which Cost falls below Worth in the low-Worth region and exceeds Worth in the high-Worth region. IT is a case in which the manufacturing cost necessary to provide value is relatively high. This is often the case for digital consumer products. For instance, the customer would like to have a notebook PC with a high-performance CPU. Generally, high-performance CPUs are costly, and thus correspond to this case. Therefore, reliance on the outsourcing of parts is unavoidable in order to enhance the customer satisfaction, and consequently, the price increases owing to the relation between supply and demand and because the manufacturer cannot control Cost.

Category 5 is a case in which Cost exceeds Worth in the low-Worth region and falls below Worth in the high-Worth region. There is a possibility of the profit being diminished by the reduction of the manufacturing Cost and the addition of Worth, though the profit is not diminished in the low-Worth region. That is, the manufacturing Cost necessary to provide Worth is relatively low. For instance, if memory capacity is doubled, Worth is thought to be double or more. On the other hand, Cost is reduced to half or less by the productivity gain, which can be attained by self-help. The Worth that the customer demands can be achieved by technology that other companies cannot offer, hence the importance of how Worth is designed.

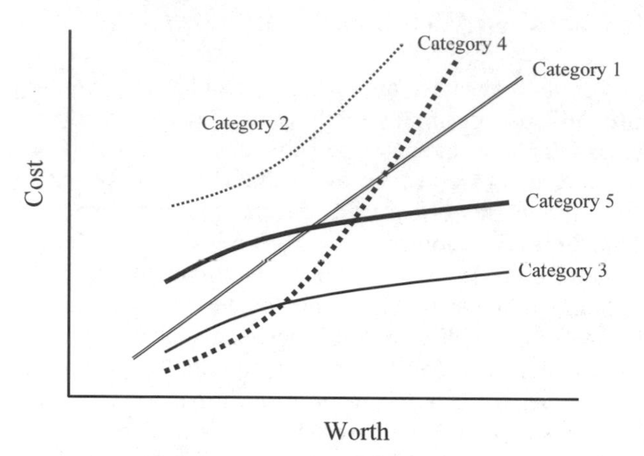

Figure 44.4 Relation between Worth and Cost

44.5 Proposal of Extended DFX Methodology

We propose Extended DfX methodology that enhances the DfX design procedure for digital consumer products development. DfX is a philosophy and practice advocated by Gatenby of Bell Laboratories, of AT&T, in 1990 [6] that ensures quality products and services, reduces the time to market for a product, and minimizes life-cycle cost. It advocates evaluating various problems throughout the life cycle at an early stage of product development, and decreasing the redesign in the latter half of product development as much as possible. In practice, the design method/tool shown in Figure 44.2 is systematically applied according to the DfX methodology. It is relatively easy to apply the DfX methodology to large-scale product development, but for digital consumer products a more concrete way of focusing on Worth is required. Therefore, the DfX methodology is expanded to include the design of Worth as shown in Figure 44.5. Worth is set first, and Cost is derived through functional design and structural design. Worth becomes the target for the customer and Cost becomes the target for the manufacturer loading to a trade-off between Worth and Cost.

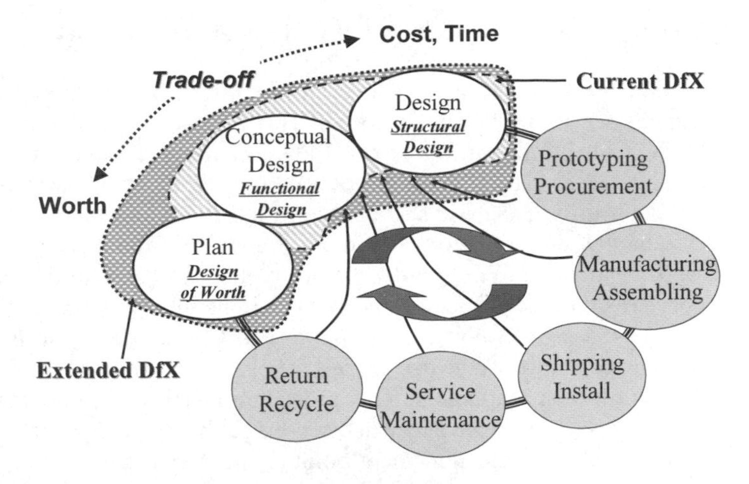

Figure 44.5 Concept of extended DfX methodology

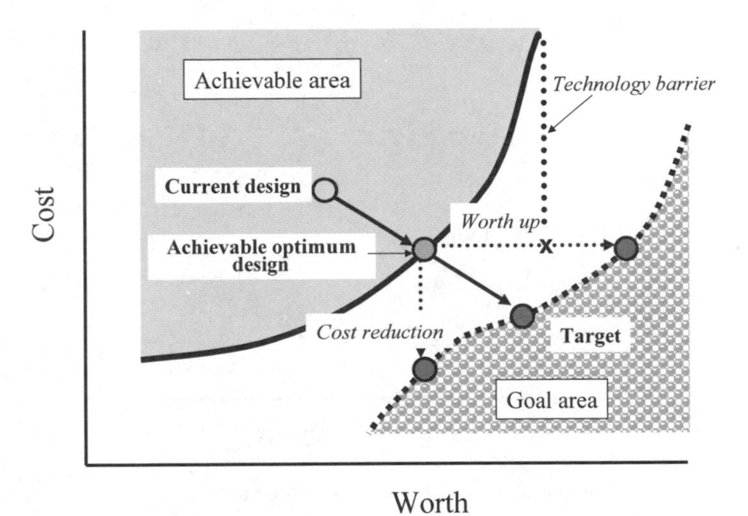

Figure 44.6 Worth/Cost map

In general, the relationship between Worth and Cost is mapped onto the Worth/Cost graph as shown in Figure 44.6. An achievable area is obtained by trade-off analysis, but generally the achievable area and the goal area do not correspond. This is a kind of trade-off. A trade-off analysis method that uses GA has recently been established and can be applied. Thus, the problem becomes clear by plotting current design on the Worth/Cost graph. For instance, the cooling method becomes a problem when the generated heat grows by advancing CPU performance as shown in Figure 44.7 in notebook PC design. If we introduce a large fan system to remove the generated heat from notebook PC with a high-performance CPU, the entire PC becomes large, and its overall Worth for customer decreases overall. Therefore, a technical breakthrough for heat decapitation is required.

We explain the procedure of the extended DfX by referring to Figure 44.8. First of all, the target is set on the Worth/Cost map. This is done at the planning stage. For example, PC with Worth equivalent to $4000 is developed at a Cost of $2000 for the power PC user. Next, “Design of Worth”, “Functional design”, and “Structural design” performed are in accordance with the DfX methodology. Worth is obtained from “Design of Worth”; Cost is assessed from “Design of the Structure”; and, as a result, Worth and Cost are plotted on the Worth/Cost map. In general, because the design achievable area doesn't satisfy the target at this stage, we need to redesign to obtain new Worth and Cost close to the target by controlling design parameters and design restrictions. New Worth and Cost are plotted on the Worth/Cost map again. This procedure enables us to approach the target. An initial target is re-evaluated when we judge that the achievement of an initial target is difficult, and the agreement point of the design feasible region and the design target is set. In practice, this design process is executed by using the Worth/Function/Structure relation graph shown in Figure 44.9.

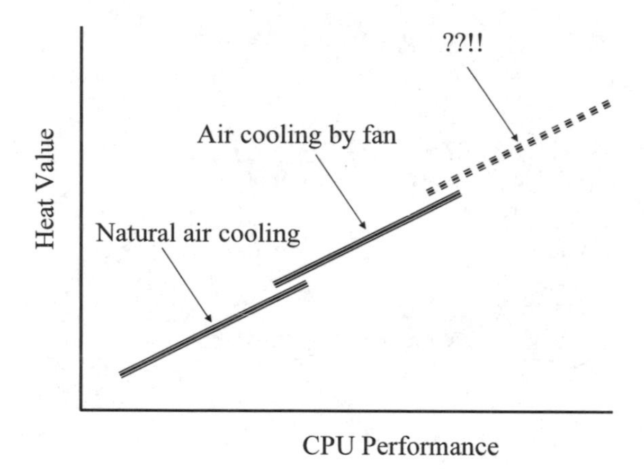

Figure 44.7 Need for Break-through technology

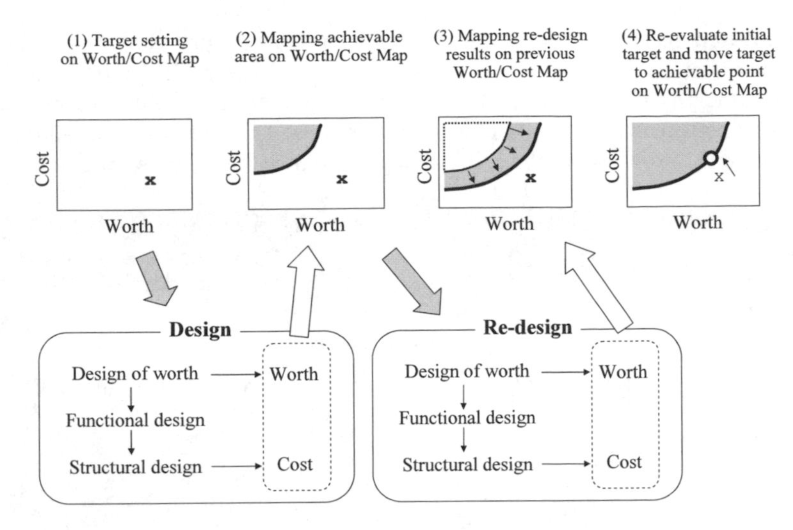

Figure 44.8 Procedure of extended DfX

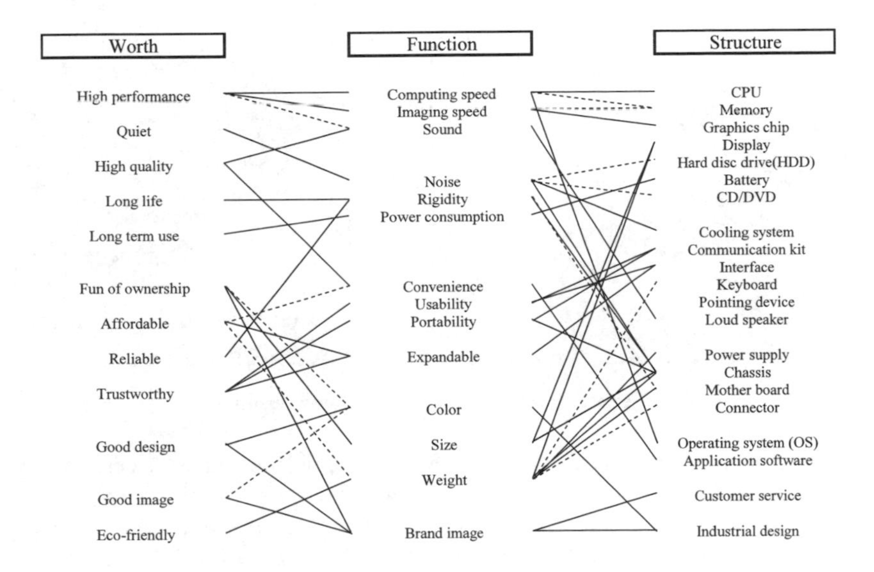

Figure 44.9 Worth/Function/Structure relation graph

44.6 Example of Trade-off Analysis and Satisfying Design

When Extended DfX is applied to practical design, two methods are necessary. One is trade-off analysis, and the other is satisfying design. Here, we introduce the example of trade-off analysis and satisfying design. Recently, the multi-objective optimization technique, for example, MOGA (Multi-Objective optimization by applying Genetic Algorism) has reached a practical stage [7]. By using MOGA, it has become possible to efficiently obtain the Pareto optimal solution that shows the trade-off between two or more objective functions. This solution corresponds to the optimal design solution of the achievable area shown in Figure 44.6.

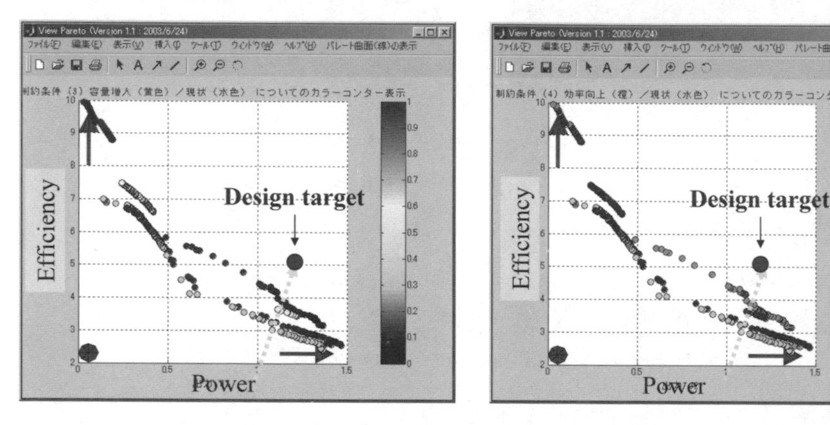

Figure 44.10 Example of trade-off analysis

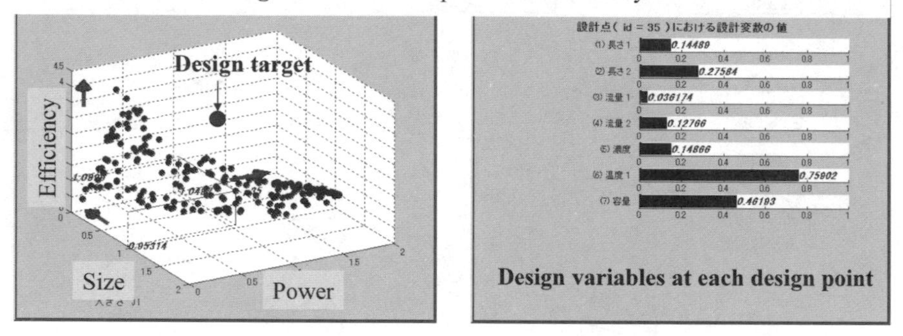

Figure 44.11 Process of satisfying design

Figure 44.10 shows an example of the conceptual design of a certain power plant. This example has 9 design variables such as length, flow rate and temperature, and three objective functions (power output, size, and efficiency). Figure 44.10 shows the Pareto optimal solution obtained by MOGA, which can be expressed not only as a three-dimensional map, but also by a bar chart referring to the design variables of the value of the design trade-off of the arbitrary objective function. However, the Pareto optimal solution did not satisfy the design target, and, in this case, a satisfying design is needed. The design constraint for each design variable finally affects the Pareto optimal solution, and therefore, in order to obtain a satisfying design, we should control the design constraints to satisfy the design target. Figure 44.10 shows the process of putting the Pareto solution close to the design target for each design variable. We determined the effect of the design variable on the Pareto solution for each design variable as shown in Figure 44.11. The designer can examine this process by trial and error easily because he/she knows the way in which the Pareto solution is affected from his/her design experience. The consequence that the Pareto solution corresponds to the design target is finally shown in Figure 44.12. In this case, it was judged to be attainable although it was a rather challenging design target.

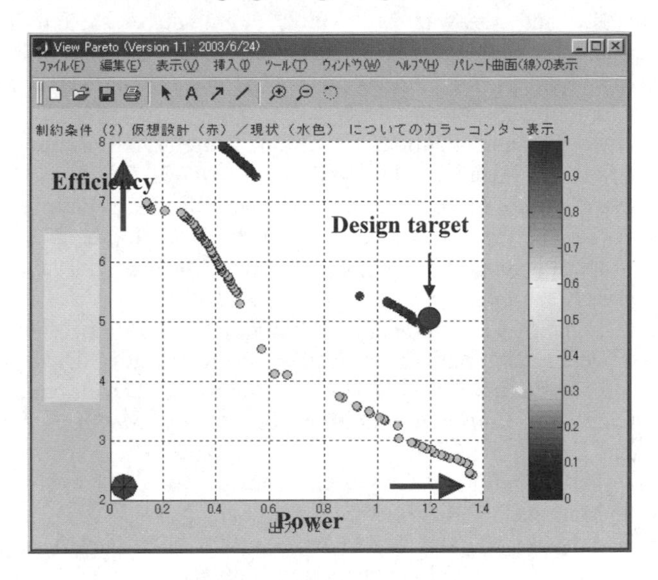

Figure 44.12 Design solution

44.7 Conclusion and Future Works

In this paper, features of digital consumer product design were first described from the perspective of the design of Worth. Next, an Extended DfX methodology was introduced to enhance DfX (Design for X) that was already established for digital consumer products. Moreover, a practical example was introduced to demonstrate the trade-off analysis and the satisfying design procedure, which is the key technology when Extended DfX is applied.

The idea of the Extended DfX methodology and a part of its practice were described. However, in order to apply the Extended DfX more concretely, we need to research the following:

- Definition of Worth and its quantitative prediction method
- Efficient algorithm of satisfying design
- Efficient prediction method of cost at initial designing stage.

We are sure that design engineering essentially concentrates not only on the accumulation of technology but also on a strategic approach to product development, and therefore, we intend to accomplish this step by step.

44.8 References

[1] Ulrich, K.T., and Eppinger, S.D., *Product Design and Development* (Third Edition), ISBN 0-07-247146-8.

[2] Ohtomi, K., and Ozawa, M., 2002, "Innovative Design Process and Information Technology for Electromechanical Product Development," *Concurrent Engineering: Research and Application*, SAGE Publications, Vol.10, No.4, pp.335-340.

[3] Yanagisawa, H., and Fukuda, S., 2003, "Interactive Reduct Evolutional Computation for Aesthetic Design," *Proc. of DETC'03: ASME Int. 23rd Computers and Info. in Engineering*, Chicago, Illinois, September 2-6, 2003.

[4] Yanagisawa, H., and Fukuda, S., 2004, "Development of Interactive Industrial Design support System Considering Customer's Evaluation," *JSME International Journal, Series C.* (Accepted, 2004).

[5] Dong, C., Zhang, C., and Wang, B., 2003, "Integration of Green Quality Function Deployment and Fuzzy Multi-Attribute Utility Theory-Based Cost Estimation for Environmentally Conscious Product Development," *Int. J. of Environmentally Conscious Design & Manufacturing*, Vol.11, No.1, pp.12-28.

[6] Gatenby, D.A., and Foo, G., 1990, "Design for X: Key to Completive, Profitable Markets," *AT&T Technical Journal,* pp. 2-13.

[7] Hiroyasu, T., Miki, M., and Watanabe, S., 2000, "The New Model of Parallel Generic Algorithm in Multi-Objective Optimization Problem – Divided Range Multi-Objective Genetic Algorithms," *IEEE Proceedings of the 2000 Congress on Evolutionary Computation*, pp. 333-340.

45

Development of Integrated Design System for Structural Design of Machine Tools

Myon-Woong Park, and Young-Tae Sohn

Abstract: The design process of machine tools is regarded as a sequential, discrete and inefficient process, as it requires various kinds of design tools and many working hours. This paper describes an integrated design system, embedding a design methodology that can support systematically the structural design and analysis of machine tools. The system is a knowledge-based design system and has three machine-tool-specific functional modules, including: configuration design and analysis, structural element design, and structural analysis support module. A machine configuration appropriate for design requirements is selected using the configuration design and analysis module. The arrangement of ribs for each structural part is then decided in the structural element design module. The structural analysis support module converts the design result into script file which is used to evaluate the designed structure by utilizing FEA software "ANSYS." The system is applied to the design of a tapping machine, and shows that the machine structure can be designed quickly and conveniently with minimum dependency on the capability of a design engineer.

Keywords: Integrated Design System, Machine Tools, Structural Design, Structural Analysis, Knowledge–based Design System

45.1 Introduction

The machine tools dominate over the quality of machine elements and, consequently, affect the development of other machinery. The design technology of a machine tool requires complicated and diversified design knowledge. Also, it evolves over a long period by accumulating designers' experiences and knowledge, rather than by revolutionary advances in a short period.

The machine tools are required to have properties of high speed and accuracy. To realize these requirements, light weight and high stiffness of structural parts, such as bed and column, are essential along with the quality function of each unit part [6, 7]. CAE analysis is indispensable to estimate the structural stability and dynamic behaviour of a machine tool. Much research to integrate design and analysis has been carried out actively [3, 7, 8, 9], and commercial CAD/CAE systems such as I-DEAS, and Pro/E have been developed. However, these are general-purpose systems, and thus exclusive functions for machine tool design are feeble. Also, in order to get reasonable results from CAE analysis, a design engineer must establish proper finite element model and boundary conditions, and therefore it is very difficult for a design engineer to use CAE functions without any help of CAE engineers.

In this research, an integrated design system, ICAD/TM, has been developed by applying a knowledge-based system [1, 2, 4, 5], by which a design engineer can design and analyze structural parts of a machining centre easily and quickly with no help from CAE experts. The system has three functional modules - the configuration design and analysis; the structural element design; and the structural analysis support module. The function of the configuration design and analysis module is to select the appropriate machine configuration and decide key outer dimensions. The selection is carried out by knowledge inference. The structural element design module is to design the internal structure of bed and column with the arrangement of ribs for reinforcement. The function of the structural analysis support module is to generate a finite element model including boundary conditions of the designed machine structure for ANSYS software with a simple interactive procedure. The system enables a designer to evaluate the alternatives swiftly and to select an optimum design in the early stage of development due to its simplified process of integrating design and analysis.

45.2 Integrated Design of Machine Tool Structure

The Machining Center and the Design Process

Generally, a machining centre for the representative machine tool is composed of several unit parts, such as a supporting unit including bed and column, a main spindle, a feed-drive unit for positioning, and accessory parts including the magazine, tool changer and controller. The main spindle unit is attached on the column, and the workpiece table and column are put on the bed as shown in Figure 45.1. The machining centres are roughly classified into horizontal and vertical styles of machines according to the direction of the spindle axis, and various detail types are possible according to the feeding direction of the column and whether or not the column is fed. The spindle direction of the horizontal machine tool is horizontal, and chip removal is easy since most cutting chips are dropped right under the cutting tool. On the other hand, the vertical machine tool has a vertical spindle, and moving the cutting-tool to a machining position and loading/unloading of the workpiece are easy. Eight types of machine tools - four horizontal and four vertical types as shown in Figures 45.2(a) and 45.2(b) - can be designed and evaluated by using the developed system.

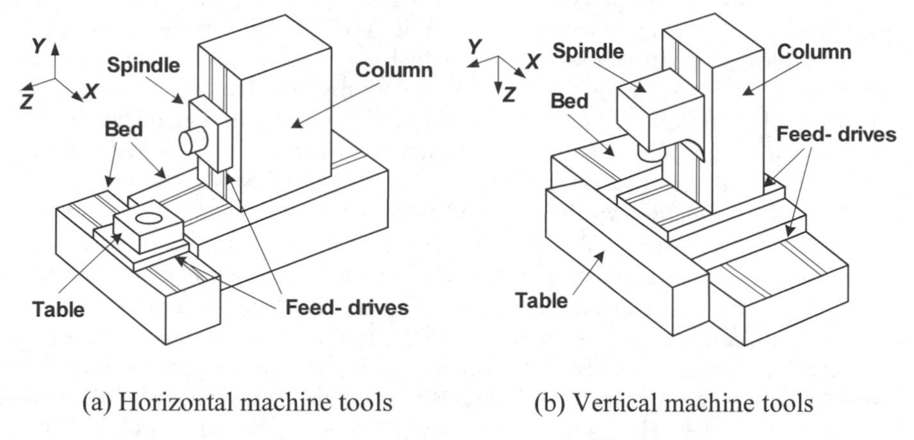

(a) Horizontal machine tools (b) Vertical machine tools

Figure 45.1 Typical configuration of machine tools

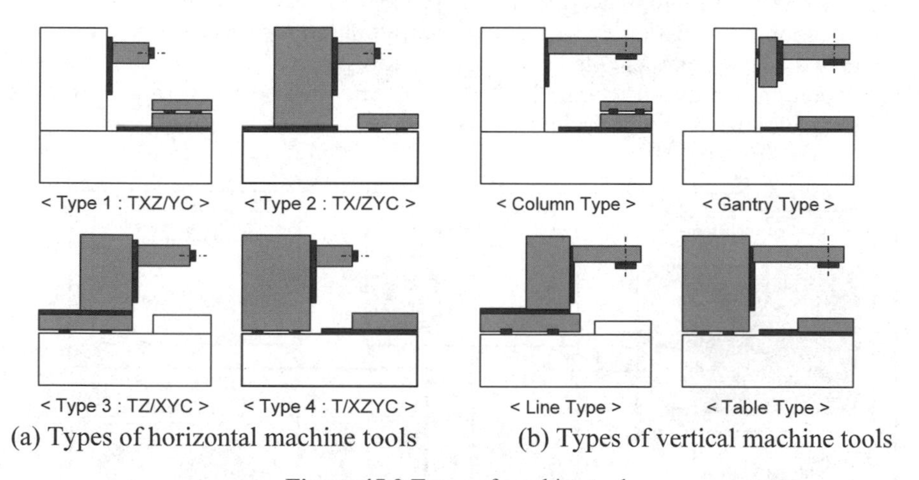

(a) Types of horizontal machine tools (b) Types of vertical machine tools

Figure 45.2 Types of machine tools

Usually, the design of a machining centre starts with the selection of the appropriate style and type based on customer requirements or goal specifications, and then basic configuration of the machine tool and principal dimensions of structural parts are decided and evaluated by static a and dynamic analysis program. Following this, each unit part of the machine tool is designed and evaluated based on the acceptable configuration of machine tool. Once the design of each unit part has been completed, analysis of the machine configuration is carried out once more in order to ensure the static and dynamic stability. In this paper, the design of the spindle and feed drive unit are excluded, and the part related to the design and analysis of the machine structure only is described.

Integrated Design of Machine Structure

Figure 45.3 shows the process of design of the basic structure and supporting elements of the machining centres by using ICAD/TM. The developed system has the functions for configuration design, modal analysis on machine configuration, rib design, and interfacing for FEA. These tightly interfaced function modules enable a coherent flow of information during the process of design and analysis.

The basic machine structure is designed based on design knowledge in accordance with the process defined in the configuration design module. The modal analysis module, which is the dedicated analysis program, is used to assess dynamic characteristics of the structure. If the structure is judged to be inappropriate, the location, size, and type of elements are rearranged interactively according to the result of the assessment. The design change is repeated until the structure is configured satisfactorily. Then, the ribs are allocated inside the structural elements, like bed and column, using the rib design module. Upon completion of the rib design, the finite element model is generated by the module for model interface. The model is fed to ANSYS where the structural analysis by the finite element method is carried out. If the designed machine is considered to be structurally inappropriate, the design process is reactivated so that the goal specification is satisfied by changing configuration or inside structure. The integrated design system is intended to design and analyze the machine configuration and the structural elements following the consistent flow during the early stage of machine tool design.

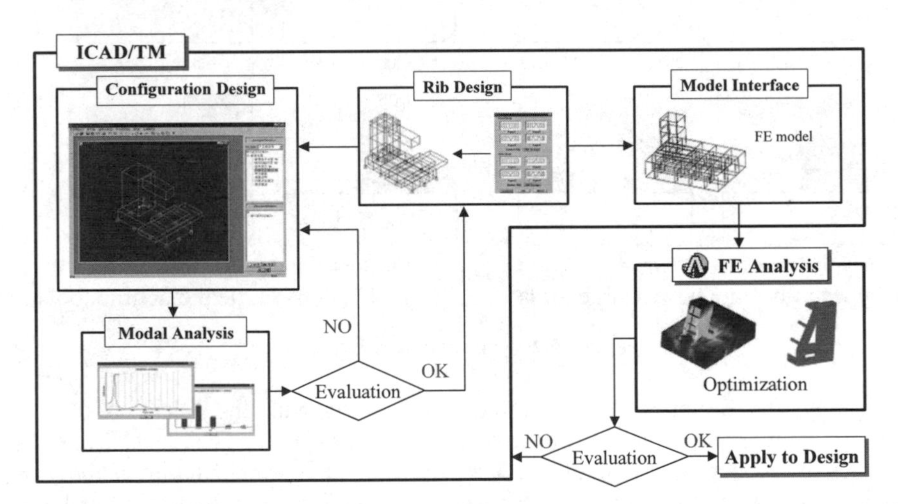

Figure 45.3 Integrated design and analysis flow of the system

45.3 Structural Design and Analysis of Machining Centers

Design of Machine Configuration

The goal of the design of machine configuration is to set up an appropriate machine structure for goal specifications with a decision for the machine type, and the sizes, positions, approximate weight, and materials of the structural components, as well as a structural evaluation of the machine structure. Through the interview with experts on machine tool design, the design process for the machine configuration has been analyzed and modelled using a structural analysis and design tool, called IDEF0 [10]. The design process embedded in the system is composed of several steps such as machine type selection, configuration design, and configuration analysis, and is managed by a design manager according to the sequence of steps to determine whether or not the design is changed.

In the machine type selection step, an appropriate type of machine for the customer requirement is selected first, based on the characteristics of each type of machine, and then the specific machine type is selected by inferring design knowledge on the required machine properties such as stiffness, cutting capability, occupying space and manufacturing cost, *etc.*.

The configuration design step is the process in which the sizes and positions of the structural component parts are determined, as well as the strokes of the feed mechanism. Usually all the dimensions of the machine configuration are determined in accordance with the size of the workpiece table. Therefore, the size of the workpiece table is decided first based on the size and weight of the workpiece and working load. Then the sizes and positions of the each structural component part are calculated in proportion to the size of the workpiece table by inferring the design knowledge such as stroke-decision formulae, size-decision formulae, and position-decision formulae. A suitable bed type is selected, based on the overall size of the machine and workpiece table, working load, and weight of the workpiece. Since the accessory components like tool magazine, tool changer, and controller box affect the dynamic behaviour of the machine tool, those sizes and positions are determined as well. In order to evaluate the machine configuration with structural analysis, the material properties and weight of each component part, as well as joining methods among component parts, are determined in the configuration design step.

The configuration analysis step estimates the dynamic behaviour of the determined machine configuration with a modal analysis that is embedded in the system. Once the machine configuration is evaluated as not suitable, the design engineer can change the machine configuration with re-execution of the design steps or modification of the sizes and positions of the components.

Design of Structural Elements

Design of structural elements such as bed and column is an important process as the assembled structure greatly influences the machining accuracy and stiffness. Bed and column are basically shell structures with ribs inside for reinforcement.

The shape and location of the ribs are decided by the designer's intention to make the structure stiff while keeping it light.

The library of basic shapes and the arrangement of ribs were set up based on the analysis of patterns of beds and columns. The design of the ribs is carried out by selecting and combining the basic patterns according to the maximum cutting force, weight of the workpiece, and size of the machine. This enables the design of various structures of ribs to be done swiftly and efficiently. There are four basic patterns of bed and column, respectively, in Figure 45.4.

The basic patterns of bed are selected for thc shapes of the front view and the side view. The patterns for column are the basic shapes of the front and the top view. Considering the characteristics of column, types 1 and 2 for a shape of the front view, and types 3 and 4 for a shape of the top view are selected. Each basic pattern has the parameters for defining the location and number of ribs, which are assigned interactively during the design process.

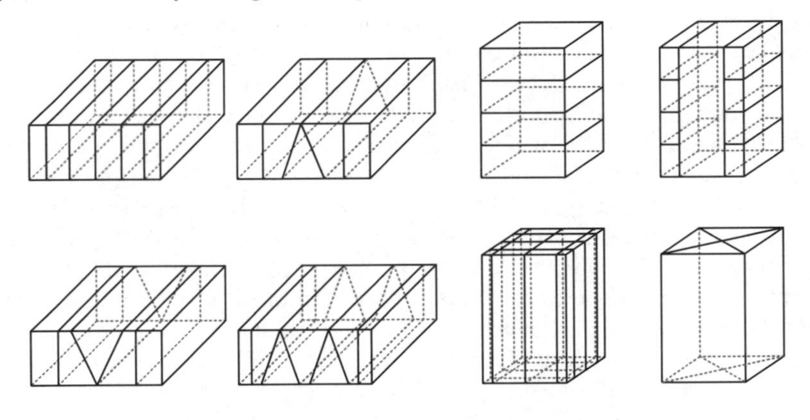

(a) Basic rib patterns for bed (b) Basic rib patterns for column

Figure 45.4 Basic shapes and arrangement of ribs

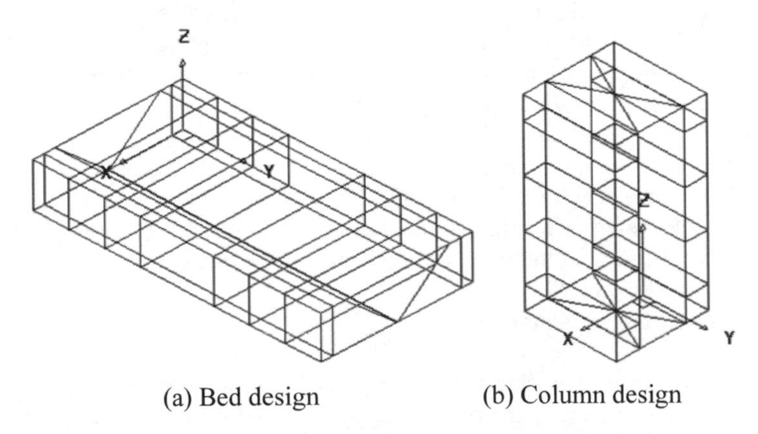

(a) Bed design (b) Column design

Figure 45.5 Design examples of bed and column by combining basic patterns

Figure 45.5(a) shows an example of the design of bed structure by combining type 3 and type 1. In the case of Figure 45.5(b), the front of the column is shaped the same as type 2 and the top view has the pattern of type 4. In order to lighten the walls and ribs, holes can be applied at the middle of those elements. The pattern, shape, number, and location of the holes are defined by assigning values interactively to the parameters representing them.

Analysis of Machine Configuration

Analysis and evaluation of the design result are necessary in order to predict the static and dynamic characteristics of the structure. The analysis module consists of configuration analysis and support for finite element analysis.

Machine configuration can be optimized through the analysis of the vibration characteristic and energy distribution. The structure is modelled as in Figure 45.6 using four simplified elements – node, beam, spring, and support. In modal analysis, the maximum amplitude of the node where the cutting force is applied at a natural frequency up to the 10th mode is predicted. Based on the modal analysis, potential and kinetic energy of each element are calculated, which show the degree of deformation and dynamic appliance.

Therefore, the configuration analysis is used to estimate the maximum dynamic deformation and the degree of displacement of each component part of the machine. Natural frequency of the structure predicted from the analysis is used to select the stable range of spindle speed so that resonance may not occur. The machine configuration is modified in order for each component to have uniform energy distribution by changing the position and size of the component, because the more a component has potential and kinetic energy, the higher the displacement and dynamic compliance of the component.

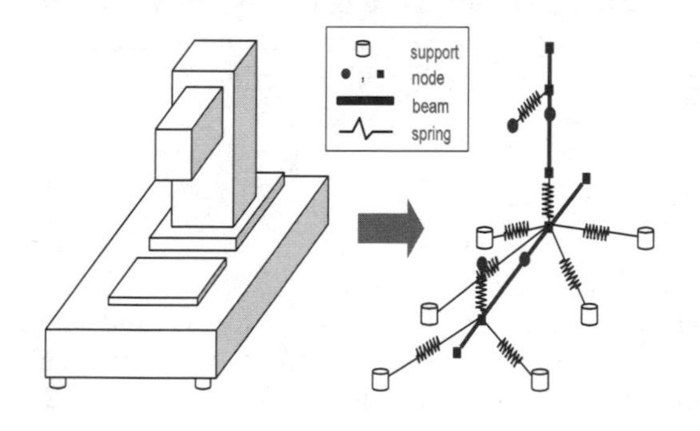

Figure 45.6 Modelling for modal analysis

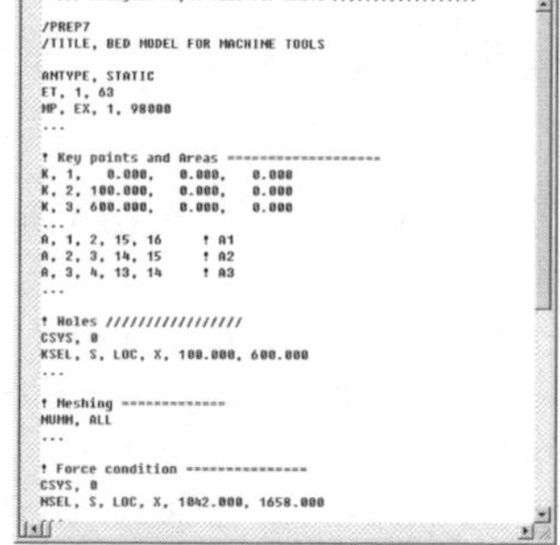

```
TMV1000AA3Bed.AAM
! /// Analysis input file for ANSYS //////////////////

/PREP7
/TITLE, BED MODEL FOR MACHINE TOOLS

ANTYPE, STATIC
ET, 1, 63
MP, EX, 1, 98000
...

! Key points and Areas ------------------
K, 1,   0.000,   0.000,   0.000
K, 2, 100.000,   0.000,   0.000
K, 3, 600.000,   0.000,   0.000
...
A, 1, 2, 15, 16     ! A1
A, 2, 3, 14, 15     ! A2
A, 3, 4, 13, 14     ! A3
...

! Holes /////////////////
CSYS, 0
KSEL, S, LOC, X, 100.000, 600.000
...

! Meshing -------------
NUMM, ALL
...

! Force condition ---------------
CSYS, 0
NSEL, S, LOC, X, 1042.000, 1658.000
```

Figure 45.7 Contents of the script file

Support Function for Finite Element Analysis

The function of the supporting FEA is to generate an analysis model of the structural elements for evaluation of static stability through finite element analysis. FEA is regarded as the most appropriate method to evaluate the stability and dynamic behaviour of machine structure, and many proven commercial systems are available. Therefore, it is sensible to utilize the commercial system for reliable and useful results.

However, to evaluate the structural elements through the commercial system, the assistances of FEA experts is necessary because the structural elements should be modelled interactively in the system, and the information on load, boundary condition, and method for mesh generation should be defined as well in accordance with the characteristics of structural elements. It is not easy for a designer with little experience in FEA to conduct the process of design optimization, which is the iteration of the evaluation and the modification of the design with an evaluation result.

Therefore, the function of interfacing design with analysis has been developed to support FEA on the design result. The analysis model including information relevant to the structural analysis of the designed machine tool is created as a script file for ANSYS. The interface enables less experienced engineers to carry out analysis using ANSYS by providing the script file with no special manipulation. The analysis model created by the system consists of shell elements since the machine tool is modelled as in the form of a plate structure. Analysis models for bed and column are created separately.

The model file includes information on shape, load, boundary condition, and generation of finite elements as shown in Figure 45.7. The shape information of the design structure is created by defining the vertices and cross points of intersection lines among the ribs as sequential key points. The areas are defined by linking the numbers assigned to the key points, and then the shape is defined using those areas. Loading information is defined through calculation of self-weight and machining force according to the type of machining centre. The boundary condition is given depending on the method for combining structural elements. Then, the commands for generating finite elements, executing analysis, and issuing results are written in the analysis model file. The structural analysis can be started by simply feeding the file to ANSYS, and the result of the analysis is available with little delay.

45.4 Implementation of the System

The architecture of the design system developed is as shown in Figure 45.8. The knowledge base established through analysis of the design process and the interview with design experts is systematically managed by the knowledge manager. The design manager controls the design process according to precedence, activates the inference engine, visualizes the design result by Parasolid modeller, and stores the result as design history. The design analyzer evaluates the dynamic characteristic of the machine configuration. The model interface generates a geometric model of the designed machine tool and the analysis model file for ANSYS.

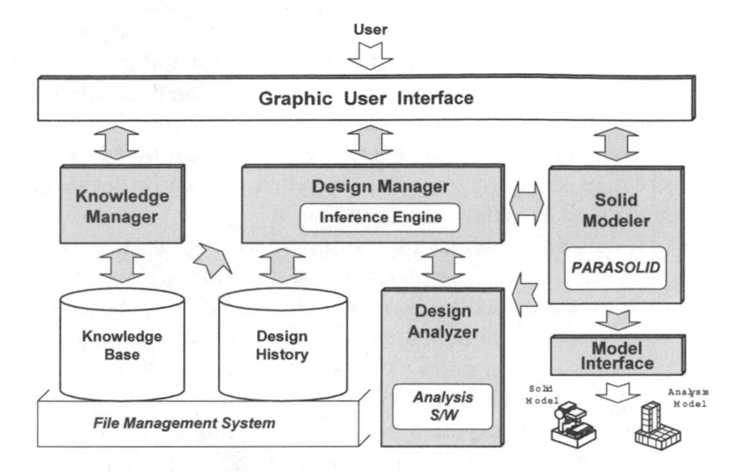

Figure 45.8 Functional architecture of the system

Figure 45.9 shows the main screen of the ICAD/TM system implemented by using Visual C++ 6.0 and Open GL. The menu for supporting functions like the management of the design project, design knowledge, and the modeller are allocated at the upper end of the screen. The area for the management of the design process and design history are on the right, and the functions for producing results through a geometric model and analysis model are on the left. The design process of each module is selected at the design management area, and the current status is shown by the icon as the design progresses. The design version and the composition part of the version can be inquired about through the history manager. A change of design is possible through reprocessing the pertinent stages. It is also possible, using the correction function of the solid modeller, but in that case, the processes following the changed part must be made invalid and the icon representing the design status must be changed interactively.

45.5 Design Instance

The system was applied to the design of the machine configuration and structural elements of a tapping machine. The design started with a selection of the type of machine. Inputting design requirements through the dialog box, the system recommends the most appropriate type of configuration. In the example of Figure 45.10, the 'line type' was selected when stiffness, cutting capacity, and manufacturing cost were given priority. When maximum size and weight of the workpiece, and maximum cutting force are given as 800 (mm) x 1000 (mm), 200 (kg), and 50 (kg), respectively, the table size and key outer dimensions of the machine are automatically decided by the system, as visualized in Figure 45.10. According to the modal analysis of the basic structure, maximum amplitude is expected to be 1.58 (mm) at 5Hz, as in Figure 45.11. Therefore, the designer can reduce maximum compliance, or adjust the location where the maximum compliance occurs, by changing the design based on the result of analysis.

Resonance can also be avoided, reflecting the result of the analysis, during the design of the spindle and considering the elements related to vibration. Energy distribution among the units is checked and the size and location of the unit are modified so that potential and kinetic energy are not concentrated in a certain part. The optimal structure satisfying the design requirement can be decided at the early stage of design through this analysis process.

Figure 45.12 shows the process of rib design where a basic pattern is chosen and design parameters for the arrangement of the ribs are input. Figure 45.13 shows the result of the structural elements design. In this example, type 1 is adopted as the basic pattern for both the front and side view of the bed. Type 2 and type 3 were taken for front and top views of the column. By the combination of those patterns, the internal holes are also designed with a rectangular shape.

The design result can be exported to the analysis model file through the model interface function. By reading the file with ANSYS, without any additional modelling or parameter setting, structural analysis can be carried out as shown in Figure 45.14. A designer learns that the maximum displacement of the design column is 103 (μm), and changes the configuration or geometry of the rib if reinforcement is required. Therefore, even a designer not familiar with structural analysis can evaluate the stability of his design quickly and conveniently, and reflect the evaluation of the design.

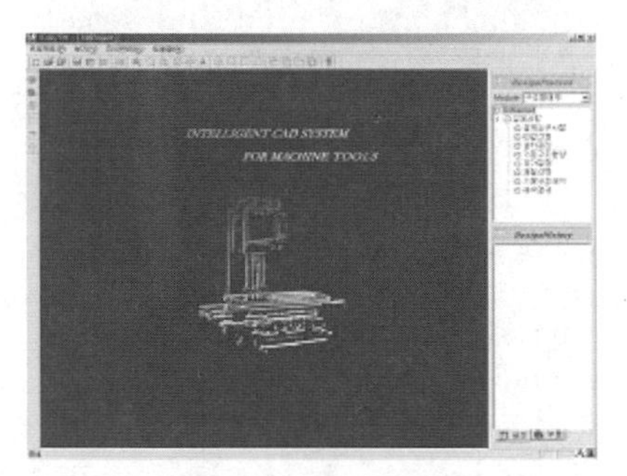

Figure 45.9 Main window of the system

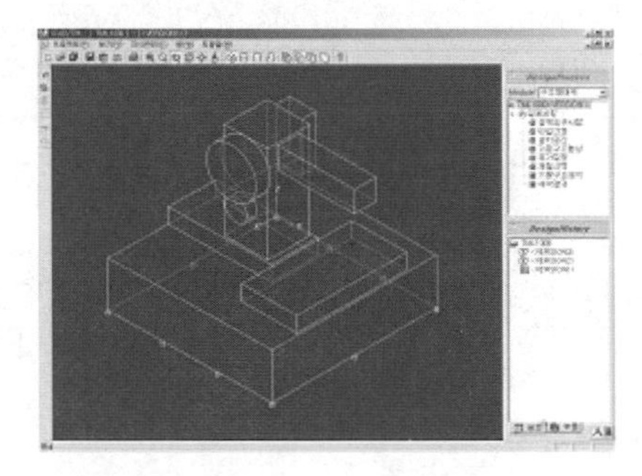

Figure 45.10 Design of configuration

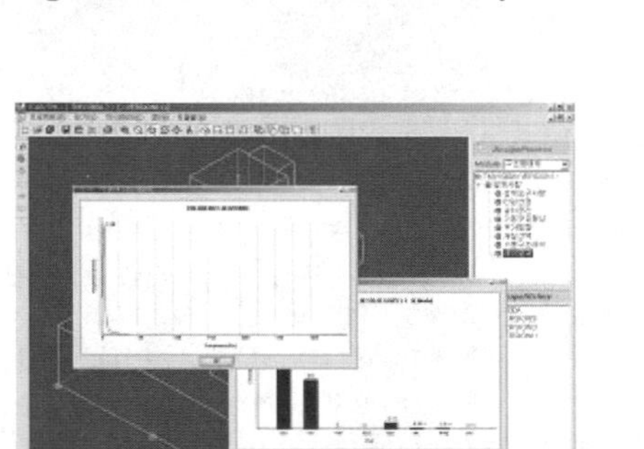

Figure 45.11 Results of modal analysis

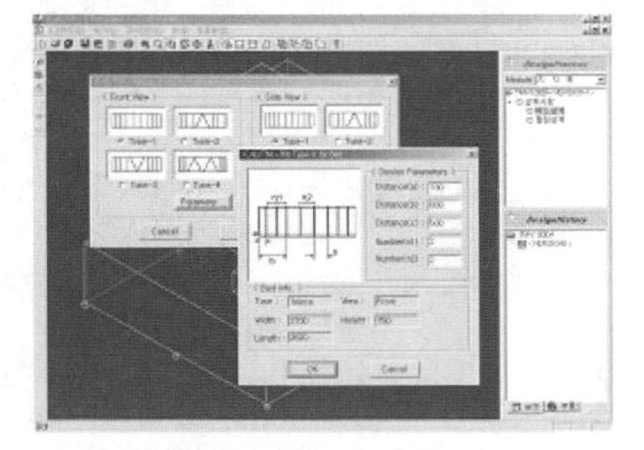

Figure 45.12 Process of rib design

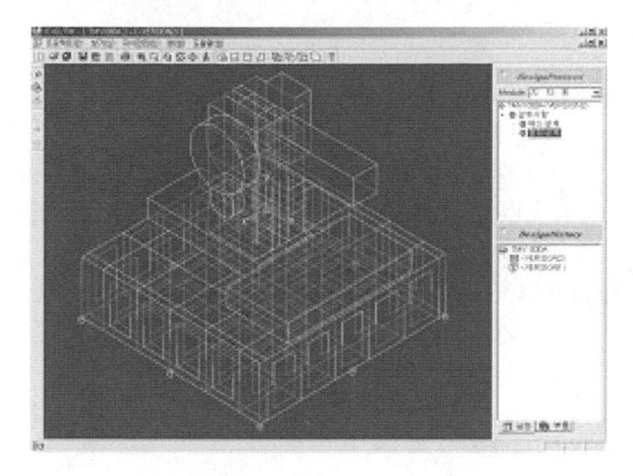

Figure 45.13 Structural element design

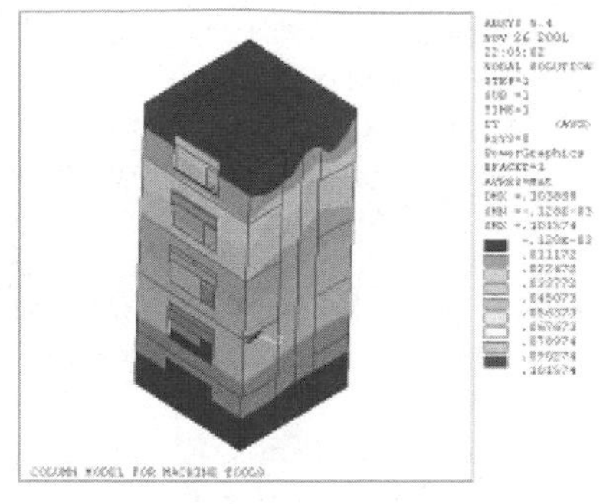

Figure 45.14 Result of FEA of column

45.6 Conclusion

An integrated design system enabling efficient design and swift evaluation of machine tools structures has been developed. The system has been implemented by integrating configuration design and analysis, and structural element design and analysis. During the process of design and analysis, human dependency could be minimized by establishing a knowledge base and decision making algorithm for design, analysis, and preparation for analysis. The implemented system was proven efficient and convenient while the system was used by design engineers developing a new model of tapping machine.

45.7 References

[1] Corbett, J., Woodward, J., 1991, "A CAD Integrated 'Knowledge-Based System' for Design of Die Cast Components," *CIRP Annals,* Vol. 40/1, pp. 103-106.

[2] Xinan, F., Zhang, Y., 1992, "Intelligent CAD for Box-Type Machine Components," *Annals of the CIRP*, Vol. 41/1, pp. 217-220.

[3] Yoshikawa, H., Tomiyama, T., Kiriyama, T., 1994, "An Integrated Modelling Environment Using the Metamodel," *CIRP Annals*, Vol. 43/1, pp. 121-124.

[4] Kimura, F., Suzuki, H., 1995, "Representing Back-ground Information for Product Description to Support Product Development Process," *Annals of the CIRP*, Vol. 44/1, pp. 113-116.

[5] Dixon, J.R., 1995, "Knowledge-Based Systems for Design," *Transactions of the ASME*, Vol. 117, pp. 11-16.

[6] Park, M.W., Cha, J.H., Park, J.H., Kang, M., 1999, "Development of an intelligent design system for embodiment design of machine tools," *Annals of the CIRP*, Vol., 48/1, pp. 329-333.

[7] Wu, B.C., Young, G.S., and Huang, T.Y., 2000, "Application of a two-level

optimization process to conceptual structural design of a machine tool," *Int. J. Machine Tools & Manufacture*, Vol. 40, pp. 783-794.
[8] Cirak, F., Scott, M.J., Antonsson, E.K., Ortiz, M., Schroder, P., 2002, "Integrated modeling, finite-element analysis, and engineering design for thin-shell structure using subdivision," *Computer Aided Design*, Vol. 34, No. 2, pp. 137-148.
[9] Hicks, B.J., Cullry, S.J., 2002, "An Integrated modeling environment for the embodiment of mechanical systems," *Computer Aided Design*, Vol. 34, No. 6, pp. 435-451.
[10] Colquhoun, G., Bains, R., and Crossly, R., 1993, "A state of the art review of IDEF0," *Int. J. CIM*, Vol. 6, No. 4, pp. 252-264.

46

The Structured Design of a Reconfigurable Control Process

ElSayed ElBeheiry, Waguih ElMaraghy, and Hoda ElMaraghy

Abstract: This paper presents the design of a control process that intelligently manages and unifies the reconfigured operations of individual manufacturing physical systems like robotic and CNC systems. The goal is to develop a Unified, Reconfigurable Open Control Architecture (UROCA) system that represents a control level higher than the open architecture controllers but utilizes their powerful features. The unifying architecture UROCA is designed based on a new controller approach inspired by the concept of human's left/right brain and whole brain intelligence, which assures traceability between the requirements and capabilities of the controller and a high degree of operational flexibility. The UROCA design process follows the guidelines of a methodology called the Design Approach for Real-Time Systems (DARTS). This DARTS method proceeds from and builds on the application results of a specification methodology called Real-Time Structured Analysis (RTSA). The outcome of this design is a three-layer (bidirectional) hierarchical control architecture having deliberative left brain for normal operations and reactive right brain for contingent operations or navigation control. A third hybrid mode is also enabled.

Keywords: control architecture design, hierarchical control, structured design, deliberative and reactive planning, intelligence

46.1 Introduction

Open architecture controllers have been proven essential for all aspects of reconfiguration in future manufacturing systems [1-4]. They provide a powerful open software/hardware environment for users to concentrate on the application at hand rather than porting them to different machines of different proprietary controllers. An open architecture controller is built from multi-vendor, plug-compatible modules and component parts, which allows the integration of off-the-shelf hardware and software components into a controller infrastructure that supports a "de facto" standard environment, *i.e.*, buy rather than develop components.

The control architecture, which is a system platform encapsulating software and hardware objects, represents the backbone of any reconfigurable control process. Its design has been the topic of intensive research in the past decade. The result is a wide variety of machine control architectures, which diverge in the kernel structures and differ in the abstraction levels, and in terms of a global solution, lead to application problems regarding communication, standardization, and operation [5]. Our research project "reconfigurable control process for manufacturing" aims at solving this problem before it arises. The availability of such a unified reconfigurable control in industry is expected to have a significant impact from the economy, productivity, and technology points of view.

The schematics of the proposed three-layer architecture are shown in Figure 46.1. The UROCA control process we propose is inspired by the human learning principles and the right brain, left brain and whole brain design methodologies.

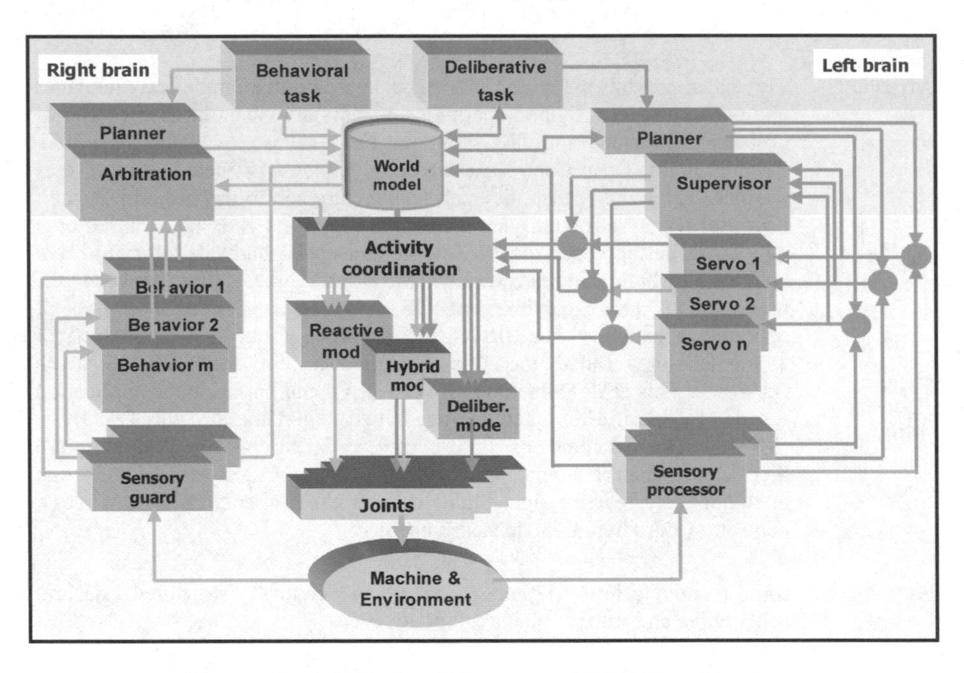

Figure 46.1 Schematics of the proposed UROCA architecture

The topmost layer in UROCA (Figure 46.1) is thought to be organizational reproduction of human knowledge handling and management functions, *e.g.* planning, learning, knowledge and data acquisition, and tasks coordination. The intermediate layer contains two deliberate planners one for each brain and each one receives separate task description from the upper layer. This will enhance the overall architecture modularity by decoupling the two brains to the lowest degree. Thus changes made to one brain will not significantly affect the other. For robotic application, the left planner, using a world model, performs tasks related to command decoding, path generation, trajectory planning, kinematics inversion, *etc.*. The right planner, using the same world model, would possibly generate navigation plans, gripping plans, or sensor-based task control plans, and compliance control depending on the design and goals of the behaviours in the left physical layer. The combination of the deliberative and reactive planning in the right brain will provide this hybrid brain with means to be more oriented towards the design and task goals without affecting its reactive features. In the proposed design, the commands generated by the two brains are not allowed to reach the servo controllers in the physical layer unless they all pass through an activity coordinator. This coordinator, aided by information provided by the planners and the world model, decides one of the following three operational modes: (i) deliberative mode in which only control commands generated by the left brain are allowed to pass, (ii) reactive mode in which only control commands generated by the right brain are allowed to pass, and (iii) hybrid mode in which the coordinator would link one or more of the deliberative and reactive commands. In the deliberative mode, owing to the modularity offered by the coordinator, the servo and supervisory deliberative commands can be added to form a hierarchical controller. Alternatively, the coordinator can only pass the supervisory commands in order to form a centralized sub-mode of operation or it can only pass the servo commands for forming a decentralized sub-mode of operation.

UROCA includes two main sensory processes called the sensory processor and the sensory guard. The sensory processor considers low-level sensor data for building various world representations based on processes like filtering, observation, fusion, *etc.*. In other words, the sensory processor only considers processing of signals that can be manipulated by observation, control and signal processing algorithms. The outcomes of the sensory processor are mainly dedicated for the operation of the left-brain planner. The sensory guard detects environmental states for which the left brain cannot compensate. Two main concepts would govern the operation of this sensory guard. The first concept is based on the recognition of certain values in the low-level sensor data that is originally needed for the left-brain control. This requires high-level or advanced sensor data processing. The other concept is to let the sensory guard has its own sensors.

46.2 UROCA Design Goals and Specification

Design methodologies can be classified as right-brain, left-brain and whole-brain as shown in Figure 46.2. For UROCA, we apply DARTS which is a structured and whole-brain design methodology. The procedural design approach, starting from requirements and ending by verification, is depicted in Figure 46.3.

UROCA design goals are summarized in Table 46.1. Opening system architecture provides it with reconfiguration abilities that are not affordable by closed architectures. The degree of hardware vendor-neutrality, the level of software/hardware component integrity, and the extent of components accessibility by users are all important when the system openness is judged. These measures, in addition to a modular architecture design, enhance reconfigurability requirements like flexibility, portability, scalability, reusability, re-planning, interchangeability, interoperability, responsiveness, reflexiveness, and learning.

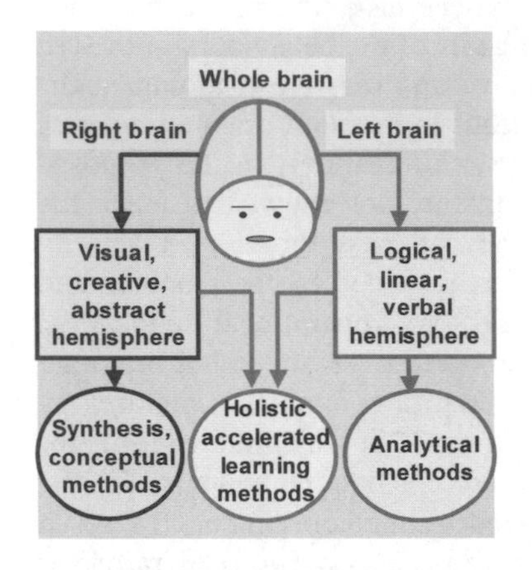

Figure 46.2 Design methodologies

Figure 46.3 UROCA design approach

Table 46.1 UROCA design goals and their definitions

Design goal	Definition
Learning	Ability to learn new tasks
Portability	Porting to new platforms
Interoperability	Good use of resources and data exchange
Exchangeability	Replacing modules of same functionalities
Responsiveness	Fast response to changes
Completeness	Capturing all relevant characteristics
Scalability	Adding or removing modules
Flexibility	Adapting system functionality
Optimality	Ultimate performance potentials
Goal-orientation	Clarity of means for achieving goals
Reflexiveness	Tightness of control loops around sensors and actuators
Planning	Sufficiency of generated plans and actions
Certainty	Data and information reliability
Simplicity	Usability, readability and user-friendliness

Basically, UROCA reconfigurability deals with both control and software reconfigurations as well as searching for every possible approach for unifying both types of reconfigurations. For unifying the UROCA reconfiguration process, it will follow similar patterns: (i) it exploits commonalities among robotic and CNC systems and avoids reinventing solutions to problems that have been tackled before and (ii) it employs software modules that are highly reusable within groups or families of similar patterns. Moreover, UROCA will favour the design of highly modular modules, which may be combined with each other to produce new system, probably in environments quite different from the one for which they were originally developed. Design specifications of published architectures made for industrial robotic and CNC system are shown in Table 46.2.

Table 46.2 Control architecture publications versus specification

Specification	Publication	Brooks [7], 1986	Albus and Quintero [8], 1990	Gat [9], 1992	Heikkila and Roning [10], 1992	Simon et al. [11], 1993	Yen et al. [12], 1993	Altintas et al. [13], 1993	Teltz and Elbestawy [14], 1993	Tunstell and Jamashidi [15], 1994	Chatila [16], 1995	Park et al. [17], 1995	Rober and Shin [18], 1995	Thieleman et al. [19], 1996	Michaud et al. [20], 1996	Khatib et al. [21], 1997	Fiedler and Schilb [22], 1997	Landers and Ulsoy [23], 1997	Hu et al. [24], 1998	Borrelly et al. [25], 1998	Huff and Edwards [26], 1999	Nilsson and Johansson [27], 1999	Fernandez and Gonzalez [28], 1999	Ulmer and Kurffess [29], 1999	Park et al. [30], 1999	Surya et al. [31], 2000	Wills et al. [32], 2000	Erol and Altantas [33], 2000	Hong et al. [34], 2000	Heilala [35], 2000	Volpe et al. [36], 2001	Yang and Hong [37], 2001	Oldknow and Yellowly [38], 2001	Wu et al. [39], 2001	Yavuz and Bradshow [40], 2002	Tan et al. [41], 2002	Freire, et al. [42], 2003	Proposed UROCA system
Structure	Open					√		√	√			√	√	√			√	√	√	√	√	√	√	√		√	√	√	√				√			√		√
	Closed	√	√	√	√		√			√	√				√	√									√						√	√		√	√		√	
Reasoning	Deliberative		√					√	√			√		√			√	√		√	√	√		√		√	√	√	√			√	√			√		
	Reactive	√					√																											√				
	Heterogen.			√	√	√				√	√				√	√	√		√	√	√		√		√						√				√			√
Control design	Top-down		√	√				√	√			√		√			√	√		√	√	√		√		√	√	√	√		√	√	√	√		√		√
	Bottom-up	√																																			√	
	Hybrid					√		√							√	√			√				√		√										√			√
Processing	Centralized				√						√																		√					√				√
	Distributed			√		√							√	√	√	√	√	√	√				√	√		√		√				√	√			√		√
	Hierarchical		√					√	√			√				√				√	√	√			√		√								√			√
Integrity	Sensor fusion		√		√	√		√	√		√	√		√	√	√	√		√	√	√	√	√				√	√	√		√							√
	Command fusion			√		√										√							√		√						√			√	√		√	√
	arbitration scheme	√		√											√										√										√			

UROCA is intended for use with different industrial machines like robotic and CNC systems, the ultimate goal is to have a machine-independent, application-independent architecture. In other words, we emphasize the transition from the technology of open controllers to the technology of universal controllers. This challenging goal cannot be accomplished without knowing the state-of-the-art of building up control architectures for industrial machines. The wide range of terminology makes it difficult to understand the results from the development and application of other researchers' architectures [6]. In spite of this fact, we have made an effort to provide the reader with an insight into the scope of the existing control architectures in terms of specification and requirements. Tables 46.2 and 46.3 provide the reader with this knowledge.

Table 46.3 QFD analysis of specification against design goals

Specification \ Design goal		Responsiveness	Scalability	Simplicity	Certainty	Reflexiveness	Interoperability	Flexibility	Learning	planning	Goal-oriented	Optimality	Completeness	Exchangability	Portability
Structure	Open	□	□	○	Δ	□	□	□	□	□	Δ	○	○	□	□
	Closed	○	▲	□	□	○	Δ	▲	○	○	□	□	□	▲	▲
Reasoning	Deliberative	▲	▲	▲	▲	▲	Δ	Δ	□	□	□	□	○	Δ	Δ
	Reactive	□	□	□	□	□	□	□	Δ	Δ	Δ	▲	○	□	□
	Hetrogeneous	○	○	○	○	○	▫	▫	▫	▫	▫	○	○	○	○
Control design	Top-down	Δ	Δ	Δ	Δ	Δ	Δ	Δ	□	□	□	□	○	Δ	Δ
	Bottom-up	□	□	□	□	□	□	□	Δ	Δ	Δ	Δ	Δ	□	□
	Hybrid	▫	▫	▫	▫	▫	▫	▫	▫	▫	▫	▫	Δ	○	○
Processing	Central	○	○	Δ	Δ	Δ	Δ	Δ	▫	▫	▫	□	□	Δ	Δ
	Distributed	□	□	□	□	□	□	□	Δ	Δ	Δ	Δ	Δ	□	□
	Hierarchical	▫	▫	▫	▫	▫	▫	▫	○	○	○	▫	▫	▫	▫
Integrity	Sensor fusion	Δ	Δ	Δ	Δ	Δ	▲	▲	▲	▫	▫	▫	□	▲	▲
	Command fusion	○	○	○	○	○	▫	▫	▲	▫	▫	□	○	▫	□
	arbitration scheme	□	□	▫	□	□	□	□	▲	▲	▲	○	▲	▲	Δ

Satisfaction improves ←

Legend	◉	○	○	Δ	▲
	very good	good	moderate	not good	very bad

The chosen architecture specification for comparing those architectures is based on structuring, reasoning, control design, processing, and integration. Table 46.3 summarizes the results of a Quality Function Deployment (QFD) analysis of the architectural specification against the (design goals) requirements. This primary step is important for us in order to form bases for having the design modularity emphasized and the reconfiguration aspects of the UROCA architecture unified.

The control design within any architecture falls into one of the following three categories: (i) centralized control, (ii) decentralized control that is sometimes, absolutely decoupled, and (iii) hierarchical control that is sometimes distributed. In a centralized control, the data from the entire system is fused, processed centrally and the control commands are the ones that are processed to all the actuators. This centralized scheme is hard to design and can be tricky if reconfiguration is demanded. However, it is the best for optimizing system performance. Conversely, the decentralized controllers are simpler to design and easier to reconfigure but at the expense of lower performance potentials. We have chosen UROCA motion controllers to be hierarchical because they represent a good compromise between the above two types of control and effectively resolve the trade-off that exists between the reconfigurability and the performance. Moreover, the design is more modular via the use of the activity coordinator that can add or decouple the supervisory signals from the servo signals in the left brain in order to provide us with either hierarchical or centralized or decoupled control modes.

46.3 UROCA Structured Design

The DARTS method supports the decomposition of a real-time system into concurrent tasks and defining the interfaces between these tasks. The DARTS method, like the Object Oriented Design (OOD) method, proceeds from and builds on a Real-Time Structured Analysis (RTSA) method. The RTSA is an important analysis and specification step that precedes the design phase. Recently, the method has been more elaborated and automated for generating auto-designs [43].

The design example that is presented here only focuses on one of the UROCA design concepts in which the right brain is intended for supervising the operation of the deliberative, left brain. The reactive, right brain accommodates changes in the world caused by either stationary objects changing their status, or moving objects altering their status, or new (objects) comers inserting themselves into the world. The stationary objects would change their status as they change their orientation, come to motion from rest, update their geometrical configuration, *etc.*. Changing speed, acceleration, and directivity leads to changing the status of a moving object. All those kinds of changes need sensor guards that detect the change and allow the world to perceive it and re-structure itself correctly. The planners in both brains continuously receive data from the world. The left brain continues to work and reconfigure itself against small changes in order to keep on the targeted performance potentials. Our design perspective here is that we consider, for instance, increasing the contact force in a robotic welding process can be accommodated by the left brain via the supervisory controller by switching from a position control to a force control. That is an internal reconfiguration that the left brain can afford. Another resolution is that the position control left brain can link to a right- brain, reactive force controller to form a whole-brain, hybrid control process. But the reconfiguration that the left brain cannot afford is in the case where contingent events come through the world and reactivity is then superimposed by the right brain.

Figure 46.4 is the first result that comes from the application of the design methodology. It is the UROCA context diagram that defines the boundary of the reconfigurable process to be developed and the external environment. It shows all the inputs to the process and outputs from the process. It explicitly shows the terminators sources and sinks of data. Figure 46.5 functionally decomposes the reconfigurable control process into four main processes or transformations. Figure

46.6 shows the Data Flow Diagram (DFD) in which some details of operations management within a single brain and among the two brains are explicitly explored. Figures 46.6 represents the major step in the RTSA specification methodology, while Figure 46.7 shows the results of one important step in DARTS methodology, where the whole UROCA architecture is structured into concurrent tasks in both brains. This requires grouping processes that could form one complete task provided that no common concurrency exists inside the same task [43].

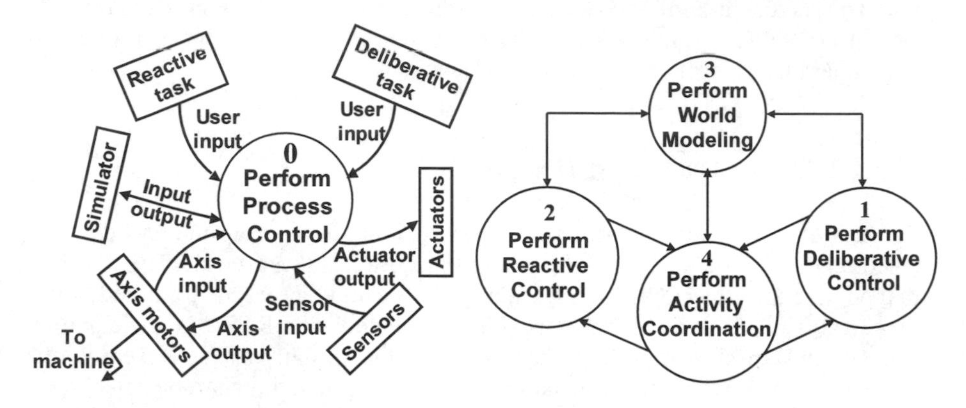

Figure 46.4 Context diagram

Figure 46.5 Functional decomposition

46.4 Conclusions

This paper introduces the design of the new UROCA architecture and explores the possibilities and usefulness of combining reactivity, planning, and deliberation in a two-brain control architecture for industrial purposes. A structured design methodology called DARTS preceded by a real-time specification methodology called RTSA was applied for analysis, specification and design of the UROCA architecture.

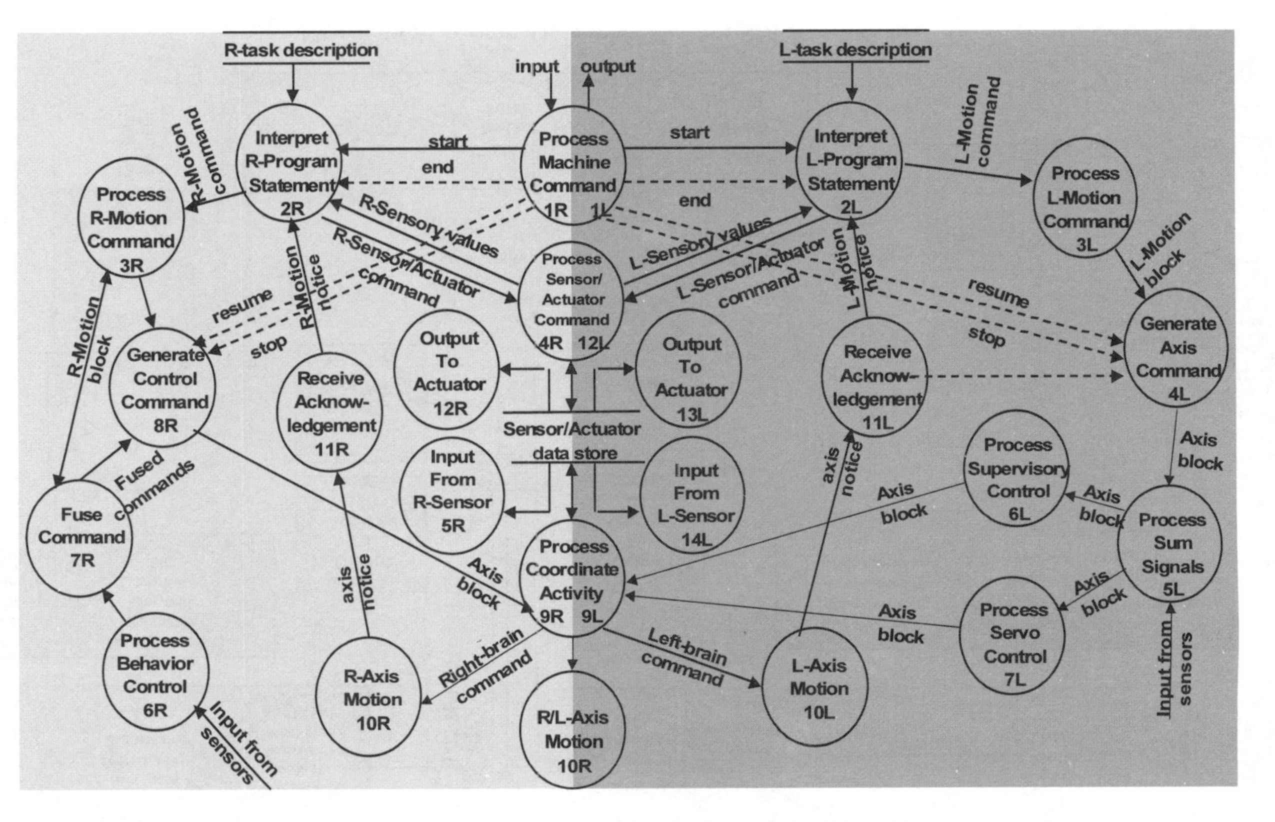

Figure 46.6 Data Flow Diagram (DFD) for the UROCA architecture

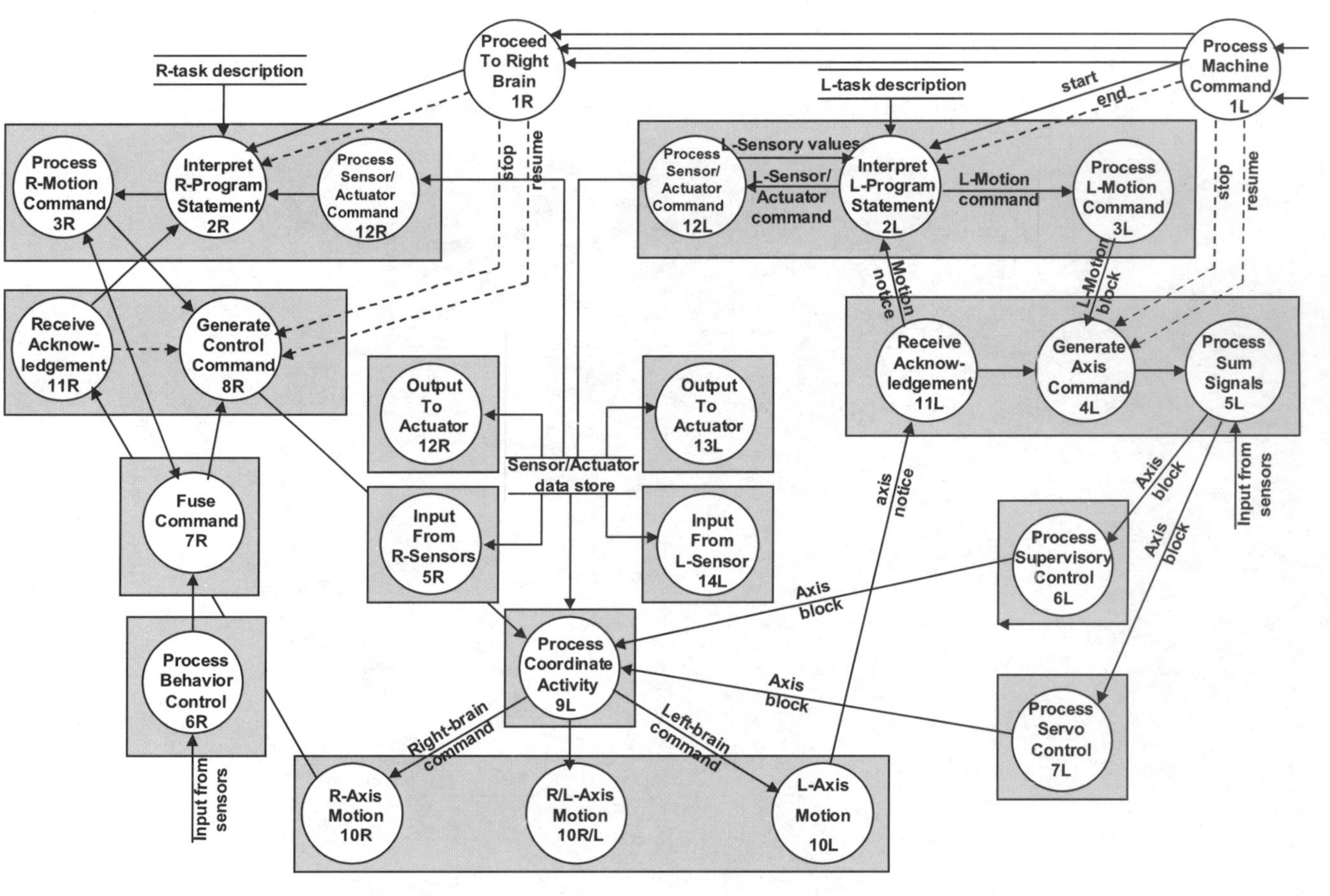

Figure 46.7 Concurrent task structuring in UROCA architecture

46.5 Acknowledgment

The Natural Science and Engineering Council of Canada (NSERC) under a Strategic Projects Grant in Value Added Products and Processes supported the work reported in this paper.

46.6 References

[1] Pritschow, G., Altintas, Y., and Jovane, F., and Koren, Y., 2001, "Open controller architecture- Past, Present and future," *CIRP Annals, Manufacturing Technology,* Vol. 50(2), pp. 463-470.

[2] Proctor, F., 1998, "Practical open architecture controllers for manufacturing application," In: *Open Architecture Control Systems* (Koren, Y., *et al.* editors), IITA Publ., pp. 103-113.

[3] Koren, Y., 1999, "Reconfigurable manufacturing systems," *Annals of the CIRP,* Vol. 48, pp. 2-13.

[4] Scheifele, D., 2001, "Reconfigurable open control systems," *CIRP 1st Int. Conf. on Reconfigurable Manufacturing,* Ann Arbor, Michigan.

[5] Coste-Maniere, E., and Simmons, R., 2000, "Architecture, the backbone of robotic systems," *Proc. of the IEEE Int. Conf. on Robot. & Automation,* San-Francisco, pp. 67-72.

[6] Senehi, M. K., and Kramer; T. R., 1998, "A framework for control architectures," *Int. J. of Computer Integrated Manufacturing,* Vol. 11(4), pp. 347 - 363.

[7] Brooks, R.A., 1986, "A robust layered control system for mobile robot," *IEEE Journal of Robotic Automation,* Vol., 2(1), pp 14-23.

[8] Albus, J., McCain, H., and Lumina, R., 1987, "NASA/NSB standard reference model for telerobot control systems architecture (NASREM)," *NBS Technical Report*, No. 1235, NBS, Gaithersburg, MD, USA.

[9] Gat, E., 1992, "Integrating planning and reacting in heterogeneous asynchronous architecture for controlling real-world mobile robots," *Proceedings of the 10th Conference on Artificial Intelligence,* pp. 1-7.

[10] Heikkila, T., and Roning, J., 1992, "PEM-modelling: A framework for designing intelligent robot control," *Journal of Robots Mechatronics,* Vol. 4(5), pp. 437–444.

[11] Simon, D., *et al..*, 1993, "Computer-aided design of a generic robot controller handling reactivity and real time control issues," *IEEE Trans. Cont. Sys. Tech.,* Vol. 1(4), pp. 213-229.

[12] Yen, J., *et al.*, 1993, "Employing fuzzy logic for navigation and control in an autonomous mobile system," *Proc. of the ACC.*, San Francisco, USA, pp. 1850–1854.

[13] Altintas, Y., Newell, N., and Ito, M., 1993, "A hierarchical open architecture multiprocessing CNC system for motion and process control," *Manufacturing Science Eng., ASME PED,* Vol. 64, pp. 195-205.

[14] Teltz, R., Urbasik, K., Elbestawi, M.A., 1994, "Intelligent, open architecture control for machining systems," *ASME-PED Publ., Manufacturing Sci. Eng.,* Vol. 68-2, pp. 851-864.

[15] Tunstel, E. and Jamshidi, M., 1994, "Fuzzy logic and behavior control strategy for autonomous mobile robot mapping," *3rd IEEE Conf. on Fuzzy Systems*, Vol. 1, pp. 514–517.

[16] Chatila, R., 1995, "Deliberation and reactivity in autonomous mobile robots," *Rob. Auton. Systems,* Vol. 16, pp. 197-211.

[17] Park, J., *et al.*, 1995, "An open architecture real-time controller for machining processes," *CIRP 27,* May 21-23, 1995, Ann Arbor, Michigan.

[18] Rober, S.J., and Shin, Y.C., 1995, "Modeling and control of CNC machines using a PC-based open architecture controller," *Mechatronics*, Vol. 5(4), pp. 401-420.

[19] Thielemans, H., Demestere, L., and Van Brussel, H., 1996, "HEDRA: Heterogeneous distributed real-time architecture," *Contr. Eng. Pract.,* Vol., 4(2), pp. 187-193.

[20] Michaud, F., Lachiver, G., and Dinh, C.T.L., 1996, "A new control architecture combining reactivity, planning, deliberation and motivation for situated autonomous agents," *Proc. Int. Conf. Sim. Adapt. Behaviour,* pp. 1-10.

[21] Khatib, O., Quinlan, S., Williams, D., 1997, "Robot planning and control," *Robot. & Autonomous Systems,* Vol., 21, pp. 349-261

[22] Fiedler, P.J. and Schilb, C.J., 1997, "Open architecture systems for robotic workcells," *IWACT 1997 Conference,* Columbia, Ohio.

[23] Landers, R.G. and Ulsoy, A.G., 1997, "Supervisory machining control on an open architecture platform," *CIRP Int. Conference & Exhibition on Design and Production of Dies and Molds,* Istanbul, Turkey, pp. 97-104.

[24] Hu, H., Gu, D., and Brady, M., 1998, "A modular computing architecture for autonomous robots," *Microprocessors and Microsystems*, Vol. 21, pp. 349-361.

[25] Borrelly, J.J., *et al.*, 1998, "The ORCCAD Architecture," *Int. J. Rob. Res.,* Vol. 17(4), pp. 338-359.

[26] Huff, B.L., and Edwards, C.R., 1999, "Layered supervisory control architecture for reconfigurable automation," *Prod. Plann. Control,* Vol. 10(7), pp. 659-670.

[27] Nilsson, K., and Johansson, R., 1998, "Integrated architecture for industrial robot programming and control," *Robotics and Autonomous Systems,* Vol. 29 (1999) pp. 205-226.

[28] Fernandez, J.A., Gonzales, J., 1999, "The NEXUS open system for integrating robotics software," *Robotics and Computer Integrated Manufacturing,* Vol. 15, pp. 431-440.

[29] Ulmer Jr., B.C., and Kurfess, T.R., 1999, "Integration of an open architecture controller with a diamond turning machine," *Mechatronics,* pp. 349-361.

[30] Park, J.M., *et al.*, 1999, "A hybrid control architecture using sequencing strategy for mobile manipulator," *Proc. IEEE/RSJ Int. Conf. Intell. Rob. Systems,* pp. 1279-1284.

[31] Surya, K., Yamazaki, K., Kawaga, Y., 2000, "PC-based open architecture servo controller for CNC machining," *Workshop on Real Time Operating Systems and application and Second Real Time Linux Workshop*, 2000.

[32] Wills, L., *et al.*, 2000, "An open platform for reconfigurable distributed, hierarchical control system," *Proc. Of the Digital Avionics Sys. Conf.,* October 2000, Philadelphia, PA.

[33] Erol, N.A., Altintas, Y., Ito, M.R., 2000, "Open system architecture modular tool kit for motion and machining process control," *IEEE/ASME Mechatronics,* Vol. 5(3), pp. 281-291.

[34] Hong, K.S., Choi, K.H., Kim, J.G., and Lee, S., 2000, "A PC-based open robot control system: PC-ORC," *Rob. & Comp. Integr. Manufacturing,* Vol. 17, pp. 355-365.

[35] Heilala, J., 2001, "Open real-time robotics control- PC hardware, Windows/VxWorks operating systems and communication," *http://aut-pc29.hut.fi/kurssit/as-116-140/sem_s01/heilala.pdf.*

[36] Volpe, R., *et al.*, 2001, "The CLARAty Architecture for robotic autonomy," *Proc. of the 2001 IEEE Aerospace Conf.,* Big Sky, Montana.

[37] Yang, M.Y., Hong, W.P., 2001, "A PC – NC milling machine with new simultaneous 3-axis control algorithm," *The International Journal of Machine Tool and Manufacture,* Vol. 41, pp. 555-566.

[38] Oldknow, K.D. and Yellowley, I., 2001, "Design, implementation and validation of a system for the dynamic reconfiguration of an open architecture machine tool controls," *Int. J. Mach. Tools & Manufacture,* Vol. 41, pp. 795-808.

[39] Wu, S.F., Mei, J.S., and Niu, P.Y., 2001, "Path guidance and control of a guided wheeled mobile robot," *Control Engrg. Practice,* Vol. 9(1), pp. 97–105.

[40] Yavuz, H. and Bradshaw, A., 2002, "A new conceptual approach to the design of hybrid control architectures for autonomous mobile robots," *J. Intell. Rob. Systems,* Vol. 34, pp. 1-26.

[41] Tan, K.K., *et al.*, 2002, "Development of an integrated and open-architecture precision motion control System," *Control Engineering Practice,* Vol. 10, pp. 757-772.

[42] Freire, E., *et al.*, 2003, "A new mobile robot control approach via fusion of control signals," *IEEE Trans. Sys., man, & Cyber., Part B: Cyber.,* pp. 1-11.

[43] Mills, K.L. and Gomaa, H., 2002, "Knowledge-based automation of a design method for concurrent systems," *IEEE Trans. Soft. Eng.,* Vol. 28(3), pp. 228-255.

Subject Index

W